数学家的智慧

——胡和生文集

胡和生 著

上海教育出版社
SHANGHAI EDUCATIONAL PUBLISHING HOUSE

图书在版编目(CIP)数据

数学家的智慧:胡和生文集 / 胡和生著. —上海:上海教育出版社,2017.6(2018.1 重印)
ISBN 978-7-5444-7568-6

Ⅰ.①数… Ⅱ.①胡… Ⅲ.①数学—文集 Ⅳ.①01-53

中国版本图书馆CIP数据核字(2017)第109419号

责任编辑 赵海燕 蒋徐巍 张莹莹
特约编辑 虞 彬
书籍设计 陆 弦
技术编辑 李伟克
印装监制 朱国范

数学家的智慧——胡和生文集
Shuxuejia de Zhihui——Hu Hesheng Wenji
胡和生 著

出版发行 上海教育出版社有限公司
官 网 www.seph.com.cn
地 址 上海市永福路 123 号
邮 编 200031
印 刷 上海中华商务联合印刷有限公司
开 本 787×1092 1/16 印张 34.5 插页 17
版 次 2017 年 6 月第 1 版
印 次 2018 年 1 月第 2 次印刷
书 号 ISBN 978-7-5444-7568-6/G·6232
定 价 168.00 元

如发现质量问题,请向本社调换 电话 021-64377165

专心致志，刻苦钻研，持之以恒，不受干扰

胡和生

胡和生在复旦大学数学研究所办公室（2005年）

胡和生与祖父名画家胡郯卿（1932年）

胡和生与堂妹胡韵生（左一）在复兴公园（1933年）

胡和生与弟弟胡庚生在复兴公园（1936年）

胡和生与大姐胡冠琛（左一）、叔叔胡伯洲（左三）、大哥胡东初（左四）、弟弟胡庚生（右二）及祖父胡郯卿（1936年）

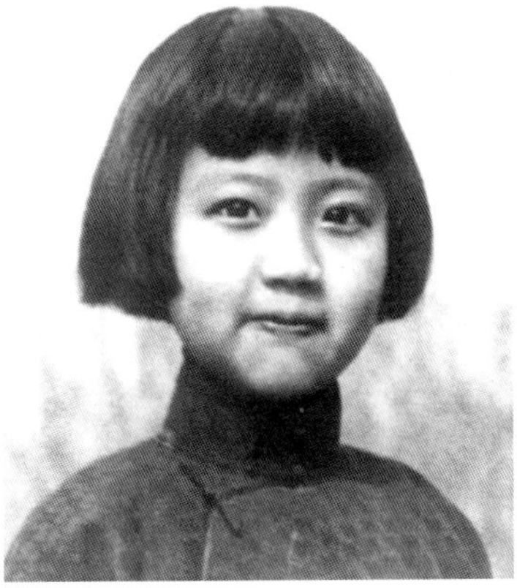

胡和生小学二年级（1935年）

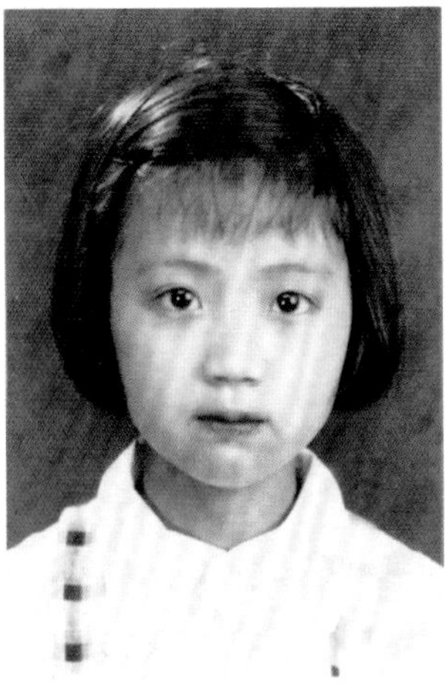

胡和生小学四年级（1937年）

胡和生初中一年级（1939年）

胡和生（1948年）

胡和生与小妹胡美琛（左二）、大姐胡冠琛（左三）、二姐胡润生（左四）在杭州（1952年）

胡和生与谷超豪结婚照（1957年夏）

胡和生与苏步青、谷超豪、卢景义（后排左一）、姜国英（后排左二）、张国樑（后排左三）、黄宣国（后排右二）、陈咸平（后排右一）（1980年）

胡和生与部分家人（左一弟弟胡庚生、左二大姐胡冠琛、左三四妹胡华琛、左四小妹胡美琛、左六父亲胡伯翔、左七外甥女冯如英、右六大姐夫冯大墉、右五谷超豪、右四侄子胡晓松、右三小妹夫欧阳光中、右二外甥欧阳村茂、右一儿子谷晓明）（1984年）

胡和生与黄宣国（前排左一）、忻元龙（前排左二）、潘养廉（前排左四）、沈纯理（前排右二）、徐幼于（前排右一）及国内微分几何进修教师班学员（1986年）

谷超豪、胡和生夫妇和学生们在复旦大学数学系图书资料室，左起：成庆明、周子翔、胡和生、丁青、谷超豪、林峻岷（1987年）

谷超豪、胡和生夫妇与独子谷晓明在复旦大学第九宿舍（1988年）

胡和生与周子翔（左一）、林峻岷（右一）在复旦大学数学系图书资料室（1988年）

胡和生与时任京都大学数理解析研究所所长Araki（1990年）

胡和生与陈省身、郑士宁夫妇、陈恕行（1991年）

胡和生与谷超豪（1991年）

胡和生、谷超豪与苏步青（1991年）

中国科学院数学物理学部数学学科学部委员合影（1992年）

胡和生、谷超豪与洪家兴（1995年6月17日）

胡和生与陈省身、郑士宁夫妇（1995年）

胡和生、谷超豪与王红（左一）、许洪伟（右二）、东瑜昕（右一）在杭州大学数学系（1997年）

胡和生、谷超豪与忻元龙（右二）、林峻岷（左一）、周汝光（左二）、于祖焕（左四）、乔志军（右三）、黎镇琦（右一）（1997年）

胡和生在上课（1997年）

胡和生出席授予李骏为复旦大学教授仪式（1999年）

胡和生与Michael Francis Atiyah、葛墨林（2000年）

胡和生、谷超豪与李大潜（2001年9月21日）

胡和生在世界数学家大会上作诺特报告（2002年8月24日）

谷超豪、胡和生夫妇与Vladimir Matveev夫妇、范恩贵在上海市政协文化俱乐部（2003年7月）

胡和生、谷超豪与Yvonne Choquet-Bruhat访问浙江大学数学科学中心，左一刘克峰、右一许洪伟（2003年11月）

谷超豪、胡和生夫妇与吴文俊、陈丕和夫妇、吴天骄（2004年5月26日）

胡和生、谷超豪与葛墨林（后左）、孙义燧（后右）在天台山（2004年）

谷超豪、胡和生夫妇与杨振宁、翁帆夫妇（2005年5月15日）

胡和生、谷超豪与谢纳庆

（2005年5月15日）

胡和生与时任教育部副部长吴启迪

（2005年5月15日）

胡和生与丁青在黄山
（2005年10月）

胡和生与潘养廉（左一）、忻元龙（左三）、黄宣国（右一）在上海瑞泰宾馆（2008年1月）

胡和生、张淌（周元燊夫人）与母校清心女中（现上海市第八中学）女生班同学（2007年3月）

胡和生与复旦大学副校长金力（2010年1月）

胡和生与时任上海交通大学校长张杰（2011年4月）

胡和生与复旦大学党委副书记陈立民（2016年6月）

胡和生与傅吉祥（2017年5月）

胡和生与嵇庆春（2017年5月）

目录

前言

寒来暑往韶华过，春华秋实梦依在.2017 年 6 月 20 日是我的九十岁生日，我的朋友们、学生们提议和协助我，编撰了这本文集.

这本文集包含了两个部分：第一部分是我的生平记述，对母校、老师的纪念文章等；第二部分是我自 1953 年起发表的部分学术论文，涉及微分几何、数学物理等研究领域.

20 世纪 50 年代初，我开始微分几何研究，最初研究仿射联络空间的几何学.我将苏联几何学家 Norden 的共轭仿射联络对推广为多个共轭联络，并证明了几个定理，完成了我的第一篇论文.这篇论文后来发表在《数学学报》，得到 Norden 的重视和肯定.在这之后，我对于理解黎曼流形的等距同构群做了许多工作."文化大革命"使我的科研一度中断，后来杨振宁教授访问复旦大学.在他的鼓励下，我开始对 Yang-Mills 场的理论进行研究，取得了一些很有意思的成果，特别是关于有限能量或者缓增能量条件下 Schwarzschild 黑洞背景中的 Yang-Mills 场.之后，我把相当部分的精力用于研究 Darboux 变换和 Backlund 变换在孤子解构造的应用上.

我与很多国家的数学家进行过科学交流与合作.法国数学家的成就是非常伟大的，在世界上有着非常重要的影响.我年轻的时候学习了 E. Cartan 的许多作品，他的想法和方法使我能做一系列微分几何的工作.1980 年以来，我有很多机会直接从法国数学家那里学习.Yvonne Choquet - Bruhat 教授就曾建议和鼓励我考虑数学物理中的一些问题.我想借此机会向所有的外国朋友表达我的感谢.

附录中收录了两篇文章.一篇是 Yvonne Choquet - Bruhat 教授与我在南京师范大学附属中学与该校师生座谈的实录;另一篇则是在众多中央和地方的报刊上有关我的通讯、报道中我非常喜欢的一篇.

自 1979 年国家恢复研究生制度以来,我一直承担研究生学位基础课的教学,并开设了现代微分几何、数学物理、孤立子理论、外微分形式、规范场理论等专题讨论班,培养了硕士生 30 多名,博士生 20 多人.他们中的很多人已成为极其优秀的数学家或其他行业的翘楚.

谢纳庆教授是我与谷超豪的关门弟子.他在繁忙的科研之余,为我收集了文集中的所有学术论文,他还以极大的热情为文集的出版做了大量的具体工作,付出了艰辛的劳动,我对他表示谢意.

虞彬同志为文集出版提供了很多资料,同时还做了很多联络工作,我也对他表示谢意.

对于文集的整理、出版,洪家兴院士、张伟平院士、丁青教授、许洪伟教授助力甚多;中国科学院院士工作局、浙江大学、复旦大学、上海市第八中学的许多同志也为文集的出版做了很多工作.文集中的一些照片是我的父亲胡伯翔先生拍摄的,他是中国近代著名的画家、摄影家.另有部分照片是复旦大学的刘畅、文汇报的臧志成、复旦大学附属中学的樊留凤以及许多不知名的朋友拍摄的.特向他们表示感谢.

我于 1987 年 2 月加入中国民主同盟.文集的出版得到了民盟上海市委专职副主委方荣同志的关心.尤其要感谢全国人大常委会副委员长、民盟中央主席张宝文先生在文集付梓之时题写了书名.

如果没有来自友人上海教育出版社资深编辑施桂芬编审的一再鼓励,以及她和赵海燕女士为文集所做出的不懈努力,那么本书的出版大概是不可能的.我在此谨对她们两位表示深深的谢意.

最后我要由衷地感谢上海教育出版社王耀东副总编辑对我这本文集给予了高度重视和巨大支持,使本文集能在我九十岁生日之际出现在大家面前.

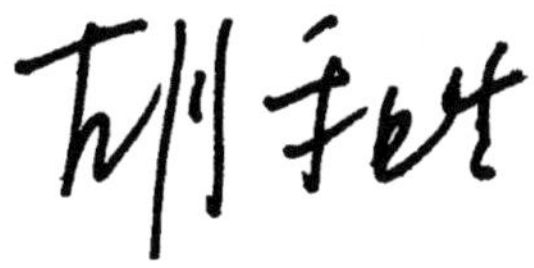

2017 年 5 月

第一部分

奋斗的历程

在数学的征途上

胡和生

(一) 女儿当自强

我出生于一个艺术的家庭，祖父胡郯卿是很有名的画家，是自学成才的.父亲胡伯翔成名很早，国画自成一家，将西方画法融入国画之中，很有创新，他又是中国摄影艺术的最初倡导者之一.他们都是依靠自己的艰苦奋斗而成为社会精英的.我家有兄弟姐妹七人，女孩五人，我居中.父亲时常教育我们要爱国，要学好本事，有一技之长立足社会，要有奋斗自强的精神，不断进步，取得成功.

在小学里，我各门功课成绩优秀，老师们称赞我聪明好学，很喜欢我.五年级时，日本侵略中国，抗战期间，国家遭受浩劫.在南京，祖父家中珍藏的古画及本人珍品被日寇抢劫一空，差一点丢了性命.祖父在南京享有盛名，为了躲避日伪对他的拉拢与迫害，他逃到了上海和我们住在一起，隐居起来，不再对外作画.当时有些日本人和汉奸要父亲为他们作画，父亲拒绝了，并停止出售作品.祖父与父亲这种爱国行为给我留下深刻的印象.

在我六年级时，大姐肺结核病重，肺部双侧均有空洞，入医院手术，并切除了四根肋骨，为她治病，家产花尽.那时上海孤岛，粮食短缺，米价很贵，我跟二姐常到粮店去排长队挤买碎米、苞米面.由于营养严重不足，原来身体就瘦弱的我，变得更为体弱多病.接着，我考进了著名的清心女中，学校离家远，我坚持每天快步行走近半小时上学.夏季雷雨天和台风之后，马路上常常积满了水，几乎到达膝盖，就蹚水过去；冬季脚后跟冻疮严重溃烂，我忍痛拖着鞋去上学.当时我求学的决心很大，从不缺课，这种艰难条件更锻炼了我的意志.

高中时期，太平洋战争已经爆发了，上海市区租界变为日寇的占领地，随时传来日军横行霸道残害中国同胞的消息.我们到校读书，也常面临日军戒严、交通断绝的威胁，我就读的是女子中学，尤其感到紧张.我家在环龙路（现在南昌路）沿马路，对面法国总会（现在的科学会堂）也被日本侵略军占用，他们站在平台上，就可看到我家，我们非常害怕，白天黑夜都拉着窗帘，生怕被日军看到我们家里有好几个女孩子，真是心惊肉跳地过日子.这些都使我认识到，国家不富强，就要受欺侮，人民要遭殃，我长大一定要为国出力，为国效劳.我很快就接受了"科学救国"的思想，立志努力读书，报效祖国.

（二）踏上数学之路

中学毕业后，我选择数学作为自己的专业方向，先后在上海交通大学数学系和大夏大学数理系学习，得到老师们很好的训练.大学毕业时，上海已经解放，老师推荐我到同济大学担任助教，但我还想继续读书，投考研究生，北京大学和浙江大学都录取了我，父母担心我从小体弱多病，不能适应北方的气候，又听说浙大苏步青教授是几何学大家，于是我就到浙大报到.

一到浙大，就为那里浓厚的学术空气所感染，系里老师除了认真教学外，还提倡科学研究，许多教师都有自己的研究方向，力求在数学上有所创造，经常出论文.这一片学术繁荣情景我从来未曾见到过.

苏先生是有名的严师，那学期他招了三名研究生.我们一进校，他马上为我们开设讨论班，要我们读书和做报告，并安排我做第一个报告.我认真做了准备，在讨论班上仔仔细细地报告了书中的内容，对苏先生的提问都能正确回答，苏先生那张严肃的面孔也泛起了笑容，称赞我"讲得很好".我听说过苏先生对学生很严，总爱挑报告中各种各样的毛病，这次我能顺利过关，感到很高兴.苏先生看到我读书理解深入，经得起他的提问，便要我读国外数学刊物上最新发表的论文，有英文的，有德文的，后来还有俄文的，往往很长，从二三十页到上百页都有，内容有关高维欧氏空间的子流形方面，特别是关于超曲面的变形理论.当时没有复印机,单单抄下来也得花许多时间,在这样的重压下,我感到只有硬拼.我体力差,白天工作之后到晚上就没有力气了,只好先睡觉,到半夜里再起来接着读.凭着数学的推理和反复体会,把论文读懂,也从中享受了读书的苦与乐,对数学的兴趣进一步增强了.由于学习的工作量很大,使我不得不抓紧时间加倍努力,又因为总有一些干扰来影响自己,我当时为自己制订了16字的座右铭"专心致志,刻苦钻研,持之以恒,不受干扰",天天要看一遍.

苏先生的严格要求,使我懂得做学问的人必须具备的素质.为了对我进一步培养,1951年暑假前,苏先生与中国科学院联系,把我收为中国科学院的研究实习员,因而我多了一重身份,也有了一份收入,又和中国科学院数学研究所有了正式的联系.苏先生帮我制订每一年度的研究计划,要我每季度写季度工作报告,寄到北京.当时华罗庚先生担任数学研究所所长,他仔细审阅了我的报告.这种写季度报告的方式,帮助我有效地克服惰性,增强了我的责任感,我受益匪浅.

1952年院系调整,我随苏先生到了复旦大学,苏先生在这里开始了重新创业的奋斗,我也开始在微分几何的研究中,获得一系列的成果.我最初研究仿射联络空间的几何学,其中第一篇论文是将苏联几何学家诺尔琴的共轭仿射联络对推广为 n 个共轭联络,这篇论文交给苏先生后,他很高兴,仔细地看了我的稿子,在进行了一些文字修改后,很快就把它推荐到《数学学报》发表,陈建功先生知道后鼓励我说"你有了第一篇,就会有第二篇、第三篇……",文章发表后得到诺尔琴的重视和肯定,在苏联《数学评论》作了详细介绍.接着我又较快地得出仿射联络空间方面的一些成果,写成两篇论文.后来我研究高维欧氏空间与常曲率空间中超曲面的变形理论、常曲率流形的结构等问题,发表了十多篇论文,这些工作改进了著名几何学家 E. Cartan, T. Y. Thomas 和苏联通讯院士 Yanenko 的研究成果.陈省身教授在美国的《数学评论》中介绍了我的成果.由于做出了一系列评价相当好的成果,在1956年我被评为中国科学院数学研究所的先进工作者,这是数学所的第一次评奖活动,华罗庚教授很重视,亲自写信给我表示祝贺和鼓励.苏步青教授的指导,陈建功、华罗庚、陈省身等老一辈大数学家的鼓励,使我很受感动,信心越来越足,决心在数学研究的道路上继续奋斗,一定要向更高的目标前进.

(三) 坚强自信,力克困难

紧接着1957年的"反右运动",1958年开始的一股否定与批判基础研究的"左"倾思潮干扰了数学界,复旦数学系陈建功先生与我两人成为批判对象.什么"理论脱离实际"、"英才教育"等帽子向我们头上抛来,来势之凶猛令人难以置信.由于大家对1957年"反右运动"记忆犹新,十分恐惧,都吓得不敢与我讲话了,只有陈建功先生偷偷地对我说"彼此彼此".我相信自己选定的、从事基础数学研究和教学的事业是正确的,是国家所需要的.遭批判我感到很委屈,但在压力面前我不灰心、不退缩,以更加积极、更加进取的态度来对待科研.说我理论脱离实际,我就努力去学习实际知识,学习了弹性力学、量子力学及广义相对论等方面的知识.在下工厂解决实际问题时,我得到了锻炼,做出了成绩,并且和原子能系的几位教师合作,开展了群论和核谱的研究.在这段时间里,我坚持数学基础理论的研究,我又学习齐性空间几何学和群表示论,完成了黎曼空间运动群空隙性的研究,意大利著名数学家Fubini在1903年首先发现了黎曼流形运动群的参数数目有空隙,提出空隙性问题,引起了很大的重视,成为当时几何方面的一个热门的重要课题,20世纪40年代至60年代王宪钟、Yano(日)、Vrancenu(罗)、Teleman(罗)、Egorov(苏)、Wakakuwa(日)等人都投入了对这个问题的研究,虽有很多进展,但不理想,他们确定了第一、第二空隙,但第二空隙是在空间维数大于248时才能得到证明.1959年至1966年间,我研究齐性空间微分几何理论,深入钻研探讨迷向群与运动群之间的关系,并通过对正交群及其子群的研究,确定了正交群的最大不可约子群的维数,经过复杂的论证和计算,得到确定所有空隙的一般方法,同时也确定了有关的黎曼度量,解决了这个持续六十多年的重要问题.苏先生十分高兴,在会上称赞说这是别开生面的工作,远远超越了前人.十多年后,改革开放,著名美籍日本几何学家Kobayashi来华参加国际会议,一见面就说起我的这项工作,赞扬我的这项成果.由于日本几何学家当时也在研究这个热门问题,我的这项工作给日本数学家留下深刻的印象.从1983年起直到21世纪初,我多次访日,日本数学家还继续多次提到这项成果.这一段经历使我深深感到,把困难与挫折视为机遇对待，是走向成功之路.四十年后的今天，我回忆起这段往事，仍然激动不已.在困难和挫折面前，一定要坚强自信，一定要继续艰苦奋斗，发扬拼搏精神.只有这样，才能不会被困难所压倒，才能把自己的工作推向新的高度.使我感到非常惋惜的是，当我的科研教学正处于黄金时期，“文化大革命”像暴风骤雨那样席卷而来，进一步迈向高峰的征途被堵死了.不断地受批判，不断地劳动，研究工作已经无法进行了，但我并未失去信心，我相信科学的春天终究会来临.

（四）继续奋斗，走向世界

到了 1974 年，还在“文革”期间，杨振宁教授访问上海，建议和复旦大学的教师开展与规范场有关的数学问题的研究，成立了一个科研小组，谷超豪是组长，成员都是数学与物理方面的精干，我是研究组成员之一.

合作持续了几年，是卓有成效的，得到很有意义的成果，完成了一系列的合作论文，我也作了实质性的贡献.这时“文革”已经结束，改革开放的划时代变革开始了，这使我获得了无穷动力，快马加鞭，决心追回失去的年华.那时国际交流的环境开始形成，要用国际水平来衡量我们的工作，这又是新的挑战，当时所完成的关于规范场方面的一系列工作，有一部分是走在当时国际前列的，为国际学术界所重视，面临这样的形势，我只有努力拼搏，赶快跟上世界潮流，把失去的时间抓回来.我感到自己数学生涯中的第二个春天也将来临.

我在有质量规范场、引力场中规范场的静态解和规范场的团块现象等方面创造性的研究成果在国际交流中受到注目.例如 1979 年去美国访问时，我的关于有质量规范场的研究就很快地引起重视.我深入探讨了静态解的存在性问题，发现质量 $m=0$ 和 $m\neq 0$ 两种情况的重大差异，因而发现了质量 $m\to 0$ 时的不连续性. 对这一事实，美国著名物理学家 S. Deser 在他本人发表的论文和给杨振宁的信中称“胡是第一个给出了经典场论中质量 $m\to 0$ 时不连续性的显式事例”，“成果十分有意义”.接着我又进一步发展了这方面的研究，就 Yang－Mills 场的团块现象和黑洞外 Yang－Mills 场是否存在的问题，进行了深入的研究，得到了法国科学院院士 Lichnerowicz 和 Choquet - Bruhat 的高度赞赏，多次请我到国际学术会议作特邀报告，又多次请我到法兰西学院作学术演讲.1983 年中国数学家派出了以苏步青教授为首，由王元和我组成的代表团，去参加日本数学会年会，王元和我在会上作了特邀学术报告.随后我又多次到法、德、意、瑞、日等国进行研究和讲学活动，并十多次在国际学术会议上作大会特邀报告，1990 年还作为中国数学会代表团三位成员之一到日本首次正式参加世界数学联盟代表大会，会后又参加世界数学家大会.这段时期前后我担任过中国数学会副理事长、上海数学会理事长和《中国数学学报》副主编，1991 年我被选为中国科学院院士.

新的困难又向我袭来，1995 年我因患结肠癌在中山医院进行手术，著名外科专家王承培教授成功地为我做了手术，我的冷静、坚强和自信性格，使我较快地恢复健康.

过了一年多，我又投入到紧张的科研和教学工作中，在可积系统与微分几何方面取得了新的进展.从 20 世纪 80 年代开始，我把现代的孤立子理论和微分几何联系起来，发展了孤立子理论中的 Darboux 变换方法，并应用于调和映照和线汇论等方面的研究中去，给出了 Minkowski 空间伪球线汇的分类及构作等，特别是在最近五年，建立起射影空间的 Laplace 序列和 Toda 方程之间的联系，并给出求解方法及实例，又得到有关 Laplace 序列的嵌入性定理，还和同事们合作给出复射影空间的 Laplace 序列成为调和序列的充要条件，并给出第一个周期性调和映照序列的实例，等等.这方面的研究先后应邀在法国、日本的国际会议上作大会报告，受到很高评价，又在德国、中国香港地区等地讲学访问.2000 年我应邀请出席了法国科学院院士大会，Choquet 院士在会上将我介绍给全体院士，介绍了我的学术成就，赞扬我在规范场及可积系统方面做了在物理上和数学上都极有意义的工作.

2002 年，我荣幸地得到世界妇女数学组织的邀请，在四年一次的世界数学家大会(2002，北京)作艾米·诺特讲座(一小时)，报告中包括了我当年获得的最新成果.这个讲座是为纪念世界伟大女数学家艾米·诺特的，从 1994 年起，每四年请一位女数学家作学术演讲，能在世界数学家大会上作诺特报告，是我很大的荣誉.

2003 年，我当选为第三世界科学院院士，又获得何梁何利“科学与技术进步奖”.

从成为浙大研究生起，我就已开始做一些对工农干部和大学生的教学工作，到了复旦以后，虽然我长期属于中国科学院和复旦的科研编制，但一直从事几何方面的教学工作，开大学生的基础课、专门课程和讨论班，指导毕业论文，从帮助苏先生指导高年级学生和研究生，到后来自己招收研究生，一直到现在.20 世纪 50 年代后期起，苏先生的行政工作和社会活动一直非常繁忙，我就努力地为他分担教学任务，他对我很放心，也很放手.这样，我的教学工作就一直很繁重，贯穿到我的整个生活.我对教学工作是非常热爱的，觉得将自己学到的知识教给青年一代的工作是最为重要、最为有意义的，尤其是将难的内容教懂学生时，我感到十分快乐.我努力把教学和科研结合起来，尽可能地把自己学到的、在研究的新的重要内容和前沿的内容教给学生.我努力去了解学生的情况，全面地关心他们，并在教学上根据他们的情况和能力对他们提出不同的要求，并就如何改进学习方法进行有针对性的指导，这就使不同程度的学生都能很快进步，大家说我能做到因材施教.我还把优秀的学生推荐到国外高水平的学府去深造，有好几位已经有相当高的国际声誉.在国内的几何学界，也有我的一批好学生在努力工作，成为教学和研究的骨干.近年来我的研究生中有三位获得国家优秀博士论文奖.一面从事科研工作，一面又承担繁重的教学，当然很艰苦，运动一来，还要受委屈，但我始终忠诚于教育事业，坚持自己认为正确的方向，作不懈的努力.

回顾自己从事数学工作的历程，我深深地感到，要取得成就，就必须有长期奋斗的决心，就必须不断学习，深入思考，刻苦钻研，持之以恒.在人生的道路上也必然会遇到各种挫折和困难，这时就需要目光远大，有勇气面对困难，坚持正确的方向，化困难为机遇，并以此作为继续前进的动力.

岁月在流逝，时代在前进，我为自己能生活在祖国正在腾飞，人民生活不断地改善的时代而感到十分高兴.我也为自己能为祖国的建设事业竭尽绵薄之力而自豪.目前我已进入高龄时期，我将努力保持自己的朝气，继续发扬自强不息的精神，为数学学科的发展，为青年优秀人才的成长再贡献一份力量.

（本文发表于《科学的道路》，上海教育出版社出版，2005 年，259 - 263）

长川源自渊泉　永无疆校史绵延

——祝贺母校市八中学140周年校庆

胡和生

今天是市八中学140周年校庆，首先让我向全体师生致以热烈的祝贺.回到母校，参加隆重的庆祝活动，看到市八中学在办学方面取得了巨大的成绩，心里非常高兴，学校历届领导，各位老师为教育事业所作出的重大贡献使我深受鼓舞.

我是在1939年进入市八中学的前身清心女中，1945年毕业，受到了为期六年的高水平的中学教育，这对我的一生产生了深远的影响，回忆起来，我的心情是很激动的.

首先我感到清心女中的老师们教学有很高的水平，非常认真.清心的英语水平很高，我小学时英语学得少，一到清心，有很大差距.到了初中二年级之后，才逐步适应了，这时老师对我的进步，多方加以鼓励，对我的英文作文加以好评，这使我树立了信心，使我的英语基础学得很扎实，发音准确.后来几十年中我虽然长期不讲英语，但在改革开放之后，我接待外宾，以及到国外访问，不经训练，仍然没有困难，人们还以为我是从国外留学回来的.又如数学，老师教得很有启发性，教导我们数学的思维方法，使我很感兴趣，又有扎实的训练，我以后决定从事数学研究，和当时的训练很有关系.在学校还举办大、小楷作文比赛等，特别是对音乐艺术、唱歌、美术也重视，培养学生有美学的修养、艺术风气.学习风气浓厚、很活跃，使我获得扎实的全面的基础.

其次我感到在学校里，我们同学之间充满着友爱、互助的精神.大家学习上相互关心、相互讨论、共同得益、共同进步，产生了很深厚的感情.在毕业后，大家分开了四十多年.十年前回到母校参加130周年校庆，同班同学30多位又相见了，回忆起当年的友情，大家都十分兴奋.从此，在我们老班长周福如的主持下，我们每年都有一两次的集体活动.我们都是年逾古稀的老人了,一见面又仿佛回到青少年时代,在清心的学生生活又回到眼前.

第三在清心期间,正是民族危难的时期,开始进初一学习时,上海已沦为孤岛,在日本侵略者的包围之中.后来日本侵略者又直接统治上海市中心地带,我们随时随地都感到安全受到严重的威胁.学校里的爱国气氛很浓,刚初一时,我们学生每天捐一分钱慰劳抗日战士和救助难民,这对我们有很大的教育意义.我们去上学要走很多路,从家中到学校(学校在英租界南京西路)经常遇到戒严不能通行,学校里要我们准备炒米粉,防止断粮.在那段时间里,我们都深深地感到,一定要使国家强大起来,这成为我们学习的动力.

和我们做学生时的情况对比,现在的市八中学已得到很大的发展.刚才参观了校史展览,我感触很深.今天的同学们有非常好的学习条件,我希望大家能珍惜宝贵的时间,学会怎样做人,做一个有理想、有道德、遵纪守法的高素质的公民;要勤奋学习,提高自己的科学文化水平,从各方面打好坚实的基础,要身体和心理都很健康.我们的国家欣欣向荣,面临着建设富强、康乐的社会主义国家的伟大任务,希望大家为将来承担伟大的建设任务努力做好准备.我相信,市八中学一定能够为国家培养出一批优秀人才,成为有杰出贡献的科学家、工程师、文学家、艺术家等,希望同学们有这样的雄心壮志.

桃李芬芳

——回忆大夏大学

胡和生

离开大夏已经半个多世纪了.在这五十多年中,我长期从事教学和研究工作,到过国内外的许多高等学校,在我脑海中,始终萦回着在大夏求学时的情景.我永远记住那宁静而生机盎然的校园,一条美丽的河流,带来了都市罕见的大自然风光,使人心旷神怡.老师们诲人不倦、关怀学生的言行,对我有很大教育意义.同学们更是一个真情友爱、爱国勤学的集体,和大家在一起真正感到其乐融融.

我是1948年秋转学到大夏大学的,当时大夏大学理工学院设有土木系、化学系和数理系.数理系的学生同时学习数学和物理课程,可以偏数学或偏物理,我算是偏数学的,但也学了一些别的大学不为数学系学生专门开设的物理课程,例如电磁学、光学等.从今天的眼光来看,系里的设备是简单的,一所平房,有几个房间.当时数理系的学生人数很少,这些设备用起来倒也还宽敞,整个国家很穷,这样的条件也很不错了.我特别喜欢的是这里有一个面积很大、长长的大厅,它的大半部分是实验室,小半部分是教室,教室里的课桌很好、很大,它也可以用来做力学实验桌.这是我们每天的活动场地,有不少课程是在这里讲授的,我们做实验也在这里.另外,更重要的,这也是我们课外活动的好场所,因为这个大厅很宽敞、明亮、舒适,再加上实验设备很公开,便于同学动手、用脑、做实验,有利于锻炼动手能力,提高感性知识,因此同学们动手能力强,物理学得活.这样的条件是一般大学所缺少的.同学们课后并不离开,喜欢留在这里自学与讨论,相互切磋,互问互教,既能使学习迅速提高,又能增强同学们的凝聚力和纯真友谊.我原来体弱多病,到大夏后,健康状况也有了显著的改善.

我感到最好的条件还是老师们认真、热情的教导.他们给我很大的教育、帮助与鼓励,使我坚定地在科学的道路上前进.江仁寿教授是系主任,他教我电磁学,用 Page 与 Adams 写的厚厚一本书作教材,上课时学生就我一个,一个老师教一个学生,他还是那么认真,那么孜孜不倦地讲解,就像对一大班学生讲课一样,每堂课都要布置习题,亲自批改习题.他对我的严格要求,使我对物理产生了很大的兴趣.刘朝阳教授教我微分几何、光学等三门课程,他是同济大学物理系的教授,在大夏是兼职,他是一位儒雅书生,学风踏实、细致,物理、数学都很有水平,他很关心我的学习,给我很多鼓励,在我大学毕业时,他向同济大学推荐,聘请我担任他所讲授的热力学的助教.施孔成教授,他教我高等代数等两门课程,讲课声音嘹亮,十分生动,劲头十足,课外与学生们打成一片,关心每位学生,很受学生欢迎.一个系有这样好的几位教授在担任本科生的教学,这表明大夏确实是一所严肃认真、教学水平很高的学校.这几位教授都希望我能继续深造,在我决定要报考研究生时,他们都非常鼓励.

我是 1950 年 1 月毕业的,当时上海已经解放,学校里有了很大的变化,有了党的领导,成立了党、团组织,原有的地下党也逐步公开了.学生中政治上求进步,学习上争上游的气氛很浓.由于数理系全系只有十多名学生,但课程很多,还有实验,本校教授、讲师、助教虽有多位,还得聘请外校教师来兼职,因此学校不胜负担,为节约教育资源,领导上决定自 1950 年 2 月起,将数理系的学生转到同济大学就读,成为同济大学的学生.消息很突然,同学们在离校前,开了一个离别会,我也参加了,会上同学们表示,到了同济后一定要发扬大夏刻苦踏实的学风,齐心协力继续努力学习,力争上游,并决定暑期再聚会.果然不负众望,到了 1950 年暑期相聚,同学们高兴地说,他们在同济每个班上都是第一名,我听了十分高兴,深深为同学们的这种热情和成绩所感动,发言鼓励他们继续奋勇前进,自己也更努力地为报考研究生作准备.暑假考试后,我被浙江大学和北京大学两所大学录取.1950 年秋季新学期开始,在浙江大学开始了我向数学的前沿迈进的道路.每当回想往事,总是对大夏的老师、同学充满了感激和怀念之情.

严师的风范

胡和生

1950年9月，我到了浙江大学，成为苏步青教授的研究生.那一年，浙江大学和北京大学都在上海招收数学学科的研究生，两所学校都录取了我，北大先在报上公布录取名单.我在大学期间曾患肺结核，北大录取后先由北大驻上海招生站安排体检，结果不太好，肺部病灶虽大部分已钙化，但还有部分纤维化病灶.招生办人员对我说，到北大后再进行体检，能否通过到那时才能知道.父母亲也担心我不能适应北方的气候与环境.正在此时，我收到了浙江大学的录取通知书和同济大学数理系的聘书.我一向对几何的兴趣比较浓厚，报考前听黄正中老师说起过，浙江大学苏步青教授在几何学研究方面很有成就，这给我留下深刻印象.当时一般人认为，同济大学的助教职位也是相当不错的，能得到这样的工作是很幸运的，但是我想自己年纪还轻，应该增进知识，去浙大当研究生应是最好的选择，只是担心能否通过浙大入学的体检关.到了杭州之后，X光透视肺部发现同样的问题，但浙大的医生让我通过了，我真是欣喜万分.我很幸运能跟随苏步青教授学习，走向数学研究的道路.到浙大入学迄今已经51年了，苏先生的指导与教育使我终身受益，影响我的整个人生.

到了浙大数学系之后，我马上就被系里浓厚的学术气氛所吸引.我本来以为做研究生只是听听课、多读一点书、多长一些数学知识.现在一看，系里的许多教师都有自己的研究方向和课题，还有小型学术讨论会，在教学的同时，都在进行着研究工作，力求在数学上有所创造，这种气氛是我以前从未感受过的.

苏先生当时除了在数学系任教授外，还担任着学校的教务长，工作很忙，那学期他招了三名研究生，除我之外，另两位是来自广西大学的一对孪生姐妹.入学后他马上为我们开设讨论班，要我们读书和做报告，他要我们读的是意大利数学家 Levi - Civita 的名著“Absolute Differential Calculus”.在大学里我已有阅读英文参考书的经验，对英文和数学都不感到有特别的困难，但我还是对内容作了充分的准备，在讨论班中仔仔细细地报告了书中的内容，对苏先生的提问都能正确回答，苏先生那张严肃的面孔也泛起了笑容，称赞我“讲得很好”.我听说过苏先生对学生管得很严，总会挑报告中各种各样的毛病，这次我能顺利过关，感到很高兴.

后来的讨论班上，我经常听到苏先生对学生的报告提问，答得不好的，他会批评、训斥(被称为挂黑板)，甚至当场停止报告，责令下次重新报告，等等.这些虽然不是对我的，但他那极其响亮威严的质问声，也吓得我心里怦怦地跳，对于自己的报告也越发不敢怠慢.苏先生不仅要求我们把所读的内容(包括推理和推导)精确无误地表述出来，还要我们抓住中心，把作者的思路简明扼要地叙述出来，所以每次报告，我总是精心准备，反复思考体会，在黑板前看到他带笑点头时，我的心就放松下来了.就是通过这样的训练，我开始学会怎样去读数学专著，掌握其基本内容与方法.

苏先生的第二步训练就是读论文，在我第一次报告后，他陆续地指定了许多论文，要我读懂了报告，这样每周就添加了一次报告论文的讨论班，只有我们三个一年级的研究生参加，每次都是我报告.这些论文是发表在国际数学杂志上的最新研究成果，有英文的、德文的和俄文的，有的是近百页的长文章，当时没有复印机，单单抄下来也得花许多时间.在这样的重压下，我感到只有硬拼，因此硬着头皮复习德文，自学俄文.靠字典，靠仔细阅读和推导，一步一步弄懂文章的数学内容和难点，并力求抓住重点，终日演算和思索，还常常开夜车到深夜.有一次，第二天早上要报告，我开夜车到下半夜，实在支撑不住，伏案睡着了，到上午八时讨论班开始时还未醒.苏先生在教室等，见我未到，就匆匆忙忙地到宿舍来找我，咚咚的敲门声把我惊醒，我十分紧张，苏先生见到我一书桌的论文、辞典和讲稿，怒气就消失了，但要我马上随他到课堂去做报告.这段时间我报告的内容主要涉及欧氏空间与黎曼空间的曲面论、超曲面理论.

由于学习的工作量很大，我不得不抓紧时间加倍努力，我每天必牢记自己当时制订的座右铭“专心致志，刻苦钻研，持之以恒，不受干扰”.苏先生的严格要求，使我懂得了做学问的人必须具备的素质.经过苏先生这样反复的锻炼,我读论文的能力有很大的提高,领会到怎样才算真正读懂一篇论文,并且知道要研究前沿的课题,必须阅读最新的论文,这又使我加强了对几何的浓厚兴趣,享受着学习数学的苦与乐,就这样我度过了读书关、读论文关.

苏先生在业务上要求很严,对学生却非常关心.我到浙大刚两个月光景,生了一场疟疾,发烧超过 40 度,苏先生特地亲自到宿舍看望我,要我好好休息、养病,不要急于看书,这使我终生难忘.当时做研究生只有很少一点生活津贴,每月 12 元,只够吃饭.在第一学期近寒假时,一天,苏先生找我谈话,对我说,研究生的津贴太少了,很清苦,下学期可聘你为助教,这样待遇好多了,你考虑一下是否愿意,再告诉我.我知道老师关心我,但不懂如何处理是好,因此特地去找白正国先生商量.白先生对我说苏先生要你转为助教是对你好,但助教教学工作忙,读书的时间少些,做研究生可以有足够的时间做研究,白先生要我自己考虑决定.我因为家中对我有些补贴,经济不紧张,后来就决定还是做研究生.又过了几个月,接近 1951 年的暑期,苏先生与中国科学院联系,把我收为中国科学院的研究实习员,因而我多了一重身份,有了一份收入,又和中国科学院数学研究所有了正式的联系.苏先生帮我制订每一年度的研究计划,要我每季度写季度工作报告,寄到北京.

浙大数学系这时又达到一个高峰,除原有的徐瑞云、白正国等教授外,浙大早期毕业,后来出国留学的卢庆骏、张素诚、曹锡华等先生已从国外回来,成为数学系的教师,又有几位业务很强的青年助教,如谷超豪、张鸣镛等开始崭露头角,研究生们成长迅速、表现突出.系里是学术上一片欣欣向荣、繁荣昌盛的景象.陈建功、苏步青两位先生 20 世纪 20 年代在日本设想的建立一个现代化的数学教学科研中心的计划,经过他们 20 年的艰苦奋斗,已经开始形成了.

1951 年下半年开始了“三反”、“五反”和思想改造运动,苏先生在其中受了许多委屈,但仍然关注着学生们,关注着数学系的发展.1952 年全国高校院系调整,把苏步青、陈建功两位先生和浙大数学系的相当一批成员调入上海复旦大学,其他成员分别调到浙江师范学院、南京大学、厦门大学、华东师范大学或留在浙江大学负担工科数学的教学.苏先生从心里对这样的调整是不同意的,这个损失太大了,但这时正是思想改造运动之后,不可能有其他选择,只有服从分配,重新创业了.

那时,孪生姐妹因不能适应苏先生的训练只读了一学期就离开了浙大.1951 年暑期后从中山大学分配来一位学几何的研究生谢兰安,我有了一位师弟,苏先生称之为金童玉女.院系调整时,苏先生就带领我们到了复旦,陈先生也带着他的得意门生夏道行和龚昇(中科院研究实习员)到达复旦.浙大数学系的图书杂志也大部分转到复旦来了,他们把浙大的优秀学风也带到了复旦.

在复旦,苏先生仍然被任命为教务长,行政工作还是很忙,苏先生继续举行微分几何的讨论班.在苏先生的指导下,苏先生、谢兰安与我三人组成的讨论班也搞得十分认真,从不间断.浙大举办讨论班之风很快地在复旦数学系产生影响,几位复旦的和来自别的学校的教授都恢复了中止多年的研究工作.

1953 年,苏先生在复旦招收了两位研究生,一位是复旦毕业生(原浙大学生)叶敬棠,另一位是东北人大毕业生韦述萱.这时,从浙大到北京俄专学俄文的谷超豪也到了复旦,系里也有别的较我们年长的教师参加几何讨论班,从而使这个讨论班改变了冷冷清清的状态.谷超豪和我当时都是讨论班的骨干,讨论班的内容丰富多彩,水平也大有提高,苏先生感到非常高兴,浙大创建的微分几何学派在复旦又显出了蓬勃的生机.这时,系里的教授,如陈传璋、黄缘芳、周怀衡先生等,也都仿效陈建功先生和苏先生的做法,举办了积分方程、代数和分析方面的讨论班,作为促进学术研究、培养青年人才的有效手段.后来复旦党委非常重视,把这种讨论班称为小型科学讨论会,在学校推广,并在报纸上作了详细的报道,在全国产生影响.

苏先生一贯把培养青年教师和研究生作为己任,他一直坚持为我们开设高级的几何课程.在浙大,我听过他讲授的“一般空间微分几何学”和 E. Cartan 的“黎曼几何”这两门课.到复旦后,他又挑选了一本苏联几何学家诺尔琴的俄文专著《仿射联络空间》作为教材.这些教材使我学到很多东西.当时国内大力提倡学习苏联,苏先生知道微分几何在苏联很受重视,十分高兴.他从新中国成立初就开始学习俄文,此时已经可以达到阅读和开始翻译苏联教材和专著的程度.20 世纪 50 年代中期,苏先生一直强调要学习李群和外微分形式方法,他根据菲尼可夫写的《外微分形式方法》一书,讲授了外微分形式理论,并且边学边讲，引导我们去注意难度很高的 Cartan - Kähler 外微分形式方程的理论.他重视启发我们用外微分形式理论去解决微分几何问题，他自己也用这些理论去处理高维空间共轭网的问题，有时遇到困难也与我们讨论，他说这是教学相长，共同提高.由于我们在 Cartan - Kähler 外微分形式理论与活动标形法的学习上花了一番功夫，他又说将外微分形式课程真正学到手的只有谷超豪、胡和生两人.

苏先生原来的专长是仿射微分几何和射影微分几何，但由于微分几何的不断发展，他并不引导我从事仿射微分几何和射影微分几何方面的研究，他认为要学习国际上最新发展的内容，一定要看新的好书.他非但自己读，而且读了就教大家，往往在读了新书的同时将书中内容逐字逐句地完整地翻译下来，然后向学生讲授.苏先生讲课声音洪亮，黑板书写清晰，因此记笔记容易，每听完一门课，笔记就是原著的翻译本，可以做到原著无遗漏，不失真.但是，他的课并不容易懂，因为他不作解释，要我们自己去体会，这往往要花许多时间，程度差的学生难以适应.他启发与督促我们去从事研究，但并不提出具体的问题让我们去做，这样的好处是能训练我们自己提出问题的能力，能寻找适合于自己兴趣和水平的问题进行研究.拿我的第一篇论文为例，他在讲授诺尔琴的仿射联络空间时，详细地介绍诺尔琴引进的一对共轭仿射联络，这就使我想到为什么只有一对联络呢？是否可以引进多个共轭的仿射联络呢？于是我就提出多重共轭的仿射联络的概念，并给出了几个定理，完成了第一篇论文.这篇论文交给苏先生后，他仔细看了我的稿子，在进行了一些文字修改后，就把它推荐到《数学学报》发表.后来诺尔琴为这篇论文在苏联评论性杂志上写了较详细的评价，加以肯定.苏先生鼓励我继续努力，我又较快地得出关于仿射联络空间的一些成果，写成两篇论文.

苏先生在查文献时看到重要文章常提醒我注意，例如，苏联通讯院士雅宁柯的两本长篇著作《高维欧氏空间超曲面的变形理论》、《子流形的变形理论》就是他推荐给我阅读的，这使我对变形理论有了较深入的了解，产生了很大的兴趣，并进一步提高了我运用外微分形式方法的技巧，在此基础上我做了几篇有关黎曼空间超曲面变形的文章，陈省身先生等在美国的《数学评论》上介绍了我的工作.这样，在苏先生的教育和熏陶下，我走上了科学研究之路.上述的一些工作以及关于常曲率空间特征的研究，使我在 1956 年被评为中国科学院数学研究所的先进工作者，这是新中国成立后科学院的第一次评奖.华罗庚先生(所长)特地写信通知我并向我祝贺，这是对我的鼓励和鞭策，苏先生对此感到非常高兴.

20世纪60年代初，当我在数学研究上陆续取得成果的时候，我觉得应对自己有更高的要求，要努力扩大知识领域，要有好的工具，为此我阅读了李群、群表示论、广义相对论等方面的许多论著，希望能抓住重要问题做出重要成果.正在此时，苏先生在讨论班上介绍黎曼空间运动群的空隙性问题，包括Fubini的早期发现，王宪钟和Yano的工作，等等，我感到这正是我所希望研究的重要问题，也正是当时微分几何研究的一个热点，我继续查阅了不少文献，利用齐性空间理论，李群及其表示论等工具，给出了正交群的最大不可约子群维数的最佳公式，建立了确定空隙的一般方法，由此可得出黎曼空间运动群的所有空隙，苏先生大为高兴，表扬了我.

1956年，苏先生和陈建功先生去北京参加了全国12年科学发展规划的制订工作，对于如何发展中国数学及复旦如何发挥作用等有了许多新的想法，他们根据国家需要和科学发展的趋势，提出要在复旦大力开创和发展计算数学、微分方程、概率论和力学等学科，也提倡对原有学科内容进行更新.苏先生同意谷超豪把主要精力转到偏微分方程，并决定让他去苏联进修两年.但苏先生对几何组还是紧抓不放，这样，我便成了他在复旦建设微分几何的主要助手.

苏先生时常用“亲如家人”这句话来形容他同谷超豪和我的关系.除了苏先生对我的关怀和培养之外，我还应该感谢苏师母(原籍日本，名松木米子，后改名苏松本)，她慈祥、贤惠、勤劳，也把我们当子女看待，对我们非常好也非常客气.从我在浙大做研究生开始，她一直称我为胡小姐，她在生活上多方面关照我、帮助我.她知道我工作忙，不善于整理家中的杂物，就亲自到我家帮我整理，我心里既非常感激，又实在过意不去.苏师母不幸于1986年病故，我与谷超豪曾多次到苏州，在她的墓前献花，怀念我们敬爱的师母.

值苏师百岁华诞之际，特将苏师指导我走上科研之路的经历写成此文，表示我衷心的敬意和感谢，敬祝苏师健康、长寿、快乐.

严格要求　诲人不倦　鼓励创造

——追思苏步青老师

胡和生

1950年9月,我到浙江大学做苏步青教授的研究生.1952年暑期院系调整,我随苏先生到复旦大学.53年来,苏先生的指导和教育,使我终生受益.苏先生的逝世,使我十分悲痛,也十分怀念.他一生致力于教育事业,始终忠诚于教育和科研,经受过许多困难挫折,历尽艰辛,取得成功.苏先生的一生是光辉的一生.他的创业精神,他的高尚品德和战胜困难的毅力,永远是我们学习的榜样.

苏先生一贯把培养人才当作最重要的工作,下面我谈谈自己的感受.

(一) 严格要求.研究生(或青年教授)必须读懂他所指定的数学名著,在讨论班上做报告,报告时不仅要把内容、计算、推理说清楚,还要说出作者的主要思路.在讨论班上,他要随时提问,当场作答,答得不好,他会批评、训斥,甚至当场停止报告,责令下次重新报告.在学会怎样读书之后,他根据学生的不同情况,布置阅读论文,当时他给我阅读国际数学刊物上最新的论文,有英文、德文、法文和俄文的,有的近百页,每个星期都要做报告,他不断压重担,促使我们尽快成长,我们深深地感到这是“大松博文式”的训练.

(二) 诲人不倦.从20世纪40年代到60年代,他在从事本科教学和繁重的行政工作的同时,一直为研究生和青年教师开设很前沿的课程,每周讲两三个小时.过去他自己主要的研究方向是射影微分几何,但讲授的却是黎曼几何、李群、外微分形式等更现代化的内容,边学边教,他把很难读、很前沿的名著一丝不苟地翻译下来,在课堂上宣讲,让我们自己去钻研和体会国际大师们很深刻、很困难的著作,从中吸收思想和方法.

(三) 鼓励创造.读书、读论文到一定阶段,他会提出一些值得研究的方向要大家去做创造性的论文，具体题目由自己考虑，有了成果苏先生就很高兴，帮你仔仔细细地修改，推荐到数学杂志上发表，推荐到全国性学术会议上去作报告，作得好，他显得非常高兴，并且希望你能进一步提高质量，做到“青出于蓝而胜于蓝”、“别开生面”.他不将学生限制在他所从事的方向，而是不断地要求学生们拓广视野，博采众长，有自己的新的研究方向和课题.

（四）亲切关怀.他对学生非常关切.我进浙江大学不久，发了疟疾，他亲自到研究生宿舍来看我，叫我暂时不要看书，养好身体.当时当研究生只有很少的津贴，他还为我安排了中国科学院数学研究所研究实习员的职位，使我经济比较宽裕，并有很好的研究交流条件.后来，他曾为几位患重病的学生争取到很好的医疗条件，使他们得以转危为安.

（五）持之以恒.他的教学、研究，包括讨论班活动，从不间断，院系调整到复旦后，一度参加几何讨论班的只有我一个人，他照样每周举行，从不间断.20世纪70年代，有一次下大雨，水深过膝，他拎着淌水的胶鞋如期赶到课堂来主持讨论班.那时他已超过75岁.这种精神，学生们无不万分感动，也正是这种精神，使我懂得做学问的人必须具备的素质.

在苏先生的长期培育下，从浙江大学开始的中国微分几何学派得以形成和发展，研究人才不断出现，学术研究范围不但一直处于国际前沿，并且不断地扩展.苏先生一生的心血，有着丰硕成果.苏先生种种感人的事迹，牢牢地被铭记在人们的心中.苏先生，安息吧，我们会很好地继承您的遗志，继续为办好您所开创的事业而努力奋斗.

坚持真理　奋斗不息

——向周培源教授致敬

谷超豪　胡和生

周培源教授是深受我国人民敬爱的前辈科学家，我们在20世纪50年代初，也就是在刚刚开始从事教育科研工作的时候，就知道他在广义相对论和湍流理论中有重要成就，并有幸和他见面.40年来，我们知道他一直坚持不懈地为社会主义建设而努力奋斗，在教育、科研、对外交流和政治建设等领域，都作出重大的贡献.他那热爱祖国、酷爱真理、坚持原则、顽强奋斗的精神，使我们深受感动，是我们学习的典范.

下面，用我们印象最深的两件事来说明周老的这种精神.

第一件事是关于引力理论的研究.爱因斯坦广义相对论的建立是20世纪科学史上的重大事件.他建立了刻画引力场的爱因斯坦方程，对牛顿的引力理论做了革命性的变革.研究广义相对论的学者们往往把引力理解为“时空的弯曲”，在引力场出现时，时空就成为弯曲的了.这种时空和平直的时空（即闵可夫斯基时空）有着本质的区别.平直时空，也就是狭义相对论的时空，存在着惯性坐标系，参考于这种坐标系，空间是欧几里得式的，时间是实数轴，这和我们通常的时空几何性质并没有差别，只是在不同的惯性坐标系之间，它们是以洛仑兹变换相互联系的.弯曲的时空比平直的时空复杂得多，它只是在小领域中和闵可夫斯基空间相近似，区域大了，差别就显示出来了.对于它，并不存在惯性坐标系，而人们所采用的坐标系也可以是没有任何特殊意义的.几何量和物理量，只能是在坐标变换下不变或协变的量，因而在实际运用起来，会遇到许多困难.例如，两个不同的点的速度，一般就无法用统一的方法加以比较，说得专门一点，就是向量的莱维齐维塔平行移动是和路径有关的，沿不同的路径平行移动所得到的向量是不一样的.

周老作为一个造诣很深的物理学家,感到广义相对论的研究结果虽然很多,爱因斯坦方程的解也已有许多,但能够真正解决的物理问题却并不很多.例如,一根棒的引力场就没有能够算出来.他感到把时空复杂化并不是最好的办法.在一次国际会议上他提出质疑,然后就提出了自己的看法,他认为,引力场的背景时空仍然应是平直的,引力势则是由爱因斯坦方程和谐和条件(坐标条件)所决定的.这样,就能避免因坐标选取所引起的混淆.由于广义相对论的坐标无关论的解说(即时空弯曲论)已深入人心,周老的理论是不易为人们所接受的,但他毫不让步,虽已 80 高龄,仍一直写文章,作演讲,培养研究生,反复阐明自己的主张.不仅如此,他还建议做有关的实验,通过实验来核验这项理论.

最近十余年来,我们有幸和周老多次会面.我们都是学数学的,而且学的是微分几何,对于弯曲的时空先入为主,而且习以为常,所以开始时也感到周老的理论不容易理解.经过周老多次的阐述,我们逐步理解了他的想法,感到很有道理.我们认为,他的理论本身在数学上具有相容性,物理图像和概念是清晰的.最近我们通过自己的研究,进一步从数学上明确了周老的引力理论和其他物理理论一样,是洛仑兹变换下的不变理论.

我们希望周老所设计的实验能够取得成功,也希望周老的理论有利于研究自然界四种基本作用力的统一(即电磁作用、弱相互作用、强相互作用和引力相互作用的统一性).

第二件事是周老在"文革"那种极其困难情形下,仍然坚持主张自然科学基础理论的重要性,和"四人帮"一伙做坚决的斗争.

在理科的教学和科研工作中,周老一贯重视基础理论.1972 年周总理接见杨振宁教授,周老参加了这次接见.这是杨振宁教授第二次回国访问.杨振宁教授想起了访问上海、北京时所见到的一些情况,就非常坦率地说:"中国在教育科研中重视理论和实际的结合,这是很好的,在经济比较落后的条件下,这也是必需的.但是目前中国理工科大学不重视基础教学和理论研究,这是目光短浅的表现,应引起重视.在科研机关里,也存在着不重视理论的倾向……"周总理对杨振宁教授的意见非常重视.同年 7 月 14 日周总理在接见以任之恭为团长、林家翘为副团长的美国华裔科学家代表团时，当面指示周老要把北京大学的理科办好，把基础理论提高.周老根据毛主席和周总理一贯重视自然科学基础理论的精神，写了一篇关于综合大学理科教育革命的文章，提倡加强基础研究.周老这篇文章完全是针对当时“四人帮”的，针对着当时占压倒地位的极“左”思潮的.本文作者之一在那时被安排讲授线性代数，在讲到线性代数方程组理论时，就有人认为是脱离实际的，并且只许讲对称阵的情况，要讲一般阵，就要受批判，原因是在有限元计算中只出现对称阵，因而讲非对称阵的线性方程组就是脱离实际.这样的事现在可以当笑话来讲，在当时却是现实.周老写这篇文章是要有很大的勇气的.文章发表以后，基础理论的教学和科研情况果然有了一些起色，就是当时在“四人帮”严密控制的复旦大学，也有人增添了某些理论教学内容，基础理论研究也有恢复的倾向.但是，周老的文章触怒了“四人帮”一伙，他们妄图以此为借口既批判周老，又影射攻击周总理.周老在文章中以300年前的微积分的研究为例，说明对当前的生产尚无直接作用的理论课题，也应予以足够的重视.“四人帮”一伙有意歪曲，强令一些数学工作者查阅科学史资料，又组织了连篇累牍的文章，定下调子，以微积分有生产实践的来源和直接的应用为借口来批判所谓“300年后有用论”，造成一股围攻的逆流.不仅如此，这一伙人还胡说什么自然科学的基础理论就是唯物辩证法，混淆哲学和自然科学的界线(其实这是恩格斯早就区别清楚了的)，企图进一步挥舞“大批判”的大棒来攻击周老.我们当时身在上海，不得和外界联系，没有周老的信息.1974年周如玲(周老的女儿，当时她在中科院物理所工作，我们原来不认识她)来上海，希望能利用复旦大学计算机算题，到我们家里说了周老的困难处境和他坚持真理、毫不妥协的情况，给了我们很深刻的印象.当时我们为周如玲安排了上机的时间，但始终不敢透露她和周老的关系，以免惹起麻烦.“四人帮”被粉碎后，周老在《红旗》杂志上发表文章进一步阐述自己的观点并对“四人帮”进行了揭露批判.这一事件的过程充分说明了周老对自然科学基础理论的重要性有着深刻的体会，并在最困难的情况下，为维护正确的观点做不懈的斗争.这种精神是永远值得我们学习的.

周老对于基础理论研究，不仅有他的明确主张，而且有一系列高水平、高质量的研究成果.他身体力行,除引力论外,他对极其困难的湍流理论,迄今仍继续进行研究,继续培养研究生,对他们的学习和工作予以亲切的关怀,这使我们非常感动.

周老已 90 高龄,仍然担负着多方面的非常繁重的任务.我们衷心地希望他多多保重身体,健康长寿,继续为人民做出更大的贡献.

(本文曾发表于《科学巨匠　师表流芳》,中国科学技术出版社出版,1992 年,109 - 111)

在纪念华罗庚教授诞辰100周年活动上的讲话

胡和生

尊敬的各位领导、各位来宾：

大家好.我代表谷超豪与本人作一个发言.

我们怀着十分崇敬与思念的心情来参加纪念华罗庚教授诞辰100周年活动.

华罗庚教授十分热爱祖国的精神是十分可佩的,新中国成立后,他放弃了在美国的极好的科研条件,毅然回到祖国,献身于祖国的科研和教育事业.

华先生1951年回到中国,对我国的数学发展产生十分巨大而关键的影响,他在数学上的成就,为世界和中国的数学史添写了光辉的篇章.他的学术思想、治学态度,诲人不倦的精神,为人民教育事业鞠躬尽瘁的精神,永远是我们的榜样.

华先生回国后,建立新中国的中科院数学研究所,并担任所长,华先生极其重视并花了大量心血去培养年轻的数学家.由于他在数学上的深厚广博与极高水平,又全心全意地以宽广的胸襟无私地将学问传授给青年一代,他亲自培养了大量的青年数学家,其中包括王元院士、陈景润院士、潘承洞院士、陆启铿院士、万哲先院士等五名院士,他们分别属于数论方向、多复变函数论方向及代数方向.中国科技大学是华先生亲自关怀、培育和成立的一所学校,华先生一开始就担任副校长和亲自担任数学系主任.他主张树立超越式的学风,培养了一大批杰出的优秀人才.我们对华先生对中国数学事业的伟大贡献由衷地佩服与崇敬.

谷超豪在1951年就认识华老了,几十年中,他虽然没有直接在华老身边工作,但有许多和他见面的机会,不断地得到他的关怀、鼓励和教导.20世纪50年代,谷超豪刚刚踏入数学领域时,华老就给他许多鼓励,教导他要学习数学中最深刻、最有意义的成果、方法和思想.见面时,华老总是喜欢向谷超豪提数学问题,了解谷超豪理解问题的深入程度.到了20世纪60年代,谷超豪已经30多岁了,华老就提醒谷超豪要抓紧时间,要努力去创造有特点的数学工具和方法.在华老去世前的十年中,华老多次和谷超豪谈到应该努力使数学为国民经济服务,同时又要努力发展数学的基础理论.虽然谷超豪做得很不够,但这些教导使谷超豪得到了很大的益处,并永远铭记在他的心中,谷超豪自己所未能做到的,会让同事们、学生们去做.另外在参加全国人民代表大会期间,华老也往往带着谷超豪去其他代表处交流,谈天,十分亲切.这些都让谷超豪印象深刻.

下面谈谈华先生给我印象最深的几件事.

1951年,在我跟随苏步青教授做研究生时,经苏先生推荐进入了当时刚成立的以华罗庚教授为所长的中科院数学所工作,仍留在苏步青教授身边,继续在他的指导下从事研究工作.数学所规定我每一季度要书面向华罗庚所长汇报自己的学习和研究,这件事对我的成长起着很大的鞭策和鼓励作用.

1954年我第一次见到华先生,那是中科院举行院士会议,华先生批准我的申请,让我去北京数学所看看,去工作一个月,并批准我列席部分院士会议,这样我才有机会见到华先生,聆听教导及汇报自己的学习和研究.这次华先生和华师母还请我到他家吃饭,第一次认识了华师母,后来我还替华苏、华密拍了照.

另一次,1956年中科院数学所在华先生主持下举行了建所第一次评奖.我后来收到了华罗庚所长亲笔写的一张明信片,通告我这次评奖结果,吴文俊一等奖,冯康二等奖,王光寅与我得到三等奖,并向我祝贺,要我继续努力,这件事使我非常感动,也受到极大的鼓励,鞭策我一辈子努力工作.

1983年,为加强中日数学界的交流,应日方邀请,华先生组织一个中国数学会代表团,赴日本参加日本数学会年会并作特别报告,苏步青教授为团长,王元与我为团员,苏先生说我是华先生提名的,这是华先生对我的信任和培养.

1985年2月22日,我和谷超豪为去欧洲进行学术交流而经过广州,林伟同志告诉我们说华老在广州,我们又得到了一次和他见面的机会,我们去看他时,他兴致很高,说了他最近的工作,又要我们代他向法国的许多学者问好,又教导我们,要努力学习人家的长处,同时也要敢于在强手面前拿出自己的东西进行交流.他一直要说下去,我们怕他太累.当时我们看他精神很好,身体还不错,都为他高兴.但是想不到,这是我们最后一次见到他了.

今天在纪念华罗庚教授诞辰一百周年之时,我们深深地怀念他,他的重大贡献将永留人间,他的伟大精神激励着大家前进.年轻的一代数学家正一批一批地成长.大家正继承华老的遗愿,把发展数学、振兴中华的重任很好地承担起来,正向着把我国建成数学强国的方向前进.

谢谢大家.

怀念徐瑞云先生

胡和生

1950年，全国有9人报考浙江大学苏步青先生、陈建功先生的研究生.最后，我和一对来自广西大学的双胞胎姐妹考取了苏先生的研究生，夏道行考取了陈先生的研究生.

1950年9月我考入浙大后，就有人说我长得很像徐瑞云先生.徐瑞云先生是从德国留学回来的，是中国第一位女数学博士，她在浙大数学系地位很高.徐先生专长函数论，是正教授.当时浙大数学系共有4位正教授，除徐先生外，还有苏步青先生、陈建功先生、卢庆骏先生.我记得我进浙大时，第一年是卢先生任系主任，到了第二年，徐先生任系主任了.

徐先生课教得很好，外文也好，她翻译了很多书，其中有一本《实变函数论》，作者是那汤松.这本书被徐先生翻译后，就变成大家通用的教材.我认为当时连北大都没有像徐先生这样的人才.她教的课也是比较深的，学生要拿到好成绩非常困难.

徐先生看过我入学前写的小传以后，很是欣赏.她是上海人，她先生是江希明教授，是当时浙大生物系系主任.江先生也是留德的博士，人非常好，对我也非常好.他们结婚后没有孩子，徐先生每到周末便会邀请我去她家做客，她说她没有孩子，也没有妹妹，就把我当作自己的妹妹.徐先生家的房子是在浙大外边的，也是浙大最好的，需要爬上山去.当时我自己的二姐在杭州工作，所以我逢周末看过二姐后，就会去徐先生家.每次都会被徐先生留下来，在她家吃午饭，并呆上半天，她会和我谈对数学的看法以及系里的事情.徐先生是一个非常健谈的人，话很多，我想这可能也是徐先生很早就做数学系领导的原因之一.徐先生对每个学生都很热情，都很关心，学生们都很喜欢她.

1952年全国高校院系调整，我随苏先生到了复旦大学，陈先生和浙大数学系的一小部分成员也调入复旦大学，徐先生则被调到浙江师范学院，做系主任.当时正是思想改造运动之后，大家不可能有其他选择，只有服从分配.

我到了上海复旦大学后，经常给徐先生写信或者打电话，每到春节都会给徐先生寄贺年卡.

1957年，高教部在复旦大学试点新中国成立后的第一次博士招考，我和谷超豪、夏道行、龚昇报名了考试.这次考试除了要考业务课，还要考政治、哲学和外文.我为了静心准备，不想呆在上海，于是便给徐先生打电话，询问可否去杭州她家住一段时间？徐先生同意我的请求，热情欢迎我去她家复习、备考.之后，我就去杭州，在徐先生家住了差不多十天.在这段时间里，我得到了徐先生的热情招待和多番鼓励.在我自己的努力和徐先生的关心下，我在那次考试中取得了非常优异的成绩，后来高教部没有同意复旦举行博士论文答辩，我最后也没有取得博士学位.但我永远忘不了徐先生对我的帮助.

1958年开始，一股否定与批判基础研究的左倾思潮干扰了数学界，在复旦数学系，陈建功先生与我成为了批判对象.陈先生后来被任命为杭州大学副校长，徐先生那时是杭州大学数学系系主任.1966年，"文化大革命"象暴风骤雨那样席卷而来，我在复旦大学受到批斗，同时听说陈先生和徐先生在杭州大学不断受批判，徐先生不仅被打成了反动学术权威，还背上了走资派的罪名.后来我听人说陈先生解放了，徐先生还未解放.我心里想徐先生应该也快解放了吧，但不久噩耗传来，徐先生白天在校内被反绑着双手批斗，晚上回到家后，跟随了她几十年的老保姆也批斗她，徐先生不堪如此凌辱，愤而自杀.我深为震惊和哀痛.

徐瑞云先生是对我国数学发展有重要贡献的前辈学者，在她诞辰一百周年之际，我深切地怀念这位热爱祖国、品德高尚的女数学家.现在浙大数学系人才辈出，学术上一片欣欣向荣、繁荣昌盛，这是对徐先生最好的纪念.

（本文曾发表于《中国数学会通讯》2015年第2期）

为什么要好好学习数学

胡和生

我们在日常生活中处处需要数学.有时用的是很简单的数学,如加减乘除、比较大小等.我们还会遇到一些复杂的数学问题.譬如理财,你必须考虑收益和风险,往往预期收益越高风险越大,哪个方案对你合适,就要有概率论的思想.又譬如,你在做家务时先做什么后做什么,哪些事情可以同时进行,必须统筹安排才能够节省时间,这是数学中运筹学的内容.现在生产自动化的程度越来越高,计算机用得越来越多,没有一定的数学水平就不能适应这个时代.

数学是一门非常重要的科学,而且越来越重要.它是各门学科的基础和工具.就工程科学而言,我们知道桥梁、汽车和飞机的设计必须进行很复杂的计算.我国载人航天飞船上天,从外形设计、轨道计算到飞行控制都会遇到许多复杂的数学问题,需要用各种数学工具去解决.现代医学也非常需要数学,譬如CT、B超、核磁共振等先进的医学仪器就涉及许多数学问题.物理学、化学、生物学等基础学科更是完全离不开数学.

社会科学里也要用到数学.要学好经济学就要用到高深的数学.诺贝尔经济学奖得主往往是很好的数学家.2002年来中国访问的纳什就是一个例子.他是美国科学院院士,在基础数学研究中有很大的贡献,他同时也致力于研究经济学,终于得到了诺贝尔经济学奖.美国电影《美丽心灵》就是以纳什为原型拍摄的.

总之,在现代社会中,无论从事哪一种行业的工作都需要大量的深入的数学知识.对中学生来说,学好数学是非常必要的,对于今后的学习和工作必然大有益处.

中国这几十年来数学有很大的进步.研究数学的人数比以前大大增加了,水平也有很大的提高,但是和美国、法国、德国、俄罗斯相比还有很大的差距.过去我们常说要成为一个数学大国,也就是说要有很多高水平的数学家出现.我国人口很多,现在的数学水平也不错,大家重视起来努力一下,要达到这个目标还是相当有把握的.现在我们又提出要成为一个数学强国,这是更艰巨的任务.要成为一个数学强国就要有相当多的、水平非常高的、非常突出的世界公认的大数学家,他们有相当多的创造性很强的研究成果,能带动世界数学的发展和帮助其他各门学科进步.这个任务就落在青年人的身上.

我小时候是抗日战争时期,国家受侵略,社会不安定,基本生活甚至连生命安全都没有保障.新中国成立以后,我到浙江大学做了研究生,努力发奋学习.我的老师是苏步青教授,他认为我是个可造就的人才,十分重视对我的培养.苏先生对我的要求非常严格,也非常爱护我,给我做很多事情,而且放手让我做,这也是培养我的一种方式.当时国家经济不发达,很少对外交流,基本上是关起门来自己干.现在大家的条件优越得多了,每年有很多水平很高的专家来讲学,取得研究成绩就会有各种形式的奖励,成才后到国外交流的机会很多,但关键还在于自己的努力.那么怎样才能学好数学呢?

第一点就是要对数学产生兴趣.有人认为数学难,所以没兴趣.实际上数学是一门很有趣味的学科,你解决一个个困难,对数学内容有越来越多的体会时,你的兴趣就一步步提高了.所以要钻进去不要怕难,才会产生兴趣.

第二点就是要肯吃苦.深入钻研的确是非常苦的.你不仅仅要会代公式会算题,更重要的是自己能理解、会思考,能抓住最根本的东西.做研究就是要做前人没有做到的事.有时候做一项研究三个月或者半年都没有进展,这是非常艰苦的阶段,我们一定要不怕困难,坚持下去,深入地弄清楚问题的实质和关键,然后就能想办法解决.

第三点就是要自信,不要看到困难就退缩了.攀登科学的高峰会碰到一个又一个困难,我们就是在克服困难中走向成功之路.要相信自己是有能力的,一定能够克服困难,这样就能全力以赴.

最后要看到自己的弱点或者最需要改进的地方,坚决下决心改进.我做研究生的时候,自己立下 16 个字的座右铭:“专心致志,刻苦钻研,持之以恒,不受干扰”.那个时候受到形形色色的干扰是相当多的,我依据这个座右铭完全专心地、非常刻苦地钻研数学并且不断取得进步.现在经济情况好,好玩的东西比较多,赚钱的机会也多了,这样考虑学问的时间就少了，所以一定要排除这些干扰.此外，还可能有很多别的干扰，如受到批评、办事不太顺利、考得不好等.你一定要不闹情绪，专心致志，持之以恒，好好学习，那么你一定会取得成功.

（本文曾发表于《科学画报》2013年第8期）

第二部分

数学的征途

共轭的仿射联络的扩充

胡和生

中国科学院数学研究所，复旦大学

一、绪　　言

诺而勤(Норден)[1]研究仿射联络空间 X_n 的共轭对 G^i_{jk}，Γ^i_{jk}；所用的基础是在 X_n 的各点的配极 b_{ij}，设两方向场 v^i，w^j，满足条件

$$b_{ij}v^i w^j = 0,$$

那么这两方向称为共轭. 当两共轭方向沿一曲线 L 顺次在联络 G^i_{jk}，Γ^i_{jk} 下平行移动后，共轭的性质依旧保留，称这两联络关于 b_{ij} 是共轭联络.

本文的目的是关于两共轭联络的扩充，就是把诺而勤的研究推广到 m 个联络的场合而导出可能扩充的部分.

已知 m 个联络 $\underset{1}{\Gamma}$，$\underset{2}{\Gamma}$，…，$\underset{m}{\Gamma}$ 和一个对称的拟似张量 $b_{i_1 i_2 \cdots i_m}$ 的时候；凡满足

$$b_{i_1 i_2 \cdots i_m} v^{i_1}_{(1)} v^{i_2}_{(2)} \cdots v^{i_m}_{(m)} = 0$$

的 m 个方向 $v^i_{(1)}$，…，$v^i_{(m)}$ 经过各联络 $\underset{v}{\Gamma}$ 的平行移动之后仍旧满足同一关系；这样定义的 m 个联络称为 m 重共轭的.

首先推广混合微分的概念，且应用于 m 重共轭联络. 为方便计，仅叙述 $m=3$ 的情况，必要时对一般场合也加以证明.

必须指出，关于共形对应的合成变换的诺而勤的研究能否推广到一般情况，现尚未确定.

二、混合微分的推广

假定向量 $\underset{1}{x}, \underset{2}{x}, \cdots; \underset{1}{y}, \underset{2}{y}, \cdots; \cdots; \cdots$ 是属于 n 个不同的空间,而这些空间的度数一般可以不等,则对于多重线形函数

$$w=f(\underset{1}{x}, \underset{2}{x}, \cdots; \underset{1}{y}, \underset{2}{y}, \cdots; \underset{1}{z}, \underset{2}{z}, \cdots; \cdots)$$

我们说是对应了中间张量.

对于中间张量运算的定义,是与普通运算相同,但只有在两个张量的对应指数均属于同一空间时才能相加,而对张量的缩短,对称化,反称化也只能对于同一空间中的指数才能施行.

现在来讨论混合张量的绝对微分.为方便计,就 $n=3$ 来叙述.

假定有三个空间 X_l, X_m, X_n,且假定 $l \geqq m \geqq n$,它们的点与点之间建立了可微分对应,使对 X_m 中一个点一定有 X_l 中一点与之对应.X_n 中的点有 X_m 中的点及 X_l 中的点与之对应,如果 $l=m=n$,则成立了一对一的对应.

现在我们在 X_n 中来讨论中间张量 $A_{\lambda a i \cdots}$ 之野.需注意者,以后对 X_l 中的指数以 λ, μ, ν, $\cdots$ 来表示,而 $a, b, c, \cdots$ 及 $i, j, k, \cdots$ 分别代表 X_m 及 X_n 中的指数,根据一意的对应,可将野看为仅属于 X_l,也可将它看为仅属于 X_m 或 X_n.

现在假定在 X_l, X_m, X_n 中均定义了仿射联络,对于 X_n 中的任何一条曲线来计算

$$z=A_{\lambda a i \cdots} u^{\lambda} v^{a} w^{i} \cdots$$

的微分,式中

$$u^{\lambda}, \cdots \in X_l; v^{a}, \cdots \in X_m; w^{i}, \cdots \in X_n.$$

且假定方向 u^{λ}, v^{a}, w^{i}, $\cdots$ 沿对应曲线为平行移动,假定这些平行移动的条件为

$$du^{\lambda}=-\omega_{\nu}^{\lambda} u^{\nu}; dv^{a}=-\omega_{b}^{a} v^{b}; dw^{i}=-\omega_{k}^{i} u^{k},$$

于是

$$
\begin{aligned}
dz &= dA_{\lambda a i\cdots} u^{\lambda} v^{a} w^{i} \cdots - A_{\lambda a i\cdots} \omega^{\lambda}_{\nu} u^{\nu} v^{a} w^{i} \\
&\quad - A_{\lambda a i\cdots} \omega^{a}_{b} u^{\lambda} v^{b} w^{i} - A_{\lambda a i\cdots} \omega^{i}_{j} u^{\lambda} v^{a} w^{j} - \cdots \\
&= (dA_{\lambda a i\cdots} - \omega^{\mu}_{\lambda} A_{\mu a i\cdots} - \omega^{b}_{a} A_{\lambda b i\cdots} - \omega^{j}_{i} A_{\lambda a j\cdots} - \cdots) u^{\lambda} v^{a} w^{i}.
\end{aligned}
$$

记

$$
\delta A_{\lambda a i\cdots} = dA_{\lambda a i\cdots} - \omega^{\mu}_{\lambda} A_{\mu a i\cdots} - \omega^{b}_{a} A_{\lambda b i\cdots} - \omega^{j}_{i} A_{\lambda a j\cdots} - \cdots
$$

且称为混合的绝对微分.

今导入空间的曲线坐标 $X_l: x^{\lambda}$, $X_m: y^{a}$, $X_n: z^{i}$，则

$$
x^{\lambda} = x^{\lambda}(z^1, \cdots, z^n), \quad y^{a} = y^{a}(z^1, \cdots, z^n).
$$

记

$$
\frac{\partial x^{\lambda}}{\partial z^{i}} = x^{\lambda}_{i}, \quad \frac{\partial y^{a}}{\partial z^{i}} = y^{a}_{i},
$$

且

$$
\begin{gathered}
\omega^{\mu}_{\lambda} = H^{\mu}_{\nu\lambda} dx^{\nu}; \quad \omega^{b}_{a} = G^{b}_{ca} dy^{c}; \quad \omega^{j}_{i} = \Gamma^{j}_{ki} dz^{k}; \\
\delta A_{\lambda a i\cdots} = (\partial_k A_{\lambda a i\cdots} - H^{\mu}_{\nu\lambda} X^{\nu}_{k} A_{\mu a i\cdots} - G^{b}_{ca} y^{c}_{k} A_{\lambda b i\cdots} - \Gamma^{j}_{ki} A_{\lambda a j\cdots} - \cdots) dz^{k}; \\
\nabla_k A_{\lambda a i\cdots} = \partial_k A_{\lambda a i\cdots} - H^{\mu}_{\nu\lambda} X^{\nu}_{k} A_{\mu a i\cdots} - G^{b}_{ca} y^{c}_{k} A_{\lambda b i\cdots} - \Gamma^{j}_{ki} A_{\lambda a j\cdots} - \cdots.
\end{gathered}
$$

称为 $A_{\lambda a i\cdots}$ 的混合共变微分.

此处 k 是属于度数最低的空间 X_n 的共变微分，由混合微分的定义可知运算是仍保留着普通共变微分的法则.

当 X_l, X_m, X_n 度数相同而空间合而为一时，

$$
\nabla_k A_{\lambda a i\cdots} = \partial_k A_{\lambda a i\cdots} - H^{\mu}_{k\lambda} A_{\mu a i\cdots} - G^{b}_{ka} A_{\lambda b i\cdots} - \Gamma^{j}_{ki} A_{\lambda a j\cdots} - \cdots.
$$

下面为明显起见，不以文字来区别不同的联络，而以对联络 G^{i}_{jk} 的共变微分用(.) 记之，对联络 Γ^{i}_{jk} 的共变微分在其下加一划对 H^{i}_{jk} 的共变微分不加记号，则得

$$
\nabla_k (A^{\cdots\cdots}_{i\cdots} B^{i\cdots}_{\cdots\cdots}) = \nabla_k (A^{\cdots\cdots}_{(i)\cdots} B^{(i)\cdots}_{\cdots\cdots}) = \nabla_k (A^{i\cdots}_{\cdots\cdots} B^{\cdots\cdots}_{\underline{i}\cdots}).
$$

即在合并的时候,缩短的指数无论把它看作属于哪个联络,仍不会影响到微分的结果.

同时应用这些符号来表示关于不同联络的共变微分,例如,

$$\nabla_s b_{ijk} = \partial_s b_{ijk} - H^m_{si} b_{mjk} - H^m_{sj} b_{imk} - H^m_{sk} b_{ijm},$$

$$\nabla_s b_{i(j)(k)} = \partial_s b_{ijk} - H^m_{si} b_{mjk} - G^m_{sj} b_{imk} - G^m_{sk} b_{ijm},$$

$$\nabla_s b_{i(j)\underline{k}} = \partial_s b_{ijk} - H^m_{si} b_{mjk} - G^m_{sj} b_{imk} - \Gamma^m_{sk} b_{ijm},$$

等等.

此处再导入 H, G, Γ 的平均联络,它的系数为

$$z^k_{ij} = \frac{1}{3}(H^k_{ij} + G^k_{ij} + \Gamma^k_{ij}),$$

而对于平均联络的共变微分,则采用下面的写法:

$$\overset{c}{\nabla}_s b_{ijk} = \partial_s b_{ijk} - z^m_{si} b_{mjk} - z^m_{sj} b_{imk} - z^m_{sk} b_{ijm}.$$

显然的,对于平均联络,我们没有区分指数所在区域的必要,并且可知

$$\begin{aligned}\overset{c}{\nabla}_s b_{ijk} &= \frac{1}{3}(\nabla_s b_{i(j)\underline{k}} + \nabla_s b_{j(k)\underline{i}} + \nabla_s b_{k(i)\underline{j}})\\ &= \frac{1}{3}(\nabla_s b_{ijk} + \nabla_s b_{(i)(j)(k)} + \nabla_s b_{\underline{ijk}}).\end{aligned}$$

三、共 轭 联 络

在仿射联络空间 X_n 中已知对称的拟似张量 b_{ijk} 的一场,如果在这空间的每点,对于此点发生的三个方向满足条件

$$b_{ijk} u^i v^j w^k = 0, \tag{1}$$

则称为 X_n 中已决定了方向有关的三重配极,而 u^i, v^j, w^k 三个方向谓之三重共轭的.

现在再来定义共轭联络:在空间 X_n 中已知三个联络 H, G, Γ,联络系数为 H^i_{jk}, G^i_{jk}, Γ^i_{jk}.如果有三个方向 u^i, v^j, w^k 关于已知的 b_{ijk} 是共轭的,且 u^i, v^j, w^k 的方向沿

一条曲线L顺次在联络H，G，Γ下平行移动，而这三重配极的性质依旧成立，则这三联络称为关于b_{ijk}是三重共轭的联络.

首先讨论三重共轭联络的条件，对于方向u^i，v^j，w^k的平行移动有如下的关系：

$$\delta u^i = \lambda u^i, \quad \delta v^{(j)} = \mu v^j, \quad \delta w^{\underline{k}} = v w^k. \tag{2}$$

由于经过方向的平行移动后(1)仍保留，我们得到

$$\delta b_{i(j)\underline{k}} u^i v^j w^k + b_{ijk} \delta u^i v^j w^k + b_{ijk} u^i \delta v^{(j)} w^k + b_{ijk} u^i v^j \delta w^{\underline{k}} = 0.$$

由(2)得到

$$\delta b_{i(j)\underline{k}} u^i v^j w^k = 0.$$

此为(1)的必然结论，但u^i，v^j，w^k是任意的，只要满足(1).故知

$$\delta b_{i(j)\underline{k}} = \alpha b_{ijk},$$

亦即

$$\nabla_s b_{i(j)\underline{k}} d u^s = \alpha b_{ijk}, \quad \alpha = 2\omega_s d u^s,$$

ω_s称为辅助张量.另一方面，由于对于任何道路L是无关的，于是得到由联络H，G，Γ所定义的共轭联络的特征为

$$\nabla_s b_{i(j)\underline{k}} = 2\omega_s b_{ijk}. \tag{3}$$

由于b_{ijk}是拟似张量，$b_{ijk}u^i v^j w^k = 0$决定了一个配极，同时$\lambda^2 b_{ijk} u^i v^j w^k = 0$也决定了一个配极.现在来讨论对于规范化$\check{b}_{ijk} = \lambda^2 b_{ijk}$，$\omega_s$起了什么变化.

$$\nabla_s \check{b}_{i(j)\underline{k}} = \nabla_s (\lambda^2 b_{ijk}) = 2\check{b}_{ijk} (\omega_s + \partial_s \log \lambda) = 2\check{\omega}_s \check{b}_{ijk}.$$

故经规范化后辅助张量受到下列变换：

$$\check{\omega}_s = \omega_s \omega_s + \partial_s \log \lambda, \tag{4}$$

即规范化前后辅助张量之差为斜度.

现在作变形张量

$$H_{ij}^{k}-\Gamma_{ij}^{k}=\underset{1}{t}{}_{ij}^{k},\quad G_{ij}^{k}-\Gamma_{ij}^{k}=\underset{2}{t}{}_{ij}^{k}. \tag{5}$$

因此 $H_{ij}^{k}-G_{ij}^{k}=\underset{1}{t}{}_{ij}^{k}-\underset{2}{t}{}_{ij}^{k}$.以下方便上记$\underset{1}{t}{}_{ij}^{k}-\underset{2}{t}{}_{ij}^{k}=(\underset{1}{t}-\underset{2}{t})_{ij}^{k}$. 利用基本关系(3),易得

$$\begin{aligned}\nabla_{s}b_{ijk}-\nabla_{s}b_{i(j)\underline{k}}&=\nabla_{s}b_{ijk}-2\omega_{s}b_{ijk}\\&=(G_{sj}^{m}-H_{sj}^{m})b_{imk}+(\Gamma_{sk}^{m}-H_{sk}^{m})b_{ijm}\\&=(\underset{2}{t}-\underset{1}{t})_{sj}^{m}b_{imk}-\underset{1}{t}{}_{sk}^{m}b_{ijm}.\end{aligned}$$

同样的获得

$$\nabla_{s}b_{(i)(j)(k)}-2\omega_{s}b_{ijk}=(\underset{1}{t}-\underset{2}{t})_{si}^{m}b_{mjk}-\underset{2}{t}{}_{sk}^{m}b_{ijm},$$

$$\nabla_{s}b_{\underline{i}\underline{j}\underline{k}}-2\omega_{s}b_{ijk}=\underset{1}{t}{}_{si}^{m}b_{mjk}+\underset{2}{t}{}_{sj}^{m}b_{imk}.$$

此处必需注意,由于对称的性质及不同联络的关系,属于同一联络的指数对调是无影响的,而属于不同联络的指数不能对调.

$$\begin{aligned}\nabla_{s}b_{ijk}-2\omega_{s}b_{ijk}&=(\underset{2}{t}-\underset{1}{t})_{sj}^{m}b_{imk}-\underset{1}{t}{}_{sk}^{m}b_{ijm},\\k\longleftrightarrow i\qquad&=(\underset{2}{t}-\underset{1}{t})_{sj}^{m}b_{imk}-\underset{1}{t}{}_{si}^{m}b_{kjm},\\j\longleftrightarrow k\qquad&=(\underset{2}{t}-\underset{1}{t})_{sk}^{m}b_{imj}-\underset{1}{t}{}_{si}^{m}b_{kjm},\\i\longleftrightarrow j\qquad&=(\underset{2}{t}-\underset{1}{t})_{sk}^{m}b_{imj}-\underset{1}{t}{}_{sj}^{m}b_{kim}.\end{aligned}$$

比较上列四式的结果,

$$\left.\begin{aligned}\underset{1}{t}{}_{si}^{m}b_{mjk}=\underset{1}{t}{}_{sj}^{m}b_{imk}=\underset{1}{t}{}_{sk}^{m}b_{ijm},\\\underset{2}{t}{}_{si}^{m}b_{mjk}=\underset{2}{t}{}_{sj}^{m}b_{imk}=\underset{2}{t}{}_{sk}^{m}b_{ijm}.\end{aligned}\right\} \tag{6}$$

于是得到

$$\left.\begin{aligned}\nabla_{s}b_{ijk}-2\omega_{s}b_{ijk}&=(\underset{2}{t}-2\,\underset{1}{t})_{si}^{m}b_{mjk},\\\nabla_{s}b_{(i)(j)(k)}-2\omega_{s}b_{ijk}&=(\underset{1}{t}-2\,\underset{2}{t})_{si}^{m}b_{mjk},\\\nabla_{s}b_{\underline{i}\underline{j}\underline{k}}-2\omega_{s}b_{ijk}&=(\underset{1}{t}+\underset{2}{t})_{si}^{m}b_{mjk}.\end{aligned}\right\} \tag{7}$$

一般讨论 m 重共轭的时候,m 个方向$v_{(1)}^{i}$, …, $v_{(m)}^{i}$ 在联络 $\underset{1}{\Gamma}$, $\underset{2}{\Gamma}$, …, $\underset{m}{\Gamma}$下平行移动

保存下列关系：

$$b_{i_1 i_2 \cdots i_m} v^{i_1}_{(1)} v^{i_2}_{(2)} \cdots v^{i_m}_{(m)} = 0;$$

我们以 $m-1$ 个变形张量

$$\underset{1}{t}{}^k_{ij} = \underset{1}{\Gamma}{}^k_{ij} - \underset{m}{\Gamma}{}^k_{ij},\ \underset{2}{t}{}^k_{ij} = \underset{2}{\Gamma}{}^k_{ij} - \underset{m}{\Gamma}{}^k_{ij},\ \cdots,\ \underset{m-1}{t}{}^k_{ij} = \underset{m-1}{\Gamma}{}^k_{ij} - \underset{m}{\Gamma}{}^k_{ij}$$

来进行讨论，得着

$$\begin{aligned}&\nabla_s b_{(i_1 i_2 \cdots i_m)_r} - 2\omega_s b_{i_1 i_2 \cdots i_m} \\ &= (\underset{1}{t} + \underset{2}{t} + \cdots + \underset{r-1}{t} + \underset{r+1}{t} + \cdots + \underset{m-1}{t} - (m-1)\underset{r}{t})^m_{si_1} b_{mi_2 \cdots i_m}.\ (r = 1,\ 2,\ \cdots,\ m-1)\end{aligned} \tag{7'}$$

此处 $\nabla_s b_{(i_1 i_2 \cdots i_m)_r}$ 表示对第 r 个联络 $\underset{r}{\Gamma}{}^k_{ij}$ 的共变微分.

又得

$$\nabla_s b_{(i_1 i_2 \cdots i_m)_m} - 2\omega_s b_{i_1 i_2 \cdots i_m} = (\underset{1}{t} + \underset{2}{t} + \cdots + \underset{m-1}{t})^m_{si_1} b_{mi_2 \cdots i_m}. \tag{7''}$$

现在回到三个联络的情形.由于 b_{ijk} 是对称的，从(6)得到

$$\nabla_s b_{i(j)\underline{k}} = \nabla_s b_{j(k)\underline{i}} = \nabla_s b_{k(i)\underline{j}} = 2\omega_s b_{ijk}.$$

因为对于上节所讨论的平均联络成立

$$\overset{c}{\nabla}_s b_{ijk} = 2\omega_s b_{ijk}, \tag{8}$$

所以三重共轭的平均联络与共轭联络有相同的基本张量及辅助张量.

现在要解出 $\underset{1}{t}{}^k_{ij}$，$\underset{2}{t}{}^k_{ij}$.为此，首先固定 b_{ijk} 中的一个指数，例如 $k=1$，且假定行列式 $|b_{ij1}| \neq 0$，于是求得逆张量 $\tilde{b}^{ij1}$.由于 b_{ijk} 是对称的，故 $\tilde{b}^{ij1}$ 也是对称的，而且

$$\nabla_s b_{ij1} - 2\omega_s b_{ij1} = (\underset{2}{t} - 2\underset{1}{t})^m_{sj} b_{mi1}.$$

两边乘以 $\tilde{b}^{ir1}$，则得

$$\left.\begin{aligned}&\tilde{b}^{mr1} \nabla_s b_{mj1} - 2\omega_s \delta^r_j = (\underset{2}{t} - 2\underset{1}{t})^r_{sj}, &(\mathrm{I})\\ &\tilde{b}^{mr1} \nabla_s b_{(m)(j)(1)} - 2\omega_s \delta^r_j = (\underset{1}{t} - 2\underset{2}{t})^r_{sj}. &(\mathrm{II})\end{aligned}\right\} \tag{9}$$

同样固定 b_{ijk} 中的 k，且 $k=2, 3, \cdots, n$，可以得到：

$$\left.\begin{array}{l}\left.\begin{array}{l}\tilde{b}^{mr2}\nabla_s b_{mj2}-2\omega_s\delta_j^r=(\underset{2}{t}-2\underset{1}{t})^r_{sj},\\ \cdots\cdots,\\ \tilde{b}^{mrn}\nabla_s b_{mjn}-2\omega_s\delta_j^r=(\underset{2}{t}-2\underset{1}{t})^r_{sj};\end{array}\right\}(\mathrm{I})\\ \left.\begin{array}{l}\tilde{b}^{mr2}\nabla_s b_{(m)(j)(2)}-2\omega_s\delta_j^r=(\underset{1}{t}-2\underset{2}{t})^r_{sj},\\ \cdots\cdots,\\ \tilde{b}^{mrn}\nabla_s b_{(m)(j)(n)}-2\omega_s\delta_j^r=(\underset{1}{t}-2\underset{2}{t})^r_{sj};\end{array}\right\}(\mathrm{II})\end{array}\right\}\tag{9'}$$

将(9, Ⅰ)与(9′, Ⅰ)各式比较，得

$$\tilde{b}^{mr1}\nabla_s b_{mj1}=\tilde{b}^{mr2}\nabla_s b_{mj2}=\cdots=\tilde{b}^{mrn}\nabla_s b_{mjn},$$

同样由(9, Ⅱ)及(9′, Ⅱ)

$$\tilde{b}^{mr1}\nabla_s b_{(m)(j)(1)}=\tilde{b}^{mr2}\nabla_s b_{(m)(j)(2)}=\cdots=\tilde{b}^{mrn}\nabla_s b_{(m)(j)(n)}.$$

为了保证 $\underset{1}{t}$，$\underset{2}{t}$ 有解，b 必须满足上列两套条件.

将(9, Ⅰ)及(9′, Ⅰ)的各式相加，再除以 n，可得

$$\left.\begin{array}{ll}\frac{1}{n}\tilde{b}^{mrh}\nabla_s b_{mjh}-2\omega_s\delta_j^r=(\underset{1}{t}-2\underset{2}{t})^r_{sj}, & (\mathrm{I})\\ \frac{1}{n}\tilde{b}^{mrh}\nabla_s b_{(m)(j)(h)}-2\omega_s\delta_j^r=(\underset{1}{t}-2\underset{2}{t})^r_{sj}. & (\mathrm{II})\end{array}\right\}\tag{10}$$

但 h 是哑指数.

由(10)可解 $\underset{1}{t}$，$\underset{2}{t}$，实际上，除了辅助张量 w_k 以外，完全可以决定这些量. 为此，添加下列条件：

$$\underset{1}{t}{}^k_{ki}=\underset{1}{t}{}^k_{ik},\quad \underset{2}{t}{}^k_{ki}=\underset{2}{t}{}^k_{ik}.\tag{11}$$

于是获得

$$\omega_s=\frac{1}{n(n-1)}\tilde{b}^{mih}\nabla_{[s}b_{i]mh},\tag{12}$$

但

$$\nabla_{[s} b_{i]mh} = \nabla_s b_{imh} - \nabla_i b_{smh}.$$

同样将(11)代入(7)可得

$$\nabla_{[s} b_{i]jk} = \nabla_{[s} b_{(i)](j)(k)} = \nabla_{[s} b_{\underline{i}]\underline{j}\underline{k}} = 2\omega_{[s} b_{i]jk}. \tag{13}$$

故当给予条件(11)时，三重共轭联络必须满足(13).这些都可以看成柯达齐(Codazzi)普遍方程在所论情形的推广.

现将(12)代入(10)，得

$$\left.\begin{aligned}
(\underset{2}{t} - 2\,\underset{1}{t})^r_{sj} &= \frac{1}{n}\,\tilde{b}^{pqh}\left\{\delta^r_p\,\nabla_s b_{qjh} - \frac{2}{n-1}\,\nabla_{[s} b_{p]qh}\delta^r_j\right\}, && (\mathrm{I})\\
(\underset{1}{t} - 2\,\underset{2}{t})^r_{sj} &= \frac{1}{n}\,\tilde{b}^{pqh}\left\{\delta^r_p\,\nabla_s b_{(q)(j)(h)} - \frac{2}{n-1}\,\nabla_{[s} b_{(p)](q)(h)}\delta^r_j\right\}, && (\mathrm{II})\\
(\underset{1}{t} + \underset{2}{t})^r_{sj} &= \frac{1}{n}\,\tilde{b}^{pqh}\left\{\delta^r_p\,\nabla_s b_{\underline{q}\underline{j}\underline{h}} - \frac{2}{n-1}\,\nabla_{[s} b_{\underline{p}]\underline{q}\underline{h}}\delta^r_j\right\}. && (\mathrm{III})
\end{aligned}\right\} \tag{14}$$

利用(13)及平均联络的关系(8)，我们可以证明(14，Ⅲ)是(14，Ⅰ)和(14，Ⅱ)的推论.由(14)得

$$\left.\begin{aligned}
\underset{2}{t}{}^r_{sj} &= \frac{1}{3n}\,\tilde{b}^{pqh}\left\{-\delta^r_p(\nabla_s b_{qjh} + 2\,\nabla_s b_{(q)(j)(h)}) + \frac{6}{n-1}\,\nabla_{[s} b_{p]qh}\delta^r_j\right\}, && (\mathrm{I})\\
\underset{1}{t}{}^r_{sj} &= \frac{1}{3n}\,\tilde{b}^{pqh}\left\{-\delta^r_p(\nabla_s b_{(q)(j)(h)} + 2\,\nabla_s b_{qjh}) + \frac{6}{n-1}\,\nabla_{[s} b_{(p)](q)(h)}\delta^r_j\right\}, && (\mathrm{II})\\
\underset{2}{t}{}^r_{sj} &= \frac{1}{3n}\,\tilde{b}^{pqh}\left\{\delta^r_p(\nabla_s b_{qjh} + 2\,\nabla_s b_{\underline{q}\underline{j}\underline{h}}) - \frac{6}{n-1}\,\nabla_{[s} b_{p]qh}\delta^r_j\right\}, && (\mathrm{III})\\
\underset{1}{t}{}^r_{sj} &= \frac{1}{3n}\,\tilde{b}^{pqh}\left\{\delta^r_p(\nabla_s b_{(q)(j)(h)} + 2\,\nabla_s b_{\underline{q}\underline{j}\underline{h}}) - \frac{6}{n-1}\,\nabla_{[s} b_{(p)](q)(h)}\delta^r_j\right\} && (\mathrm{IV})
\end{aligned}\right\} \tag{15}$$

显然，从这些方程式可以看出：

在三重共轭联络时，知道二个联络及已给 b_{ijk} 就可以得到其他一个联络，并且可以证明$\underset{1}{t}{}^r_{sj}$，$\underset{2}{t}{}^r_{sj}$ 与规范化无关.

在 m 重共轭联络时，由于固定 $b_{i_1 i_2 \cdots i_m}$ 中的$m-2$个指数，我们应用与三重共轭时同

样的办法,可以得到

$$\left.\begin{aligned}&\tilde{b}^{mra\cdots b}\ \nabla_s b_{(mja\cdots b)_1}-2\omega_s\delta^r_j=(\underset{2}{t}+\underset{3}{t}+\cdots+\underset{m-1}{t}-(m-1)\underset{1}{t})^r_{sj},\\&\cdots\cdots\\&\tilde{b}^{mra\cdots b}\ \nabla_s b_{(mja\cdots b)_{m-1}}-2\omega_s\delta^r_j=(\underset{1}{t}+\underset{2}{t}+\cdots+\underset{m-2}{t}-(m-1)\underset{m-1}{t})^r_{sj}.\end{aligned}\right\}\tag{16}$$

此处 $a\cdots b$ 是1, 2, …, n 中任意 $m-2$ 个指数，而(16)中 $a\cdots b$ 不是哑指数,要保证 $\underset{1}{t}$, $\underset{2}{t}$ 有解,b 必须满足

$$\begin{aligned}&\tilde{b}^{ija\cdots b}\ \nabla_s b_{(ija\cdots b)_1}=\tilde{b}^{ija'\cdots b'}\ \nabla_s b_{(ija'\cdots b')_1},\\&\cdots\cdots\\&\tilde{b}^{ija\cdots b}\ \nabla_s b_{(ija\cdots b)_{m-1}}=\tilde{b}^{ija'\cdots b'}\ \nabla_s b_{(ija'\cdots b')_{m-1}}.\end{aligned}$$

而 $a'\cdots b'$ 亦为1, 2, …, n 中任意 $m-2$ 个指数,于是,$b_{i_1i_2\cdots i_m}$ 满足上列条件而 $b_{i_1i_2\cdots i_m}$ 有 H^n_m 个，此处

$$H^n_m=\frac{(n+m-1)!}{m!\ (n-1)!}.$$

故

$$(m-1)(H^n_{m-2}-1)\leqq H^n_m \tag{17}$$

是为了 b 是有解的必要条件.

利用三重共轭联络时同样的方法,可以得到

$$\omega_s=\frac{1}{n^{m-2}(n-1)}\tilde{b}^{i_1i_2\cdots i_m}\ \nabla_{[s}b_{i_1]i_2\cdots i_m}.$$

代入(7′)(7″)得

$$\begin{aligned}&(\underset{2}{t}+\underset{3}{t}+\cdots+\underset{m-1}{t}-(m-1)\underset{1}{t})^r_{sj}\\&=\tilde{b}^{i_1i_2\cdots i_m}\ \frac{1}{n^{m-2}}\left\{\delta^r_{i_1}\ \nabla_s b_{(i_2ji_3\cdots i_m)_1}-\frac{2}{n-1}\nabla_{[s}b_{i_1]i_2\cdots i_m}\delta^r_j\right\},\end{aligned}$$

$$(\underset{1}{t}+\underset{3}{t}+\cdots+\underset{m-1}{t}-(m-1)\underset{1}{t})^r_{sj}$$

$$=\widetilde{b}^{i_1i_2\cdots i_m}\frac{1}{n^{m-2}}\left\{\delta^r_{i_1}\nabla_s b_{(i_2ji_3\cdots i_m)_2}-\frac{2}{n-1}\nabla_{[s}b_{i_1]i_2\cdots i_m}\delta^r_j\right\},$$

……

$$(\underset{1}{t}+\underset{2}{t}+\cdots+\underset{m-2}{t}-(m-1)\underset{m-1}{t})^r_{sj}$$

$$=\widetilde{b}^{i_1i_2\cdots i_m}\frac{1}{n^{m-2}}\left\{\delta^r_{i_1}\nabla_s b_{(i_2ji_3\cdots i_m)_{m-1}}-\frac{2}{n-1}\nabla_{[s}b_{i_1]i_2\cdots i_m}\delta^r_j\right\}.$$

相加的结果：

$$(\underset{1}{t}+\underset{2}{t}+\cdots+\underset{m-1}{t})^r_{sj}=-\widetilde{b}^{i_1i_2\cdots i_m}\frac{1}{n^{m-2}}\{\delta^r_{i_1}(\nabla_s b_{(i_2ji_3\cdots i_m})_1+\cdots$$
$$+\nabla_s b_{(i_2ji_3\cdots i_m)_{n-1}}-2\nabla_{[s}b_{i_1]i_2\cdots i_m}\delta^r_j\}.$$

将最后一式减去上面任何一式，可以从已知 $m-1$ 个联络及 $b_{i_1i_2\cdots i_m}$ 求得第 m 个共轭联络，故得：

在仿射联络空间 X_n 中，对于 $b_{i_1i_2\cdots i_m}$ 为共轭的 m 个联络 $\underset{1}{\Gamma}$，$\underset{2}{\Gamma}$，…，$\underset{m}{\Gamma}$. 经平行移动后，仍保存其共轭的性质时，当 m 满足

$$(m-1)(H^n_{m-2}-1)\leqq H^n_m$$

时，从已知 $m-1$ 个联络及 $b_{i_1i_2\cdots i_m}$ 可求得与之共轭的第 m 个联络.

四、无挠率的联络

现在讨论 H，G，Γ 均无挠率的情形：

$$H^k_{ij}=H^k_{ji},\quad G^k_{ij}=G^k_{ji},\quad \Gamma^k_{ij}=\Gamma^k_{ji},$$

于是成立

$$\underset{1}{t}{}^k_{ij}=\underset{1}{t}{}^k_{ji},\quad \underset{2}{t}{}^k_{ij}=\underset{2}{t}{}^k_{ji}. \tag{1}$$

从(3.7)容易明了,在已知的无挠率的 H, G, Γ 之下 b_{ijk} 是不能任意选定的,实际上必须成立

$$\left.\begin{aligned}&\nabla_{[s}b_{i]jk}=2\omega_{[s}b_{i]jk},\\&\nabla_{[s}b_{(i)](j)(k)}=2\omega_{[s}b_{i]jk},\\&\nabla_{[s}b_{\underline{i}]\underline{j}\underline{k}}=2\omega_{[i}b_{i]jk}.\end{aligned}\right\}\tag{2}$$

故得:

若三重共轭联络是无挠率的,那么,b_{ijk} 所满足的条件和有挠率共轭联络加上条件(3.11)时是一样的,就是满足柯达齐普遍方程的推广.

关于 s, i 反称化

$$\nabla_s b_{i(j)\underline{k}}-\nabla_s b_{(i)(j)\underline{k}}=(\underset{2}{t}-\underset{1}{t})^m_{si}b_{mjk};$$

因为无挠率,故得

$$\nabla_{[s}b_{(i)](j)\underline{k}}=2\omega_{[s}b_{i]jk}.$$

同样可得

$$\nabla_{[s}b_{i](j)(k)}=\nabla_{[s}b_{i]\underline{j}\underline{k}}=\nabla_{[s}b_{i](j)k}=\nabla_{[s}b_{i]j\underline{k}}=\nabla_{[s}b_{(i)](j)\underline{k}}=\nabla_{[s}b_{\underline{i}](j)\underline{k}}=2\omega_{[s}b_{i]jk}.$$

利用 b_{ijk} 的对称性质,及(3.6)

$$\nabla_s b_{i(j)\underline{k}}=\nabla_s b_{i(k)j}=2\omega_s b_{ijk}.$$

可是

$$\nabla_s b_{i(k)j}-\nabla_s b_{i\underline{j}(k)}=2\omega_s b_{ijk}-\nabla_s b_{i\underline{j}(k)}=-\underset{1}{t}{}^m_{si}b_{mjk},$$

所以

$$\nabla_{[s}b_{\underline{i}]\underline{j}(k)}=2\omega_{[s}b_{i]jk}.$$

其余类推.因此得下列结果:

无挠率的共轭联络中,b_{ijk} 的共变微分,对于有两个联络相同时经反称化后均为相

等，即 b_{ijk} 的共变微分，不论其三个联络是否相同，（包括三个相同，二个相同，均不同）经反称化后皆相等，即满足柯达齐方程的推广.

现在来研究为联络 H，G，Γ 经射影变换后要保存共轭的性质，则辅助张量的变化如何.

射影变换的关系式为：

$$'H_{ij}^{k}-H_{ij}^{k}=\underset{1}{T}{}_{ij}^{k}=\delta_{i}^{k}\sigma_{j}+\delta_{j}^{k}\sigma_{i},$$
$$'G_{ij}^{k}-G_{ij}^{k}=\underset{2}{T}{}_{ij}^{k}=\delta_{i}^{k}p_{j}+\delta_{j}^{k}p_{i},$$
$$'\Gamma_{ij}^{k}-\Gamma_{ij}^{k}=\underset{3}{T}{}_{ij}^{k}=\delta_{i}^{k}\pi_{j}+\delta_{j}^{k}\pi_{i},$$

但 $\sigma_{i}=\frac{1}{n+1}\underset{1}{T}{}_{mi}^{m}$，其余类推.

在变换下，我们得到

$$'\nabla_{[s}b_{i](j)\underline{k}}=\nabla_{[s}b_{i](j)\underline{k}}-T_{j[s}^{m}b_{i]mk}-T_{k[s}^{m}b_{i]jm}$$
$$=2(\omega_{[s}-\frac{1}{2}p_{[s}-\frac{1}{2}\pi_{[s})b_{i]jk},$$

于是

$$'\omega_{s}=\omega_{s}-\frac{1}{2}p_{s}-\pi_{s}.$$

现在我们再来讨论在零挠率时，曲率张量，辅助张量及基本张量间的关系.作

$$z=b_{ijk}u^{i}v^{j}w^{k}$$

微分后得下列形式：

$$\nabla_{s}z=\nabla_{s}b_{i(j)\underline{k}}u^{i}v^{j}w^{k}+b_{ijk}\nabla_{s}u^{i}v^{j}w^{k}+b_{ijk}u^{i}\nabla_{s}v^{(j)}w^{k}+b_{ijk}u^{i}v^{j}\nabla_{s}w^{\underline{k}},$$

上式的右方经微分后为

$$\nabla_{r}\nabla_{s}b_{i(j)\underline{k}}u^{i}v^{j}w^{k}+\nabla_{s}b_{i(j)\underline{k}}\nabla_{r}u^{i}v^{j}w^{k}+\nabla_{s}b_{i(j)\underline{k}}u^{i}\nabla_{r}v^{(j)}w^{k}+\nabla_{s}b_{i(j)\underline{k}}u^{i}v^{j}\nabla_{r}w^{\underline{k}}$$
$$+\nabla_{r}b_{i(j)\underline{k}}\nabla_{s}u^{i}v^{j}w^{k}+b_{ijk}\nabla_{r}\nabla_{s}u^{i}v^{j}w^{k}+b_{ijk}\nabla_{s}u^{i}\nabla_{r}v^{(j)}w^{k}+b_{ijk}\nabla_{s}u^{i}v^{j}\nabla_{r}w^{\underline{k}}$$

$$+\nabla_r b_{i(j)\underline{k}} u^i \nabla_s v^{(j)} w^k + b_{ijk} \nabla_r u^i \nabla_s v^{(j)} w^k + b_{ijk} u^i \nabla_r \nabla_s v^{(j)} w^k + b_{ijk} u^i \nabla_s v^{(j)} \nabla_r w^{\underline{k}}$$

$$+\nabla_r b_{i(j)\underline{k}} u^i v^j \nabla_s w^{\underline{k}} + b_{ijk} \nabla_r u^i v^j \nabla_s w^{\underline{k}} + b_{ijk} u^i \nabla_r v^{(j)} \nabla_s w^{\underline{k}} + b_{ijk} u^i v^j \nabla_r \nabla_s w^{\underline{k}}.$$

将上式对 r, s 反称化,并由于

$$\nabla_{[r} \nabla_{s]} z = \nabla_{[r} \nabla_{(s)]} z = \nabla_{[r} \nabla_{\underline{s}]} z,$$

以及

$$\nabla_{[r} \nabla_{s]} z = 0.$$

故得到

$$\nabla_{[r} \nabla_{s]} b_{i(j)\underline{k}} u^i v^j w^k + b_{ijk} \nabla_{[r} \nabla_{s]} u^i v^j w^k$$
$$+ b_{ijk} u^i \nabla_{[r} \nabla_{s]} v^{(j)} w^k + b_{ijk} u^i v^j \nabla_{[r} \nabla_{s]} w^{\underline{k}} = 0. \tag{3}$$

由于

$$\nabla_{[r} \nabla_{s]} u^m = \frac{1}{2} R_{rsi}{}^m u^i,$$

$$\nabla_{[r} \nabla_{s]} v^{(m)} = \frac{1}{2} \rho_{rsj}{}^m v^j,$$

$$\nabla_{[r} \nabla_{s]} w^{\underline{k}} = \frac{1}{2} \sigma_{rsk}{}^m w^k.$$

此处 $R_{rsi}{}^m$, $\rho_{rsj}{}^m$, $\sigma_{rsk}{}^m$ 分别表示联络 H, G, Γ 的曲率张量,而

$$\nabla_{[r} \nabla_{s]} b_{i(j)k} = 2 \nabla_{[r} (\omega_{s]} b_{i(j)\underline{k}}) = 2 \nabla_{[r} \omega_{s]} b_{ijk} + 2\omega_{[s} \omega_{r]} b_{ijk} = 2 \nabla_{[r} \omega_{s]} b_{ijk}.$$

代入(3),因为 u^i, v^j, w^k 是任意的,于是

$$R_{rsi}{}^m b_{mjk} + \rho_{rsj}{}^m b_{imk} + \sigma_{rsk}{}^m b_{ijm} = -4 \nabla_{[r} \omega_{s]} b_{ijk}. \tag{4}$$

当 m 个无挠率联络时,有如下关系:

$$\underset{1}{R}{}_{rsi_1}{}^h b_{hi_2\cdots i_m} + \underset{2}{R}{}_{rsi_2}{}^h b_{i_1 h i_3\cdots i_m} + \cdots + \underset{m}{R}{}_{rsi_m}{}^h b_{i_1\cdots i_{m-1} h} = -4 \nabla_{[r} \omega_{s]} b_{i_1 i_2\cdots i_m}. \tag{5}$$

但 $\underset{1}{R}{}_{rsi_1}{}^h$, $\underset{2}{R}{}_{rsi_2}{}^h$, $\cdots$, $\underset{m}{R}{}_{rsi_m}{}^h$ 表示关于无挠率共轭联络 Γ_1, Γ_2, $\cdots$, Γ_m 的曲率张量.

现在再讨论三个联络的情形，当 H，G，Γ 相同时，则曲率张量 $R_{rsi}{}^{m}$，$\rho_{rsj}{}^{m}$，$\sigma_{rsk}{}^{m}$ 亦相同，这时方程(5)变为

$$R_{rst}{}^{m}b_{mjk}+R_{rsj}{}^{m}b_{imk}+R_{rsk}{}^{m}b_{ijm}=-4\nabla_{[r}\omega_{s]}b_{ijk}.$$

此方程式亦可由另法求得，记

$$\nabla_{s}z_{ijk}=B_{sijk},$$

将它展开

$$\partial_{s}z_{ijk}-H^{h}_{si}z_{hjk}-H^{h}_{sj}z_{ihk}-H^{h}_{sk}z_{ijh}=B_{sijk},$$

而

$$\nabla_{r}\nabla_{s}z_{ijk}=\nabla_{r}B_{sijk},\tag{6}$$

展开即得

$$\partial_{r}\nabla_{s}z_{ijk}-H^{m}_{sr}\nabla_{m}z_{ijk}-H^{m}_{si}\nabla_{s}z_{mjk}-H^{m}_{rj}\nabla_{r}z_{imk}-H^{m}_{rk}\nabla_{s}z_{ijm}$$
$$=\partial_{r}B_{sijk}-H^{m}_{rs}B_{mijk}-H^{m}_{ri}B_{smjk}-H^{m}_{rj}B_{simk}-H^{m}_{rk}B_{sijm}.\tag{7}$$

所以(6)，(7)是等价的，于是可积条件成为

$$\nabla_{[r}\nabla_{s]}z_{ijk}=\nabla_{[r}B_{s]ijk},$$

即

$$-R_{rsi}{}^{m}z_{mjk}-R_{rsj}{}^{m}z_{imk}-R_{rsk}{}^{m}z_{ijm}=2\nabla_{[r}B_{s]ijk}.$$

当 $\nabla_{s}b_{ijk}=2\omega_{s}b_{ijk}$ 时，可积条件为

$$R_{rsi}{}^{m}b_{mjk}+R_{rsj}{}^{m}b_{imk}+R_{rsk}{}^{m}b_{ijm}=-4\nabla_{[r}\omega_{s]}b_{ijk}.\tag{8}$$

这相当于 $\underset{1}{t}{}^{i}_{jk}=\underset{2}{t}{}^{i}_{jk}=0$ 的场合，和普遍的伟尔(Weyl)联络颇类似.

当 H^{i}_{ik}，Γ^{i}_{jk} 均为欧氏联络时，由(4)我们得到

$$\rho_{rsj}{}^{m}b_{imk}=-4\nabla_{[r}\omega_{s]}b_{ijk}.$$

固定 k，且假设 $|b_{ijk}|\neq 0$.上式乘以 b^{hik}，得

$$\rho_{rsj}{}^{m}=-4\nabla_{[r}\omega_{s]}\delta_{j}^{m}.$$

依照在无挠率空间的皮安基-派德华恒等式(Bianchi-Podova)

$$\rho_{rsj}{}^{m}+\rho_{sjr}{}^{m}+\rho_{jrs}{}^{m}=-4(\nabla_{[r}\omega_{s]}\delta_{j}^{m}+\nabla_{[s}\omega_{j]}\delta_{r}^{m}+\nabla_{[j}\omega_{r]}\delta_{r}^{m})=0.$$

缩短 m，j 得

$$(n-2)\nabla_{[r}\omega_{s]}=0.$$

所以，当 $n>2$ 时，三个无挠率的共轭联络有二个是欧氏的，其他一个一定也是欧氏的.同样获得：

m 个无挠率的共轭联络，有 $m-1$ 个是欧氏联络，则当 $n>2$ 时，其他一个联络一定也是欧氏的.

参考文献

[1] Порден，А.П.，Пространства Аффинной Связности，1950.

An Extension of Conjugate Affine Connections

Hu Hou-Sung
(Institute of Mathematics，Academia Sinica and Fu-Tan University)

Abstract

Norgen has given a definition of conjugate pairs of connections G_{jk}^{i}，Γ_{jk}^{i} in an affinely connected space X_n，based upon a given polarity b_{ij} at any point in X_n. Two directions v^i and w^j are said to be conjugate if they satisfy the following condition：

$$b_{ij}v^{i}w^{j}=0.$$

If these conjugate directions, when parallelly displaced along a curve L under the connections G^i_{jk}, Γ^i_{jk} respectively, remain to be conjugate, then G^i_{jk} and Γ^i_{jk} are defined as conjugate connections with respect to the tensor b_{ij}.

The aim of this paper is to extend Norgen's results to a set of m connections and then to establish some analogous theorems.

Given m connections $\underset{1}{\Gamma}$, $\underset{2}{\Gamma}$, $\cdots$, $\underset{m}{\Gamma}$ and a symmetric psendo-tensor $b_{i_1 i_2 \cdots i_m}$ in an affinely connected space X_n, we define that m directions $v^i_{(1)}$, $\cdots$, $v^i_{(m)}$ are m-conjugate if they satisfy the condition

$$b_{i_1 i_2 \cdots i_m} v^{i_1}_{(1)} v^{i_2}_{(2)} \cdots v^{i_m}_{(m)} = 0.$$

Now let each direction $v^i_{(k)}$ be parallelly displaced under the connection $\underset{k}{\Gamma}$ $(k=1, 2, \cdots, m)$ and along a common curve L; these connections $\underset{1}{\Gamma}$, $\cdots$, $\underset{m}{\Gamma}$ are defined to be m-conjugate if the displaced directions of $v^i_{(1)}$, $\cdots$, $v^i_{(m)}$ thus obtained remain to be m-conjugate.

The notion of mixed covariant diffrentiation is generalized to the case of m-conjugate connections.

The main results of the present paper are as follows:

1. Given $(m-1)$ connections $\underset{1}{\Gamma}$, $\cdots$, $\underset{m-1}{\Gamma}$ and a symmetric pseudotensor $b_{i_1 i_2 \cdots i_m}$; if we can determine the mth connection $\underset{m}{\Gamma}$ such that $\underset{1}{\Gamma}$, $\cdots$, $\underset{m}{\Gamma}$ are m-conjugate with respect to $b_{i_1 i_2 \cdots i_m}$, then the integer m is to satisfy the condition

$$(m-1)(H^n_{m-2}-1) \leqq H^n_m.$$

2. The tensor $b_{i_1 i_2 \cdots i_m}$ must satisfy the generalized Codazzi equations

$$\nabla_{[s} b_{(i_1] i_2 \cdots i_m)_r} = 2 w_{[s} b_{i_1] i_2 \cdots i_m},$$

where $\nabla_s b_{(i_1 i_2 \cdots i_m)_r}$ denotes the covariant differentiation with regard to connection Γ, and [ ] the sign of antisymmetrization.

3. If $\underset{1}{\Gamma}$, $\cdots$, $\underset{m}{\Gamma}$ are connections without torsion, then the Riemannian curvature tensor $R_{ijk}{}^{l}$ and $b_{i_1 i_2 \cdots i_m}$ obey the relations

$$\underset{1}{R}{}_{rsi_1}{}^{h} b_{hi_2\cdots i_m} + \underset{2}{R}{}_{rsi_2}{}^{h} b_{i_1 h i_3 \cdots i_m} + \cdots + \underset{m}{R}{}_{rsi_m}{}^{h} b_{i_1\cdots i_{m-1}h} = -4\,\nabla_{[r} w_{s]} b_{i_2 i_2 \cdots i_m}.$$

Especially we obtain:

When the assigned $(m-1)$ connections in X_n $(n>2)$ are euclidean, then also is the mth connection which forms an m-conjugate set with the assigned.

(本文曾发表于《数学学报》1953 年第 3 卷第 4 期,343 - 357)

特殊的仿射联络空间*

胡和生
中国科学院数学研究所，复旦大学

§1. 在 n 维的仿射联络空间中，一对称张量 b_{ij} 及二个仿射联络 G^i_{ik}，Γ^i_{ik} 如满足

$$\nabla_k b_{i(j)} = \frac{\partial b_{ij}}{\partial x^k} - G^m_{ki} b_{mj} - \Gamma^m_{kj} b_{im} = 2\omega_k b_{ij},$$

$$(i, j = 1, 2, \cdots, n)$$

依据 A.П.诺尔勤[1]的说法，联络偶 G^i_{ik}，Γ^i_{ik} 关于张量 b_{ij} 是共轭的. 作者[2]曾经扩充这个思想而定义 m 个联络关于一个 m 阶的对称共变张量是共轭的场合，当时曾提出由 $m-1$ 个联络如何决定第 m 个联络的问题，从而得到一系列的方程. 这些方程的共存性与它的解的唯一性，则还需对 $m-1$ 个联络或张量作出另外的规定.

本文特选 n 阶的反称张量 $e_{i_1 i_2 \cdots i_n}$，而考察关于这 n 阶张量为共轭的 n 个联络的全体. 这时可完全决定全体联络，由合于一定条件的 $n-1$ 个联络也能唯一地决定第 n 个联络，并且联系于等价仿射联络能保持空间的"体积"这一性质，对我们的结果形成了一个几何的解释.

§2. 由于本文中所讨论的联络与共同拟似的联络及等价仿射的联络都有关系，我们首先介绍它的意义，并扩充等价仿射联络到有挠率的情形.

设 Γ^k_{ij}，$\breve{\Gamma}^k_{ij}$ 是两个仿射联络的支量；如果存在关系式

$$\Gamma^k_{ij} = \breve{\Gamma}^k_{ij} + q_i \delta^k_j, \tag{2.1}$$

---

* 1954 年 8 月 20 日收到.

则称这两联络为相互共同拟似的.所谓两个联络的共同拟似,其意义如次：同一方向的向量依这两联络沿任何曲线平行移动之后,它们仍有相同的方向.当 $\breve{\Gamma}_{ij}^{k}$ 是无挠率时,则 Γ_{ij}^{k} 称为半对称联络的支量,而且 $\breve{\Gamma}_{ij}^{k}$ 可用 Γ_{ij}^{k} 完全决定.

共同拟似联络的曲率张量的支量 $R_{rsh}{}^{i}$ 与 $\breve{R}_{rsh}{}^{i}$ 之间存在关系

$$\breve{R}_{rsh}{}^{i}=R_{rsh}{}^{i}-(\partial_r q_s-\partial_s q_r)\delta_h^i, \tag{2.2}$$

它们相等的充要条件是 q_i 变成斜度.

§3. 如所周知,如果在无挠率的仿射联络空间一个体积对向量的平行移动是不变的,则称这空间为等价仿射的并且成立

1. $\Gamma_{ks}^{s}=\partial_k \log e$,

2. $R_{ij}=R_{ji}$,

3. $R_{rsm}{}^{m}=0$.

这里 e 是空间的密度.

现在来讨论有挠率的空间的场合.假设存在对向量的平行移动是不变的体积,联络支量应有如何性质呢?

由向量 $\underset{1}{v}^{i_1}$, $\underset{2}{v}^{i_2}$, …, $\underset{n}{v}^{i_n}$ 所作的体积

$$V=e_{i_1 i_2\cdots i_n}\underset{1}{v}^{i_1}\underset{2}{v}^{i_2}\cdots\underset{n}{v}^{i_n} \tag{3.1}$$

经过向量的平行移动之后,如果 $dV=0$, 则

$$\nabla_k e_{i_1 i_2\cdots i_n}=0.$$

改写之

$$\partial_k e_{i_1 i_2\cdots i_n}-\Gamma_{k i_1}^{s}e_{s i_2\cdots i_n}-\cdots-\Gamma_{k i_n}^{s}e_{i_1 i_2\cdots i_{n-1} s}=0.$$

由于 $i_p=i_q$ 时,上式为恒等式,只需考虑指数是 1, 2, …, n 的情形,记 $e_{12\cdots n}=\varepsilon$ 而得

$$\Gamma_{ks}^{s}=\partial_k \log \varepsilon.$$

(3.1)的可积条件是

$$\nabla_{[l}\nabla_{k]}e_{i_1i_2\cdots i_n}=0.$$

计算之,得到

$$R_{ekm}{}^{m}=0.$$

所以条件1,3,仍然成立.

现设Γ_{ij}^{k}与一等价仿射空间$\breve{\Gamma}_{ij}^{k}$是半对称的.由(2.2)可得

$$\partial_r q_s-\partial_s q_r=0,$$

即q_s为斜度,所以曲率张量与李契张量与$\breve{\Gamma}_{ij}^{k}$所作的曲率张量和李契张量各相等,因之,李契张量也是对称的,即2亦成立.

在上述假定中,设

$$q_r=\partial_r\log\varphi,$$

则由(2.1)可得

$$q_r=\partial_r\log\left(\frac{\varepsilon}{e}\right)^{1/n}.$$

§4. 如上所言,在n维仿射联络空间中n个向量$\underset{1}{v}^{i_1},\cdots,\underset{n}{v}^{i_n}$的体积用

$$V=e_{i_1i_2\cdots i_n}\underset{1}{v}^{i_1}\cdots\underset{n}{v}^{i_n}$$

来定义,这里$e_{i_1i_2\cdots i_n}$是反称张量.

设空间有n个联络其支量记为$\underset{1}{\Gamma}_{jk}^{i},\underset{2}{\Gamma}_{jk}^{i},\cdots,\underset{n}{\Gamma}_{jk}^{i}$,如果向量$\underset{1}{v}^{i_1},\underset{2}{v}^{i_2},\cdots,\underset{n}{v}^{i_n}$分别依$\underset{1}{\Gamma},\underset{2}{\Gamma},\cdots,\underset{n}{\Gamma}$平行移动,而体积保持不变,就需成立

$$dV=0,$$

或

$$\nabla_k e_{(i_1)_1(i_2)_2\cdots(i_n)_n}=0.$$

这里$(\ )_i$表示依$\underset{i}{\Gamma}$所作的共变微分,就是说,

$$\nabla_k e_{(i_1)_1(i_2)_2\cdots(i_n)_n} = \partial_k e_{i_1 i_2\cdots i_n} - \underset{1}{\Gamma}{}^m_{ki_1} e_{mi_2\cdots i_n} - \underset{2}{\Gamma}{}^m_{ki_2} e_{i_1 m i_3\cdots i_n} - \cdots$$
$$- \underset{n}{\Gamma}{}^m_{ki_n} e_{i_1 i_2\cdots i_{n-1} m} = 0, \tag{4.1}$$

固定 $i_1, i_2, \cdots, i_n$ 为 $1, 2, \cdots, n$，就有

$$\partial_k e - \underset{1}{\Gamma}{}^1_{k1} e - \underset{2}{\Gamma}{}^2_{k2} e - \cdots - \underset{n}{\Gamma}{}^n_{kn} e = 0,$$

或

$$\underset{1}{\Gamma}{}^1_{k1} + \underset{2}{\Gamma}{}^2_{k2} + \cdots + \underset{n}{\Gamma}{}^n_{kn} = \partial_k \log e, \tag{4.2}$$

轮换地给定指标 $i_1, i_2, \cdots, i_n$ 的数值，就再得到

$$\left.\begin{array}{l} \underset{1}{\Gamma}{}^2_{k2} + \underset{2}{\Gamma}{}^3_{k3} + \cdots + \underset{n}{\Gamma}{}^1_{k1} = \partial_k \log e, \\ \cdots\cdots \\ \underset{1}{\Gamma}{}^n_{kn} + \underset{2}{\Gamma}{}^1_{k1} + \cdots + \underset{n}{\Gamma}{}^{n-1}_{kn-1} = \partial_k \log e, \end{array}\right\} \tag{4.2$'$}$$

相加得到

$$\underset{1}{\Gamma}{}^s_{ks} + \underset{2}{\Gamma}{}^s_{ks} + \cdots + \underset{n}{\Gamma}{}^s_{ks} = n\partial_k \log e. \tag{4.3}$$

所以，当 $n-1$ 个联络是等价仿射的，则第 n 个联络亦为等价仿射的.

在(4.1)中置 $i_1 = i_2 = i$，就有

$$\underset{1}{\Gamma}{}^m_{ki} e_{mii_3\cdots i_n} + \underset{2}{\Gamma}{}^m_{ki} e_{imi_3\cdots i_n} = 0,$$

或

$$\underset{1}{\Gamma}{}^m_{ki} = \underset{2}{\Gamma}{}^m_{ki} \quad (m \neq i).$$

一般就有

$$\underset{\alpha}{\Gamma}{}^m_{ki} = \underset{\beta}{\Gamma}{}^m_{ki} \quad (m \neq i) \quad (\alpha, \beta = 1, 2, \cdots, n), \tag{4.4}$$

又置 $i_1 = 2, i_2 = 1, i_3 = 3, \cdots, i_n = n$，利用(4.1)，就得

$$\partial_k \log e = \underset{1}{\Gamma}{}^2_{k2} + \underset{2}{\Gamma}{}^1_{k1} + \underset{3}{\Gamma}{}^3_{k3} + \cdots + \underset{n}{\Gamma}{}^n_{kn},$$

与(4.2)相减得

$$\underset{1}{\Gamma}{}^{1}_{k1}-\underset{2}{\Gamma}{}^{1}_{k1}=\underset{1}{\Gamma}{}^{2}_{k2}-\underset{2}{\Gamma}{}^{2}_{k2};$$

一般就有

$$\underset{\alpha}{\Gamma}{}^{i}_{ki}-\underset{\beta}{\Gamma}{}^{i}_{ki}=\underset{\alpha}{\Gamma}{}^{j}_{kj}-\underset{\beta}{\Gamma}{}^{j}_{kj}\quad(\alpha,\ \beta=1,\ 2,\ \cdots,\ n).\tag{4.5}$$

但 i，j 并不表示和式.置

$$\underset{\sigma}{\Gamma}{}^{i}_{kj}-\underset{n}{\Gamma}{}^{i}_{kj}=\underset{\sigma}{t}{}^{i}_{kj}\quad(\sigma=1,\ 2,\ \cdots,\ n-1),$$

则(4.4)(4.5)可写为

$$\underset{\sigma}{t}{}^{m}_{ki}=0\quad(m\neq i),\tag{4.4$'$}$$

$$\underset{\sigma}{t}{}^{i}_{ki}=\underset{\sigma}{t}{}^{j}_{kj}\quad(i,\ j\ \text{并非和式});\tag{4.5$'$}$$

从(4.4)$'$及(4.5)$'$可置

$$\underset{\sigma}{t}{}^{m}_{ki}=q_k\delta^m_i.$$

从此已可见 $n-1$ 个联络 $\underset{\sigma}{\Gamma}$ 与 $\underset{n}{\Gamma}$ 为共同拟似，因之也互相共同拟似，再将

$$\underset{\sigma}{\Gamma}{}^{i}_{jk}=\underset{n}{\Gamma}{}^{i}_{jk}+\underset{\sigma}{q}\ j\delta^i_k\tag{4.6}$$

代入(4.1)，得到

$$\partial_k e_{i_1i_2\cdots i_n}-\underset{n}{\Gamma}{}^{m}_{ki_1}e_{mi_2\cdots i_n}-\underset{n}{\Gamma}{}^{m}_{ki_2}e_{i_1mi_3\cdots i_n}-\cdots$$

$$-\underset{n}{\Gamma}{}^{m}_{ki_n}e_{i_1i_2\cdots i_{n-1}m}-(\underset{1}{q}_k+\underset{2}{q}_k+\cdots+\underset{n-1}{q}_k)e_{i_1i_2\cdots i_n}=0;$$

如 i_1，i_2，…，i_n 至少有一对相等，这式恒成立，如 i_1，i_2，…，i_n 均不相等，就得唯一的式子

$$\underset{1}{q}_k+\underset{2}{q}_k+\cdots+\underset{n-1}{q}_k=\partial_k\log e-\underset{n}{\Gamma}{}^{s}_{ks}.\tag{4.7}$$

定理 若 n 个联络关一 n 阶反称张量 $e_{i_1i_2\cdots i_n}$ 是共轭的，则

(i) 可以任意选择其中的一个（例如 $\underset{n}{\Gamma}$）；

(ii) 其余的联络与 $\underset{n}{\Gamma}$ 是共同拟似的(也互相共同拟似的);

(iii) 共同拟似向量 $\underset{1}{q}_k$, $\underset{2}{q}_k$, …, $\underset{n-1}{q}_k$ 必须满足(4.7).

于此可见如果互相共同拟似的 $n-1$ 个联络是已知的话,则可以唯一地确定第 n 个联络.

现在将结果给以几何解释:

假如 G^i_{jk} 为我们所讨论空间的仿射联络支量,就可以在空间找到与 G^i_{jk} 相共同拟似的 n 个联络,使空间中的某一个已给由 n 个向量所作成的体积当其各个向量分别依这 n 个联络平行移动时,体积为不变.

假如 G^i_{jk} 为使所给体积不变的等价仿射空间的联络支量,而规定 n 个联络中的任意一个与 G^i_{jk} 共同拟似(即依据它向量的平行移动与这向量依据 G^i_{jk} 平行移动有相同的方向),则其余的向量依其余的联络平行移动时,其方向也与依 G^i_{jk} 平行移动时相同.因此同一体积沿 G^i_{jk} 平行移动与沿 $\underset{1}{\Gamma}$, …, $\underset{n}{\Gamma}$, 平行移动后只可能组成体积的向量长度有所不同.

§5. 现在再讨论几个特殊的情形.

(i) 如果要求所有的联络都是无挠率的,则由(4.6)得着

$$\underset{\sigma}{q}_{[j}\delta^i_{k]}=0.$$

缩短 i, k,

$$\underset{\sigma}{q}_j=0 \quad (\sigma=1, 2, \cdots, n-1),$$

即 n 个联络完全相同且

$$\underset{s}{\Gamma}{}^s_{ks}=\partial_k \log e \quad (s=1, 2, \cdots, n).$$

这就是说,n 个联络就是使体积不变的等价的仿射联络.

(ii) 很显然的,如果 $\underset{\sigma}{q}_j=0$ $(\sigma=1, 2, \cdots, n-1)$ 而联络是有挠率的,则得 n 个相等的有挠率联络,这归入§3 中所讨论的有挠率的等价仿射联络的情形.

(iii) 如果 n 个联络中有一个是半对称的,那么全体联络都是半对称的,并且与同一

联络是半对称的.

(iv) 当 $\sum_{\sigma=1}^{n-1} \underset{\sigma}{q}_k$ 是一斜度时，$\underset{n}{\Gamma}$ 就是一般的等价仿射的联络，曲率张量间存在关系

$$\underset{\sigma}{R}_{rsh}{}^{i} = \underset{n}{R}_{rsh}{}^{i} - \delta_h^i(\partial_r \underset{\sigma}{q}_s - \partial_s \underset{\sigma}{q}_r) \quad (\sigma = 1,\ 2,\ \cdots,\ n-1),$$

将此 $n-1$ 式相加，得到

$$\begin{aligned}\underset{n}{R}_{rsh}{}^{i} &= \frac{\underset{1}{R}_{rsh}{}^{i} + \cdots + \underset{n-1}{R}_{rsh}{}^{i}}{n-1} + \delta_h^i\left(\partial_r\left(\sum \underset{\sigma}{q}_s\right) - \partial_s\left(\sum \underset{\sigma}{q}_r\right)\right)\\ &= \frac{\underset{1}{R}_{rsh}{}^{i} + \cdots + \underset{n-1}{R}_{rsh}{}^{i}}{n-1}.\end{aligned}$$

这就是说，第 n 个联络的曲率张量恰为前面 $n-1$ 个联络曲率张量的算术平均.

又当每个 $\underset{\sigma}{q}_r$ 均为斜度时，则

$$\underset{1}{R}_{rsh}{}^{i} = \underset{2}{R}_{rsh}{}^{i} = \cdots = \underset{n}{R}_{rsh}{}^{i}.$$

最后再导入 n 个联络 $\underset{1}{\Gamma},\ \underset{2}{\Gamma},\ \cdots,\ \underset{n}{\Gamma}$ 的平均联络，它的系数为

$$l_{ij}^k = \frac{1}{n}(\underset{1}{\Gamma}_{ij}^k + \underset{2}{\Gamma}_{ij}^k + \cdots + \underset{n}{\Gamma}_{ij}^k),$$

于是对平均联络的共变微分

$$\begin{aligned}\overset{c}{\nabla}_k e_{i_1 i_2 \cdots i_n} &= \frac{1}{n}(\nabla_k e_{(i_1 i_2 \cdots i_n)_1} + \nabla_k e_{(i_1 i_2 \cdots i_n)_2} + \cdots + \nabla_k e_{(i_1 i_2 \cdots i_n)_n})\\ &= \frac{1}{n}(\nabla_k e_{(i_1)_1 (i_2)_2 \cdots (i_n)_n} + \nabla_k e_{(i_2)_1 (i_3)_2 \cdots (i_n)_{n-1} (i_1)_n} + \cdots\\ &\quad + \nabla_k e_{(i_n)_1 (i_1)_2 (i_2)_3 \cdots (i_{n-1})_n}) = 0.\end{aligned}$$

所以如果 n 个向量分别依 $\underset{1}{\Gamma},\ \underset{2}{\Gamma},\ \cdots,\ \underset{n}{\Gamma}$，平行移动时体积不变，则当各个向量均依其平均联络平行移动后，体积亦不变.

利用(4.3)可得

$$l^s_{ks} = \partial_k \log e.$$

参考文献

[1] Норден, А.П., Пространства Аффинной Связности, 1950.

[2] 胡和生,共轭的仿射联络的扩充,数学学报 4(1953),343-357.

Some Special Affinely Connected Spaces

Hu Hou-Sung

(Institute of Mathematics, Academia Sinica and Fu-Tan University)

Abstract

Norgen has defined a pair of affine connections G^i_{jk}, Γ^i_{jk} as conjugate connections with respect to a symmetric tensor b_{ij}, when they satisfy the condition

$$\nabla_k b_{i(j)} \equiv \frac{\partial b_{ij}}{\partial x^k} - G^m_{ki} b_{mj} - \Gamma^m_{kj} b_{im} = 2\omega_k b_{ij}.$$

In a recent paper the author has generalized this notion to the case of a symmetric tensor of order m, and has solved the problem of determing the mth connection from the given $m-1$ connections. We have system of equations and for the consistency and uniqueness of the solution of the system of equations we obtained some restrictions which must be added to the tensor or the $m-1$ connections.

In the present note we consider n conjugate connections with respect to an antisymmetric tensor $e_{i_1 i_2 \cdots i_n}$ and prove the following theorem:

If n connections are conjugate with respect to an antisymmetric tensor $e_{i_1 i_2 \cdots i_n}$, then

(i) one of them for example $\underset{n}{\Gamma}$, can be choosen arbitrarily,

(ii) the remaining connections $\underset{1}{\Gamma}$, $\underset{2}{\Gamma}$, $\cdots$, $\underset{n-1}{\Gamma}$ are common-pseudo with $\underset{n}{\Gamma}$ (hence mutually common-pseudo),

(iii) the common-pseudo vectors $\underset{1}{q}_k$, $\underset{2}{q}_k$, $\cdots$, $\underset{n-1}{q}_k$ must satisfy the following condition:

$$\underset{1}{q}_k+\underset{2}{q}_k+\cdots+\underset{n-1}{q}_k=\partial_k \log e-\underset{n}{\Gamma}^s_{ks}.$$

As a geometrical interpretation of these connections we have: If there exist a volume in an affinely connected space formed by n vectors such that it remains unaltered when these vectors undergo parallel displacements respectively with regard to the connections $\underset{1}{\Gamma}$, $\underset{2}{\Gamma}$, $\cdots$, $\underset{n}{\Gamma}$, then these connections must satisfy the conditions (i), (ii), (iii).

If G^i_{jk} is such an affine connection in our space, then we can always find n connections common-pseudo with G^i_{jk} such that the volume formed by certain n vectors remains unaltered when the forming vectors undergo parallel displacements with regard to $\underset{1}{\Gamma}$, $\underset{2}{\Gamma}$, $\cdots$, $\underset{n}{\Gamma}$ respectively.

If the connection G^i_{jk} is equi-affine and any one of the n connections is common-pseudo with it, then the directions of the vectors remain the same when they are parallelly displaced under $\underset{1}{\Gamma}$, $\underset{2}{\Gamma}$, $\cdots$, $\underset{n}{\Gamma}$ respectively.

Especially, when the n connections are all symmetric, then they become one and the same.

(本文曾发表于《数学学报》1955 年第 5 卷第 3 期,325 - 332)

德沙格定理在射影空间超曲面论上的推广

胡和生

中国科学院数学研究所，复旦大学

本文的目的是利用诺尔勤[1]（А. П. Норден）装配超曲面的方法，推广德沙格(Desargues)定理.

诺尔勤在研究射影空间 P_{n+1} 的超曲面 X_n

$$\boldsymbol{x}=\boldsymbol{x}(u^1,\ u^2,\ \cdots,\ u^n)$$

时，在 X_n 的每点 x 添上一过 $\boldsymbol{x}$ 而不在 X_n 的切平面上的直线，称为第一法线，它可以由点 $\boldsymbol{X}$（不在 X_n 的切平面上）与 $\boldsymbol{x}$ 的联线来决定，$\boldsymbol{X}$ 称为第一法点.又对 X_n 的每点 $\boldsymbol{x}$，安排一个在 X_n 的切平面上的 $n-1$ 维平面与它对应，这 $n-1$ 维平面称为第二法集，它是由 n 个点

$$\boldsymbol{y}_i=\partial_i\boldsymbol{x}-l_i\boldsymbol{x}\quad (i=1,\ 2,\ \cdots,\ n)$$

所决定，点 $\boldsymbol{y}_i$ 称为第二法集支点，有了这样装配的超曲面，就称为归范化的超曲面，由此可以决定超曲面的内在联络 Γ^i_{jk}，而能够把超曲面的基本方程写成如下形式：

$$\partial_i\boldsymbol{x}=\boldsymbol{y}_i+l_i\boldsymbol{x},$$

$$\nabla_j\boldsymbol{y}_i=l_j\boldsymbol{y}_i+p_{ji}\boldsymbol{x}+b_{ij}\boldsymbol{x},$$

$$\partial_j\boldsymbol{X}=m^k_j\boldsymbol{y}_k+m_j\boldsymbol{x}-\lambda_j\boldsymbol{X}.$$

式中 ∇_j 表示依联络 G^i_{jk} 的共变导微，p_{ji}，b_{ij}，m^k_j，m_j，λ_i 是 u^1，…，u^n 的适当的函数，∂_i 表示依 u_i 的普通偏导微[2].

依诺尔勤的定义，P_{n+1} 中的一对超曲面 X_n，X'_n，点与点之间有一一对应，且在对应点处第一法线及第二法集都是共同的，如此的两超曲面称为有共同归范化.这里首先讨论这两超曲面联络间的关系.

两超曲面的方程为

$$\boldsymbol{x}=\boldsymbol{x}(u^1, \cdots, u^n),$$

$$\bar{\boldsymbol{x}}=\bar{\boldsymbol{x}}(u^1, \cdots, u^n).$$

每对对应点 $\boldsymbol{x}$，$\bar{\boldsymbol{x}}$ 对应同一组 u^1，…，u^n 的数值.记第二法集支点分别为 $\boldsymbol{y}_i$，$\bar{\boldsymbol{y}}_i$.联络分别为 Γ，$\check{\Gamma}$. 由共同归范化的定义，必须成立

$$\bar{\boldsymbol{x}}=\boldsymbol{X}+\sigma\boldsymbol{x}, \tag{1}$$

$$\bar{\boldsymbol{y}}_i=a_i^k\,\bar{\boldsymbol{y}}_k. \tag{2}$$

而 X_n 与$\overline{X}_n$ 的基本方程中的第二式给出：

$$\nabla_j\boldsymbol{y}_i=l_j\boldsymbol{y}_i+p_{ji}\boldsymbol{x}+b_{ji}\boldsymbol{X}, \tag{3}$$

$$\overline{\nabla}_j\,\bar{\boldsymbol{y}}_i=\bar{l}_j\,\bar{\boldsymbol{y}}_i+\bar{p}_{ji}\,\bar{\boldsymbol{x}}+\bar{b}_{ji}\boldsymbol{X}. \tag{4}$$

将(1)(2)代入(4)得

$$\partial_j(a_i^l\boldsymbol{y}_l)-\check{\Gamma}_{ji}^k a_k^l\boldsymbol{y}_l=\bar{l}_j a_i^k\boldsymbol{y}_k+\bar{p}_{ji}(\boldsymbol{X}+\sigma\boldsymbol{x})+\bar{b}_{ji}\boldsymbol{X},$$

或

$$\partial_j a_i^l\boldsymbol{y}_l+a_i^l\partial_j\boldsymbol{y}_l-\check{\Gamma}_{ji}^k a_k^l\boldsymbol{y}_l=\bar{l}_j a_i^k\boldsymbol{y}_k+\sigma\,\bar{p}_{ji}\boldsymbol{x}+(\bar{p}_{ji}+\bar{b}_{ji})\boldsymbol{X}.$$

利用

$$a_i^l\,\tilde{a}_l^j=\delta_j^i$$

定义函数 $\tilde{a}_l^j$，将上式乘以$\tilde{a}_m^i$，得

$$\tilde{a}_m^i\partial_j a_i^l\boldsymbol{y}_l+\partial_j\boldsymbol{y}_m-\check{\Gamma}_{ji}^k a_k^l\boldsymbol{y}_l\,\tilde{a}_m^i=\bar{l}_j\boldsymbol{y}_m+\sigma\,\bar{p}_{ji}\,\tilde{a}_m^i\boldsymbol{x}+(\bar{p}_{ji}+\bar{b}_{ji})\,\tilde{a}_m^i\boldsymbol{X}.$$

将此式减(3)式，且比较系数，得

$$\sigma \bar{p}_{ji} \tilde{a}^i_m - p_{jm} = 0,$$

$$(\bar{p}_{ji} + \bar{b}_{ji}) \tilde{a}^i_m - b_{jm} = 0,$$

$$\Gamma^l_{jm} + \tilde{a}^i_m \partial_j a^l_i - \check{\Gamma}^k_{ji} a^l_k \tilde{a}^i_m = (\bar{l}_j - l_j)\delta^l_m. \qquad (5)$$

将(5)式乘以 a^m_s 得

$$\partial_j a^l_s - \check{\Gamma}^k_{js} a^l_k + \Gamma^l_{jm} a^m_s = a^l_s(\bar{l}_j - l_j). \qquad (6)$$

此式左边,可记为 $\nabla_j a^l_{(s)}$,而右边,利用了诺尔勤所得到的关系式 $\bar{l}_j = -\lambda_j$[3] 可书写

$$\nabla_j a^l_{(s)} = a^l_s(\bar{l}_j - l_j) = -a^l_s(\lambda_j + l_j) = -2\omega_j a^l_s. \qquad (7)$$

这里

$$\omega_j = \frac{1}{2}(\lambda_j + l_j)$$

称为辅助张量,而(7)式就表明着:两超曲面上的一对联络是关于 a^l_s 共轭的.

如果要两个联络相同,则 a^l_s 要满足

$$\nabla_j a^l_s = -2\omega_j a^l_s.$$

特别讨论当二组第二法点是符合的情形,此时

$$\bar{\mathbf{y}}_i = \underset{(i)}{\lambda} \mathbf{y}_i, \qquad ((i)\text{ 不是和式}) \qquad (8)$$

第二法点符合,表示对应方向的切线相交,这个关系不因坐标变换而改变,所以经坐标变换后

$$\bar{\mathbf{y}}'_i = \underset{(i)}{\lambda}' \mathbf{y}'_i,$$

于是 $\underset{(i)}{\lambda}'$ 与 $\underset{(i)}{\lambda}$ 之间应满足下列关系

$$\sum \underset{(i)}{\lambda}' \frac{\partial u^l}{\partial \bar{u}^i} \frac{\partial \bar{u}^i}{\partial u^m} = \delta^m_l \underset{(m)}{\lambda}.$$

考察方程组

$$\xi_i^l \frac{\partial \bar{u}^i}{\partial u^m} = \delta_l^m \underset{(m)}{\lambda}$$

而知有唯一的解 $\xi_i^l = \underset{(m)}{\lambda} \frac{\partial u^l}{\partial \bar{u}^i}$，因之

$$\underset{(i)}{\lambda}' \frac{\partial u^l}{\partial \bar{u}^i} = \underset{(m)}{\lambda} \frac{\partial u^l}{\partial \bar{u}^i}.$$

于是

$$\underset{(i)}{\lambda}' = \underset{(m)}{\lambda} \quad (i,\ m=1,\ \cdots,\ n),$$

固定 i，变动 m 就得到

$$\underset{(1)}{\lambda} = \underset{(2)}{\lambda} = \cdots = \underset{(n)}{\lambda}.$$

(8)式就化为

$$\bar{\boldsymbol{y}}_i = \lambda \boldsymbol{y}_i.$$

故经适当的规范化后得

$$\bar{\boldsymbol{y}}_i = \boldsymbol{y}_i. \tag{9}$$

于是

$$a_k^i = \delta_k^i.$$

代入(6)，就有

$$\Gamma_{jm}^l - \check{\Gamma}_{jm}^l = \delta_m^l (\bar{l}_j - l_j).$$

由于内在联络 Γ, $\check{\Gamma}$ 是无挠率的，就成立

$$\delta_m^l (\bar{l}_j - l_j) = \delta_j^l (\bar{l}_m - l_m).$$

令 $m = l$, $j \neq l$, 就得到

$$\bar{l}_j = l_j.$$

由此也得到 $\Gamma=\check{\Gamma}$，两个联络相同.

由于

$$\boldsymbol{y}_i=\partial_i\bar{\boldsymbol{x}}-l_i\bar{\boldsymbol{x}}=\partial_i\boldsymbol{x}-l_i\boldsymbol{x},$$

利用(1)得到

$$\partial_i\boldsymbol{X}=(1-\sigma)\partial_i\boldsymbol{x}+[(\sigma-1)l_i-\partial_i\sigma]\boldsymbol{x}+l_i\boldsymbol{X},$$

求其可积条件

$$\begin{aligned}\partial_{ij}\boldsymbol{X}=&(1-\sigma)\partial_{ij}\boldsymbol{x}-\partial_j\sigma\partial_i\boldsymbol{x}+(\sigma-1)l_i\partial_j\boldsymbol{x}-\partial_i\sigma\partial_j\boldsymbol{x}\\&+\boldsymbol{x}\partial_j[(\sigma-1)l_i-\partial_i\sigma]+\partial_jl_i\boldsymbol{X}+l_i\partial_j\boldsymbol{X},\end{aligned}$$

计算后得

$$(\sigma-1)\partial_jl_i\boldsymbol{x}+\partial_jl_i\boldsymbol{X}=(\sigma-1)\partial_il_j\boldsymbol{x}+\partial_il_j\boldsymbol{X}.$$

比较系数得 l_i 是斜度，所以装配一定是调和的.

(9)式也可以写成

$$\partial_i\bar{\boldsymbol{x}}-l_i\bar{\boldsymbol{x}}=\partial_i\boldsymbol{x}-l_i\boldsymbol{x}.$$

记

$$l_i=-\partial_i\log\varphi,$$

上式即为：

$$\varphi\partial_i\bar{\boldsymbol{x}}+\partial_i\varphi\bar{\boldsymbol{x}}=\varphi\partial_i\boldsymbol{x}+\partial_i\varphi\boldsymbol{x},$$

因之

$$\partial_i(\varphi\bar{\boldsymbol{x}})=\partial_i(\varphi\boldsymbol{x}),$$

由此

$$\varphi\bar{\boldsymbol{x}}=\varphi\boldsymbol{x}+a.$$

a 为常数，因此 $\boldsymbol{x}$，$\bar{\boldsymbol{x}}$ 通过一定点，亦即第一法线通过一定点.得

定理： 设有 P_{n+1} 中二超曲面 X_n 及 X'_n，它们之间存在一对应，使得对应点的对应切线都相交，那么两个曲面的对应一定是配景的，即对应点的联线是共点的.

这个定理与德沙格定理有密切的关系，如果两超平面是配景的，由于德沙格定理；设 A，$\bar{A}$，B，$\bar{B}$ 是两对对应点，则 A，B 与 $\bar{A}$，$\bar{B}$ 相交并且交点在一 $n-1$ 维的平面上，其逆亦真.

现在把这个情形扩充到超曲面上来，此定理表示其逆定理部分且把条件减低到对应点的切线相交，至于正定理部分是容易证明的.

设 X_n，X'_n 成配景(关于 O)，A，B，$\bar{A}$，$\bar{B}$ 是两对对应点，O，A，B，$\bar{A}$，$\bar{B}$ 共平面.因之 AB，$\bar{A}\bar{B}$ 相交，令 B 趋向于 A，则 $\bar{B}$ 趋向于 $\bar{A}$，BA 趋向于 A 点的切线，$\bar{B}\bar{A}$ 趋向于 $\bar{A}$ 的对应于 AB 的切线，所以 $\bar{A}\ \bar{B}$ 与 AB 也相交.

对于这定理尚可作如下的对偶定理.

定理： 设有 P_{n+1} 中二超曲面 X_n，X'_n，它们之间存在一对应，使得二对应邻近切超平面交集在一超平面上，那么对应超平面的交集都在一超平面上.其逆亦真.

最后，作者在此对苏步青教授的指导与鼓励表示感谢.

参 考 文 献

[1] Норден，А.П.，Пространства Аффинной Связности，1950.

[2] 见前书 227 页.

[3] 见前书 246 页.

(1954 年 11 月 14 日收到)

(本文曾发表于《复旦学报》1955 年第 1 期，46－50)

论保持平均曲率的黎曼测度 V_m 在欧氏空间 E_{m+1} 的变形*

胡和生
中国科学院数学研究所,复旦大学

§1. 如所知,二维黎曼测度

$$ds^2 = E(u, v)du^2 + 2F(u, v)du\,dv + G(u, v)dv^2;$$

在 E_3 中的实现不是唯一的,而容有与单项目的两个任意函数有关的连续变形.如此,曲面的测度不能决定曲面,但不同于此,m 维黎曼测度($m \geqslant 3$)如能在 E_{m+1} 中实现,一般地说实现它的超曲面只能是唯一的,也就是说,超曲面 $V_m \subset E_{m+1}$ 一般是不能变形的,而能变形的只是狭窄的一类超曲面.

关于1级的测度的分类问题,在近一世纪几何学者的工作中得到了解决,Bianchi Sbrana, E. Cartan, Weise, T. Y. Thomas, Н. А. Розенсон, А. М. Лопшиц 等都做过这工作.Н. Н. Яненко[1]运用外微分形式的方法很详细地把这些学者的工作综合起来,并且完全给出高维欧氏空间可变形的超曲面的分类.

对于三维欧氏空间的曲面,一般不能使其保持平均曲率而变形,但是存在着这样的曲面偶,它与单项目的4个函数有关[2].本文的目的,是利用Н. Н. Яненко关于可变形超曲面的分类;来讨论 E_{m+1} 中保持平均曲率的超曲面的变形问题,并且将不可变形的超曲面的第二基本形式;用平均曲率及测度张量明确地表示出来,最后的结果也可看作 T. Y. Thomas 的研究[3]的拓广.

---

* 1955年1月28日收到.

§2. 对于曲面$V_m \subset E_{m+1}$，联上m参数标形族$(r, I_\alpha)\alpha=1, \cdots, m+1$使其始点在曲面$r=r(u^1, \cdots, u^m)$上.我们通常选择半直交的标形族，即向量$I_{m+1}$与曲面直交，而为曲面的法线.那么，曲面的导来形式为

$$dr=\omega^i I_i,$$

$$dI_\alpha=\omega_\alpha^\beta I_\beta,$$

$$I_\alpha I_\beta=g_{\alpha\beta}, \quad g_{m+1,\alpha}=\delta_{m+1,\alpha}, \quad \alpha, \beta=1, \cdots, m+1, i, j=1, \cdots, m,$$

曲面的组织方程(即 Gauss - Петерсон - Codazzi 方程)是

$$(\omega^\alpha)'=[\omega^\beta \omega_\beta^\alpha],$$

$$(\omega_\beta^\alpha)'=[\omega_\beta^\gamma \omega_\gamma^\alpha].$$

在半直交标形下，我们将形式ω_β^α分为二群：

$$\omega_j^i=\Gamma_{jk}^i \omega^k, \qquad \text{(内在形式)}$$

$$\omega_i^{m+1}=\lambda_{ij}\omega^j, \quad \lambda_{ij}=\lambda_{ji}, \quad \text{(混合形式)}$$

而

$$\omega_{m+1}^{m+1}=0.$$

不难得到，$\lambda_{ij}\omega^i\omega^j$是超曲面的第二基本形式，从而混合形式的系数$\lambda_{ij}$就是第二基本形式的系数.下面提出 R. Beez 的定理.

定理. 超曲面$V_m \subset E_{m+1}$当秩数$r>2$时，是不可变形的.就是说，如果超曲面V_m的秩数>2，而且超曲面$\bar{V}_m$与V_m有同一测度，那么这两超曲面V_m，$\bar{V}_m$一定互相符合.

这里，秩数是混合形式ω_i^{m+1}中的线性独立的数目.秩数的几何意义是：曲面的切超平面所依赖的参数的个数.另一方面，定义测度$ds^2=g_{ij}du^i du^j$的秩数为双一次形式系统$\Omega_{ij}=\sum_{k<l} R_{ijkl}[\omega^k \omega^l]$的秩数. 可以证明，如 1 级的测度有秩数$r \geqslant 2$，则测度的秩数与它所实现的曲面的秩数符合.由于 R. Beez 的定理，讨论变形问题归结到测度的秩数$r \leqslant 2$的情形，按照 H. H. Яненко 来做如下的分类.

$r=0$ 时，测度是欧氏的，即 0 级的情形.

$r=1$ 的测度是不可能的.

$r=2$ 时，以规范和乐标形为参考标形，那么测度 $ds^2=g_{ij}du^idu^j$ 所实现的曲面 $V_m\subset E_{m+1}$ 的混合形式 $\psi_i=\omega_i^{m+1}$ 满足下列条件：

$$\psi_\alpha=0,\quad \alpha=3,\cdots,m; \tag{1}$$

$$\psi_i=\lambda_{ij}du^j,\quad \lambda_{ij}=\lambda_{ji},\quad i,j=1,2. \tag{2}$$

研究超曲面的变形问题，就是考察 Gauss－Петерсон－Codazzi 方程的解 λ_{ij} 的自由度的问题.由(1)(2)把 Петерсон－Codazzi 方程写成张量形式

$$\Gamma^i_{\alpha j}\lambda_{ik}-\Gamma^i_{\alpha k}\lambda_{ij}=0, \tag{3}$$

$$\frac{\partial\lambda_{i2}}{\partial u^1}-\frac{\partial\lambda_{i1}}{\partial u^2}=\Gamma^j_{i1}\lambda_{j2}-\Gamma^j_{i2}\lambda_{j1}, \tag{4}$$

$$\frac{\partial\lambda_{ij}}{\partial u^\alpha}=\Gamma^l_{i\alpha}\lambda_{lj},\qquad i,j=1,2;\ \alpha=3,\cdots,m, \tag{5}$$

而且把 Gauss 方程写成

$$R_{1212}=\lambda_{11}\lambda_{22}-\lambda_{12}^2. \tag{6}$$

称(3)为决定系统，我们根据决定系统的分类来确定 Gauss－Петерсон－Codazzi 方程的解.

1) 如果决定系统的秩数为 2，则可得

$$\lambda_{ij}=\pm\Delta_{ij}\sqrt{\frac{R_{1212}}{\Delta_{11}\Delta_{22}-\Delta_{12}^2}},\quad i,j=1,2; \tag{7}$$

其中 Δ_{ij} 是由量 $\Gamma^i_{\alpha j}$ 所组成的某些二阶张量，在这情形之下，测度有唯一的实现.

2) 如果决定系统的秩数为 1，则

$$\frac{\Gamma^1_{\alpha2}}{\Gamma^1_2}=\frac{-\Gamma^2_{\alpha1}}{\Gamma^2_1}=\frac{\Gamma^2_{\alpha2}-\Gamma^1_{\alpha1}}{\Gamma^2_2-\Gamma^1_1}=k_\alpha.$$

决定系统还原为方程

$$\Gamma_2^1\lambda_{11}+(\Gamma_2^2-\Gamma_1^1)\lambda_{12}+\Gamma_1^2\lambda_{22}=0.$$

因而,得到下列几种支情形.

2_1:Γ_j^i 有二个不同的特征根.我们可选择标形使得

$$\lambda_{12}=0, \tag{8}$$

于是 Gauss 方程为

$$R=R_{1212}=\lambda_{11}\lambda_{22}, \tag{9}$$

而且 Петерсон－Codazzi 方程的可积分条件为

$$As^2+2Bs+C=0. \tag{10}$$

式中A,B,C是$\Gamma_{jk}^i(i,j,k=1,2)$及R_{ijkl}的函数,而且$S=\lambda_{11}^2$.此时有下列三种情形:

2_{11}. 如果(10)有二个不同的根,那么可把测度实现为二个不符合的曲面,称它为稀疏的变形.

2_{12}. 如果(10)有重根,那么获得 2_{11} 的极限情形.实现的曲面是唯一的.

2_{13}. $A=B=C=0$,则 Codazzi 方程是完全可积分的:$\lambda_{11}=\lambda_{11}(u^1u^2,\cdots,u^m c)$.测度容许有许多的实现,而实际上,被实现在单参数族而不相符合的等测度的曲面上.

2_2:如果 Γ_j^i 有一个特征根,那么可适当地选择标形,使得

$$\lambda_{11}=0, \tag{11}$$

$$R=R_{1212}=-\lambda_{12}^2,$$

于是

$$\lambda_{12}=\pm\sqrt{-R}. \tag{12}$$

这时有二种可能性.

2_{21}. $\Gamma_{11}^2\neq 0$:实现的曲面是唯一的.

2_{22}. $\Gamma_{11}^2=0$:实现是非唯一的.它依赖于单项目的一个任意函数.

3) 最后讨论决定系统的秩数等于 0 的情形.

这时可以证明能选择这样的规范和乐标形使满足 $I_i \cdot I_\alpha = 0$ ($i=1, 2$; $\alpha = 3, \cdots, m$) 在这标形下 Γ^i_{jk} ($i, j, k=1, 2$) 只依赖于参数 u^1, u^2 而 $\Gamma_\alpha = \Gamma^1_{1\alpha} = \Gamma^2_{2\alpha}$ ($\alpha=3, \cdots m$) 只依赖于参数 $u^3, \cdots, u^m$，并且

$$\lambda_{ij} = \bar{\lambda}_{ij}(u^1, u^2) e^{v(u^3, \cdots, u^m)}. \tag{13}$$

于是 Gauss－Петерсон－Codazzi 方程的求解问题归结到

$$r(u^1, u^2) = \bar{\lambda}_{11} \bar{\lambda}_{22} - \bar{\lambda}^2_{12}, \tag{14}$$

$$-\frac{\partial \bar{\lambda}_{i1}}{\partial u^2} + \frac{\partial \bar{\lambda}_{i2}}{\partial u^1} = \Gamma^j_{i1} \bar{\lambda}_{j2} - \Gamma^j_{i2} \bar{\lambda}_{j1} \tag{15}$$

的求解问题，这里

$$R = R_{1212} = e^{2v(u^3, \cdots, u^m)} r(u^1, u^2), \quad V = \int \Gamma_\alpha d u^\alpha. \tag{16}$$

可以证明 $V_m \subset E_{m+1}$ 的问题归结到 $V_2 \subset S_3$ 的问题.

当固定数值 $u^3 = u^3_0, \cdots, u^m = u^m_0$ 时，V_2 测度为

$$d\tilde{s}^2 = g_{ij}(u^1, u^2, u^3_0, \cdots, u^m_0) du^i du^j = \tilde{g}_{ij} du^i du^j$$
$$(i, j=1, 2)$$

的流形 V_2，它的曲率张量为$\tilde{R}_{1212}$，与 ds^2 的曲率张量 $R_{1212}(u^1, u^2, u^3_0, \cdots, u^m_0)$ 之间有如下的关系：

$$R_{1212} = \tilde{R}_{1212} - k^2_0. \tag{17}$$

$$k^2_0 = \sum_{\alpha=3}^{m} \Gamma^2_\alpha(u^3_0, \cdots, u^m_0). \tag{18}$$

而它们的联络系数 $\Gamma^k_{ij}(u^1, u^2)$ 是相同的.①

由于任意二维测度 $d\tilde{s}^2$ 可以安装在正常曲率 k^2_0 的 S_3 之中，我们找出函数$\tilde{\lambda}_{ij}(u^1,$

---

① 在这里，有假设$\tilde{g}_{11} \tilde{g}_{23} - \tilde{g}^2_{12} = 1$ 的必要，但由于经过一个适当的保持和乐性的坐标系统变换，一定可以达到这个要求，证明见作者的"论黎曼测度 V_m 在常曲率空间 S_{m+1} 中的变形"[4].

u^2）使满足关系式

$$\widetilde{R}_{1212}-k_0^2=R_{1212}(u^1,\ u^2,\ u_0^3,\ \cdots,\ u_0^m)=\widetilde{\lambda}_{11}\widetilde{\lambda}_{22}-\widetilde{\lambda}_{12}^2, \tag{19}$$

$$\frac{\partial\widetilde{\lambda}_{i2}}{\partial u^1}-\frac{\partial\widetilde{\lambda}_{i1}}{\partial u^2}=\Gamma_{i1}^j\widetilde{\lambda}_{j2}-\Gamma_{i2}^j\widetilde{\lambda}_{j1}, \tag{20}$$

而且函数

$$\overline{\lambda_{ij}}=\frac{\widetilde{\lambda}_{ij}(u^1,\ u^2)}{e^{v(u_0^3,\ \cdots,\ u_0^m)}} \tag{21}$$

满足(14)，(15)，于是函数

$$\lambda_{ij}(u^1,\ u^2,\ u^3,\ \cdots,\ u^m)=e^{v(u^3,\ \cdots,\ u^m)}\,\overline{\lambda_{ij}}(u^1,\ u^2) \tag{22}$$

变为 $V_m\subset E_{m+1}$ 的 Gauss－Петерсон－Codazzi 方程的解，因此解的自由度依赖于单项目的二个函数，就是曲面的实现依赖于单项目的二个函数.

特别当 $\Gamma_\alpha(\alpha=3,\ \cdots,\ m)$ 时，问题归结到 $V_2\subset E_3$，同样地，实现不是唯一的，而是与单项目的二个任意函数有关的.

§3. 现在来讨论保持平均曲率的曲面变形问题并按照测度张量及平均曲率表示 λ_{ij}.由于 $r\geqslant 2$ 时显然不能变形，我们只需按照上节的 $r=2$ 的三种情形来讨论.

首先指出，高维欧氏空间的超曲面的平均曲率是由下式定义的

$$H=g^{ij}\lambda_{ij}. \tag{23}$$

现依各种情况分别进行讨论.

1）如果决定系统秩数等于 2，那么曲面显然不能变形；而 λ_{ij} 即依(7)来表示.

2）决定系统的秩数等于 1：

2_1. $\lambda_{12}=0$，$R=\lambda_{11}\lambda_{22}$.

由于黎曼－克里斯托费尔张量变形后仍不变，且由(23)得

$$\lambda_{11}=\frac{R}{\lambda_{22}},$$

$$\lambda_{22}=\frac{H \pm \sqrt{H^2-4g^{11}g^{22}R}}{2g^{22}},$$

所以当 $H^2 \neq 4g^{11}g^{22}R$ 时,得到稀疏的实现,即实现为二个不符合的曲面这两曲面容许相互的保持平均曲率的变形.当 $H^2=4g^{11}g^{22}R$ 时,实现是唯一的,即不容许变形.

2_2. $\lambda_{11}=0$, $\lambda_{12}=\pm\sqrt{-R}$,

且由(23)

$$\lambda_{22}=\frac{H \pm 2g^{12}\sqrt{-R}}{g^{22}},$$

因此当保持平均曲率时与 2_1 情形一样,实现是稀疏的或唯一的.

3) 决定系统秩数等于0:从前面我们知道 $V_m \subset E_{m+1}$ 的变形归结到 $V_2 \subset S_3$ 的变形问题,现在我们首先要证明当 $V_m \subset E_{m+1}$ 保持平均曲率而变形时,对应的 $V_2 \subset S_3$ 也保持平均曲率而变形.由于 $V_m \subset E_{m+1}$ 的平均曲率

$$\begin{aligned} H(u^1, u^2, \cdots, u^m) &= g^{ij}(u^1, \cdots, u^m)\lambda_{ij}(u^1, \cdots, u^m) \\ &= g^{ij}(u^1, \cdots, u^m)\frac{e^{v(u^3, \cdots, u^m)}}{e^{v(u_0^3, \cdots, u_0^m)}}\tilde{\lambda}_{ij}(u^1, u^2) \quad (i, j=1, 2). \end{aligned}$$

利用

$$\frac{\partial g_{ab}}{\partial u^c}=[ac, b]+[bc, a]=g_{db}\Gamma_{ac}^d+g_{da}\Gamma_{bc}^d \quad (a, b, c, d=1, \cdots, m),$$

可得

$$\frac{\partial g_{ij}}{\partial u^\alpha}=2g_{ij}\Gamma_\alpha \quad (i, j=1, 2, \alpha=3, \cdots, m)$$

并注意到 Γ_α 为只依赖于参数 $u^3, \cdots, u^m$ 这一事实,则知

$$g_{ij}(u^1, u^2, \cdots, u^m)=\bar{g}_{ij}(u^1, u^2)e^{2v(u^3, \cdots, u^m)}, \quad v=\int\Gamma_\alpha du^\alpha.$$

由§2可知

$$\widetilde{g}_{ij}=g_{ij}(u^1,u^2,u_0^3,\cdots,u_0^m)=\bar{g}_{ij}(u^1,u^2)e^{2v(u_0^3,\cdots,u_0^m)}$$
$$=g_{ij}(u^1,\cdots,u^m)\frac{e^{2v(u_0^3,\cdots,u_0^m)}}{e^{2v(u^3,\cdots,u^m)}},$$

由于I_i, $I_\alpha=0$,即$g_{i\alpha}=0$,容易证明

$$\hat{g}^{ij}=g^{ij}\frac{e^{2v(u^3,\cdots,u^m)}}{e^{2v(u_0^3,\cdots,u_0^m)}}.$$

依据普通的定义S_3中二维黎曼流形的平均曲率为

$$\widetilde{H}(u^1,u^2)=\frac{\widetilde{g}_{ij}\widetilde{\lambda}_{ij}}{2},$$

所以

$$H(u^1,u^2,\cdots,u^m)=\frac{2e^{v(u_0^3,\cdots,u_0^m)}}{e^{v(u^3,\cdots,u^m)}}\widetilde{H}(u^1,u^2). \tag{24}$$

这样证明了我们的要求,于是研究决定系统等于0的$V_m\subset E_{m+1}$的保持平均曲率的变形问题,只要研究$V_2\subset S_3$中保持平均曲率的变形.下面,我们开始讨论这个问题.

设S_3中两曲面V_2,$\bar{V}_2$互为变形,A与$\bar{A}$为对应点,在曲面V_2的每点A联上直交三面体T,使其第三轴合于曲面法线,而且最初二轴落在切平面上:

$$dA=\omega_1 I_1+\omega_2 I_2,\quad \omega_3=0,$$
$$dI_k=\omega_{ki}I_i,\quad \omega_{ik}=-\omega_{ki}\quad (i,k=1,2,3).$$

在曲面$\bar{V}_2$的对应点$\bar{A}$,做直交三面体$\bar{T}$,使得当$A\to\bar{A}$且在对应点的切平面重合时,$\bar{T}$要 重合到T去.

$$d\bar{A}=\bar{\omega}_1\bar{I}_1+\bar{\omega}_2\bar{I}_2,$$
$$\bar{\omega}_{ik}=-\bar{\omega}_{ki}.$$
$$d\bar{I}_k=\bar{\omega}_{ki}\bar{I}_i,$$

利用双方测度 $d\bar{A}^2=dA^2$ 的性质,可得

$$\bar{\omega}_1=\omega_1,\quad \bar{\omega}_2=\omega_2.$$

外导微上式的结果,

$$\bar{\omega}_{12}=\omega_{12}.$$

又外导微关系式

$$\omega_3=0,$$

便得到

$$[\omega_{13}\omega_1]+[\omega_{23}\omega_2]=0.$$

按嘉当引理写出

$$\omega_{13}=\tilde{\lambda}_{11}\omega_1+\tilde{\lambda}_{12}\omega_2,$$
$$\omega_{23}=\tilde{\lambda}_{12}\omega_1+\tilde{\lambda}_{22}\omega_2.$$

如同 E_3 中一样,保持平均曲率即保持 $\tilde{\lambda}_{11}+\tilde{\lambda}_{22}$. 因此得到

$$[\omega_1,\bar{\omega}_{23}-\omega_{23}]-[\omega_2,\bar{\omega}_{13}-\omega_{13}]=0,$$

于是问题的所有方程是

$$\left.\begin{aligned}&\omega_3=0,\ \bar{\omega}_3=0,\\&\bar{\omega}_1-\omega_1=0,\ \bar{\omega}_2-\omega_2=0,\ \bar{\omega}_{12}-\omega_{12}=0,\\&[\omega_1,\bar{\omega}_{23}-\omega_{23}]-[\omega_2,\bar{\omega}_{13}-\omega_{13}]=0.\end{aligned}\right\}\qquad(25)$$

为了导微上列的发甫方程,首先注意在黎曼空间中的公式

$$(\omega_{ij})'=[\omega_{ik}\omega_{kj}]+\sum_{k<l}R_{ijkl}[\omega_k\omega_l].$$

对于常曲率空间现在应变成的方式

$$R_{ijkl}=k_0^2(\tilde{g}_{ik}\tilde{g}_{jl}-\tilde{g}_{il}\tilde{g}_{jk})$$

在直交标形下，

$$R_{ijkl}=k_0^2(\delta_{ik}\delta_{jl}-\delta_{il}\delta_{jk}),\quad i,j,k=1,2,3.$$

因此

$$D\omega_{12}=[\omega_{13}\omega_{32}]+R_{1212}[\omega_1\omega_2]=[\omega_{13}\omega_{32}]+k_0^2[\omega_1\omega_2].$$

于是 $\bar{\omega}_{12}-\omega_{12}=0$ 的可积条件为

$$[\bar{\omega}_{13}\bar{\omega}_{23}]-[\omega_{13}\omega_{23}]=0. \tag{26}$$

从而,系统(25)的导微方程式为(26)及下列二方程

$$\begin{aligned}&[\omega_1\omega_{13}]+[\omega_2\omega_{23}]=0,\\&[\omega_1\bar{\omega}_{13}]+[\omega_2\bar{\omega}_{23}]=0.\end{aligned} \tag{27}$$

由于问题归结到这样：求二维积分流形 m_2 使在其上$[\omega_1\ \omega_2]\neq 0$，我们用不着导微(25)的最末式.

根据嘉当的准衡得知这系统是在对合下的,且正则积分元素链的标数是 $S_1=4$，$S_2=0$. 所以解的自由度与单项目的 4 个函数有关,也就是说：$V_2\subset S_3$ 时，保持平均曲率而且互为变形的曲面偶;是与单项目的 4 个函数有关的;这和 $k_0^2=0$ 的情形也就是在§1所提到的结果相同.

因此当决定系统秩数等于 0 时，$V_m\subset E_{m+1}$ 保持平均曲率的变形问题与单项目的 4 个函数有关.也就确定了这类超曲面一般能由测度张量及平均曲率完全确定.

§4. 现在我们要将这种情形中不可变形的超曲面的第二基本形式的系数 λ_{ij} 用平均曲率及测度张量来表示,我们先述 $V_2\subset S_3$ 的情形，而此时与 T. Y. Thomas 所研究 $V_2\subset E_3$ 的情形极为类似,故较简单地叙述,而省略复杂的计算.

设 $V_2\subset S_3$，其平均曲率为

$$\widetilde{H}=\frac{\widetilde{g}^{ij}\,\widetilde{\lambda}_{ij}}{2}, \tag{28}$$

而且 Gauss 方程为

$$\widetilde{R}_{1212}=(\widetilde{\lambda}_{11}\widetilde{\lambda}_{22}-\widetilde{\lambda}_{12}^{2})+k_0^2(\widetilde{g}_{11}\widetilde{g}_{22}-\hat{g}_{12}^{2}), \tag{29}$$

按方程

$$|\widetilde{g}_{ij}|\widetilde{K}=\widetilde{\lambda}_{11}\widetilde{\lambda}_{22}-\lambda_{12}^{2} \tag{30}$$

定义总曲率 $\widetilde{K}$.

由(29)成立关系

$$\widetilde{R}_{1212}=|\widetilde{g}_{ij}|\widetilde{K}+k_0^2|\widetilde{g}_{ij}|. \tag{31}$$

在任何一点选择坐标系统使 $g_{ij}=\delta_{ij}$，则(28)，(30)可写为

$$\widetilde{H}=\frac{\widetilde{\lambda}_{11}+\widetilde{\lambda}_{22}}{2}, \tag{32}$$

$$\widetilde{K}-\widetilde{\lambda}_{11}\widetilde{\lambda}_{22}-\widetilde{\lambda}_{12}^{2}, \tag{33}$$

而 Gauss－Петерсон－Codazzi 方程为

$$\widetilde{R}_{1212}=(\widetilde{\lambda}_{11}\widetilde{\lambda}_{22}-\widetilde{\lambda}_{12}^{2})+k_0^2, \tag{34}$$

$$\widetilde{\lambda}_{ij,k}-\widetilde{\lambda}_{ik,j}=0, \tag{35}$$

在这坐标系统下

$$4(\widetilde{H}^2-\widetilde{K})=(\widetilde{\lambda}_{11}-\widetilde{\lambda}_{22})^2+4\widetilde{\lambda}_{12}^{2}\geqslant 0. \tag{36}$$

如果 $\widetilde{H}^2-\widetilde{K}=0$，则由上式可证明在任何坐标系统下成立关系式

$$\widetilde{\lambda}_{ij}=\widetilde{H}\widetilde{g}_{ij}. \tag{37}$$

由此式再利用 Петерсон－Codazzi 方程可得，如果曲面上某区域满足条件：$\widetilde{H}^2-\widetilde{K}=0$，则曲面的总曲率及平均曲率均为常数.

从(32)，(33)，(35)利用 Ricci 恒等式

$$\widetilde{\lambda}_{ij,kl}-\widetilde{\lambda}_{ij,lk}=-\widetilde{\lambda}_{ik}\widetilde{R}^{h}_{jkl}-\widetilde{\lambda}_{hj}\widetilde{R}^{h}_{ikl},$$

及(34)并经过复杂的演算(相当于 T. Y. Thomas 的原文)可得

$$\tilde{\lambda}^{\alpha\beta}Q_{\alpha\beta}=\tilde{g}^{\alpha\beta}P_{\alpha\beta}, \tag{38}$$

式中

$$Q_{\alpha\beta}=2(\tilde{H}^2-\tilde{K})\tilde{H}_{\alpha\beta}-4\tilde{H}\tilde{H}_\alpha\tilde{H}_\beta+\tilde{K}_\alpha\tilde{H}_\beta+\tilde{K}_\beta\tilde{H}_\alpha,$$

$$P_{\alpha\beta}=(\tilde{H}^2-\tilde{K})\tilde{K}_{\alpha\beta}+\tilde{K}_\alpha\tilde{K}_\beta+\tilde{H}\tilde{K}_\alpha\tilde{H}_\beta-\tilde{H}\tilde{K}_\beta\tilde{H}_\alpha-2(\tilde{K}+k_0^2)(\tilde{H}^2-\tilde{K})^2\tilde{g}_{\alpha\beta};$$

而且

$$\tilde{\lambda}^{\alpha\beta}=\frac{\tilde{\lambda}_{\alpha\beta}\text{ 在 }|\tilde{\lambda}_{\alpha\beta}|\text{ 中的余因子}}{|\tilde{g}_{\mu\nu}|}.$$

当 $\tilde{H}^2-\tilde{K}\geqslant 0$ 时，式(38)都成立.

改写(38)及(28)

$$\left.\begin{aligned}&Q_{22}\tilde{\lambda}_{11}-2Q_{12}\tilde{\lambda}_{12}+Q_{11}\tilde{\lambda}_{22}=|\tilde{g}_{\mu\nu}|\tilde{g}^{\alpha\beta}P_{\alpha\beta},\\&\tilde{g}_{22}\tilde{\lambda}_{11}-2\tilde{g}_{12}\tilde{\lambda}_{12}+\tilde{g}_{11}\tilde{\lambda}_{22}=2|\tilde{g}_{\mu\nu}|\tilde{H}.\end{aligned}\right\} \tag{39}$$

其系数矩阵为

$$\left\|\begin{matrix}Q_{22}&Q_{12}&Q_{11}\\\tilde{g}_{22}&\tilde{g}_{12}&\tilde{g}_{11}\end{matrix}\right\|.$$

一般这矩阵秩数等于 2，不失一般性我们可以假定最后二行的行列式不等于 0，于是按照(39)及(30)可在坐标系统 $\tilde{g}_{\alpha\beta}=\delta_{\alpha\beta}$ 下来计算 λ_{ij} 的数值，然后再用张量形式表达出来而得

$$\tilde{\lambda}_{ij}=RW^{-1}\tilde{g}_{ij}+2UW^{-1}Q_{ij}\pm\{[(\tilde{H}^2-\tilde{K})W-U^2]^{\frac{1}{2}}/W\}V_{ij}. \tag{40}$$

这里

$$W=(\tilde{g}^{ij}Q_{ij})^2-4|Q_{ij}|/|\tilde{g}_{ij}|,$$

$$R=\tilde{g}^{ij}R_{ij}\tilde{g}^{hk}Q_{hk}-4\tilde{H}|Q_{ij}|/|\tilde{g}_{ij}|,$$

$$U=\tilde{H}\tilde{g}^{ij}Q_{ij}-\tilde{g}^{ij}P_{ij},$$

$$V_{ij}=(Q_{ih\varepsilon jk}+Q_{jh\varepsilon ik})\tilde{g}^{hk},\quad \varepsilon_{12}=-\varepsilon_{21}=(|\tilde{g}_{ij}|)^{\frac{1}{2}},\quad \varepsilon_{11}=\varepsilon_{22}=0.$$

这样，我们已将 $V_2 \subset S_3$ 的第二基本形式系数 $\tilde{\lambda}_{ij}$ 在坐标系统 $g_{ij}=\delta_{ij}$ 下用 $\tilde{H}$ 及 $\tilde{K}$ 表示起来，而且 $V_2 \subset E_3$ 的问题恰相当于 $k_0^2=0$ 的情形.

现在回到 $V_m \subset E_{m+1}$ 的情形.当决定系统的秩数等于 0 时，对于一般的曲面利用(21)(22)，可表示其第二基本形式系数 $\lambda_{ij}(u^1, \cdots, u^m)$ 如此：

$$\lambda_{ij}(u^1, u^2, \cdots, u^m)=\frac{e^{v(u^3, \cdots, u^m)}\,\tilde{\lambda}_{ij}(u^1, u^2)}{e^{v(u_0^3, \cdots, u_0^m)}}, \tag{41}$$

而且这里的 $\tilde{\lambda}_{ij}(u^1, u^2)$ 决定于(40).而其中的 $\tilde{K}$ 为

$$|\tilde{g}_{ij}|\tilde{K}=R_{1212}(u^1, u^2, u_0^3, \cdots, u_0^m). \tag{42}$$

这样，在选定了 $u_0^3, \cdots, u_0^m$ 之后，$\lambda_{ij}(u^1, \cdots, u^m)$ 就能决定起来，从这里可以看到，除了所列举的 $\lambda_{ij}(u^1, \cdots, u^m)$ 以外，已经没有其他可能有的解，所以 $\lambda_{ij}(u^1, \cdots, u^m)$ 的决定与 $u_0^3, \cdots, u_0^m$ 的选择无关.我们可表述由 $H(u^1, \cdots, u^m)$ 与 $g_{ab}(u^1, \cdots, u^m)$ 来决定 $\lambda_{ij}(u^1, \cdots, u^m)$ 的过程如下：先任意选择一组数值 $u_0^3, \cdots, u_0^m$；依据(42)，得到 $\tilde{K}$. 按照(24)，得到 $\tilde{H}(u_1, u_2)$ 再把它们代到(40)中去，求到 $\lambda_{ij}(u^1, u^2)$. 最后由(41)就能完全决定 $\lambda_{ij}(u^1, \cdots, u^m)$.

由于主曲率是方程

$$\left|\lambda_{ij}-\frac{1}{R}g_{ij}\right|=0, \quad i, j=1, 2, \cdots, m$$

的根，在黎曼测度的秩数等于 2 的情况下，只有两个非平凡的主曲率，而它们的乘积为 $\lambda_{11}\lambda_{22}-\lambda_{12}^2$，就是 $R_{1212}(u^1, u^2, \cdots, u^m)$；它们的和是 $g^{ij}\lambda_{ij}$，就是平均曲率.因而保持主曲率的变形问题就与保持平均曲率的变形问题相同.

作者在写此文时，得苏步青教授的指导，敬致深切的感谢.

参考文献

[1] Яненко Н. Н., Некоторые вопросы теории вложения римановых метрик в евклидоы пространства,

УМН, VIII вып. 1(53)(1953), 21-100.

[2] Фиников С. П., Метод внешних форм Картана. Гостехиздат, 1948.

[3] Thomas, T. Y., Algebraic determination of the second fundamental form of a surface by its mean curveture. *Bull. Amer. Math. Soc.*, **51**(1945), 390-399.

[4] 胡和生，论黎曼测度 V_m 在常曲率空间 S_{m+1} 中的变形.(尚未发表)

On the Deformation of a Riemannian Metric V_m of Class Ⅰ Which Preserves the Mean Curvature

Hu Hou-Sung

(Institute of Mathematics, Academia Sinica and Fu-Tan University)

Abstract

It is well known that the hypersurface realizing an m-dimensional Riemannian metric of class Ⅰ in E_{m+1} is not deformable in general. The only possible deformable metric must be of rank 2. With aid of the method of exterior differential forms N. N. Yanenko has recently given a complete classification of deformable hypersurfaces.

The object of the present paper is to investigate the deformation of $V_m \subset E_{m+1}$ which preserves the mean curvature. Of course, the metric must necessarily be of rank 2. In the theory of Yanenko the system of determination which takes the form

$$\Gamma^i_{\alpha j}\lambda_{ik} - \Gamma^i_{\alpha k}\lambda_{ij} = 0, \quad \alpha = 3, \cdots, m; \quad i, j = 1, 2 \tag{S}$$

in suitable coordinates plays an important rôle. In our case we have the following result:

(i) When the system (S) is of rank 2, the hypersurface is non-deformable.

(ii) When (S) is of rank 1, the hypersurfaces is either non-deformable or deformable discretely.

(iii) When (S) is of rank 0, the problem is reduced to the deformation of V_2 in S_3

(space of positive constant curvature) which preserves the mean curvature.

In order to obtain the final result of case (iii) we use, at first, Cartan's criterion for a system of exterior differential equations and thence derive that a paris of deformable surfaces with preservation of mean curvatures depends upon four arbitrary functions of a single argument. In the next, using the method of T. Y. Thomas we give an explicit expression of the coefficient, of the second fundamental form of $V_2 \subset S_3$ in terms of the metric tensor and the mean curvature in general. In consequence, there can also be obtained the expression of the coefficients of the second fundamental form of a general V_m in the case (iii), those for the cases (i)(ii) being trivail. Thus we have generalized the results of T. Y. Thomas (1945) to the case $V_m \subset E_{m+1} (m > 2)$.

(本文曾发表于《数学学报》1956年第6卷第1期,127－137)

论黎曼测度 V_m 在常曲率空间 S_{m+1} 中的变形*

胡和生

中国科学院数学研究所,复旦大学

§1. 关于 m 维黎曼测度($m \geqslant 3$)

$$ds^2 = g_{ij} du^i du^j \quad (i, j = 1, \cdots, m)$$

在欧氏空间 E_{m+1} 中的安装及变形问题在近一世纪的几何学者的工作中得到了解决,而所确定的 $V_m \subset E_{m+1}$ 一般是不能变形的.就是说,能够变形的只是狭窄的一类超曲面. Яненко[1]运用了外微分形式的方法,很详细地综合了这些工作,并且完全地给出了高维欧氏空间的可变形的超曲面的分类.

如所知,$V_m \subset S_{m+1}$ 一般也是不能变形的[2].作者运用外微分形式和 Яненко 的方法来研究这个问题,首先定义测度的 k_0-秩数,证明 k_0-秩数>2 的测度的不可变形性并且讨论高斯-柯达齐-彼杰尔松方程的相关性.最后,确定了高维常曲率空间的可变形的超曲面的完全分类.

§2. 在 $m+1$ 维黎曼空间

$$ds^2 = g_{\alpha\beta} \omega^\alpha \omega^\beta \quad (\alpha, \beta = 1, \cdots, m+1) \tag{1}$$

的每点对于已给的形式 $\omega^1, \omega^2, \cdots, \omega^{m+1}$ 时常能建立相切的标形 $I_1, I_2, \cdots, I_{m+1}$,使得

$$dr = \omega^\alpha I_\alpha,$$

* 1955 年 10 月 10 日收到.

$$dI_\alpha = \omega_\alpha^\beta I_\beta.$$

而且

$$ds^2 = dr^2 = I_\alpha I_\beta \omega^\alpha \omega^\beta = g_{\alpha\beta}\omega^\alpha \omega^\beta,$$

我们称方程

$$\left.\begin{array}{l} dg_{\alpha\beta} = \omega_\alpha^\gamma g_{\gamma\beta} + \omega_\beta^\gamma g_{\gamma\alpha}, \\ (\omega^\alpha)' = [\omega^\beta \omega_\beta^\alpha], \\ (\omega_\alpha^\beta)' = [\omega_\alpha^\gamma \omega_\gamma^\beta] + R_{\alpha\gamma\delta}^\beta [\omega^\gamma \omega^\delta] \end{array}\right\} \tag{3}$$

为黎曼空间的组织方程，这里 $R_{\alpha\gamma\delta}^\beta$ 为曲率张量.

特别，当黎曼空间是常曲率空间 k_0 时，组织方程变为

$$\left.\begin{array}{l} dg_{\alpha\beta} = \omega_\alpha^\gamma g_{\gamma\beta} + \omega_\beta^\gamma g_{\gamma\alpha}, \\ (\omega^\alpha)' = [\omega^\beta \omega_\beta^\alpha], \\ (\omega_\alpha^\beta)' = [\omega_\alpha^\gamma \omega_\gamma^\beta] + k_0 g^{\varepsilon\beta}(g_{\varepsilon\gamma}g_{\varepsilon\delta} - g_{\varepsilon\delta}g_{\alpha\gamma})[\omega^\gamma \omega^\delta]. \end{array}\right\} \tag{4}$$

对于常曲率空间 S_{m+1} 的超曲面 V_m

$$r = r(u^1, u^2, \cdots, u^m),$$

可以选择这样的标形 $(r, I_i)(i=1, 2, \cdots, m)$ 使得每点的标形与外面空间 S_{m+1} 的相切标形符合，并且我们通常选取半直交的标形族，即向量 I_{m+1} 与曲面 V_m 相直交，而为曲面的法线.

如此，曲面的导来形式为

$$\left.\begin{array}{l} dr = \omega^i I_i \quad (i=1, 2, \cdots, m;\ \alpha = 1, 2, \cdots, m+1), \\ dI_\alpha = \omega_\alpha^\beta I_\beta, \quad I_\alpha I_\beta = g_{\alpha\beta}, \\ g_{m+1\alpha} = \delta_{m+1\alpha}, \quad \omega^{m+1} = 0. \end{array}\right\} \tag{5}$$

由(4)我们可得曲面的组织方程

$$\left.\begin{aligned}&[\omega^i\omega_i^{m+1}]=0,\\&(\omega_i^{m+1})'-[\omega_i^j\omega_j^{m+1}]=0,\\&(\omega_j^i)'=[\omega_j^k\omega_k^i]+[\omega_j^{m+1}\omega_{m+1}^i]+k_0g^{li}(g_{lh}g_{jk}-g_{lk}g_{jh})[\omega^h\,\omega^k].\end{aligned}\right\}\tag{6}$$

称后面两个方程为高斯－柯达齐－彼杰尔松方程.

在半直交标形下，我们将形式 ω_β^α 分为三群：

$$\left.\begin{aligned}&\omega_j^i=\Gamma_{jk}^i\omega^k, &&\text{（内在形式）}\\&\omega_i^{m+1}=\lambda_{ij}\omega^j,\quad\lambda_{ij}=\lambda_{ji}, &&\text{（混合形式）}\\&\omega_{m+1}^{m+1}=0. &&\text{（外在形式）}\end{aligned}\right\}\tag{7}$$

而且

$$\omega_{m+1}^i=-g^{ij}\omega_j^{m+1}.\tag{8}$$

这样可将组织方程写成张量形式

$$\left.\begin{aligned}&R_{ijkl}-k_0(g_{ik}g_{jl}-g_{il}g_{jk})=\lambda_{ik}\lambda_{jl}-\lambda_{il}\lambda_{jk},\\&\lambda_{ik,j}-\lambda_{ij,k}=0.\end{aligned}\right\}\tag{9}$$

其中 $\lambda_{ij}\omega^i\omega^j$ 是曲面的第二基本形式，因此混合形式的系数即为第二基本形式的系数.

§3. 首先导入曲面的秩数的定义.

定义. 曲面 V_m 在已给点M的秩数是混合形式系统ω_i^{m+1} 在对应点M的秩数，换句话说，就是这个系统中线性独立形式的数目.

容易看到，曲面的秩数与已给点的标形的选择无关，下面将 Beez[3]定理推广到常曲率空间.

定理. 设 $V_m\subset S_{m+1}$.当秩数 >2 时，V_m 是不可变形的，也就是说，如果 V_m 的秩数 >2，则与 V_m 等长的任何曲面$\overline{V}_m$ 一定与 V_m 符合.

对曲面 V_m 装配二阶可微分的直交标形族 I_1，…，I_{m+1} 使这些满足条件

$$dr=\omega^iI_i\quad(i=1,\,2,\,\cdots,\,m),$$

$$dI_\alpha = \omega_\alpha^\beta I_\beta \quad (\alpha, \beta = 1, \cdots, m+1),$$

$$I_\alpha I_\beta = \delta_{\alpha\beta}, \quad \omega_\alpha^\beta + \omega_\beta^\alpha = 0.$$

对于等长的曲面 $\bar{V}_m$ 选择半直交标形族$\bar{I}_1$, $\bar{I}_2$, …, $\bar{I}_{m+1}$ 使得

$$d\bar{r} = \bar{\omega}^i \bar{I}_i = \omega^i \bar{I}_i.$$

利用等长的性质可知标形 $\bar{I}_1$, …, $\bar{I}_{m+1}$ 也是直交的,且为单位的.

考察

$$\omega^i = \bar{\omega}^i \tag{10}$$

的可积条件:

$$[\omega^j \Delta_j^i] = 0,$$

此处

$$\Delta_j^i = \bar{\omega}_j^i - \omega_j^i = -\Delta_i^j.$$

利用唯一性定理,可以证明

$$\Delta_j^i = 0, \quad \bar{\omega}_j^i = \omega_j^i. \tag{11}$$

现在作出上式的可积条件.由于标形是直交的,

$$(\omega_j^i)' = [\omega_j^k \omega_k^i] - [\omega_j^{m+1} \omega_i^{m+1}] + k_0(\delta_{ik}\delta_{jl} - \delta_{il}\delta_{jk})[\omega^k \omega^l],$$

$$(\bar{\omega}_j^i)' = [\omega_j^k \omega_k^i] - [\bar{\omega}_j^{m+1} \bar{\omega}_i^{m+1}] + k_0(\delta_{ik}\delta_{jl} - \delta_{il}\delta_{jk})[\omega^k \omega^l],$$

因此可得

$$[\omega_i^{m+1} \omega_j^{m+1}] = [\bar{\omega}_i^{m+1} \bar{\omega}_j^{m+1}]. \tag{12}$$

两边乘 $\bar{\omega}_i^{m+1}$, 我们有

$$[\omega_i^{m+1} \bar{\omega}_i^{m+1} \omega_j^{m+1}] = 0. \tag{13}$$

分下列两种情形来讨论:

1. 如果至少有一个i使$[\omega_i^{m+1}\bar{\omega}_i^{m+1}]\neq 0$，则由(12)

$$\omega_j^{m+1}=0 \mod \omega_i^{m+1},\ \bar{\omega}_i^{m+1}\quad (j=1,\ 2,\ \cdots,\ m).$$

这意味着ω_i^{m+1}，$\bar{\omega}_j^{m+1}$的秩数为2，而与假设矛盾.

2. 如果对于所有指数i及任意选择直交标形I_1，…；I_{m+1}有

$$[\omega_i^{m+1}\bar{\omega}_i^{m+1}]=0,$$

那么

$$\bar{\omega}_i^{m+1}=\lambda_i\omega_i^{m+1}\quad (i=1,\ 2,\ \cdots,\ m).$$

由于ω_i^{m+1}的秩数大于2，当$m\geqslant 3$时，将此式代入(12)可以证明

$$\lambda_i=1,$$

于是得到

$$\left.\begin{aligned}&\omega^i=\bar{\omega}^i,\\&\omega_j^i=\bar{\omega}_j^i,\\&\omega_i^{m+1}=\bar{\omega}_i^{m+1}=-\omega_{m+1}^i=-\bar{\omega}_{m+1}^i,\\&I_\alpha\cdot I_\beta=\bar{I}_\alpha\cdot\bar{I}_\beta=\delta_{\alpha\beta}\quad (i,\ j=1,\ 2,\ \cdots,\ m;\ \alpha,\ \beta=1,\ 2,\ \cdots,\ m+1).\end{aligned}\right\}\tag{14}$$

因此曲面V_m与$\bar{V}_m$符合，定理证毕.

§4. 在本节我们讨论黎曼测度在常曲率空间中的安装的一些问题.

定义. m维黎曼测度$ds^2=g_{ij}\omega^i\omega^j\ (i,\ j=1,\ 2,\ \cdots,\ m)$是可以被安装到$m+1$维常曲率空间$S_{m+1}$中的，如果在$S_{m+1}$中存在三次连续可微分的曲面$V_m$，它的向径为$r=r(u^1,\ \cdots,\ u^m)$而有已给测度

$$dr^2=ds^2=g_{ij}\omega^i\omega^j,\tag{15}$$

我们称曲面$V_m\subset S_{m+1}$为测度的实现曲面.

我们容易建立m维黎曼测度能够被安装在$m+1$维常曲率空间的充要条件.

定理. 为了已知的 m 维测度 $ds^2=g_{ij}\omega^i\omega^j$ 能被安装到常曲率 k_0 空间 S_{m+1} 的充要条件是存在满足条件(6)及(8)的形式 ω_i^{m+1}, ω_{m+1}^i.

现在导入测度的 k_0-秩数的概念,在后面的变形问题的研究中,这概念极其重要.

定义. 称反称的双一次形式系统

$$k_{ij}=\sum_{k<l}(R_{ijkl}-k_0(g_{ik}g_{jl}-g_{il}g_{jk}))[\omega^k\omega^l] \tag{16}$$

的秩数为测度 $ds^2=g_{ij}\omega^i\omega^j$ 的 k_0-秩数.

可以看到,测度的 k_0-秩数等于矩阵 $R_{ijkl}-k_0(g_{ik}g_{jl}-g_{il}g_{jk})$ 的秩数,并且秩数的定义对于 ω^i 的不退缩一次变换是不变的.

我们来研究测度的 k_0-秩数与曲面的秩数间的关系.

定理. 如果能够把具有 k_0-秩数$\geqslant 2$ 的测度 $ds^2=g_{ij}\omega^i\omega^j$ 安装在常曲率 k_0 空间 S_{m+1},则 k_0-秩数与它的实现的曲面的秩数符合.

证. 设 $\omega_i^{m+1}=\varphi_i$ 为已给测度 ds^2 所实现的曲面 $V_m\subset S_{m+1}$ 的混合形式,它必须满足高斯关系式(6).按照(16)可得

$$K_{ij}=[\varphi_i\varphi_j], \tag{17}$$

因而,很明显地看出,在测度的 k_0-秩数$\geqslant 2$ 的假定下,K_{ij} 的基层符合于φ_i, φ_j 的基层;于是证明了定理.

将此定理与§3的推广的Beez's定理联系起来,可以知道,在测度 $ds^2=g_{ij}\omega^i\omega^j$ 的 k_0-秩数$\geqslant 2$时它在常曲率 k_0 空间中实现的曲面是不能变形的,所以在后文所讨论的变形问题只须考虑测度的 k_0-秩数$\leqslant 2$ 的情形.

§5. 现在我们来讨论高斯-柯达齐-彼杰尔松方程的相关性.所得到的结果与 T. Y. Thomas[4] 对欧氏空间超曲面所得到的相类似.

定理. 如果能够把测度 $ds^2=g_{ij}\omega^i\omega^j$ 安装在常曲率 k_0 空间 S_{m+1},而且测度的 k_0-秩数$\geqslant 4$,那么,柯达齐-彼杰尔松方程是高斯方程的推论.

证. 由(6)得到高斯方程:

$$(\omega_j^i)'-[\omega_j^k\omega_k^i]-k_0g^{li}(g_{lh}g_{jk}-g_{lk}g_{jh})[\omega^h\omega^k]=[\omega_j^{m+1}\omega_{m+1}^i],$$

或者，按照(8)改写做

$$(\omega_j^i)'=[\omega_j^k\omega_k^i]-2k_0g_{jk}[\omega^i\omega^k]=-g^{il}[\psi_j\psi_l], \tag{18}$$

外绝对微分上式并且用 D 表示外绝对微分得到

$$\begin{aligned}&D\Omega_j^i-2k_0[Dg_{jk}\omega^i\omega^k]-2k_0g_{jk}[D\omega^i\omega^k]+2k_0g_{jk}[\omega^iD\omega^k]\\&=-[Dg^{il}\psi_j\psi_l]-g^{il}[D\psi_j\psi_l]+g^{il}[\psi_jD\psi_l],\end{aligned} \tag{19}$$

式中

$$\Omega_j^i=R_{jkl}^i[\omega^k\omega^l]=(\omega_j^i)'-[\omega_j^k\omega_k^i]$$

利用皮安基恒等式[5]

$$D\Omega_j^i=0,$$

和

$$D\omega^i=0,\quad Dg_{ij}=0,\quad Dg^{ij}=0. \tag{20}$$

又可改写(19)：

$$[D\psi_i\psi_j]-[D\psi_j\psi_i]=0,$$

而

$$D\psi_i=d\psi_i-[\omega_i^k\psi_k].$$

以 ψ_j 乘(20)的两端得到

$$[\psi_i\psi_jD\psi_j]=0\quad(j\text{ 非和式}),$$

或者，导入记号

$$\theta=[\psi_jD\psi_j]\quad(j\text{ 非和式}),$$

则

$$[\psi_i\theta]=0. \tag{21}$$

按照假设 k_0-秩数 $\geqslant 4$，在一般情况下得到

$$[\psi_i\psi_j\psi_k\psi_l]\neq 0 \quad (i\neq j\neq k\neq l).$$

因此，从(21)导出

$$\theta=0,$$

即

$$[D\psi_j\psi_j]=0 \quad (j\text{ 非和式}).$$

由于最后关系式在任意标形之下都要成立，而且 $D\psi_j$，ψ_j 的变换是按照共变定律的，所以

$$[D\psi_i\psi_j]+[D\psi_j\psi_i]=0. \tag{22}$$

从(20)和(22)得到

$$[D\psi_i\psi_j]=0.$$

再利用秩数 $\geqslant 4$ 的假定，就可导出柯达齐-彼杰尔松方程

$$D\psi_i=d\psi_i-[\omega_i^j\psi_j]=0.$$

§6. 在变形问题的研究中要适当地选择在每点的标形来满足几何上的要求.在本节中，我们简单叙述几种特殊的标形.

1. 和乐标形.如果 ω^i 是全微分 du^i，则称标形 I_1，I_2，…，I_{m+1} 为和乐标形；相反地在其他情形称它为非和乐标形.对于和乐标形得到

$$dr=du^iI_i,$$

因此

$$I_i=\frac{\partial r}{\partial u^i},$$

应用和乐标形有它的优点,但是为此却减少标形的选择自由度,从而限制曲面上的标形族的变换集合.

2. 规范标形. 我们先证下列定理.

定理. 设测度的 k_0-秩数是 r,那么,经过适当的标形 I_1, …, I_m 的变换之后能够把 K_{ij} 化为如下的规范方式:

$$\left.\begin{aligned} &k_{i\alpha}=0 \quad (i=1, \cdots, m;\ \alpha=r+1, \cdots, m),\\ &k_{ij}=\sum_{k<l}(R_{ijkl}-k_0(g_{ik}g_{jl}-g_{il}g_{jk}))[\omega^k\omega^l] \quad (i, j, k, l=1, \cdots, r).\end{aligned}\right\} \tag{23}$$

证. 取 k_{ij} 的基层 $\bar\omega^1$, $\bar\omega^2$, …, $\bar\omega^r$ 作为新的基层 $\bar\omega^1$, …, $\bar\omega^r$, $\bar\omega^{r+1}$, …, $\bar\omega^m$ 中的最初 r 个形式. 在新的基层下,我们有

$$k_{ij}=\sum_{s<t}(R_{ijst}-k_0(g_{is}g_{jt}-g_{it}g_{js}))[\bar\omega^s\bar\omega^t] \quad (i, j=1, \cdots, m;\ s, t=1, \cdots, r);$$

也就是

$$R_{ijk\alpha}-k_0(g_{ik}g_{j\alpha}-g_{i\alpha}g_{jk})=0 \quad (i, j, k=1, \cdots, m;\ \alpha=r+1, \cdots, m).$$

由于上式对 (ij), $(k\alpha)$ 为对称的,因此

$$k_{k\alpha}=0 \quad (k=1, \cdots, m;\ \alpha=r+1, \cdots, m).$$

k_0-秩数为 r 的测度当标形变换后使得 $K_{ij}(i, j=1, \cdots, m)$ 满足(23)时称所对应的标形 I_1, …, I_m 为规范标形.

不难证明存在一些标形同时具有规范的及直交的性质,也有一些标形,同时具有规范的与和乐的性质.我们也注意到只有在欧氏空间才存在直交的和乐标形.

§7. 现在开始进行可变形的超曲面 $V_m\subset S_{m+1}$ 的分类.我们已经证明:当测度的 k_0-秩数 >2 时,实现的曲面不能变形,所以只有在测度的 k_0-秩数 $\leqslant 2$ 的时候,实现的曲面才有变形的可能;我们来分别考虑这些情形.

当 k_0-秩数 $=0$ 时,

$$k_{ij}=0,$$

于是

$$R_{ijkl}-k_0(g_{ik}g_{jl}-g_{il}g_{jk})=0.$$

这时曲面本身就是一个 m 维的常曲率 k_0 的空间，故为0级的.本文现不讨论其在 $m+1$ 维常曲率 k_0 的空间的变形问题.事实上，可以证明，它在 $m+1$ 维常曲率 k_0 的空间的变形问题与 $m+1$ 维欧氏空间中的超平面的变形问题相同，而其自由度为 m 个单参数的函数[6].

当 k_0-秩数 $=1$ 时，由于反称双一次形式的秩数常是偶数，所以这种情形不可能产生，于是剩下考虑测度的 k_0-秩数是 2 的情形，我们利用规范和乐的标形来进行讨论，由于 $I_1,\cdots,I_m$ 是和乐的，所以满足

$$\omega^\alpha=du^\alpha,\quad \alpha=1,2,\cdots,m.$$

这样一来，K_{ij} 采取下列形态：

$$k_{12}=(R_{1212}-k_0(g_{11}g_{22}-g_{12}^2))[du^1\,du^2],$$
$$k_{i\alpha}=0\quad(i=1,\cdots,m;\quad \alpha=3,\cdots,m),\tag{24}$$

按照 §4 的定理，形式系统 K_{ij} 的基层 $\omega^1=du^1$，$\omega^2=du^2$ 与实现的基层 $\psi_i=\omega_i^{m+1}$ 符合，于是

$$\psi_\alpha=0,\quad \alpha=3,\cdots,m;\tag{25}$$

$$\psi_i=\lambda_{ij}du^j,\quad \lambda_{ij}=\lambda_{ji},\quad i,j=1,2.\tag{26}$$

(25)的可积条件是

$$\Gamma^i_{\alpha k}\lambda_{ij}-\Gamma^i_{\alpha j}\lambda_{ik}=0,\tag{27}$$

$$\Gamma^i_{\alpha\beta}\lambda_{ij}=0.\tag{28}$$

利用 ψ_i 的秩数等于 2 的性质，可得

$$\Gamma^i_{\alpha\beta}=0.$$

故(26)的可积条件是

$$\frac{\partial\lambda_{ij}}{\partial u^k}-\frac{\partial\lambda_{ik}}{\partial u^j}=\Gamma^l_{ik}\lambda_{jl}-\Gamma^l_{ij}\lambda_{kl}, \tag{29}$$

$$\frac{\partial\lambda_{ij}}{\partial u^\alpha}=\Gamma^l_{i\alpha}\lambda_{jl}. \tag{30}$$

这些都是柯达齐-彼杰尔松条件，我们称(27)为决定的系统，它是关于张量 λ_{ij} 的齐一次方程组.根据决定系统的秩数，我们来做变形的分类.

1. 决定系统的秩数=2.这时测度的安装是唯一的，因为由(27) λ_{ij} 除了一个因子外可以完全决定；利用高斯方程

$$R_{1212}-k_0(g_{11}g_{22}-g_{12}^2)=\lambda_{11}\lambda_{22}-\lambda_{12}^2, \tag{31}$$

可以得到

$$\lambda_{ij}=\pm\Delta_{ij}\sqrt{\frac{R_{1212}-k_0(g_{11}g_{22}-g_{12}^2)}{\Delta_{11}\Delta_{22}-\Delta_{12}^2}}. \tag{32}$$

2. 如果决定系统秩数为 1，则

$$\frac{\Gamma^1_{\alpha2}}{\Gamma^1_2}=\frac{-\Gamma^2_{\alpha1}}{\Gamma^2_1}=\frac{\Gamma^2_{\alpha2}-\Gamma^1_{\alpha1}}{\Gamma^2_2-\Gamma^1_1}=k_\alpha$$

决定系统简化为

$$\Gamma^1_2\lambda_{11}+(\Gamma^2_2-\Gamma^1_1)\lambda_{12}+\Gamma^2_1\lambda_{22}=0,$$

因而，又可细分为下列几种情形.

2_1. 仿射子 Γ^i_j 有二个不同的特征根，我们可以选择标形使得

$$\lambda_{12}=0,$$

高斯方程变为

$$k=R_{1212}-k_0(g_{11}g_{22}-g_{12}^2)=\lambda_{11}\lambda_{22},$$

柯达齐-彼杰尔松方程的积分可能条件为

$$As^2+2Bs+C=0,$$

这里 A, B, C 是 Γ^i_{jk} 和 k 的函数,而且 $s=\lambda^2_{11}$,这时有三种可能.

2_{11}. 如果这代数方程有二个不同的根,那么可把测度实现在二个不符合的曲面,称它为稀疏的变形.

2_{12}. 如果这方程有重根,那么得到 2_{11} 的极限情形,实现的曲面是唯一的.

2_{13}. 如果 $A=B=C=0$,那么柯达齐-彼杰尔松方程是完全可积分的 $\lambda_{11}=\lambda_{11}(u^1, u^2, \cdots, u^m, c)$ 测度实现在单参数族而不相符合的等测度的曲面上.

2_2. 仿射子 Γ^i_j 有一个特征根,在适当的标形选择之下

$$\lambda_{11}=0,$$

$$k=R_{1212}-k_0(g_{11}g_{22}-g^2_{12})=-\lambda^2_{12}.$$

$$\lambda_{12}=\sqrt{-k}.$$

这时有二种可能:

2_{21}. $\Gamma^2_{11}\neq 0$ 实现的曲面是唯一的.

2_{22}. $\Gamma^2_{11}=0$ 实现不是唯一的,而依赖于单参数的一个任意函数.

3. 最后讨论决定系统的秩数是零的情形,这时

$$\Gamma^2_{\alpha 1}=\Gamma^1_{\alpha 2}=\Gamma^2_{\alpha 2}-\Gamma^1_{\alpha 1}=0,\quad \alpha=3, \cdots, m,$$

$$\Gamma^2_{\alpha 2}-\Gamma^1_{\alpha 1}=\Gamma_\alpha, \tag{33}$$

依据 Яненко 类似的方法,可以证明:这时能选取这样的规范和乐标形,它能满足等式

$$I_i\cdot I_\alpha=0,\quad i=1, 2;\quad \alpha=3, \cdots, m.$$

令

$$g=g_{11}g_{22}-g^2_{12}\quad (g>0).$$

由于 g_{ij} 的共变微分为 0,从(33)式就能得到

$$\frac{\partial g}{\partial u^\alpha}=4\Gamma_\alpha g,$$

或

$$\frac{\partial \log g}{\partial u^\alpha}=4\Gamma_\alpha.$$

利用组织方程及(33),并经过一些计算,可以证明:Γ^i_{jk} 只与 u^1, u^2 有关,Γ_α 与 u^1, u^2 无关,并且存在关系式

$$\frac{\partial \Gamma_\alpha}{\partial u^\beta}-\frac{\partial \Gamma_\beta}{\partial u^\alpha}=0.$$

这样一来,

$$g(u^1, u^2, u^3, \cdots, u^m)=g_1(u^1, u^2)e^{4v(u^3, \cdots, u^m)},$$

式中

$$v=\int \Gamma_\alpha\, du^\alpha$$

单为 u^3, …, u^m 的函数,g_1 单为 u^1, u^2 的函数且 $g_1>0$. 现在作变数的变换

$$\bar{u}^1=u^1, \quad \bar{u}^2=\varphi(u^1, u^2), \quad \bar{u}^\alpha=u^\alpha.$$

这里 $\varphi(u^1, u^2)$ 是某一待定的函数. 显而易见,经过这种方式的坐标变换之后,所引起的和乐标形的变换仍然是规范的,而且关系式 $I_i \cdot I_\alpha=0$ 仍然保留. 我们现在要决定 φ 使得 $g_1(u^1, u^2)$ 化为常数 1. 在这个变换下,

$$\bar{\Gamma}_\alpha=\bar{\Gamma}^2_{2\alpha}=\bar{\Gamma}^1_{1\alpha}=\Gamma^\nu_{\lambda\mu}\frac{\partial \bar{u}^1}{\partial u^\nu}\frac{\partial u^\lambda}{\partial \bar{u}^1}\frac{\partial u^\mu}{\partial \bar{u}^\alpha}+\frac{\partial^2 u^\nu}{\partial \bar{u}^1 \partial \bar{u}^\alpha}\frac{\partial \bar{u}^1}{\partial u^\nu}=\Gamma^1_{1\alpha}=\Gamma^2_{2\alpha}=\Gamma_\alpha$$

$$(\lambda, \mu, \nu=1, \cdots, m).$$

由此可见,$g_1(\bar{u}^1, \bar{u}^2)$ 的变换律为

$$\bar{g}_1(\bar{u}^1, \bar{u}^2)=g_1(u^1, u^2)\left|\frac{\partial u^i}{\partial \bar{u}^j}\right|^2.$$

如果选取 φ,使得

$$\left|\frac{\partial \bar{u}^j}{\partial u^i}\right| = \frac{\partial \varphi}{\partial u^2} = \sqrt{g_1(u^1, u^2)}.$$

则可取

$$\varphi = \int \sqrt{g_1(u^1, u^2)}\, du^2,$$

使得

$$\bar{g} = e^{4v(\bar{u}^3, \cdots, \bar{u}^m)},$$

它与 u^1，u^2 无关.我们就在这样选取的标形下来进行讨论.为简便计，仍以 u^1，…，u^m 来记所用的坐标系统，相应的 Γ，g 也不必加上一划，这时高斯-柯达齐-彼杰尔松方程是

$$\left.\begin{aligned} & R_{1212} - k_0 g = \lambda_{11}\lambda_{22} - \lambda_{12}^2, \\ & -\frac{\partial \lambda_{i1}}{\partial u^2} + \frac{\partial \lambda_{i2}}{\partial u^1} = \Gamma_{i1}^j \lambda_{j2} - \Gamma_{i2}^j \lambda_{j1}, \\ & \frac{\partial \lambda_{i1}}{\partial u^\alpha} = \Gamma_\alpha \lambda_{i1}, \quad \frac{\partial \lambda_{i2}}{\partial u^\alpha} = \Gamma_\alpha \lambda_{i2}. \end{aligned}\right\} \tag{34}$$

从最后两式，可表示 λ_{ij} 为

$$\lambda_{ij} = \bar{\lambda}_{ij}(u^1, u^2) e^{v(u^3, \cdots, u^m)},$$

$$R_{1212} - k_0 g = r(u^1, u^2) e^{2v(u^3, \cdots, u^m)}.$$

于是把解(34)的问题归结到解下列方程的问题：

$$r(u^1, u^2) = \bar{\lambda}_{11}\bar{\lambda}_{22} - \bar{\lambda}_{12}^2,$$

$$-\frac{\partial \bar{\lambda}_{i1}}{\partial u^2} + \frac{\partial \bar{\lambda}_{i2}}{\partial u^1} = \Gamma_{i1}^j \lambda_{j2} - \Gamma_{i2}^j \lambda_{j1}.$$

以下证明，这样的问题可化为 $V_2 \subset S_3$ 的变形问题，因为置

$$u^3 = u_0^3, \cdots, u^m = u_0^m.$$

所得到的 V_2 有测度

$$d\bar{s}^2 = g_{ij}(u^1, u^2, u_0^3, \cdots, u_0^m) du^i du^j.$$

而其联络系数即为 $\Gamma^i_{jk}(u^1, u^2)$，而且 $d\bar{s}^2$ 的曲率张量 $\bar{R}_{1212}$ 与 ds^2 的曲率张量 $R_{1212}(u^1, u^2, u_0^3, \cdots, u_0^m)$ 之间的关系是

$$R_{1212} = \bar{R}_{1212} - x^2,$$

此处

$$x^2 = \sum_{\alpha=3}^{m} \Gamma^2_\alpha(u_0^3, \cdots, u_0^m).$$

现在要找函数系统

$$\lambda_{11}(u^1, u^2), \quad \lambda_{12}(u^1, u^2), \quad \lambda_{22}(u^1, u^2),$$

使满足关系式

$$\bar{R}_{1212} - x^2 - k_0 g(u_0^3, \cdots, u_0^m) = r(u^1, u^2) e^{2v(u_0^3, \cdots, u_0^m)} = \lambda_{11}\lambda_{22} - \lambda_{12}^2,$$

$$\frac{\partial \lambda_{i2}}{\partial u^1} - \frac{\partial \lambda_{i1}}{\partial u^2} = \Gamma^j_{i1}\lambda_{j2} - \Gamma^j_{i2}\lambda_{j1}.$$

因为 $u_0^3, \cdots, u_0^m$ 都可以看作参数，所以我们的问题归入 V_2 在常曲率 $x^2 + k_0 g(u_0^3, \cdots, u_0^m)$ 的空间 S_3 中的安装问题. 如所知，它一定有解而且解的自由度依赖于二个单变数的未知函数.如果求得了解 $\lambda_{ij}(u^1, u^2)$ 后，作

$$\bar{\lambda}_{ij} = \frac{\lambda_{ij}(u^1, u^2)}{e^{v(u_0^3, \cdots, u_0^m)}},$$

就易知

$$\lambda_{ij}(u^1, \cdots, u^m) = e^{v(u^3, \cdots, u^m)} \bar{\lambda}_{ij}$$

满足高斯-彼杰尔松-柯达齐方程(34)，这相当于 V_m 的测度在常曲率空间 k_0 中的一个实现. 因而，得到：

定理. 当决定系统的秩数为 0 时，问题归结到 $V_2 \subset S_3$ 的变形问题，而解的自由度依赖于二个单变数的任意函数.

最后,作者在这里对苏步青教授的指导与鼓励表示感谢.

参 考 文 献

[1] Яненко Н. Н., Некоторые вопросы теории вложения римановых метрик в евклидовы пространства, *Успехи матем. науки*, **8**(1953), 21 - 100.

[2] Eisenhart, L. P., Riemannian Geometry (1925), 210 - 214.

[3] Beez, R., Zur Theorie des Krümmungsmasses von Mannigfoltigkeiten höherer Ordnung. *Zeitschr. für Math. u. Phys.*, **21**(1876). 373 - 401.

[4] Thomas, T. Y., Riemann Spaces of Class One and Their Characterization, *Acta Mathematica*, **67**(1936), 169 - 211.

[5] Cartan, E., Leçons sur la gémétrie des espaces de Rieman (1946), 210 - 211.

[6] 胡和生,常曲率空间常曲率超曲面的变形(尚未发表).

On the Deformation of a Riemannian Metric V_m in a Space of Constant Curvature S_{m+1}

Hu Hou-Sung
(Institute of Mathematics, Academia Sinica and Fuh-tan University)

Abstract

With the aid of the method of exterior differential forms N. N. Yanenko has recently given a complete classification of the deformation of m dimensional Riemannian metric

$$ds^2 = g_{ij} du^i du^j \quad (i, j = 1, \cdots, m)$$

in Euclidean space E_{m+1}. Here we propose to investigate the same problem in a space S_{m+1} of constant curvature $k_{0_{m+1}}$. Introducing the definition of the k_0 - rank of a metric, we obtain the following results:

1. In general, a $V_m \subset S_{m+1}$ is indeformable and the only possible deformable metric must be of k_0 - rank$\leqslant 2$ (an extension of Beez's Theorem).

2. When k_0 - rank$\geqslant 4$, the Peterson-Codazzi equations are consequences of Gauss equations (an extension of T. Y. Thomas' Theorem).

A complete classification of deformable hypersurfaces $V_m \subset S_{m+1}$ is given.

(本文曾发表于《数学学报》1956 年第 6 卷第 2 期,320 - 332)

常曲率空间中超曲面的变形与平均曲率

胡和生
中国科学院数学研究所，复旦大学

§1. 三维欧氏空间 E_3 中，任何曲面 V_2 有二维的黎曼测度 $ds^2=Edu^2+2Fdudv+Gdv^2$，这就是曲面 V_2 的第一基本形式.其逆，对于已给的黎曼测度，在 E_3 中如果找到曲面 V_2 而以此测度作为它的第一基本形式，我们称此曲面为测度的实现曲面.如所知，任何二维黎曼测度是能够在 E_3 中的曲面上实现的，而且所实现的曲面并不是唯一的，它是实现在与二个单参数的任意函数有关的曲面族上.这些具有相同测度的曲面称为互相变形的.由此可见，E_3 中曲面 V_2 的第一基本形式并不能决定曲面. 在高维欧氏空间中，情形并不相同，给了我们测度一般就能唯一的决定曲面[1][2].因而高维欧氏空间的曲面一般是不能变形的，而能变形的只是狭窄的一类曲面.

如果我们给变形加上一个限制，例如要互相变形的曲面有相同的平均曲率，那么我们就得到保持平均曲率的变形.一般而言，对于 E_3 中的任意曲面并不能找到与它有共同第一基本形式及共同平均曲率而不相符合的曲面，也就是说，曲面 $V_2\subset E_3$ 一般是不能保持平均曲率而变形的. 因此我们得到结论：给了我们第一基本形式及平均曲率一般就能完全决定曲面.于是，平均曲率一般可以起代替第二基本形式的作用，T. Y. Thomas[3] 在 $V_2\subset E_3$ 中将第二基本形式用测度张量及平均曲率表示出来.

外微分形式在研究变形问题时特别显得有力，这主要是由于它添加了标形选择的自由度.Яненко[4] 利用了外微分形式将一世纪来的几何学者在欧氏空间 E_{m+1} 的超曲面上的变形问题的工作完备地综合起来，并且完全地给出了高维欧氏空间可变形超曲面的分类.从此，对于已给的 m 维黎曼测度，如果它在 E_{m+1} 中能实现，我们就可知道它实现在多少超曲面上.作者利用了类似的方法,研究常曲率空间 S_{m+1} 中超曲面 V_m 的变形问题[5],推广了这些结果,同样的也得出了常曲率空间的可变形超曲面的分类.

利用 Яненко 关于 $V_m CE_{m+1}$ 的可变形超曲面的分类,作者研究了保持平均曲率的变形[6],而得出了,这类狭窄可变形的超曲面一般是不能保持平均曲率而变形的.并且将这类超曲面的第二基本形式用平均曲率及测度张量表示出来,而得到 T. Y. Thomas 的研究在高维空间可变形超曲面上的推广.

本文的目的是利用作者关于常曲率空间超曲面的分类来讨论上述问题,而将 T. Y. Thomas 的研究推广到高维常曲率空间的超曲面上去,但类似于以前[6]的一些计算,为节省篇幅计,不予重复.

§2. 在常曲率 K_0 的空间 S_{m+1} 中的超曲面

$$r=r(u^1,\ u^2,\ \cdots,\ u^m)$$

上每点选择标形 $I_1,\ \cdots,\ I_m,\ I_{m+1}$ 使向量 $I_1,\ \cdots,\ I_m$ 与曲面相切而向量 I_{m+1} 为曲面的法线.这样,曲面的导来形式为

$$\left.\begin{aligned}&dr=\omega^i I_i,\quad i=1,\ 2,\ \cdots,\ m,\quad \alpha=1,\ \cdots,\ m+1,\\&dI_\alpha=\omega_\alpha^\beta I_\beta,\\&I_\alpha I_\beta=g_{\alpha\beta},\quad g_{m+1\alpha}=S_{m+1\alpha},\quad \omega^m=0.\end{aligned}\right\}\tag{1}$$

曲面的组织方程为

$$\left.\begin{aligned}&[\omega^i\omega_i^{m+1}]=0,\\&(\omega_j^i)'=[\omega_j^k\omega_k^i]+[\omega_j^{m+1}\omega_{m+1}^i]+K_0 g^{li}(g_{lh}g_{jk}-g_{lk}g_{jh})[\omega^h\omega^k],\\&(\omega_i^{m+1})'-[\omega_i^j\omega_j^{m+1}]=0.\end{aligned}\right\}\tag{2}$$

而 ω_β^α 写成

$$\left.\begin{aligned}&\omega_j^i=\Gamma_{jk}^i\omega^k,\quad \Gamma_{jk}^i=\Gamma_{kj}^i,\\&\omega_i^{m+1}=\lambda_{ij}\omega^j,\quad \omega_{m+1}^i=-g^{ij}\omega_j^{m+1},\quad \lambda_{ij}=\lambda_{ij},\\&\omega_{m+1}^{m+1}=0.\end{aligned}\right\}\tag{3}$$

这样(2)的后面两式可写为张量形式

$$\left.\begin{aligned}&R_{ijkl}-K_0(g_{ik}g_{jl}-g_{il}g_{jk})=\lambda_{ik}\lambda_{jl}-\lambda_{il}\lambda_{jk},\\&\lambda_{ik,j}-\lambda_{ij,k}=0.\end{aligned}\right\}\tag{4}$$

称为高斯-柯达齐-彼杰尔松方程(Gauss - Codazzi - Петерсон)其中λ_{ij}为V_m的第二基本形式的系数.

现在首先导入K_0-秩数的定义

定义: 反称双一次形式系统

$$K_{ij}=\sum_{k<l}(R_{ijkl}-K_0(g_{ik}g_{jl}-g_{il}g_{jk}))[\omega^k\omega^l]\tag{5}$$

的秩数称为$ds^2=g_{ij}\omega^i\omega^j$的$K_0$-秩数.

可以看到测度的K_0-秩数等于$R_{ijkl}-K_0(g_{ik}g_{jl}-g_{il}g_{jk})$所做成的矩阵的秩数.我们在前篇[5]得到了如下的结果

定理: 当测度$ds^2=g_{ij}\omega^i\omega^j$的$K_0$-秩数$>2$时,它在常曲率$K_0$的空间$S_{m+1}$中实现的超曲面是不能变形的.

因此$V_m\subset S_{m+1}$只有在K_0-秩数$\leqslant 2$时才有变形的可能.这样明显地,K_0-秩数>2的曲面的第二基本形式系数λ_{ij}可用测度张量完全地表示出来. 现在我们根据前文[5]分别考虑所余下的情形(主要是K_0-秩数$=2$的情形)并求出λ_{ij}用平均曲率及测度张量的表示式.

K_0-秩数$=0$时V_m的测度是常曲率的. 这是级0的情形.这里不加讨论.

K_0-秩数$=1$时, 由于反称双一次形式的秩数总是偶数这种情形不可能发生.

K_0-秩数$=2$时, 我们以规范和乐标形为参考标形,那么$\psi_i=\omega_i^{m+1}$满足条件

$$\psi_\alpha=0,\quad \alpha=3,\cdots,m,\tag{6}$$

$$\psi_i=\lambda_{ij}du^j,\quad \lambda_{ij}=\lambda_{ji},\quad i,j=1,2.\tag{7}$$

讨论$V_m\subset S_{m+1}$的变形问题,实际上就是求高斯-柯达齐-彼杰尔松方程的解λ_{ij}的自由度的问题,这时可将高斯方程写为

$$R_{1212}-K_0(g_{11}g_{22}-g_{12}^2)=\lambda_{11}\lambda_{22}-\lambda_{12}^2. \tag{8}$$

考虑(6),(7)的可积条件得柯达齐–彼杰尔松方程

$$\Gamma_{\alpha k}^i\lambda_{ij}-\Gamma_{\alpha j}^i\lambda_{ik}=0, \tag{9}$$

$$\frac{\partial\lambda_{ij}}{\partial u^k}-\frac{\partial\lambda_{ik}}{\partial u^j}=\Gamma_{ik}^l\lambda_{jl}-\Gamma_{ij}^l\lambda_{kl}, \tag{10}$$

$$\frac{\partial\lambda_{ij}}{\partial u^\alpha}=\Gamma_{i\alpha}^l\lambda_{jl}. \tag{11}$$

我们称(9)为决定的系统.

§3. 现在我们要根据决定的系统来进行我们的讨论.

首先指出,平均曲率

$$H=g^{ij}\lambda_{ij}. \tag{12}$$

1) 如果决定系统的秩数=2,这时测度的安装是唯一的,利用高斯方程(8)可以得到

$$\lambda_{ij}=\pm\Delta_{ij}\sqrt{\frac{R_{1212}-K_0(g_{11}g_{22}-g_{12}^2)}{\Delta_{11}\Delta_{22}-\Delta_{12}}},$$

其中Δ_{ij}是由量$\Gamma_{\alpha j}^i$所组成的某些二阶张量.在这情形中,第二基本形式的系数λ_{ij}完全由测度张量决定.

2) 如果决定系统的秩数是1,则

$$\frac{\Gamma_{\alpha 2}^1}{\Gamma_2^1}=-\frac{\Gamma_{\alpha 1}^2}{\Gamma_1^2}=\frac{\Gamma_{\alpha 2}^2-\Gamma_{\alpha 1}^1}{\Gamma_2^2-\Gamma_1^1}=k_\alpha.$$

决定系统化为方程

$$\Gamma_2^1\lambda_{11}+(\Gamma_2^2-\Gamma_1^1)\lambda_{12}+\Gamma_1^2\lambda_{22}=0.$$

因而得到下列几种支情形

α_1: Γ_j^i有两个不同的特征根,取不变方向为I_1, I_2可得

$$\lambda_{12}=0,$$

$$K=R_{1212}-K_0(g_{11}g_{22}-g_{12}^2)=\lambda_{11}\lambda_{22}.$$

这时平均曲率

$$H=g^{11}\lambda_{11}+g^{22}\lambda_{22}.$$

从此,就可将 λ_{11}, λ_{22} 用 g_{ij} 及 H 表示出来.一般有二组解,因此可有两个不符合而具有相同的平均曲率的曲面来实现第一基本形式 $g_{ij}dx^idx^j$,称为稀疏的实现.在特殊情形下即方程只有一组解时,实现的曲面(除符合外)是唯一的.

α_2: Γ_j^i 有一个特征根,可选取适当的标形使

$$\lambda_{11}=0,$$

$$K=R_{1212}-K_0(g_{11}g_{22}-g_{12}^2)=-\lambda_{12}^2.$$

这时平均曲率

$$H=2g^{12}\lambda_{12}+g^{22}\lambda_{22},$$

λ_{12} 与 λ_{22} 就能用 g_{ij} 与 H 表示起来.得到 α_1 同样的结论.

3) 如果决定系统的秩数为 0.这时

$$\Gamma_{\alpha 1}^2=\Gamma_{\alpha 2}^1=\Gamma_{\alpha 2}^2-\Gamma_{\alpha 1}^1=0,\quad \alpha=3,\cdots,m,$$

记

$$\Gamma_{\alpha 2}^2=\Gamma_{\alpha 1}^1=\Gamma_\alpha.$$

我们可以证明.能选择如此的规范和乐标形使它还满足

$$\left.\begin{array}{l} I_i\cdot I_\alpha=0,\quad (i=1,2;\quad \alpha=3,\cdots,m) \\ \text{及} \\ h=g_{11}g_{22}-g_{12}^2=e^{4v(u^3,\cdots,u^m)}. \end{array}\right\}\qquad(12)$$

这时高斯-柯达-齐彼杰尔松方程是

$$\left.\begin{aligned}&R_{1212}-K_0(g_{11}g_{22}-g_{12}^2)=\lambda_{11}\lambda_{22}-\lambda_{12}^2,\\&-\frac{\partial\lambda_{i1}}{\partial u^2}+\frac{\partial\lambda_{i2}}{\partial u^1}=\Gamma_{i1}^j\lambda_{j2}-\Gamma_{i2}^j\lambda_{j1},\\&\frac{\partial\lambda_{i1}}{\partial u^\alpha}=\Gamma_\alpha\lambda_{i1},\quad\frac{\partial\lambda_{i2}}{\partial u^\alpha}=\Gamma_\alpha\lambda_{i2}.\end{aligned}\right\}\tag{13}$$

这里 Γ_{ik}^j 只依赖于 u^1, u^2, Γ_α 只依赖于 u^3, …, u^m.

可以证明,这样的问题可化为常曲率空间 S_3 的曲面 V_2 的变形问题,而 V_2 的测度为置 $u^3=u_0^3$, …, $u^m=u_0^m$ 于 $g_{ij}(u^1, u^2, \cdots, u^m)$ $(i, j=1, 2)$ 所得. 即

$$d\tilde{s}^2=g_{ij}(u^1, u^2, u_0^3, \cdots, u_0^m)du^idu^j=\tilde{g}_{ij}du^idu^j, (i, j=1, 2)\tag{14}$$

而它的曲率张量 $\tilde{R}_{1212}$ 与$R_{1212}(u^1, u^2, u_0^3, \cdots, u_0^m)$ 之间有关系

$$R_{1212}(u^1, u^2, u_0^3, \cdots, u_0^m)=\tilde{R}_{1212}-k^2,\ k^2=\sum_{\alpha=3}^m\Gamma_\alpha^2(u_0^3, \cdots, u_0^m).\tag{15}$$

我们来求 $d\tilde{s}^2$ 在常曲率 $k^2+K_0h(u_0^3, \cdots, u_0^m)$ 的空间 S_3 变形,也就是求下列方程系统的解 $\lambda_{ij}(u^1, u^2)(i, j=1, 2)$ 的自由度

$$\left.\begin{aligned}&\tilde{R}_{1212}-(k^2+K_0h)=\lambda_{11}\lambda_{22}-\lambda_{12}^2,\\&\frac{\partial\lambda_{i2}}{\partial u^1}-\frac{\partial\lambda_{i1}}{\partial u^2}=\Gamma_{i1}^j\lambda_{j2}-\Gamma_{i2}^j\lambda_{j1},\end{aligned}\right\}\tag{16}$$

如所知 $V_2\subset S_3$ 的变形依赖于两个单变数的任意函数,从此可证明了(13)的解 $\lambda_{ij}(u^1, u^2, \cdots, u^m)$ 依赖于两个单变数的任意函数,且(13)及(16)解之间存在关系

$$\lambda_{ij}(u^1, u^2, u^3, \cdots, u^m)=\frac{e^{v(u^3, \cdots, u^m)}}{e^{v(u_0^3, \cdots, u_0^m)}}\lambda_{ij}(u^1, u^2).\tag{17}$$

这样,我们确定了:当决定系统的秩数为 0 时, $V_m\subset S_{m+1}$ 的变形依赖于两个单变数的任意函数.下面我们要讨论在这种情形时保持平均曲率的变形.并且要求出第二基本形式用测度张量及平均曲率的表示式.

现在先证明,当 $V_m\subset S_{m+1}$ 保持平均曲率而变形时,对应的 $V_2\subset S_3$ 也保持平均曲

率而变形，由(17)知

$$H(u^1, u^2, \cdots, u^m)=g^{ij}(u^1, u^2, \cdots, u^m)\lambda_{ij}(u^1, u^2, \cdots, u^m)$$

$$=g^{ij}(u^1, u^2, \cdots, u^m)\frac{e^{v(u^3, \cdots, u^m)}}{e^{v(u_0^3, \cdots, u_0^m)}}\lambda_{ij}(u^1, u^2), (i, j=1, 2) \tag{18}$$

注意到 Γ_α 只依赖于参数 $u^3, \cdots, u^m$ 这一事实可证明

$$g_{ij}(u^1, u^2, \cdots, u^m)=\bar{g}_{ij}(u^1, u^2)e^{2v(u^3, \cdots, u^m)},$$

并且在我们的坐标系统下，由(12)可知

$$h=g_{11}g_{22}-g_{12}^2=e^{4v(u^3, \cdots, u^m)},$$

$$g^{11}=\frac{g_{22}}{h}, g^{22}=\frac{g_{11}}{h}, g^{12}=-\frac{g_{12}}{h}.$$

这样(18)可改写为

$$H(u^1, u^2, \cdots, u^m)=\frac{1}{e^{v(u_0^3, \cdots, u_0^m)}e^{v(u^3, \cdots, u^m)}}[\bar{g}_{22}\lambda_{11}(u^1, u^2)$$

$$-2\bar{g}_{12}\lambda_{12}(u^1, u^2)+\bar{g}_{11}\lambda_{22}(u^1, u^2)]. \tag{18}$$

由于 V_2 的测度张量为

$$\tilde{g}_{ij}=g_{ij}(u^1, u^2, u_0^3, \cdots, u_0^m)=\bar{g}_{ij}(u^1, u^2)e^{2v(u_0^3, \cdots, u_0^m)},$$

作其逆张量 $\tilde{g}^{ij}$，并利用(12)式可得

$$\tilde{g}^{11}=\frac{\bar{g}_{22}}{e^{2v(u_0^3, \cdots, u_0^m)}}, \quad \tilde{g}^{22}=\frac{\bar{g}_{11}}{e^{2v(u_0^3, \cdots, u_0^m)}}, \quad \tilde{g}^{12}=-\frac{\bar{g}_{12}}{e^{2v(u_0^3, \cdots, u_0^m)}}.$$

代入(18′)式，

$$H(u^1, u^2, \cdots, u^m)=2\frac{e^{v(u_0^3, \cdots, u_0^m)}}{e^{v(u^3, \cdots, u^m)}}H(u^1, u^2). \tag{19}$$

这里

$$H(u^1, u^2) = \frac{\tilde{g}^{ij}\lambda_{ij}(u^1, u^2)}{2}. \tag{20}$$

这样,证明了我们的结果,因此要研究决定系统秩数等于 0 的 $V_m \subset S_{m+1}$ 保持平均曲率变形问题,只需研究 $V_2 \subset S_3$ 中保持平均曲率的变形.利用作者在[6,7]中的结果得知 $V_2 \subset S_3$ 时保持平均曲率而互为变形的曲面偶是与单变数的 4 个任意函数有关,从此我们可以推出在决定系统秩数为 0 时,$V_m \subset S_{m+1}$ 一般是不能保持平均曲率而变形的,而能保持平均曲率变形的超曲面偶亦依赖于单变数的 4 个任意函数.

§4. 我们要将这种情形中一般的超曲面的第二基本形式 $\lambda_{ij}(u^1, \cdots, u^m)$ 用平均曲率及测度张量表示出来,为此首先利用[6]中结果可得曲面 V_2 在常曲率为 $k^2 + K_0 h(u_0^3, \cdots, u_0^m)$ 的空间 S_3 中的第二基本形式表示式(详细参阅[3] 及[6]).

$$\lambda_{\alpha\beta} = RW^{-1}g_{\alpha\beta} + 2UW^{-1}Q_{\alpha\beta} \pm \left\{\frac{[(H^2-K)W-U^2]^{\frac{1}{2}}}{W}\right\} V_{\alpha\beta}.$$

$$(\alpha, \beta = 1, 2) \tag{21}$$

这里

$$\left.\begin{aligned}
&V_{\alpha\beta} = (Q_{\alpha\mu}\varepsilon_{\beta\nu} + Q_{\beta\mu}\varepsilon_{\alpha\nu})g^{\mu\nu},\\
&W = (\tilde{g}^{\alpha\beta}Q_{\alpha\beta})^2 - \frac{4 \mid Q_{\alpha\beta} \mid}{\mid \tilde{g}_{\alpha\beta} \mid},\\
&R = \tilde{g}^{\alpha\beta}P_{\alpha\beta}\,\tilde{g}^{\mu\nu}Q_{\mu\nu} - \frac{4H \mid Q_{\alpha\beta} \mid}{\mid \tilde{g}_{\alpha\beta} \mid},\\
&U = H\,\tilde{g}^{\alpha\beta}Q_{\alpha\beta} - \tilde{g}^{\alpha\beta}P_{\alpha\beta}.
\end{aligned}\right\} \tag{22}$$

而

$$Q_{\alpha\beta} = 2(H^2-K)H_{\alpha\beta} - 4HH_\alpha H_\beta + K_\alpha H_\beta + K_\beta H_\alpha, \tag{22$'$}$$

$$\begin{aligned}
P_{\alpha\beta} = {} & (H^2-K)K_{\alpha\beta} + K_\alpha K_\beta + HK_\alpha H_\beta - HK_\beta H_\alpha - \\
& 2(K+k^2+K_0h)(H^2-K)^2\,\tilde{g}_{\alpha\beta}.
\end{aligned}$$

这里 K 为 V_2 的总曲率. 将(21)代入(17)就得到 $\lambda_{ij}(u^1, \cdots, u^m)$ 的表示式.

我们可以证明 V_2 的总曲率 K 为:

$$|\tilde{g}_{ij}| K = R_{1212}(u^1, u^2, u_0^3, \cdots, u_0^m). \tag{23}$$

此外,又可证明 $\lambda_{ij}(u^1, \cdots, u^m)$ 的决定与 $u_0^3, \cdots, u_0^m$ 的选择无关.因此,得到由 $H(u^1, u^2, \cdots, u^m)$ 与 $g_{ab}(u^1, \cdots, u^m)(a, b=1, \cdots, m)$ 来决定 $\lambda_{ij}(u^1, u^2, \cdots, u^m)$ 的过程如下:先选择一组数值 $u_0^3, \cdots, u_0^m$;依据(23)式得到 K,又按照(19)式得到 $H(u^1, u^2)$,再把它们代入(21),(22),(22′)得到 $\lambda_{ij}(u^1, u^2)$. 最后由(17)式就能完全决定 $\lambda_{ij}(u^1, \cdots, u^m)$.

至此,我们已把 $V_m \subset S_{m+1}$ 的一般的可变形的超曲面,用测度张量及平均曲率表示出来.

最后,作者在此对苏步青教授的指导与鼓励表示感谢.

参考文献

[1] T. Y. Thomas, Acta. Mathamatica 67(1936).

[2] C. B. Allendoeffer, American Journal of Math. 61(1939).

[3] T. Y. Thomas, Bull. A.M.S.(1945).

[4] H.H. Яненко, Y. M. H.(1953).

[5] 胡和生,论黎曼测度 V_m 在常曲率空间 S_{m+1} 的变形(尚未发表).

[6] 胡和生,论保持平均曲率的黎曼测度 V_m 在欧氏空间 E_{m+1} 的变形.数学学报 6 卷 1 期(1956).

[7] 胡和生,三维黎曼空间的第二基本形式与平均曲率(尚未发表).

(1955 年 11 月 30 日收到)

(本文曾发表于《复旦学报(自然科学)》1956 年第 1 期,99 - 105)

常曲率空间中的常曲率超曲面的变形问题

胡和生
中国科学院数学研究所，复旦大学

§1. 作者在前文[1]中，讨论了 m 维黎曼空间($m \geqslant 3$)

$$ds^2 = g_{ij} du^i du^j \quad (i, j = 1, 2, \cdots, m) \tag{1}$$

在常曲率 k_0 的空间 S_{m+1} 中的安装与变形问题，并依据 k_0-秩数给出了高维常曲率空间的可变形超曲面的完全分类.所谓 k_0-秩数就是双一次协变式

$$k_{ij} = \sum_{k<l} (R_{ijkl} - k_0(g_{ik}g_{jl} - g_{il}g_{jk}))[du^k du^l]$$

的秩数，也就是矩阵

$$\| R_{ijkl} - k_0(g_{ik}g_{jl} - g_{il}g_{jk}) \|$$

的秩数.这里 ijk 表示列的指标，l 为行的指标.而 R_{ijkl} 表示黎曼空间(1)的曲率张量，因此 k_0-秩数是由测度(1)本身所内在决定的.

我们已经证明：当 k_0-秩数 >2 时，曲面不能变形，因此就在 k_0-秩数 $\leqslant 2$ 时进行变形问题的研究. 当 k_0-秩数 $=0$ 时，由于

$$R_{ijkl} = k_0(g_{ik}g_{jl} - g_{il}g_{jk}) \tag{2}$$

黎曼测度(1)本身是有常曲率 k_0 的，也就是说，它属于 0 级的(class 0).对于这个场合我们未予讨论.然而事实上，常曲率 k_0 空间 S_{m+1} 中存在着常曲率 k_0 的超曲面 S_m，所以我们有研究这些超曲面的变形问题的必要，本文解决了这个问题而得出了如下的结论.

定理：在 $m+1$ 维常曲率 k_0 空间中，常曲率 k_0 的超曲面的变形依赖于 m 个单参数的

任意函数.

我们在这里指出，H.H. Яненко[2] 在研究欧氏空间超曲面的变形时，将欧氏空间超平面的变形也归入 0 级而未予解决，我们在这里同时解决了这个问题，而得到推论：

在 $m+1$ 维欧氏空间中超平面的变形依赖于 m 个单参数的任意函数.

§2. 如所知，曲面 $V_m \subset S_{m+1}$ 唯一地决定于第一基本形式和第二基本形式，求解已给测度(1)在 S_{m+1} 中的变形问题，就是从 Gauss - Peterson - Codazzi 的方程[3]

$$R_{ijkl} - k_0(g_{ik}g_{jl} - g_{il}g_{jk}) = \lambda_{ik}\lambda_{jl} - \lambda_{il}\lambda_{jk}, \tag{3}$$

$$\lambda_{ij,k} - \lambda_{ik,j} = 0, \tag{4}$$

求 λ_{ij} 的解的自由度的问题，这里 λ_{ij} 是 V_m 的第二基本形式的系数.

由于已给测度是有常曲率 k_0 的，它满足方程(2)，所论的问题就化为方程系统

$$\lambda_{ik}\lambda_{jl} - \lambda_{il}\lambda_{jk} = 0, \tag{5}$$

及(4)的求解问题.可以求得

$$\lambda_{ij} = \varphi_i\varphi_j. \tag{6}$$

式中 φ_i 必需满足(4)，就是

$$\varphi_i(\varphi_{j,k} - \varphi_{k,j}) + \varphi_j\varphi_{i,k} - \varphi_k\varphi_{i,j} = 0. \tag{7}$$

由于，(1)是常曲率的，不妨假定已经取好适当的坐标系统，使它的联络系数为

$$\Gamma_{ij}^k = -\delta_i^k \frac{\partial p}{\partial u^j} - \delta_j^k \frac{\partial p}{\partial u^i}, \tag{8}$$

从而改写(7)，

$$\varphi_i\left(\frac{\partial\varphi_j}{\partial u^k} - \frac{\partial\varphi_k}{\partial u^j}\right) + \varphi_j\frac{\partial\varphi_i}{\partial u^k} - \varphi_k\frac{\partial\varphi_i}{\partial u^j} + \varphi_i\varphi_j p_k - \varphi_i\varphi_k p_j = 0, \tag{9}$$

或者置

$$e^p = \sigma \tag{10}$$

时，化为下列形式：

$$\frac{\partial(\sigma\varphi_i\varphi_j)}{\partial u^k}=\frac{\partial(\sigma\varphi_i\varphi_k)}{\partial u^j}. \tag{11}$$

这样，得到

$$\sigma\varphi_i\varphi_j=\frac{\partial F_i}{\partial u^j}=\frac{\partial F_j}{\partial u^i},$$

式中

$$F_j=\frac{\partial F}{\partial x^j}, \tag{12}$$

并且

$$\frac{\partial^2 F}{\partial u^i\partial u^j}=\sigma\varphi_i\varphi_j. \tag{13}$$

因而，所论的 Peterson－Codazzi 的方程组等价于

$$\frac{\partial^2 F}{\partial u^i\partial u^j}=\varphi_i\varphi_j. \tag{14}$$

这又等价于下列法甫方程组

$$dF=F_i du^i, \tag{15}$$

$$dF_i=\varphi_i\varphi_j du^j, \tag{16}$$

式中 F，F_i，φ_j 均为未知函数.这样一来，$S_m\subset S_{m+1}$ 的变形问题归结到这样的问题：求法甫方程组(15)，(16)的积分流形 M_n 的解的自由度，而在 M_m 上 $[du^1 du^2\cdots du^m]\neq 0$.

为求(15)，(16)的协变式系统，将它们外导微一次，(15)的协变式恒等成立，而(16)的外导微的结果是：

$$\varphi_i[d\varphi_j du^j]+\varphi_j[d\varphi_i du^j]=0. \tag{17}$$

从法甫方程系统(15)，(16)可见 dF，dF_i 可用 du^i 来表示，因此积分元素在 du^i 的

任意选择下是根据

$$d\varphi_i = l_{ij} du^j \tag{18}$$

所确定.

现在按照 $[du^i]$ 作链[4]，可知 l_{i1} 是完全任意的.这样就确定了积分元素 ε_1 的自由度 $\gamma_1 = m$.

为确定积分元素 ε_2，将(18)代入(17)，并根据链的分划得到 l_{i2} 所应满足的关系式

$$\varphi_i(l_{21} - l_{12}) + l_{i1}\varphi_2 - l_{i2}\varphi_1 = 0, \tag{19}$$

从而，得到

$$l_{12} = \frac{\varphi_1 l_{21} + \varphi_2 l_{11}}{2\varphi_1},$$

$$l_{\alpha 2} = \frac{\varphi_2(l_{21} - l_{12}) + l_{\alpha 1}\varphi_2}{\varphi_1}, \quad \alpha = 2, \cdots, m. \tag{20}$$

方程组(19)的成立并未影响 l_{i1} 的任意性，而且由(20)完全确定了 l_{i2}，因此 ε_2 的自由度 $\gamma_2 = 0$.

为确定积分元素 ε_2，首先由(17)，(18)得到 l_{i3} 所必需满足的方程组

$$\varphi_i(l_{31} - l_{13}) + l_{i1}\varphi_3 - l_{i3}\varphi_1 = 0, \tag{21}$$

$$\varphi_i(l_{32} - l_{23}) + l_{i2}\varphi_3 - l_{i3}\varphi_2 = 0. \tag{22}$$

由(21)求得

$$l_{13} = \frac{\varphi_1 l_{31} + \varphi_3 l_{11}}{2\varphi_1},$$

$$l_{\alpha 3} = \frac{\varphi_\alpha(l_{31} - l_{13}) + l_{\alpha 1}\varphi_3}{\varphi_1}. \tag{23}$$

这表示了由 l_{i1} 完全确定 l_{i3}. 将(23)代入(22)中可以证明它们是恒等满足的，而且从此并未产生 l_{i1} 的任何关系式，因此所论的方程组是法式可解的，而且 ε_3 的自由度 $\gamma_3 = 0$.

同样地，对于 ε_4，…，ε_m 也可类似的证明均可由 l_{i1} 来确定 $l_{i\alpha}$（$\alpha=4$，…，m）且为法式可解.因此链是正则的，方程组是在对合下的，并且它的标数是 $s_1=m$，$s_2=0$，…，$s_m=0$.所以积分流形 M_m 依赖于 m 个单变数的任意函数，这就证明了§1 的定理.

至于欧氏空间超平面的变形的问题从证明中还可以看出它是相当于 $\sigma=1$ 的情况，但是，(13)和(14)是等价的，所以常曲率空间的常曲率超曲面的变形问题实际上已归结到欧氏空间超平面的变形问题，这样，我们可以更清楚地得出§1 的推论.

参 考 文 献

[1] 胡和生，论黎曼测度 V_m 在常曲率空间 S_{m+1} 的变形，数学学报六卷二期(1956)320－331.

[2] Н. Н. Яненко，Некоторые вопросы теории вложения римановых метрик в евклидовы пространства，Успехи матем. науки 8(1953)21－100.

[3] L. P. Eisenhart，Riemannian Geometry (1925)146－150.

[4] С. П. Фиников，Метод внешних форм картана(1948) 225－238.

The Deformation of the Hypersurfaces of Constant Curvature in Constant Curvature Space

Hu Hou-Sung

Abstract

In the preceding paper the author gave the complete classification of the deformation of the hypersurface V_m in a space S_{m+1} of constant curvature k_0 excepted the case when V_m is of the same constant curvature k_0.

The purpose of this paper is to consider the remaining case which is reduced to solve the equations of Gauss－Peterson－Codazzi

$$\lambda_{ik}\lambda_{jl}-\lambda_{il}\lambda_{jk}=0, \tag{1}$$

$$\lambda_{ij,k}-\lambda_{ik,j}=0, \tag{2}$$

where λ_{ij} is the coefficients of the second fundamental form of V_m.

From equations (1) we have

$$\lambda_{ij}=\varphi_i\varphi_j.$$

Using the fact

$$\Gamma_{ij}^k=-\delta_i^k\frac{\partial p}{\partial u^j}-\delta_j^k\frac{\partial p}{\partial u^i},$$

and writing

$$\sigma=e^p,\ \bar{\varphi}_i=\sqrt{\sigma}\varphi_i.$$

Wc transform the equations (2) to the Phaffian system

$$dF=F_i du^i,$$

$$dF_i=\bar{\varphi}_i\bar{\varphi}_j du^j,$$

where F, F_i, $\bar{\varphi}_i$ are all unknowns.

We prove that this system is in involution with Characteristic numbers $s_1=m$, $s_2=\cdots=s_m=0$ and get the conclusion.

Theorem: The hypersurface V_m of constant curvature k_0 in the space S_{m+1} of same constant curvature is always deformable and the degree of freedom of the deformation is m functions of a single argument.

This result is valid in the case $k_0=0$, i.e. The hypersurpace V_m with Enclidean metric is always deformable with degree of freedom mf (1) in Enclidean space E_{m+1}.

(1956 年 7 月 18 日收到)

(本文曾发表于《复旦学报(自然科学)》1956 年第 2 期,35 - 39)

负常曲率空间的一种特征

胡和生

中国科学院数学研究所，复旦大学

§1. 如果 N 维的黎曼空间 V_N 含有如此的 n 维子空间 V_n，它的诱导尺度具有常曲率，那么我们说：空间 V_N 含有 n 维的常曲率曲面.

如果 V_N 中的曲面 V_n 具有这样的性质，使切于 V_n 的空间测地线一定在 V_n 上，或者等价的说：曲面的法平面素是平行的，那么我们称 V_n 为全测地的曲面[1].

本文讨论那一些具有某种全测地超曲面系的黎曼空间的性质，而且从此得到负常曲率空间的一种特征：

如果 $m(m\geqslant 4)$ 维黎曼空间 V_m 含有 $m-1$ 系相互正交的全测地的常曲率超曲面，那么空间 V_m 一定有负常曲率，而且这些超曲面也都具有相同的负常曲率.

§2. 如所知，为了黎曼空间 V_m 要有一系全测地超曲面，充要条件是：在适当坐标之下，V_m 的线素可化为

$$ds^2=A(dx^1)^2+g_{\alpha\beta}dx^\alpha dx^\beta \quad (\alpha,\ \beta=2,\ \cdots,\ m), \tag{1}$$

这里 $x^1=\text{const.}$是那一系全测地超曲面，A 是$x^i(i=1,\ 2,\ \cdots,\ m)$的任意函数，而且 $g_{\alpha\beta}$ 不依赖于 x^1 [2].

我们容易导出这样的事实：设黎曼空间的线素可以写成

$$ds^2=g_{11}(dx^1)^2+g_{\alpha\beta}dx^\alpha dx^\beta$$

的形式；如果 $x^1=\text{const.}$是一系全测地的超曲面，那么 $g_{\alpha\beta}$ 一定与 x^1 无关.

实际上，因 $x^1=\text{const.}$ 的单位法向量是 $\left(\frac{1}{\sqrt{g_{11}}}, 0, 0, \cdots, 0\right)$ 并且沿着 $x^1=\text{const.}$ 是平行的，所以它沿 $x^1=\text{const.}$ 上任何道路的绝对微分等于 0，从而

$$\Gamma_{\beta 1}^{\alpha}=0 \quad (\alpha, \beta=2, \cdots, m),$$

即 $[\beta 1, \gamma]=0$. 又利用 $g_{1\beta}=0$，就得出所要的结果：

$$\frac{\partial g_{\beta\gamma}}{\partial x^1}=0.$$

根据这个事实，我们来证：如果 V_m 含有 m 重正交全测地的超曲面系统，那么它一定是欧氏空间.因为这时我们可以选择坐标，使这 m 重正交系统成为坐标超曲面系统，所以 V_m 的线素可以写成

$$ds^2=g_{11}(dx^1)^2+g_{22}(dx^2)+\cdots+{}^2g_{mm}(dx^m)^2,$$

而且由于超曲面系统 $x^i=\text{const.}$ 都是全测地的，g_{ii} 只包含 x^i，因而空间线素是欧氏的.

在叙述负常曲率空间的一种特征之前，我们先证明一个结果：如果 V_m 中含有 $m-1$ 系相互正交的全测地超曲面，那么一定还含有另一系超曲面与它们共同组成一个 m 重正交超曲面系统.

证明：在空间 V_m 的每点取正交标形 $I_i(i=1, 2, \cdots, m)$，使 $I_\alpha(\alpha=1, 2, \cdots, m-1)$ 是第 α 系超曲面的单位法线向量，于是 V_m 的导来方程是

$$\left.\begin{aligned} &dr=\omega_i I_i, \\ &dI_i=\omega_{ij}I_j, \quad \omega_{ij}+\omega_{ji}=0 \quad (i, j=1, 2, \cdots, m). \end{aligned}\right\} \tag{2}$$

它的组织方程是

$$\left.\begin{aligned} &(\omega_i)'=[\omega_j\omega_{ji}], \\ &(\omega_{ij})'=[\omega_{ik}\omega_{kj}]+\sum_{k<l}R_{ijkl}[\omega_k\omega_l] \end{aligned}\right. \quad (i, j, k, l=1, 2, \cdots, m). \tag{3}$$

由于 I_α 垂直于第 α 系曲面，第 α 系超曲面的微分方程应当是

$$\omega_\alpha = 0 \quad (\alpha = 1, 2, \cdots, m).$$

又因为它们是全测地超曲面，法线向量 I_α 沿着各超曲面 $\omega_\alpha = 0$ 都是平行的，

$$dI_\alpha \equiv 0 \quad (\mathrm{mod}\ \omega_\alpha). \tag{4}$$

记 $\omega_{ij} = \lambda_{ijk}\omega_k$，由于标形是正交的，$\lambda_{ijk} + \lambda_{jik} = 0$. 这样从(2)及(4)就有

$$\lambda_{\alpha jk}\omega_k \equiv 0 \quad (\mathrm{mod}\ \omega_\alpha).$$

由此获得

$$\lambda_{\alpha jk} = 0 \quad (\alpha = 1, 2, \cdots, m-1;\ j, k = 1, \cdots, m;\ k \neq \alpha).$$

于是

$$\omega_{\alpha\beta} = 0, \tag{5}$$

$$\omega_{\alpha m} = \lambda_{\alpha m\alpha}\omega_\alpha \quad (\alpha\beta = 1, \cdots, m-1; \alpha \text{ 非和式}).$$

由(3)和(5)容易导出

$$(\omega_m)' = 0,$$

这就是说，$\omega_m = 0$ 代表与前面 $m-1$ 系正交的一系超曲面.

§3. 我们提出一个问题：如果黎曼空间 V_m 含有全测地的常曲率 $k_\alpha(\alpha = 1, 2, \cdots, m-1)$ 的超曲面的 $m-1$ 正交系统，那么 V_m 应该有什么特性.

根据上节的结果首先知道，一定存在另外一系超曲面同那些 $m-1$ 系超曲面组成 m 重正交系统.现在取这 m 系的超曲面作为坐标超曲面，那么 V_m 的线素就采取下列方式：

$$ds^2 = g_{11}(dx^1)^2 + \cdots + g_{mm}(dx^m)^2.$$

由于超曲面系 $x^\alpha = \mathrm{const.}(\alpha = 1, 2, \cdots, m-1)$都是全测地的，$g_{\alpha\alpha}$ 只是x^α 和x^m 的函数而且 g_{mm} 单是 x^m 的函数，这样 V_m 的线素又可化为

$$ds^2 = g_{11}(x^1, x^m)(dx^1)^2 + \cdots + g_{m-1m-1}(x^{m-1}, x^m)(dx^{m-1})^2 + (dx^m)^2. \tag{6}$$

又因为超曲面 $x^1 = \mathrm{const.}$的线素是

$$ds^2 = g_{22}(x^2, x^m)(dx^2)^2 + \cdots + g_{m-1m-1}(x^{m-1}, x^m)(dx^{m-1})^2 + (dx^m)^2. \quad (7)$$

这系 $x^1=\text{const.}$中的所有超曲面都是等长的，因而它们有相同的曲率张量 $A_{\alpha\beta\gamma\delta}(\alpha, \beta, \gamma, \delta=2, \cdots, m)$. 实际上

$$A_{\alpha\beta\beta\alpha} = \frac{1}{4}\frac{\partial g_{\alpha\alpha}}{\partial x^m}\frac{\partial g_{\beta\beta}}{\partial x^m} \quad (\alpha, \beta=2, \cdots, m-1;\ \alpha \neq \beta),$$

$$A_{\alpha m m\alpha} = \frac{1}{2}\frac{\partial^2 g_{\alpha\alpha}}{\partial x^m} - \frac{1}{4}\frac{1}{g_{\alpha\alpha}}\left(\frac{\partial g_{\alpha\alpha}}{\partial x^m}\right)^2 \quad (\alpha\ \text{非和式};\alpha=2, \cdots, m-1),$$

且曲率张量的其余的支量都是 0.则由于 $x^1=\text{const.}$有相同的常曲率 k_1，因此

$$A_{\alpha\beta\gamma\delta} = k_1(g_{\alpha\gamma}g_{\beta\delta} - g_{\alpha\delta}g_{\beta\gamma}) \quad (\alpha, \beta, \gamma, \delta=2, \cdots, m),$$

这样就成立

$$\left.\begin{aligned}
&\frac{1}{4}\frac{\partial g_{\alpha\alpha}}{\partial x^m}\frac{\partial g_{\beta\beta}}{\partial x^m} = -k_1 g_{\alpha\alpha}g_{\beta\beta} \quad (\alpha, \beta=2, \cdots, m-1;\ \alpha \neq \beta),\\
&\frac{1}{2}\frac{\partial^2 g_{\alpha\alpha}}{\partial x^m} - \frac{1}{4}\frac{1}{g_{\alpha\alpha}}\left(\frac{\partial g_{\alpha\alpha}}{\partial x^m}\right)^2 = -k_1 g_{\alpha\alpha} \quad (\alpha=2, \cdots, m-1;\ \alpha\ \text{非和式}).
\end{aligned}\right\} \quad (8)$$

同样地对于超曲面系 $u^\gamma=\text{const}$ $(\gamma=2, \cdots, m-1)$ 类似的可以得到

$$\frac{1}{4}\frac{\partial g_{\alpha\alpha}}{\partial x^m}\frac{\partial g_{\beta\beta}}{\partial x^m} = -k_\gamma g_{\alpha\alpha}g_{\beta\beta} \quad (\alpha, \beta, \gamma=1, \cdots, m-1;\ \alpha, \beta, \gamma \neq), \quad (9)$$

$$\frac{1}{2}\frac{\partial^2 g_{\alpha\alpha}}{\partial x^m} - \frac{1}{4}\frac{1}{g_{\alpha\alpha}}\left(\frac{\partial g_{\alpha\alpha}}{\partial x^m}\right)^2 = -k_\gamma g_{\alpha\alpha} \quad (\alpha, \gamma=1, \cdots, m-1;\ \alpha \neq \gamma). \quad (10)$$

当 $m \geqslant 4$ 时，由(10)可得 $k_1=k_2=\cdots=k_{m-1}$，就是 $m-1$ 系超曲面有相同的常曲率 k_0.

改写(9)式

$$\frac{\partial}{\partial x^m}\log g_{\alpha\alpha} = -4k_0\left(\frac{\partial}{\partial x^m}\log g_{\beta\beta}\right)^{-1}, \quad (9')$$

并且注意到 $g_{\gamma\gamma}$ 只能含有 x^γ, x^m 的事实，就可知(9′)式的两端只能和 x^m 有关.积分之后得出

$$g_{\alpha\alpha}=a_{\alpha}(x^{m})\xi_{\alpha}(x^{\alpha})\quad(\alpha=1,2,\cdots,m-1;\ \alpha\text{ 非和式}).\tag{11}$$

从而(9)和(10)变成 a_{α} 所满足的方程：

$$\frac{1}{4}\frac{da_{\alpha}}{dx^{m}}\frac{da_{\beta}}{dx^{m}}=-k_{0}a_{\alpha}a_{\beta}\quad(\alpha,\beta=1,2,\cdots,m-1;\ \alpha\neq\beta),\tag{12}$$

$$\frac{1}{2}\frac{d^{2}a_{\alpha}}{(dx^{m})^{2}}-\frac{1}{4a_{\alpha}}\left(\frac{da_{\alpha}}{dx^{m}}\right)^{2}=-k_{0}a_{\alpha}\quad(\alpha=1,2,\cdots,m-1).\tag{13}$$

这样 V_m 的线素(6)经适当坐标变换之后可化为下列形式：

$$ds^{2}=\sum_{\alpha-1}^{m-1}a_{\alpha}(x^{m})(dx^{\alpha})^{2}+(dx^{m})^{2}.\tag{14}$$

从此可得曲率张量的支量：

$$\left.\begin{aligned}&R_{\alpha mm\alpha}=\frac{1}{2}\frac{d^{2}a_{\alpha}}{(dx^{m})^{2}}-\frac{1}{4a_{\alpha}}\left(\frac{da_{\alpha}}{dx^{m}}\right)^{2}\quad(\alpha=1,2,\cdots,m-1),\\&R_{\alpha\beta\beta\alpha}=\frac{1}{4}\frac{da_{\alpha}}{dx^{m}}\frac{da_{\beta}}{dx^{m}}\quad(\alpha,\beta=1,2,\cdots,m-1;\ \alpha\neq\beta),\end{aligned}\right\}\tag{15}$$

而且其余的支量都等于 0.按照(12)和(13)改写(14)，

$$R_{\alpha44\alpha}=-k_{0}a_{\alpha},$$

$$R_{\alpha\beta\beta\alpha}=-k_{0}a_{\alpha}a_{\beta}.$$

因而，

$$R_{hijk}=k_{0}(g_{hj}g_{ik}-g_{hk}g_{ij})\quad(h,i,j,k=1,2,\cdots,m).\tag{16}$$

这表示了 V_m 有常曲率 k_0，又由于 $m\geqslant 4$，从(12)式得到

$$\frac{1}{4}\left(\frac{da_{\alpha}}{dx^{m}}\right)^{2}=-k_{0}(a_{\alpha})^{2}\quad(\alpha=1,2,\cdots,m-1).$$

所以 k_0 必须是负常数.这样，就证明了下列定理：

定理： 如果 m 维黎曼空间 $V_m(m\geqslant 4)$ 含有全测地的常曲率超曲面的 $m-1$ 正交系

统,那么空间 V_m 一定有负的常曲率，而且这些超曲面也都具有相同的常曲率.

这个定理是负常曲率空间的一个特征,因为在适当坐标下,可化负常曲率 $k_0(<0)$ 的空间 V_m 的线素为

$$ds^2=e^{\sqrt{-4k_0}x^m}[(dx^1)^2+(dx^2)^2+\cdots+(dx^{m-1})^2]+(dx^m)^2,$$

而且由前节可知这里 $x^\alpha=\text{const.}$ $(\alpha=1, 2, \cdots, m-1)$ 做成全测地超曲面的 $m-1$ 正交系统.又计算它的曲率张量时容易知道,各超曲面都具有负常曲率 k_0.

参 考 文 献

[1] Cartan, E., Leçons sur la géométrie des espaces de Riemann (1946).

[2] Eisenhart, L. P., Riemannian geometry (1926).

A Characteristic Property of the Space of Negative Constant Curvature

Hu Hou-Sung

Abstract

In this paper we get the following characteristic property of the space of negative constant curvature.

If an $m(\geqslant 4)$ dimensional Riemannian space V_m contains $m-1$ families of mutually orthorgonal totally geodesic hypersurfaces of constant curvature, then V_m is a space of negative constant curvature, and all these hypersurfaces are of the same negative constant curvature.

(本文曾发表于《复旦学报(自然科学)》1957 年第 1 期,175－180)

关于黎曼空间的两种秩数

胡和生

中国科学院复旦大学数学研究室

§1. 引言.本文的目的是找出黎曼空间两种秩数的几何意义.这两种秩数特别是在黎曼空间到常曲率空间的安装与变形问题上,具有很重要意义[1].这里所得到的结果是 H. H. Яненко[2]及陈省身与 N. H. Kuiper[3]的定理的推广.在§3 中我们作出秩数几何意义的一个应用.它与 C. Tompkins[4]的一个安装定理是密切联系着的.

§2. 常曲率空间中的曲面秩数的几何意义.对于常曲率 k_0 的空间 S_{m+q}:

$$ds^2 = g_{\alpha\beta} dx^\alpha dx^\beta \quad (\alpha, \beta = 1, \cdots, m+q)$$

中的曲面 V_m: $M = M(u^1, u^2, \cdots, u^m)$,选择这样的标形$(M, I_\alpha)$,使得$(I_i)(i=1, \cdots, m)$与 V_m 相切,而$(I_{m+s})(s=1, \cdots, q)$为曲面 V_m 的相互垂直的法向量,那么对于 V_m 成立

$$\left.\begin{aligned} &dM = \omega^i I_i, \quad \omega^{m+1} = \cdots = \omega^{m+q}, \\ &dI_\alpha = \omega_\alpha^\beta I_\beta \quad (\alpha, \beta = 1, \cdots, m+q), \end{aligned}\right\} \tag{1}$$

$$(\omega_i^j)' = [\omega_i^\alpha \omega_\alpha^j] + k_0 g_{ik} [\omega^j \omega^k] \quad (i, j, k = 1, \cdots, m+q), \tag{2}$$

$$(\omega_i^{m+s})' = [\omega_i^\alpha \omega_\alpha^{m+s}], \tag{3}$$

$$\omega_{m+s}^i = -g^{ij} \omega_j^{m+s}, \tag{4}$$

而

$$\omega_j^i = \Gamma_{jk}^i \omega^k, \tag{5}$$

$$\omega_i^{m+s} = \lambda_{ij}^s \omega^j. \tag{6}$$

这里 Γ_{ik}^j 是曲面的联络系数，λ_{ij}^s 是曲面的第二基本形式的系数. 我们首先导入曲面的秩数的概念.

定义 1. 曲面 $V_m \subset S_{m+q}$ 在已给点 M 的秩数是那一些 ω_i^{m+s} 在 M 点的独立的个数. 当曲面 V_m 在每点有相同的秩数 ρ 时，我们称曲面的秩数为 ρ.

我们现在来作它的几何解释. 设曲面 V_m 的秩数为 ρ，由(3)可知方程系统 $\omega_i^{m+s}=0$ 是完全可积的. 变换相切标形 I_i 可使 $\omega^1, \cdots, \omega^\rho$ 成为 ω_i^{m+s} 的 ρ 个基底于是 $\omega^1=\omega^2=\cdots=\omega^\rho=0$ 为完全可积. V_m 可分层为 ∞^ρ $V_{m-\rho}$，而沿着每个 $V_{m-\rho}$ $\omega^1=\omega^2=\cdots=\omega^\rho=0$.

由于 $\omega^1, \cdots, \omega^\rho$ 是 ω_i^{m+s} 的基层，从(6)可得出

$$\omega_a^{m+s}=0 \quad (a=\rho+1, \cdots, m), \tag{7}$$

作(7)的可积条件，利用(5)，(6)及(7)本身，以及曲面秩数为 ρ 的性质我们获得

$$\Gamma_{ab}^\sigma=0 \quad (\sigma=1, \cdots, \rho;\ a, b=\rho+1, \cdots, m). \tag{8}$$

因此

$$\omega_a^\sigma=0 \quad (\operatorname{mod} \omega^1, \cdots, \omega^\rho). \tag{9}$$

从(7)，(9)就可得出沿着 $V_{m-\rho}$ 成立

$$dI_a=\omega_a^b I_b \quad (a, b=\rho+1, \cdots, m),$$

这表示了，$V_{m-\rho}$ 的切向量经平行移动后还是与 $V_{m-\rho}$ 相切. 所以它是全测地的.

另一方面，容易证明

$$dI_{m+s}=\omega_{m+s}^{m+t} I_{m+t}. \tag{10}$$

我们在 $V_{m-\rho}$ 上取一点 P，作全测地曲面 S_m 使它过点 P 而且与 V_m 相切. 由于 $V_{m-\rho}$ 是全测地的，$V_{m-\rho} \subset S_m$. 设 P' 为同一 $V_{m-\rho}$ 上的另一点，由性质(10)可知 S_m 在 P' 的法向量亦就是 V_m 在 P' 的法向量. 所以 S_m 也是 V_m 在 P' 点的相切全测地曲面，由此导出；沿

着$V_{m-\rho}$相切全测地曲面S_m是相同的.这样我们就得出常曲率空间曲面秩数的几何解释.

定理1. 设V_m为安装在常曲率空间S_{m+q}中的m维黎曼空间,如果V_m的曲面秩数为ρ,则V_m可分层为$\infty^\rho V_{m-\rho}$,使得每个$V_{m-\rho}$均为S_{m+q}中的全测地曲面,且沿每个$V_{m-\rho}$,V_m的相切全测地曲面S_m是相同的.

这是Яненко关于欧氏空间曲面秩数定理的推广.

§3. k_0-秩数的几何意义.我们导入m维黎曼空间V_m的测度($ds^2=g_{ij}\omega_i\omega_j$)的$k_0$-秩数的概念.

定义2. 反称双一次形式系统

$$K_{ij}=\frac{1}{2}K_{ijkl}[\omega_k\omega_l]=\frac{1}{2}(R_{ijkl}-k_0(g_{ik}g_{jl}-g_{il}g_{jk}))[\omega_k\omega_l]$$

$$(i,\ j,\ k,\ l=1,\ 2,\ \cdots,\ m) \tag{11}$$

的秩数是测度$ds^2=g_{ij}\omega_i\omega_j$的$k_0$-秩数.

由此可见测度的k_0-秩数等于矩阵$\|R_{ijkl}-k_0(g_{ik}g_{jl}-g_{il}g_{jk})\|$的秩数,这里$i$,$j$,$k$表示行的指标,$l$表示列的指标,现在要作出$k_0$-秩数的几何意义.

设V_m的k_0-秩数为P,我们不难证明:能选取这样的直交标形[1],使

$$K_{\alpha j}=(\omega_{\alpha j})'-[\omega_{\alpha k}\omega_{kj}]-k_0[\omega_\alpha\omega_j]=0\ (\alpha=\rho+1,\ \cdots,\ m;\ j=1,\ \cdots,\ m), \tag{12}$$

外导微上式并利用黎曼空间的组织方程

$$\left.\begin{aligned}(\omega_i)'&=[\omega_j\omega_{ji}],\\(\omega_{ij})'&=[\omega_{ik}\omega_{kj}]+\Omega_{ij}=[\omega_{ik}\omega_{kj}]+\frac{1}{2}R_{ijkl}[\omega_k\omega_l],\end{aligned}\right\} \tag{13}$$

我们得到

$$[\omega_{\alpha a}K_{ab}]=0\quad(a,\ b=1,\ \cdots,\ \rho). \tag{14}$$

也就是

$$K_{abcd}[\omega_{\alpha a}\omega_c\omega_d]=0, \tag{15}$$

对固定一套 a，c，d 乘以 $\rho-2$ 个 $\omega_e(e=1, \cdots, \rho)$ 使 e，$c\neq d$ 我们可推出

$$K_{abcd}\omega_{\alpha a}=0 \quad (\mathrm{mod}\ \omega_1, \cdots, \omega_\rho). \tag{16}$$

由于 K_{abcd} 为 ρ 秩，对于固定的 a，由上式得

$$\omega_{\alpha a}=0 \quad (\mathrm{mod}\ \omega_1, \cdots, \omega_\rho), \tag{17}$$

利用(17)，我们考察方程系统 $\omega_a=0(a=1, \cdots, \rho)$，可证明它是完全可积的.于是 V_m 可分层为 $\infty^\rho\ V_{m-\rho}$，又由(17)，我们可知在 $V_{m-\rho}$ 上

$$dI_\alpha=\omega_{\alpha\beta}I_\beta \quad (\alpha, \beta=\rho+1, \cdots, m), \tag{18}$$

这表示了，$V_{m-\rho}$ 是全测地的.

设 $\bar{\Omega}_{\alpha\beta}$ 是 $V_{m-\rho}$ 的曲率形式，由于 $V_{m-\rho}$ 是全测地的，

$$\Omega_{\alpha\beta}=\bar{\Omega}_{\alpha\beta}. \tag{19}$$

从(12)

$$0=K_{\alpha\beta}=\bar{\Omega}_{\alpha\beta}-k_0[\omega_\alpha\omega_\beta], \tag{20}$$

这表示了 $V_{m-\rho}$ 是常曲率的.这样就得到

定理 2. 设 V_m 是 m 维黎曼空间，如果它的 k_0-秩数等于 ρ，则 V_m 可分层为 $\infty^\rho\ V_{m-\rho}$ 使得每个 $V_{m-\rho}$ 均为 V_m 中常曲率 k_0 的全测地曲面.

这个结果推广了陈省身与 Kuiper[3] 及 Яненко[2] 关于测度的秩数的一个定理.这里的证明方法是较类似于前者.

§4. 应用.设 S_m 为常曲率 k_0 空间 S_{m+q} 中的常曲率 k_0 的曲面，则它必须满足关系式(1)—(4)，由于 S_m 是常曲率的

$$(\omega_i^j)'-[\omega_i^k\omega_k^j]=k_0g_{ik}[\omega^j\omega^k],$$

把它代到(2)并采取直交标形，则 $S_m\subset S_{m+q}$ 就必须满足

$$\sum_{s=1}^{q}[\omega_i^{m+s}\omega_j^{m+s}]=0. \tag{21}$$

如果对于某一个 i，ω_i^{m+1}，…，ω_i^{m+q} 的秩数为 q，则根据嘉当引理可知 ω_k^{m+s} 的秩数为 q，也就是曲面的秩数为 q.

如果并没有任一个 i 使 ω_i^{m+1}，…，ω_i^{m+q} 的秩数为 q，则可以证明：适当的变换标形，可使对每个 i，ω_i^{m+1}，…，ω_i^{m+q} 的秩数 ρ_i 均相等，则 $\rho_i=\rho<q$，从而也可证明曲面的秩数 $=\rho<q$（详细参阅[5]），再利用§2关于曲面秩数的几何意义就得如下定理.

定理 3. 设 S_m 为安装在 $(m+q)$ 维常曲率 k_0 空间的常曲率 k_0 的曲面，那么 S_m 的曲面秩数不大于 q. 并且 $q\leqslant m-1$ 时，过 S_m 的每点必有空间 S_{m+q} 中的测地线.

从这里我们可得

推论. $2m-1$ 维罗巴切夫斯基空间（曲率 k_0）必不含有 m 维局部常曲率 k_0 的紧致解析流形.

特别当 $k_0=0$ 时，所得的欧氏空间的结果是属于 Tompkins 的，但他那时并无解析的条件. 我们这里提供了一个从局部黎曼几何来导出这个整体结果的方法.

参 考 文 献

[1] 胡和生，1956. 论黎曼测度 V_m 在常曲率空间 S_{m+1} 中的变形. 数学学报，6，320-331.

[2] Яненко，Н. Н.，1953. Некоторые вопросы теории вложения Римановых метрик к евклидовы пространства. УМН. 8，320-331.

[3] Chern，S. S.（陈省身），and Kuiper，N. H.，1952，Some theorems on the isometric imbedding of compact Riemannian manifolds in Enclidean space. Annals of Math. 56，422-430.

[4] Tompkins，C.，1939，Isometric embedding of flat manifolds in Enclidean space，Duke Math. J. 5，58-61.

[5] Яненко，Н. Н.，1954. К теории вложения поверхпостей многомерного евклидова пространства，труды Моск，матем，о—ва，III，89-180.

（1957 年 10 月 21 日收到）

On the Ranks of Two Kinds of a Riemannian Space

Hu Hou-sung

The purpose of this paper is to give several geometrical interpretations of ranks of two kinds of a Riemannian space. These ranks are of special importance in the theory of imbeddings and deformations of an m-dimensional Riemannian space V_m into an $(m+q)$ dimensional space S_{m+q} of constant curvature. The results obtained here are the generalization of the theorems of Yanenko, Chern and Kuiper. Furthermore, an application is found to the proof of a result which is closely related to an imbedding theorem of Tompkins.

Let an m-dimensional Riemannian space V_m be imbedded in an $(m+q)$-dimensional space S_{m+q} of constant curvature k_0; let λ^s_{ij} be the coefficients of the second fundamental forms of V_m, then the rank of the matrix $\| \lambda^s_{ij} \|$ $(i, j=1, \cdots, m; s=1, \cdots, q)$ is defined to be the rank of V_m.

Theorem 1. If $V_m \subset S_{m+q}$ is of rank ρ, then V_m may be stratified into $\infty^\rho (m-\rho)$-dimensional totally geodesic surfaces $V_{m-\rho}$ in S_{m+q}, such that the totally geodesic surfaces S_m tangent to V_m along every $V_{m-\rho}$ are the same.

Let an m-dimensional Riemannian space V_m be of the metric $ds^2=g_{ij}dx^i dx^j$ and R_{ijkl} its curvature tensor, then the rank of the matrix $\| R_{ijkl}-k_0(g_{ik}g_{jl}-g_{il}g_{jk}) \|$ is defined to be the k_0-rank of the metric $ds^2=g_{ij}dx^i dx^j$ where the indices $(i\,j\,k)$ and l denote the rows and columns of the matrix respectively.

Theorem 2. If a Riemanian space V_m is of rank ρ, then V_m may be stratified into ∞^ρ totally geodesic surfaces $V_{m-\rho}$ of constant curvature k_0 in V_m.

Theorem 3. If S_m is a space of constant curvature k_0 immersed in a space S_{m+q} of the same constant curvature k_0, then the rank of S_m is not greater than q. Especially, when

$q \leqslant m-1$, there are geodesics of S_{m+q} passing through every point of S_m.

Corollary. In a $(2m-1)$ - dimensional Lobatchewski space of curvature k_0 none m - dimensional compact analytic manifolds can realize a Riemannian metric of constant curvature k_0.

（本文曾发表于《复旦学报（自然科学）》1957 年第 2 期，346 - 351）

论射影平坦空间的一个特征*

胡和生
复旦大学及中国科学院数学研究所

§1. 如所周知,黎曼空间中关于平面公理的嘉当(E. Cartan)定理[1]可以拓广到更一般的空间中去[2,3],满足平面公理的 m 维黎曼空间在它的每点容有 ∞^{m-1} 张全测地超曲面.

柏尔特拉米(Beltrami)给出常曲率空间的另一特征,只有常曲率空间才能与欧氏空间建立点的对应使它的测地线对应于欧氏空间的直线,在这个对应之下所有全测地超曲面与欧氏空间中的全体超平面相对应.

嘉当与柏尔特拉米的结果都可拓广到仿射联络空间,这时很自然地用射影平坦空间代替了常曲率空间.本文的目的在于作出射影平坦空间的一个新的特征.

定理: 如果 m 维仿射联络空间在它的每点容有这样的 $m+2$ 张全测地超曲面,使它们在仿射空间的映像构成一般位置的 $m+2$ 系超平面,那么原空间必须是射影平坦的**.

推论: 如果 m 维黎曼空间在它的每点容有这样的 $m+2$ 张全测地超曲面使它们在欧氏空间的映像构成一般位置的 $m+2$ 系超平面,那么原空间必须是常曲率的.

在§3 中我们证明定理中 $m+2$ 系不能减少了.如 $m+2$ 系的映像有 $m+1$ 系为超平面,而另一系不为超平面,则空间也不见得为射影平坦.

§2. 设 m 维仿射联络空间 V_m 中容有 $m+2$ 系全测地超曲面,并且能建立 V_m 与仿射空间 A_m 的如此的对应使这 $m+2$ 系超曲面的映像构成一般位置的 $m+2$ 系超平面***,那

* 1957 年 8 月 19 日收到.
** 作者在此感谢审查者的意见,使定理的条件得到精简.
*** 这里的一般位置是指满足条件(4)及 $B_1C_1 \neq 0$ 者.

么适当的选取坐标可使 $m+2$ 系全测地超曲面的方程顺次是

$$u^i = \text{const} \quad (\text{对固定的 } i, j=1, \cdots, m), \tag{1}$$

$$B_i(t)u^i - B(t) = 0, \tag{2}$$

$$C_i(\sigma)u^i - C(\sigma) = 0; \tag{3}$$

且对通过同一点的超平面

$$\begin{vmatrix} B_i & C_i \\ B_j & C_j \end{vmatrix} \neq 0 \quad (i \neq j). \tag{4}$$

V_m 的变位方程在和乐标形下是

$$dM = du^i I_i,$$

$$dI_i = \omega_i^j I_j, \quad \omega_i^j = \Gamma_{i\,k}^{\,j} du^k. \quad (i, j, k = 1, \cdots, m) \tag{5}$$

而 $\Gamma_{i\,k}^{\,j}$ 是 V_m 的联络系数.

由于第 i 系超曲面 $u^i = \text{const}$(对固定的 i, $i=1, \cdots, m$) 是全测地的,因此切于超曲面的向量经平行移动后还是它的切向量,这样由(5)就可得到

$$\Gamma_{j\,k}^{\,i} = 0 \ (i \neq j, i \neq k; i, j, k = 1, \cdots, m). \tag{6}$$

对于第 $m+1$ 系超曲面(2),设 $B_1 \neq 0$,则可改写为

$$du^1 = -b_\alpha du^\alpha \quad (\alpha = 2, \cdots, m),$$

这里 $b_\alpha = \dfrac{B_\alpha}{B_1}$,于是在这系超曲面上

$$dM = du^\alpha (I_\alpha - b_\alpha I_1).$$

由于它是全测地的,$d(I_\alpha - b_\alpha I_1)$ 必须是 $I_\alpha - b_\alpha I_1$ 的线性组合,从而得到

$$\omega_\alpha^1 - b_\alpha \omega_1^1 = -(\omega_\alpha^\beta - b_\alpha \omega_1^\beta) b_\beta \quad (\text{mod } du^1 + b_\alpha du^\alpha).$$

由此导出

$$\Gamma_{a a}^{a}+b_{a} \Gamma_{11}^{1}-2 \Gamma_{1 a}^{1}-2 b_{a} \Gamma_{1 a}^{1}=0, \tag{7}$$

$$-b_{a} \Gamma_{1 s}^{1}-b_{a} b_{s} \Gamma_{1 s}^{s}+b_{a} \Gamma_{a s}^{a}+b_{s} \Gamma_{s a}^{s}-b_{s} \Gamma_{a 1}^{1}+b_{a} b_{s} \Gamma_{11}^{1}-b_{a} b_{s} \Gamma_{a 1}^{a}=0,$$
$$(\text{对 } a, s \text{ 均非作和}; a, s=2, \cdots, m; a \neq s). \tag{8}$$

同样地对于第 $m+2$ 系超曲面(3)也可得到相应的关系式:

$$\Gamma_{a a}^{a}+c_{a} \Gamma_{11}^{1}-2 \Gamma_{1 a}^{1}-2 c_{a} \Gamma_{1 a}^{1}=0, \tag{9}$$

$$-c_{a} \Gamma_{1 s}^{1}-c_{a} c_{s} \Gamma_{1 s}^{s}+c_{a} \Gamma_{a s}^{a}+c_{s} \Gamma_{s a}^{s}-c_{s} \Gamma_{a 1}^{1}+c_{a} c_{s} \Gamma_{11}^{1}-c_{a} c_{s} \Gamma_{a 1}^{a}=0 \tag{10}$$

$$(\text{对 } a, s \text{ 均非作和}; a, s=2, \cdots, m; a \neq s);$$

这里 $c_{a}=\frac{C_{a}}{c_{1}}$. 记

$$\Gamma_{11}^{1}=2 p_{1}, \Gamma_{22}^{2}=2 p_{2}, \cdots, \Gamma_{m m}^{m}=2 p_{m}. \tag{11}$$

由于(4),将(11)代入(7),(9)就可得

$$\Gamma_{a 1}^{a}=p_{1}, \Gamma_{1 a}^{1}=p_{a}. \tag{12}$$

又从(8),(10)利用(4)得出

$$\Gamma_{a s}^{a}=p_{s}. \tag{13}$$

这里从(6)及(11)—(13)就可将 V_m 的联络系数写成

$$\Gamma_{j k}^{i}=\delta_{j}^{i} p_{k}+\delta_{k}^{i} p_{j} \quad(i, j, k=1, \cdots, m). \tag{14}$$

因此 V_m 是射影平坦的. 这样就证明了定理.反过来说射影平坦空间满足定理的条件.所以说这个定理是射影平坦空间的一个特征.

特别对于黎曼空间就成立引理中所提到的推论.

§3. 现在我们来讨论前述定理中的 $m+2$ 系全测地超曲面的有关条件是否能减少的问题.首先注意到,如果空间 V_m 有一般位置的 m 系全测地超曲面,就可选取坐标系统使它们的方程为 $u^k=\text{const}$,那么它们一定可以映像到仿射空间成为 m 系平行超平面,因此要减少条件,我们应该假定其余二系超曲面中有一系的方程为 $g(u^1, \cdots, u^m)=$

const,其中 g 并非 $u^1, \cdots, u^m$ 的线性函数.我们能够证明这样的空间一般并非射影平坦的.为简便计,在 $m=3$ 时讨论. 设 4,5 两系的方程是 $u^1+u^2+u^3=\text{const}$, $(u^1)^2+(u^2)^2+(u^3)^2=\text{const}$,利用它们是全测地的性质,并经过一定的计算可得

$$\Gamma_{ij}^{i}=\frac{1}{2}\left(\Gamma_{jj}^{j}+\frac{(u^i)^2+(u^j)^2}{u^i u^j(u^j-u^i)}\right)\quad (i,\ j=1,\ 2,\ 3;\ i\neq j;\text{依 } i,\ j \text{ 均非作和}).$$

外尔(Weyl)张量的支量[4]

$$W_{121}^{1}=\frac{(u^1)^4[(u^2)^3-(u^2)^2u^3-u^2(u^3)^2]+\cdots}{(u^1)^2(u^2)^2(u^3)^2(u^1-u^2)(u^2-u^3)(u^3-u^1)},$$

分子中未写出的项所含 u^1 的次数不超过三次,所以不恒等于 0,因此这个空间一般不是射影平坦的,也就是定理的条件再不能减弱了.

参考文献

[1] Cartan, E. (1946): Leçons sur la géometrie des espaces de Riemann.

[2] 苏步青(1950):科学记录,3,7-16.

[3] 谷超豪(1950):科学记录,3,53-59.

[4] Eisenhart, L. P. (1925): Riemannian geometry.

A Characterization of a Projective Flat Space

Hu Hoo-Sung

(Institute of Mathematics, Academia Sinica and Fuh-tan University)

Abstract

The aim of the present paper is to give a characterization of a projective flat space. It has an intimate connection with Cartan's theorem on the axiom of plane and Beltrami's on the geodesic correspondence of a Riemannian space and a flat space. The main result is as follows:

Theorem: If an m - dimensional affine connected space without torsion contains $m+2$ families of totally geodesic hypersurfaces, such that there exists a mapping of V_m into affine space A_m which brings the $m+2$ families of hypersurfaces onto $m+2$ families of planes in general position, then the space must be projective flat and conversely.

Especially, in the case of a Riemannian space, we obtain a characterization of a space of constant curvature.

It is also shown that the number $m+2$ cannot be replaced by a smaller one. Even when the images of $m+1$ families among $m+2$ families are hyperplanes and the remaining one is not, the space is not always projectively flat.

（本文曾发表于《数学学报》1958 年第 8 卷第 2 期，269 - 271）

常曲率空间的直交全测地超曲面系*

胡和生
中国科学院数学研究所;复旦大学
(中国科学院学部委员苏步青教授推荐)

§1. 如所知,m 维负常曲率 $k_0(<0)$ 空间的线素可以写成

$$ds^2 = e^{2\sqrt{-k_0}\, x^m}\left[(dx^1)^2 + (dx^2)^2 + \cdots + (dx^{m-1})^2\right] + (dx^m)^2,$$

从而在负常曲率空间里,存在 $m-1$ 系相互直交的全测地超曲面,并且这些超曲面显然都具有同一负曲率 k_0.

关于正常曲率空间我们得到

定理 1. m 维正常曲率空间至多只能容有 $\left[\dfrac{m+1}{2}\right]$ 系相互直交的全测地超曲面.

由此可见,虽然所有常曲率空间的全测地超曲面的分布从射影的观点来看是相同的(因为所有的常曲率空间都是射影欧氏的),但从度量的观点来看,正负常曲率空间之间却有区别.

在§3 中证明了下列定理:

定理 2. 如果黎曼空间 $V_m(m>3)$ 容有 3 系相互直交的常曲率的全测地超曲面,那么这些超曲面的常曲率都等于 k_0,并且空间 V_m 也具有同一的常曲率 k_0.

我们还证明:容有 2 系相互直交常曲率的全测地超曲面的黎曼空间 V_m 不一定是常曲率的.因而定理 2 条件中的 3 系不能减低了.

---

* 1957 年 9 月 13 日收到.

由上述两定理可知：当$m\geqslant 5$时，定理2是一切常曲率空间的特征，而当$m=4$时这性质只有负常曲率空间的特征，因为正常曲率空间V_4不能容有3系相互直交的全测地超曲面.

§2. 我们首先证明常曲率空间V_m如果存在p系相互直交的全测地超曲面，那么在适当的坐标系统下可以化V_m的线素为

$$ds^2=g_{11}(x^1,\ x^a)(dx^1)^2+\cdots+g_{pp}(x^p,\ x^a)(dx^p)^2+\frac{(dx^{p+1})^2+\cdots+(dx^m)^2}{\left\{1+\frac{k_0}{4}[(x^{p+1})^2+\cdots+(x^m)^2]\right\}^2}\quad (a=p+1,\ \cdots,\ m). \tag{1}$$

证明： 在V_m的各点装上直交标形$I_i(i=1,\ 2,\ \cdots,\ m)$，使$I_a(a=1,\ 2,\ \cdots,\ p)$各为第$a$系全测地超曲面的法线；空间的导来方程是

$$dM=\omega_i I_i,$$
$$dI_i=\omega_{ij}I_j,\ \omega_{ij}+\omega_{ji}=0,\ \omega_{ij}=\lambda_{ijh}\omega_k. \tag{2}$$

它的组织方程是

$$(\omega_i)'=[\omega_j\omega_{ji}],$$
$$(\omega_{ij})'=[\omega_{ik}\omega_{kj}]+K[\omega_i\omega_j]. \tag{3}$$

而且V_m的线素是

$$ds^2=\sum_{a=1}^{p}(\omega_a)^2+\sum_{a=p+1}^{m}(\omega_a)^2. \tag{4}$$

由于I_a垂直于第a系全测地超曲面，第a系全测地超曲面的微分方程为$\omega_a=0\ (a=1,\ 2,\ \cdots,\ p)$，从全测地的性质，得到

$$\lambda_{ajk}=0\quad (a=1,\ 2,\ \cdots,\ p;\ j,\ k=1,\ \cdots,\ m;\ k\neq a). \tag{5}$$

$$\omega_{a\alpha}=-\omega_{\alpha a}=\lambda_{a\alpha a}\omega_a. \tag{6}$$

现在考虑发甫方程系统$\omega_\alpha=0\ (\alpha=p+1,\ \cdots,\ m)$；从(3)及(6)可知

$$(\omega_\alpha)'=0 \quad (\mathrm{mod}\,\omega_{p+1},\ \cdots,\ \omega_m).$$

从而 $\omega_\alpha=0$ 是完全可积的，就是

$$\omega_\alpha=C_{\alpha\beta}\,d\,f^\beta. \tag{7}$$

又因 $\omega_1=0$ 代表一系超曲面，因此它也是完全可积的，我们获得

$$\omega_a=b_a d\,f^a \quad (a=1,\ 2,\ \cdots,\ p;\ a\ \text{非和式}), \tag{8}$$

将(7)，(8)代入(4)，并变换坐标使 $\bar{x}^i=f^i$ 且仍以 x^i 记坐标系统，则线素化为

$$ds^2=g_{11}(dx^1)^2+\cdots+g_{pp}(dx^p)^2+g_{\alpha\beta}\,dx^\alpha dx^\beta.$$

由(8)可知 $dx^a=0$ 代表 p 系全测地超曲面，所以除 g_{aa} 之外其他 g_{ij} 中不包含 x^a.再注意到：$x^a=\mathrm{const.}$ 定义曲面 V_{m-p}，而它的线素是 $g_{\alpha\beta}dx^\alpha dx^\beta$.如所知，$V_{m-p}$ 上的诱导联络一定是常曲率的，所以 V_m 的线素可化为(1).

为了具有形式(1)的线素确乎是常曲率空间的线素，充要条件是这样：

$$\frac{\partial^2\sqrt{g_{aa}}}{(\partial x^\alpha)^2}+\frac{1}{2A}\sum_{\beta=p+1}^{m}\frac{\partial\sqrt{g_{aa}}}{\partial x^m}\frac{\partial A}{\partial x^\beta}-\frac{1}{A}\frac{\partial\sqrt{g_{aa}}}{\partial x^\alpha}\frac{\partial A}{\partial x^\alpha}=-K_0\sqrt{g_{aa}}A, \tag{9}$$

$$\frac{1}{A}\sum_{\alpha=p+1}^{m}\frac{\partial\sqrt{g_{aa}}}{\partial x^\alpha}\frac{\partial\sqrt{g_{bb}}}{\partial x^\alpha}=-K_0\sqrt{g_{aa}}\sqrt{g_{bb}} \quad (a\neq b), \tag{10}$$

$$\frac{\partial^2\sqrt{g_{aa}}}{\partial x^\alpha\partial x^\beta}+\frac{1}{2}\sqrt{A}K_0\left(\frac{\partial\sqrt{g_{aa}}}{\partial x^\alpha}x^\beta+\frac{\partial\sqrt{g_{aa}}}{\partial x^\beta}x^\alpha\right)=0 \quad (\alpha\neq\beta), \tag{11}$$

其中

$$A=\frac{1}{\left(1+\frac{K_0}{4}[(x^{p+1})^2+\cdots+(x^m)^2]\right)^2}. \tag{12}$$

固定 a，令 $\sqrt{g_{aa}}=\psi(x^a,\ x^\alpha)$ 并改写(11)为

$$\frac{\partial B_\alpha}{\partial x^\beta}+\frac{K_0}{2}\sqrt{A}x^\beta B_\alpha=0 \quad (\alpha\neq\beta), \tag{13}$$

这里

$$B_\alpha = \frac{\partial \psi}{\partial x^\alpha} + \frac{K_0}{2}\sqrt{A}\, x^\alpha \psi. \tag{14}$$

解(13)，

$$B_\alpha = \rho_\alpha(x^1, \cdots, x^p, x^\alpha)\sqrt{A}. \tag{15}$$

代入(14)，并解 ψ

$$\psi = \sqrt{g_{aa}} = \sqrt{A}\sum_{\alpha=p+1}^{m} G_{\alpha a}(x^1, \cdots, x^p, x^\alpha). \tag{16}$$

利用(12)和(16)来化简(9)经相当计算后得

$$\frac{1}{\sqrt{A}}\frac{\partial^2 G_{\alpha a}}{(\partial x^\alpha)^2} - \frac{1}{2}K_0\sum_{\beta=p+1}^{m}\left(\frac{\partial G_{\beta a}}{\partial x^\beta}x^\beta - G_{\beta a}\right) = 0. \tag{17}$$

调换指数 α 为 γ，并以(17)减它，就得

$$\frac{\partial^2 G_{\alpha a}(x^1, \cdots, x^p, x^\alpha)}{(\partial x^\alpha)^2} = \frac{\partial^2 G_{\gamma a}(x^1, \cdots, x^p, x^\alpha)}{(\partial x^\gamma)^2}, \tag{18}$$

从而各边应与 x^α，x^γ无关.令 $\frac{\partial^2 G_{\alpha a}}{(\partial x^\alpha)^2} = 2g_a(x^1, \cdots, x^p)$，而且积分起来，

$$G_{\alpha a} = g_a(x^1, \cdots, x^p)(x^\alpha)^2 + h_{\alpha a}(x^1, \cdots, x^p)x^\alpha + k_{\alpha a}(x^1, \cdots, x^p)(\alpha\ \text{非和式}), \tag{19}$$

把(19)代入(17)得到 g_a 及 $k_{\beta a}$ 所满足的条件:

$$-\frac{K_0}{2}\sum_{\beta=p+1}^{m} k_{\beta a} = 2g_a. \tag{20}$$

这样，由(19)及(16)得到

$$\sqrt{g_{aa}} = \Phi_a + \left(\sum_{\beta=p+1}^{m} h_{\beta a}x^\beta - 2\Phi_a\right)\sqrt{A}, \tag{21}$$

式中

$$\Phi_a(x^1,\ \cdots,\ x^p)=\frac{4g_a}{K_0}. \tag{22}$$

令 $x^\beta=0$，代入(21)，可知 $\sqrt{g_{aa}}=-\Phi_a$ 因而成立

$$\Phi_a<0. \tag{23}$$

由(21)求得 $\frac{\partial\sqrt{g_{aa}}}{\partial x^\alpha}$，代入(10)，再经相当的合并与消去手续后得

$$\sum_{\alpha=p+1}^m h_{\alpha a}h_{\alpha b}=-K_0\Phi_a\Phi_b<0, \tag{24}$$

从这里可以看到：为了 m 维正常曲率空间容有 p 系相互直交的全测地超曲面，必须成立

$$\sum_{\alpha=p+1}^m h_{\alpha a}h_{\alpha b}<0\quad(\alpha=p+1,\ \cdots,\ m;\ a=1,\ \cdots,\ p). \tag{25}$$

问题就归结到这样：$m-p$ 维欧氏空间中能否容有 p 个向量，使任何一对的数量积取负值.根据数学归纳法可证一个

引理： 在 r 维欧氏空间中，至多只能选取 $r+1$ 个向量使每对的数量积取负值.

从此立刻得出结论：为使方程(24)容有实解，必须 $p\leqslant m-p+1$ 或 $p\leqslant\frac{m+1}{2}$.

现在我们要证明 $p\leqslant\frac{m+1}{2}$ 时，方程(24)的解确乎存在.首先取

$$\Phi_1=\Phi_2=\cdots=\Phi_m=-\frac{1}{\sqrt{K_0}},$$

从而化(24)为

$$\sum_{\alpha=p+1}^m h_{\alpha a}h_{\alpha b}=-1\quad(a,\ b=1,\ \cdots,\ p;\quad \alpha=p+1,\ \cdots,\ m). \tag{26}$$

解上述方程我们取 $h_{\alpha 1}$ 的支量为(1, 0, …, 0)为使 $h_{\alpha 2}$ 与 $h_{\alpha 1}$ 的数量积等于-1起见，必须取 $h_{\alpha 2}$ 的第一个支量为-1，我们取 $h_{\alpha 2}$：(-1, 1, 0, …, 0)同样地对于 $h_{\alpha 3}$ 它与

$h_{\alpha 1}$及$h_{\alpha 2}$的数量积为负,因此它的第一个、第二个支量必为-1及-2,因此我们选取$h_{\alpha 3}$:$(-1, -2, 1, 0, \cdots, 0)$,同样地对于$h_{\alpha 4}$我们决定它的第三个支量为$-6$,而$h_{\alpha 4}$采取$(-1, -2, -6, 1, 0, \cdots, 0)$.由于假设了$p \leqslant \frac{m+1}{2}$亦即$m-p \geqslant p-1$,因此可以类似地继续进行最后就能决定$h_{\alpha p}$的第$p-1$个支量,这样我们就得到(26)的解.

我们现在把所定的Φ_a及$h_{\alpha a}$的值代入(21),借以作出线素

$$ds^2 = \sum_{a=1}^{p} g_{aa}(x^a, x^\alpha)(dx^a)^2 + \frac{\sum_{\alpha=p+1}^{m}(dx^\alpha)^2}{\left[1+\frac{K_0}{4}\sum_{\alpha=p+1}^{m}(x^\alpha)^2\right]^2}. \tag{27}$$

根据作法以及一些直接的验算可以证明:这时,

$$R_{hijk} = K_0(g_{hj}g_{ik} - g_{hk}g_{ij}),$$

所以线素(27)定义一个正常曲率的空间.由于具有同一的常曲率的空间都是等价的,我们证完了定理1.

§3. 在证明定理2之前我们先证明两个有用的事实.

(i) 如果黎曼空间V_m容有2系直交常曲率的全测地超曲面,那么它们的交集也是常曲率的曲面且这些曲率相等.

实际上,第1系中的任一个超曲面$\overset{(1)}{V}_{m-1}$和第2系中的任一个超曲面$\overset{(2)}{V}_{m-1}$的交集$\overset{(1)}{V}_{m-1} \cap \overset{(2)}{V}_{m-1}$是$m-2$维曲面;它对于$\overset{(1)}{V}_{m-1}$和$\overset{(2)}{V}_{m-1}$都是全测地曲面,所以它和$\overset{(1)}{V}_{m-1}$,$\overset{(2)}{V}_{m-1}$都有相同的常曲率.

附带指出,从这事实及在§2的开始部分所述可以断定:如果V_m容有2系直交常曲率的全测地超曲面,那么它们必容有m重直交系统.

(ii) 设点的坐标为$y^1, \cdots, y^m$的空间V_m容有全测地超曲面V_{m-1}:$y^m = \text{const.}$,那么从高斯(Gauss)方程就可得到

$$R_{ijkl} = \underset{1}{R}_{ijkl} \quad (i, j, k, l = 1, \cdots, m-1), \tag{28}$$

这里 R_{ijkl} 和 $\underset{1}{R}_{ijkl}$ 顺次的是 V_m 和 V_{m-1} 的曲率张量的支量.

现在转到定理 2 的证明.设空间 V_m 容许 3 系直交全测地常曲率超曲面,那么必定能够选取适当坐标系统使 V_m 的线素化为

$$ds^2 = g_{11}(x^1, x^\alpha)(dx^1)^2 + g_{22}(x^2, x^\alpha)(dx^2)^2 + g_{33}(x^3, x^\alpha)(dx^3)^2 + \frac{(dx^4)^2 + \cdots + (dx^m)^2}{\left\{1 + \frac{K_0}{4}[(x^{p+1})^2 + \cdots + (x^m)^2]\right\}^2} \quad (\alpha = 4, \cdots, m),$$

可是按假设超曲面系 $x^1 = \text{const.}$ 是全测地常曲率的,从(28)得到

$$R_{\alpha\beta\gamma\delta} = k(g_{\alpha\gamma}g_{\beta\delta} - g_{\alpha\delta}g_{\beta\gamma}) \quad (\alpha, \beta, \gamma, \delta \neq 1).$$

特别是

$$R_{2323} = k g_{22} g_{33},$$

同样:

$$R_{\alpha\beta\gamma\delta} = k(g_{\alpha\gamma}g_{\beta\delta} - g_{\alpha\delta}g_{\beta\gamma}) \quad (\alpha, \beta, \gamma, \delta \neq 1, 2, 3),$$

及

$$R_{1313} = k g_{11} g_{33}, \; R_{1212} = k g_{11} g_{22}.$$

从直接计算 $R_{abac}(a, b, c = 1, 2, 3, a, b, c \neq)$ 及曲率张量其他支量均为 0,因而有

$$R_{\alpha\beta\gamma\delta} = k(g_{\alpha\gamma}g_{\beta\delta} - g_{\alpha\delta}g_{\beta\gamma})$$

就是说空间是常曲率的.定理证毕.

§4. 现在要证明,单容有 2 系直交全测地常曲率的超曲面的 V_m 不一定是常曲率的.为此,设我们的线素为

$$ds^2 = g_{11}(x^1, x^\alpha)(dx^1)^2 + g_{22}(x^2, x^\alpha)(dx^2)^2 + \frac{(dx^3)^2 + \cdots + (dx^m)^2}{\left\{1 + \frac{K_0}{4}[(x^{p+1})^2 + \cdots + (x^m)^2]\right\}^2} \quad (\alpha = 3, \cdots, m),$$

式中函数 g_{11}，g_{22}满足(9)和(11)，即 g_{11}，g_{22}具有形式(21)，但 $h_{\alpha a}$($\alpha=3$，…，m；$a=1$，2) 不满足(24).由§2 的计算过程中可见 $x^1=$const.与 $x^2=$const. 是两系相互直交的全测地超曲面且为常曲率，可是

$$R_{1212}=kg_{11}g_{22}$$

不成立，所以 V_m不是常曲率的.

(本文曾发表于《科学记录(新辑)》1958 年第 2 卷第 1 期，1－5)

K 展空间的一种新几何学*

胡和生
中国科学院复旦大学数学研究室

§1. 引言. 自从 1931 年 J. Douglas[1]研究 K 展空间的几何学以来，许多几何学家继续作出这方面的发展[2,3,4].Douglas 根据 K 展空间的坐标变换及 K 展上的参数变换定义了画法几何学、仿射几何学和体积式几何学.在本文里，将得出一种新的几何学，即 K 展的射影几何学，就是对于任意的坐标变换及参数的射影变换

$$u^\alpha = \frac{a^\alpha_\beta v^\beta + b^\alpha}{c_r v^\gamma + d} \quad (a^\alpha_\beta,\ b^\alpha,\ c_r,\ d \text{ 常数})$$

为不变的几何学.很显然，这种几何学是介于 K 展的画法几何学与仿射几何学之间而不是体积几何学.我们容易看出，如果 K 展的几何学是体积式的，而同时又是射影式的，那么它必须是仿射几何学.在§3 中我们要导出射影的基本不变量和 K 展空间为射影平坦的充要条件等等.

需要指出的是，这里的射影几何学与画法几何学有所不同，尽管有些几何学家在 K 展空间的研究中称画法几何学为射影几何学.

§2. K 展射影几何学的导出. 设 N 维空间(x^i)的 K 展的微分方程是

$$\frac{\partial^2 x^i}{\partial u^\alpha \partial u^\beta} + H^i_{\alpha\beta}(x,\ p) = 0 \quad \left(p^j_\alpha = \frac{\partial x^j}{\partial u^\alpha}\right), \tag{1}$$

式中 $(x^i)(i,\ j,\ k=1,\ \cdots,\ N)$ 是空间一点的坐标，$(u^\alpha)(\alpha,\ \beta,\ \gamma=\dot{1},\ \dot{2},\ \cdots,\ \dot{K})$ 是 K

* 1959 年 1 月 28 日收到.

展上的一点的参数，$H^i_{\alpha\beta}$是x，p 的函数，关于 p 是广义齐次的并且Γ^i_{jk}是仿射联络系数：

$$\Gamma^i_{jk}=\frac{1}{K(K+1)}\frac{\partial^2 H^i_{\alpha\beta}}{\partial p^j_\alpha\partial p^k_\beta}.$$

这些联络系数对于参数变换

$$u^a=u^a(v;\ a) \tag{2}$$

应受到下列变换：

$$\Gamma^i_{jk}=\Gamma^i_{jk}+\delta^i_jA_k+\delta^i_kA_i+p^i_\lambda B^\lambda_{jk}, \tag{3}$$

式中

$$A_k=\frac{1}{K(K+1)}\frac{\partial G^\lambda_\lambda\beta}{\partial p^k_\beta},\quad B^\gamma_{jk}=\frac{1}{K(K+1)}\frac{\partial^2 G^\gamma_{\alpha\beta}}{\partial p^j_\alpha\partial p^k_\beta}. \tag{4}$$

而且沿着每个 K 展流形成立

$$G^\gamma_{\alpha\beta}=-\frac{\partial v^\gamma}{\partial u^\rho}\frac{\partial^2 u^\rho}{\partial\gamma^a\partial v^\beta}. \tag{5}$$

在坐标变换 $x^i=x^i(\bar{x})$ 之下，$\Gamma^i_{jk}(x,\ p)$的变换规律是

$$\frac{\partial^2\ \bar{x}^i}{\partial x^j\partial x^k}+\bar{\Gamma}^i_{ab}(\bar{x},\ \bar{p})\frac{\partial\bar{x}^a}{\partial x^i}\frac{\partial\bar{x}^b}{\partial x^k}=\Gamma^l_{jk}(x,\ p)\frac{\partial\bar{x}^l}{\partial x^l}. \tag{6}$$

我们将考察在怎样的参数之下 $B^\lambda_{jk}=0$. 利用 $G^\gamma_{\alpha\beta}$关于指标的广义齐次性成立

$$G^\gamma_{\alpha\beta}=\delta^\gamma_\alpha p^k_\beta A_k+\delta^\gamma_\beta p^k_\alpha A_k. \tag{7}$$

即

$$(K+1)\frac{\partial v^\gamma}{\partial u^\rho}\frac{\partial^2 u^\rho}{\partial v^\alpha\partial v^\beta}=\delta^\gamma_\alpha\frac{\partial v^\mu}{\partial u^\rho}\frac{\partial^2 u^\rho}{\partial v^\mu\partial v^\beta}+\delta^\gamma_\beta\frac{\partial v^\mu}{\partial u^\rho}\frac{\partial^2 u^\rho}{\partial v^\mu\partial v^\alpha}. \tag{8}$$

为解出这组微分方程将它乘以 $\dfrac{\partial u^\sigma}{\partial v^\gamma}$，

$$\frac{\partial^2 u^\sigma}{\partial v^\alpha \partial v^\beta}=\frac{\partial u^\sigma}{\partial v^\alpha}E_\beta+\frac{\partial u^\sigma}{\partial v^\beta}E_\alpha, \tag{9}$$

式中

$$E_\beta=\frac{1}{K+1}\frac{\partial v^\mu}{\partial u^\rho}\frac{\partial^2 u^\rho}{\partial v^\mu \partial v^\beta}. \tag{10}$$

另一方面,作出(9)的可积条件,

$$\frac{\partial u^\sigma}{\partial v^\beta}\left(\frac{\partial E_\alpha}{\partial v^\gamma}-E_\alpha E_\gamma\right)-\frac{\partial u^\sigma}{\partial v^\gamma}\left(\frac{\partial E_\alpha}{\partial v^\beta}-E_\alpha E_\beta\right)+\frac{\partial u^\sigma}{\partial v^\alpha}\left(\frac{\partial E_\beta}{\partial v^\gamma}-\frac{\partial E_\gamma}{\partial v^\beta}\right)=0, \tag{11}$$

并且乘以 $\frac{\partial v^\delta}{\partial u^\sigma}$,便得到

$$\delta_\beta^\delta\left(\frac{\partial E_\alpha}{\partial v^\gamma}-E_\alpha E_\gamma\right)-\delta_\gamma^\delta\left(\frac{\partial E_\alpha}{\partial v^\beta}-E_\alpha E_\beta\right)+\delta_\alpha^\delta\left(\frac{\partial E_\beta}{\partial v^\gamma}-\frac{\partial E_\gamma}{\partial v^\beta}\right)=0. \tag{12}$$

令 $\delta=\alpha\neq\beta$, γ,得到$\frac{\partial E_\beta}{\partial v^\gamma}-\frac{\partial E_\gamma}{\partial v^\beta}=0$,因此 $E_\beta=\frac{\partial E}{\partial v^\beta}$.又当 $K>1$ 时,由(12)可得

$$\frac{\partial^2 E}{\partial v^\alpha \partial v^\beta}-\frac{\partial E}{\partial v^\alpha}\frac{\partial E}{\partial v^\beta}=0. \tag{13}$$

积分之,

$$e^{-E}=c_\gamma v^\gamma+d, \tag{14}$$

其中 c_γ, d 都是任意常数.这样一来,(9)式化为

$$\frac{\partial}{\partial v^\alpha}\left((c_\gamma v^\gamma+d)\frac{\partial u^\sigma}{\partial v^\beta}\right)+\frac{\partial u^\sigma}{\partial v^\alpha}c_\beta=0.$$

积分一次,

$$(c_\gamma v^\gamma+d)\frac{\partial u^\sigma}{\partial v^\beta}+c_\beta u^\sigma=a_\beta^\sigma,$$

再积分一次,就可断定:(8)的解必须具有下列形式:

$$u^{\alpha}=\frac{a_{\beta}^{\alpha}v^{\beta}+b^{\alpha}}{c_{\gamma}v^{\gamma}+d}. \tag{15}$$

反过来,(15)确是(8)的解.实际上,可以证明在变换(15)之下

$$\frac{\partial^{2}u^{\rho}}{\partial v^{\alpha}\partial v^{\beta}}=-\frac{c_{\beta}}{A}\frac{\partial u^{\rho}}{\partial v^{\alpha}}-\frac{c_{\alpha}}{A}\frac{\partial u^{\rho}}{\partial v^{\beta}}. \tag{16}$$

这里 $A=c_{\gamma}v^{\gamma}+d$,从这关系容易证明(15)是微分方程(8)的解.

在这种变换下,联络的变换规律是

$$'\Gamma_{jk}^{i}=\Gamma_{jk}^{i}+\delta_{j}^{i}A_{k}+\delta_{k}^{i}A_{j}. \tag{17}$$

我们称 $\Gamma_{jk}^{i}(x, p)$关于(6)及(17)的不变量论为 K 展空间的射影几何学.

我们也可用另一方法来验证微分方程(8)的解是(15).把$(u', \cdots, u^{k})$看作 K 维普通仿射空间的仿射坐标,那么在坐标(u)下,这空间的联络系数 $\Gamma_{uv}^{\lambda}=0$,因此在坐标 v 下

$$\bar{\Gamma}_{\alpha\beta}^{\gamma}\frac{\partial u^{\rho}}{\partial v^{\gamma}}=\frac{\partial^{2}u^{\rho}}{\partial v^{\alpha}\partial v^{\beta}} \quad 或 \quad \bar{\Gamma}_{\alpha\beta}^{\gamma}=\frac{\partial^{2}u^{\rho}}{\partial v^{\alpha}\partial v^{\beta}}\frac{\partial v^{\gamma}}{\partial u^{\rho}}. \tag{18}$$

如果 $u=u(v)$ 满足(8),那么在坐标系数(v)之下,仿射联络一定要具形状

$$\bar{\Gamma}_{\alpha\beta}^{\gamma}=\delta_{\alpha}^{\gamma}\psi_{\beta}+\delta_{\beta}^{\gamma}\psi_{\alpha}. \tag{19}$$

如所知,为使一个坐标下空间的测地线由 $K-1$ 个线性方程所表示,其充要条件是空间联络在这坐标下具有形式(19);所以经过坐标变换 $u^{\alpha}=u^{\alpha}(v)$ 之后,直线的方程仍旧是 $K-1$ 个线性方程.如果把这坐标变换看成仿射空间自身到自身的对应,那么它必须使直线变到直线.因而,它必须是射影变换(15).

§3. 射影的基本不变量和空间射影平坦性.我们容易证明,两个 K 展空间互为射影等价的充要条件是:

$$'z_{jk}^{i}=z_{jk}^{i}, \tag{20}$$

其中

$$z_{jk}^{i}=\Gamma_{jk}^{i}-\frac{\delta_{j}^{i}}{N+1}\Gamma_{\alpha k}^{\alpha}-\frac{\delta_{k}^{i}}{N+1}\Gamma_{j\alpha}^{\alpha},\tag{21}$$

这个射影的基本不变量 z_{jk}^{i} 在坐标变换下与 Π_{jk}^{i} 有一样的变换规则.

我们还可证明,K 展空间的射影的微分不变量的最一般形式是

$$F_{b_1\cdots b_s}^{a_1\cdots a_r}\left(p,\ z,\ \frac{\partial z}{\partial p},\ \frac{\partial z}{\partial x},\ \frac{\partial^2 z}{\partial x^2},\ \frac{\partial^2 z}{\partial x\partial p},\ \cdots\right).\tag{22}$$

如 K 展空间经坐标变换 $x^i=x^i(\bar{x})$ 及射影变换(15)后,K 展的方程能化为

$$\frac{\partial^2 x^i}{\partial u^\alpha \partial u^\beta}=0,\tag{23}$$

或

$$x^i=A_\alpha^i u^\alpha+c^i.\tag{24}$$

那么,同其他几何学一样,称它为射影平坦的 K 展空间.

最后,我们来证实关于 K 展空间的射影平坦性的定理.

定理: 要使 K 展空间($K>1$)为射影平坦的空间,其充要条件是:

$$z_{jk}^{i}\Big|_{a}^{a}=0.\tag{25}$$

必要性是显然的.实际上设 K 展空间是射影平坦的,那么经过坐标变换及射影变换之后,$\Gamma_{jk}^{i}=0$,因而 $z_{jk}^{i}\Big|_{a}^{a}=0$.

现在,来证条件的充分性.关于 i, a 缩短(25),

$$\Gamma_{jk}^{a}\Big|_{a}^{a}-\frac{1}{N+1}\Gamma_{ak}^{a}\Big|_{j}^{a}-\frac{1}{N+1}\Gamma_{aj}^{a}\Big|_{k}^{a}=0,$$

从而 $\Pi_{jk}^{i}=z_{jk}^{i}$, $\Pi_{jk}^{i}\Big|_{a}^{a}=0$.这表示了空间是画法平坦的,所以存在坐标系统使 $\Pi_{jk}^{i}(x)=0$.这样 $z_{jk}^{i}=0$ 或 $\Gamma_{jk}^{i}=\delta_{j}^{i}p_{k}+\delta_{k}^{i}\psi_{i}\left(\psi_{k}=\dfrac{1}{N+1}\Gamma_{ak}^{a}\right)$,所以在这个坐标下,$K$ 展的微分方程变为

$$\frac{\partial^2 x^i}{\partial u^\alpha \partial u^\beta}+(\delta^i_j\psi_k+\delta^i_k\psi_i)p^j_\alpha p^k_\beta=0. \tag{26}$$

(26)既为画法平坦,在所用的坐标系统下 $\Pi^i_{jk}=0$,所以它与

$$\frac{\partial^2 x^i}{\partial v^\alpha \partial v^\beta}=0 \tag{27}$$

成画法对应,因此(26)的每个解都是从(27)的解经变换 $v^\alpha=v^\alpha(u)$ 化出来的.现设 $x^i=x^i(u^\alpha)$ 为(26)的一组解,$x^i=x^i(v^\alpha)$ 为对应的(27)的一解,由(26),(27)可知变换 $u^\alpha=u^\alpha(v)$ 满足

$$\frac{\partial x^i}{\partial u^\alpha}\frac{\partial^2 u^\alpha}{\partial v^\beta\partial v^\gamma}=(\delta^i_j\psi_k+\delta^i_k\psi_i)\frac{\partial x^j}{\partial u^\alpha}\frac{\partial x^k}{\partial u^\gamma}\frac{\partial u^\alpha}{\partial v^\beta}\frac{\partial u^\gamma}{\partial v^\delta},$$

或

$$\frac{\partial x^i}{\partial u^\alpha}\left[\frac{\partial^2 u^\alpha}{\partial v^\beta\partial v^\delta}-\psi_k\frac{\partial x^k}{\partial u^\gamma}\left(\frac{\partial u^\alpha}{\partial v^\beta}\frac{\partial u^\gamma}{\partial v^\delta}+\frac{\partial u^\gamma}{\partial v^\beta}\frac{\partial u^\alpha}{\partial v^\delta}\right)\right]=0.$$

因 $\left\|\frac{\partial x^i}{\partial u^\alpha}\right\|$ 的序数为 K,这时方程个数(β, γ 固定)为 N,所以得到

$$\frac{\partial^2 u^\alpha}{\partial v^\beta\partial v^\delta}-\Phi_\delta\frac{\partial u^\alpha}{\partial v^\beta}-\Phi_\beta\frac{\partial u^\alpha}{\partial v^\delta}=0. \tag{28}$$

在这里 $\Phi_\delta=\psi_k\frac{\partial x^k}{\partial v^\delta}$.乘以$\frac{\partial v^\beta}{\partial u^\alpha}$,并关于 α, β 作和,便得出

$$(K+1)\Phi_\delta=\frac{\partial^2 u^\alpha}{\partial v^\beta\partial v^\delta}\frac{\partial v^\beta}{\partial u^\alpha}.$$

把它代入(28),就会看到 u^α 适合微分方程组(8),也就是说:u^α 与 v^α 之间的关系是射影变换.

我们指出,最后的定理也可以用苏步青教授在[3]文中相同的方法予以证明.

参 考 文 献

[1] Douglas, J. 1931 *Math. Ann.* **105**, 707 - 733.

[2] Su, Buchin 1947 Descriptive collineations in spaces of *K*-spread, *Trans. Amer. Math. Soc.* **61**, 495 - 507.

[3] Su, Buchin 1957 *Math. Nachr.* **16**,215 - 226.

[4] 谷超豪 1951 中国科学,Ⅱ(1),1 - 19.

(本文曾发表于《科学记录(新辑)》1959 年第 3 卷第 3 期,84 - 87)

黎曼空间的芬斯拉乘积*

胡和生

中国科学院上海数学研究所

1. 如果在一个 n 维的芬斯拉空间 F_n 里能选取这样的坐标,使空间线素化为形式

$$ds^2 = 2\mathscr{F}(a_{ab}(x^c)dx^a dx^b,\ a_{ap}(x^r)dx^p dx^q)\begin{pmatrix}a,\ b,\ c=1,\ \cdots,\ m\\ p,\ q,\ r=m+1,\ \cdots,\ n\end{pmatrix}, \quad (1)$$

就称它为线素分别是 $a_{ab}(x^c)dx^a dx^b$ 和 $a_{pq}(x^r)dx^p dx^q$ 的两个黎曼空间 R_m 和 R_{n-m} 的芬斯拉乘积.

在研究容许运动群的芬斯拉空间的问题上,这样类型的空间,具有一定的重要性,谷超豪[1]在研究容有最大阶数运动群的芬斯拉空间时,得到了其中一类具有线素 $ds^2 = 2\mathscr{F}((dx^1)^2,\ a_{\alpha\beta}(x^\gamma)dx^\alpha dx^\beta)(\alpha,\ \beta=2,\ \cdots,\ n)$.

置 $p_1 = a_{ab}dx^a dx^b$, $p_2 = a_{pq}dx^p dx^q$,改写(1)为

$$ds^2 = 2\mathscr{F}(p_1,\ p_2), \quad (2)$$

$\mathscr{F}$ 作为 dx^1, $\cdots$, dx^n 的函数正齐二次的,$\mathscr{F}$ 作为 p_1, p_2 的函数是正齐一次的.我们假定除 $p_1 = p_2 = 0$ 外,函数 $\mathscr{F}(p_1,\ p_2)$ 和其第三阶为止的所有偏导数都是 p_1 和 p_2 的连续函数.在这里我们要证明

定理. 线素能化为形式(1)的芬斯拉空间 F_n 一定容有两系全测地曲面 V_m 与 V_{n-m},它们分别和黎曼空间 R_m, R_{n-m} 有相同的线素,并且当我们取 V_m 或 V_{n-m} 上的元素为支

* 1959 年 7 月 23 日收到.

持元素时，它们是互相直交的.

这个结果是黎曼空间的乘积空间已知结果的扩充.

2. 芬斯拉空间 F_n 的尺度张量

$$g_{ij}=\frac{\partial^2\mathscr{F}}{\partial(dx^i)\partial(dx^j)}\quad(i,\ j=1,\ 2,\ \cdots,\ n),\tag{3}$$

经计算可得

$$g_{ap}=4a_{pq}dx^q a_{ab}dx^b\ \frac{\partial^2\mathscr{F}}{\partial p_1\partial p_2};\tag{4}$$

$$g_{pq}=4\ \frac{\partial^2\mathscr{F}}{\partial p_2\partial p_2}a_{rq}dx^r a_{tp}dx^t+2\ \frac{\partial\mathscr{F}}{\partial p_2}a_{pq};\tag{5}$$

$$g_{ab}=4\ \frac{\partial^2\mathscr{F}}{\partial p_1\partial p_1}a_{ad}dx^d a_{cb}dx^c+2\ \frac{\partial\mathscr{F}}{\partial p_1}a_{ab}.\tag{6}$$

由于

$$g_{ap}(x;\ dx^1,\ dx^2,\ \cdots,\ dx^m,\ 0,\ \cdots,\ 0)=0,$$

$$g_{ap}(x;\ 0,\ 0,\ \cdots,\ 0,\ dx^{m+1},\ \cdots,\ dx^n)=0.$$

因此，曲面 $V_m(x^p=\text{const.},\ p=m+1,\ \cdots,\ n)$ 和 $V_{n-m}(x^a=\text{const.},\ a=1,\ \cdots,\ m)$ 以其一方的切向量为支持元素时是互相直交的.

又在 V_m 与 V_{n-m} 上所诱导的线素分别为

$$ds_1^2=2\mathscr{F}(p_1,\ 0),\ ds_2^2=2\mathscr{F}(0,\ p_2).$$

$\mathscr{F}(p_1,\ 0)$ 为一个变量 p_1 的齐一次函数，因此 $2\mathscr{F}(p_1,\ 0)=k_1p_1$. 同样地，$2\mathscr{F}(0,\ p_2)=k_2p_2$，$k_1$，$k_2$ 为常数，变动 p_1 与 p_2 的倍数后，k_1，k_2 均可取为 1，把 p_1，p_2 适当变动一个倍数且把所得结果仍记为 p_1，p_2，则 F_n 中这二系曲面 $V_m(x^p=\text{const.})$ 与 $V_{n-m}(x^a=\text{const.})$ 上线素平方分别为 $a_{ab}dx^a dx^b$ 与 $a_{pq}dx^p dx^q$.

为证明 $V_m(x^p=\text{const.})$ 是全测地曲面，只要证明

$$\Gamma^{p}_{ab}(x^{c},\ x^{p},\ dx^{c},\ 0)=0. \tag{7}$$

事实上，F_n 的测地线方程是

$$\frac{d^2x^i}{ds^2}+\overset{*}{\Gamma}{}^{i}_{jk}(x,\ dx)\frac{dx^j}{ds}\frac{dx^k}{ds}=0$$

或

$$\frac{d^2x^i}{ds^2}+\Gamma^{i}_{jk}(x,\ dx)\frac{dx^j}{ds}\frac{dx^k}{ds}=0. \tag{8}$$

于是 $\overset{*}{\Gamma}{}^{i}_{jk}$ 为嘉当(E. Cartan)联络[2]，而且

$$\Gamma^{i}_{jk}=\overset{*}{\Gamma}{}^{i}_{jk}+C^{i}_{jh}dx^{l}\Gamma^{h}_{lk},\ C^{i}_{ih}=g^{li}C_{jlh}=\frac{1}{2}g^{li}\frac{\partial g_{jl}}{\partial(dx^h)}. \tag{9}$$

当(7)成立时，(8)的以 x^{i}_{0}，$\left(\frac{dx^a}{ds}\right)_0$，$\left(\frac{dx^p}{ds}\right)_0=0$ 为初始条件的解可由微分方程

$$\frac{d^2x^a}{ds^2}+\Gamma^{a}_{bc}(x^{d},\ x^{p}_{0},\ dx^{d},\ 0)\frac{dx^b}{ds}\frac{dx^c}{ds}=0 \tag{10}$$

及有限方程

$$x^{p}(s)=x^{p}_{0}$$

所决定.这是因为方程(10)在初始条件 $s=s_0$，$x^a=x^a_0$，$\left(\frac{dx^a}{ds}\right)=\left(\frac{dx^a}{ds}\right)_0$ 之下有唯一解 $x^a=x^a(s)$；把它与 $x^p(s)=x^p_0$ 联系起来，便得到(8)的解，并且这个解满足原来的初始条件，从而是(8)的唯一解.这就是说，与 V_m 相切的测地线必在 V_m 上，所以 V_m 是全测地的，因而要证明 V_m 是全测地曲面，就只须证明(7)式就可以了.

由于当 $dx^p=0$ 时，$g^{pa}=0$，$\Gamma^{p}_{ab}=g^{pq}\Gamma_{aqb}$，因而只要证明 $\Gamma_{aqb}(x^c,\ x^p,\ dx^c,\ 0)=0$.这里

$$\Gamma_{aqb}=\frac{1}{2}\left(\frac{\partial g_{aq}}{\partial x^b}+\frac{\partial g_{bq}}{\partial x^a}-\frac{\partial g_{ab}}{\partial x^q}\right)+C_{abc}\frac{\partial G^c}{\partial(dx^q)}+C_{abp}\frac{\partial G^p}{\partial(dx^q)}-$$

$$C_{qbc}\frac{\partial G^{c}}{\partial(dx^{a})}-C_{qbp}\frac{\partial G^{p}}{\partial(dx^{a})},\tag{11}$$

$$2G_{i}=\frac{\partial^{2}\mathscr{F}}{\partial(dx^{i})\partial x^{k}}dx^{k}-\frac{\partial\mathscr{F}}{\partial x^{i}},\ G^{p}=g^{pi}G_{i}.$$

利用$\mathscr{F}$的齐次性,可知在$dx^{p}=0$时,

$$\frac{\partial^{2}\mathscr{F}}{\partial p_{1}^{2}}=0,\ \frac{\partial^{2}\mathscr{F}}{\partial p_{1}\partial p_{2}}=0,\ \frac{\partial^{3}\mathscr{F}}{\partial p_{1}^{3}}=0.$$

从此,容易证明,当$dx^{p}=0$时,

$$\frac{\partial g_{aq}}{\partial x^{b}}=\frac{\partial g_{ab}}{\partial x^{q}}=0,\ C_{abc}=C_{abp}=C_{qbc}=C_{qbp}=0,$$

因而$\Gamma_{aqb}(x^{i},\ dx^{c},\ 0)=0$,所以$V_{m}$是全测地的.同样地,我们可证明$V_{n-m}$也是全测地的.

参考文献

[1] 谷超豪 1957 容有最大阶数运动群的芬斯拉空间,科学记录(新辑)1,197－200.

[2] Cartan, E. 1934 Les espaces de Finsler. *Actualite's scientifiques et industrielles*, 79(或参阅苏步青 1958 一般空间微分几何学第一章).

(本文曾发表于《科学记录(新辑)》1959 年第 3 卷第 10 期,363－365)

同曲率曲面

胡和生
中国科学院上海数学研究所

§1. 黎曼空间 V_N 的全测地面曲 V_m 具有这样的性质：V_m 关于其上任意两方向的黎曼曲率等于外界空间 V_N 关于这两方向的黎曼曲率，特别欧氏空间 E_N 的全测地曲面就是平面，而且平面 E_m 的变形曲面 V_m 也具有上述性质，但它并非全测地的，从这个事实看来，我们有可能推广全测地曲面概念来研究一种特殊类型的曲面称为同曲率曲面.它与 P 型次全测地曲面也有密切关系，[1] 本文的目的是研究黎曼空间同曲率曲面的一些性质，在最后一节中，讨论容有同曲率超曲面系的黎曼空间的线素形式.

§2. 我们首先给出下列定义，设黎曼空间 V_N 里的曲面 V_m 关于其上任意二维平面素的黎曼曲率常等于 V_N 关于同一平面素的黎曼曲率，我们称这种 V_m 为同曲率曲面.

设 y^α $(\alpha=1, \cdots, N)$ 是 V_N 的空标，x^j $(j=1, 2, \cdots, m)$ 是 V_m 的坐标，且 λ_1^i，λ_2^i 是确定一个平面素的二单位向量，它们在 V_N 中的支量是

$$\xi_1^\alpha=y_{,i}^\alpha\lambda_1^i,\ \xi_2^\alpha=y_{,i}^\alpha\lambda_2^i.$$

这里，$y_{,i}^\alpha=\dfrac{\partial y^\alpha}{\partial x^i}$. 由同曲率曲面的定义可知

$$\frac{R_{hijk}\lambda_1^h\lambda_2^i\lambda_1^j\lambda_2^k}{(g_{hj}g_{ik}-g_{hk}g_{ij})\lambda_1^h\lambda_2^i\lambda_1^j\lambda_2^k}=\frac{\bar{R}_{\alpha\beta\gamma\delta}\xi_1^\alpha\xi_2^\beta\xi_1^\gamma\xi_2^\delta}{(a_{\alpha\gamma}a_{\beta\delta}-a_{\alpha\delta}a_{\beta\gamma})\xi_1^\alpha\xi_2^\beta\xi_1^\gamma\xi_2^\delta}, \tag{1}$$

这里 $a_{\alpha\beta}$，g_{ij} 各为 V_N 与 V_m 的基本张量，由于 λ_1^i，λ_2^j 均为单位的，且上式关于曲面上任何方向成立，因此得到

$$R_{hijk}=\bar{R}_{\alpha\beta\gamma\delta}y^{\alpha}_{,h}y^{\beta}_{,i}y^{\gamma}_{,j}y^{\delta}_{,k}.$$

另一方面 $V_m\subset V_N$ 的 Gauss 方程为

$$R_{hijk}=\sum_{\sigma=m+1}^{N}(\Omega_{\sigma 1hj}\Omega_{\sigma 1ik}-\Omega_{\sigma 1hk}\Omega_{\sigma 1ij})+\bar{R}_{\alpha\beta\gamma\delta}y^{\alpha}_{,h}y^{\beta}_{,i}y^{\gamma}_{,j}y^{\delta}_{,k}.$$

这样就得到

定理 1. V_m 为同曲率曲面的充要条件是：它的第二基本形式的系数满足

$$\sum_{\sigma}(\Omega_{\sigma 1hj}\Omega_{\sigma 1ik}-\Omega_{\sigma 1hk}\Omega_{\sigma 1ij})=0. \tag{2}$$

全测地曲面显然是同曲率曲面，但全脐点曲面(非全测地的)却不是同曲率的.事实上容易证明：如果 V_m 同时具有全脐点和同曲率的性质，就必须是全测地的，这是因为，将 $\Omega_{\sigma 1ij}=\frac{\Omega_{\sigma}}{n}g_{ij}$ 代入(2)式并利用 $|g_{ij}|\neq 0$，就可得出 $\Omega_{\sigma}=0$.

现在来考察这类曲面的范围，当 V_N 是欧氏空间 E_N 时，V_m 也一定是欧氏式的，因此 E_N 中 m 维的同曲率曲面就是与其中 m 维平面互为变形的全体曲面，并且此外没有了，又当 V_N 是常曲率 k_0 的空间时，V_m 一定也是常曲率 k_0 的，所以这类曲面也就是与全测地曲面 V_m 互为变形的全体曲面.

§3. 特别讨论同曲率超曲面 $V_m\subset V_{m+1}$ 的情形，这时由(2)可知

$$\Omega_{ij}=\varphi_i\varphi_j, \tag{3}$$

就是 $\|\Omega_{ij}\|$ 秩为 1 或 0，当秩是 0 时 $\varphi_i=0$ 且所论的是全测地超曲面.

我们容易证明同曲率超曲面的一个几何特征

定理 2. 设 $V_m\subset V_N$，且 $|\Omega_{ij}-\lambda g_{ij}|=0$ 的初级因子为简单的，则 V_m 为同曲率曲面的充要条件是有 $m-1$ 个等于 0 的主法曲率.

我们再给出同曲率超曲面的一个性质.对 $V_m\subset V_{m+1}$，取曲面的和乐标形 $\{P, I_1, I_2, \cdots, I_{m+1}\}$，使 $I_1, \cdots, I_m$ 在切平面上，并且 I_{m+1} 是曲面的法线，这时

$$dI_{m+1}=\omega^{i}_{m+1}I_i=-g^{ij}\omega^{m+1}_{j}I_i. \tag{4}$$

由(3)得到 $\omega_j^{m+1}=\Omega_{jk}dx^k=\varphi_j\varphi_k dx^k$，从而

$$dI_{m+1}=-g^{ij}\varphi_j\varphi_k dx^k I_i. \tag{5}$$

现在分作二种情况来讨论：

i) $\varphi_j dx^j=0$ 完全可积的情况：这时 $V_m=\infty^1 V_{m-1}$ 并且沿着其中任一 V_{m-1} 上的方向 dx^i 满足 $\varphi_j dx^j=0$.所以，$dI_{m+1}=0$，即超曲面的法线沿 V_{m-1} 上的任意方向是平行的，但这时 V_{m-1} 的切平面素沿 V_{m-1} 不一定平行，而 V_m 的切平面素沿 V_{m-1} 上任何方向是平行的，V_m 的切平面包含 V_{m-1} 的切平面，故 V_{m-1} 是Ⅰ型次全测地曲面[1]在常曲率空间 S_{m+1}（包括欧氏空间）中这些 V_{m-1} 是 S_{m+1} 中的全测地曲面且沿每个 V_{m-1}，S_m 的相切全测地曲面是相同的[2].

ii) $\varphi_j dx^j=0$ 非完全可积的情况，这时我们称方程 $\varphi_j dx^j=0$ 在 V_m 上定义了 $m-1$ 维的非完整流形[3],[4]，此时 V_m 的切平面素沿非完整流形 $\varphi_i dx^i=0$ 的方向平行，如下定义称这样的非完整流形 V_{m-1} 为Ⅰ型次全测地的，那么 V_m 便是容有Ⅰ型次全测地非完整的 V_{m-1}.

在常曲率空间(包括欧氏空间) $\varphi_j dx^j=0$ 一定是完全可积的，但在一般的黎曼空间却不一定是这样，综合起来得到

定理 3. 设 V_m 为 V_{m+1} 中的同曲率超曲面，那么它的第二基本形式系数具有形式 $\Omega_{ij}=\varphi_i\varphi_j$，并且 i) 当 $\varphi_i dx^i=0$ 是完全可积时，$V_m=\infty^1 V_{m-1}$；而 V_{m-1} 是Ⅰ型次全测地的，ii) 当 $\varphi_i dx^i=0$ 不是完全可积时，那么在 V_m 上，存在Ⅰ型次全测地的 $m-1$ 维非完整流形，特别当 V_m 是常曲率空间情形 ii)不会发生而在情形 i)中 V_{m-1} 是全测地的.

当同曲率曲面 $V_m\subset V_{m+q}$ 时，关系(2)成立，根据 Яненко 的定理[5]可知 $\Omega_{\sigma 1ij}$ 的秩数 $\leqslant q$.现设 $\Omega_{\sigma 1ij}$ 的秩数为 t，那么 $\Omega_{\sigma 1ij}dx^j=0$ 在 V_m 上定义 $m-t$ 维平面素，我们得到同超曲面的情形相类似的结果.

定理 4. 设 V_m 为 V_{m+q} 中的同曲率曲面，那么 $\Omega_{\sigma 1ij}$ 的秩数 $t\leqslant q$，并且 i) 当 $\Omega_{\sigma 1ij}dx^j=0$ 是完全可积时，$V_m=\infty^t V_{m-t}$，其中 V_{m-t} 是 t 型次全测地曲面，如果 V_{m+q} 是常曲率空间，那么 V_{m-t} 是全测地的，ii) 当 $\Omega_{\sigma 1ij}dx^j=0$ 不是完全可积时，在 V_m 上存在

$m-t$ 维的非完整流形，它是 t 型次全测地的，且当 V_{m+q} 是常曲率时，这种情形不会发生.

§4. 设 V_{m+1} 容有一系同曲率超曲面，我们要确定 V_{m+1} 的线素所采取的形式，为此，选坐标系统使得这系超曲面的方程是 $y^{n+1}=\text{const.}$ 而且 y^{n+1} 曲线垂直于这系曲面，这时

$$\Omega_{ij}=-\frac{1}{2}\frac{\partial a_{ij}}{\partial y^{n+1}}\frac{1}{\sqrt{a_{n+1n+1}}},$$

将它代入 $\Omega_{ik}\Omega_{jl}-\Omega_{il}\Omega_{jk}=0$，就能获得

$$\frac{\partial a_{ik}}{\partial y^{n+1}}=\psi_i(y^1,\cdots,y^{n+1})\psi_j(y^1,\cdots,y^{n+1}).$$

再由(3)得到

$$a_{ik}=-2\int\sqrt{a_{n+1\,n+1}}\,\varphi_i\varphi_j d y^{n+1}+b_{ik}(y^1,\cdots,y^n), \tag{6}$$

从而成立

定理 5. V_{n+1} 容许一系同曲率超曲面的充要条件是：它的线素化为形式

$$ds^2=a_{ik}dy^i dy^k+a_{n+1\,n+1}(dy^{n+1})^2, \tag{7}$$

式中 $a_{n+1\,n+1}$ 是 $y^1,\cdots,y^{n+1}$ 的任意函数且 a_{ik} 是按(6)定义的.这时 $y^{n+1}=\text{const.}$ 为同曲率超曲面系的方程，特别当这些超曲面是全测地时，空间的线素化为[6]

$$ds^2=b_{ik}(y^1,\cdots,y^n)dy^i dy^k+a_{n+1\,n+1}(y^1,\cdots,y^{n+1})(dy^{n+1})^2. \tag{8}$$

如所知黎曼空间如果容有 m 重直交测地超曲面系统，那么它一定是欧氏空间.现在我们来研究容有 m 重直交同曲率超曲面系统的黎曼空间 V_m 的线素应该是怎样的问题.

设所论空间的线素是

$$ds^2=g_{11}(dx^1)^2+\cdots+g_{mm}(dx^m)^2,$$

其中 $x^i=\text{const.}\ (i=1,\cdots,m)$ 构成 m 重直交的同曲率超曲面系统，曲面 $x^1=\text{const.}$ 的线素是

$$ds^2 = g_{22}(dx^2)^2 + \cdots + g_{mm}(dx^m)^2,\ \underset{1}{\Omega}_{ij} = -\frac{1}{2}\frac{\partial g_{ij}}{\partial x^1}\frac{1}{\sqrt{g_{11}}} \quad (i,\ j = 2,\ \cdots,\ m),$$

因而

$$\underset{1}{\Omega}_{ij} = 0 \quad (i \neq j;\ i,\ j = 2,\ \cdots,\ m). \tag{9}$$

另一方面，利用关系(9)及同曲率的性质知

$$\underset{1}{\Omega}_{ii}\underset{1}{\Omega}_{jj} = 0,$$

我们假定每系同曲率超曲面都不是全测地的，从而不妨假定 $\underset{1}{\Omega}_{22} \neq 0,\ \underset{1}{\Omega}_{33} = \cdots = \underset{1}{\Omega}_{mm} = 0$，这样就可得出 $\frac{\partial g_{33}}{\partial x^1} = 0,\ \cdots,\ \frac{\partial g_{mm}}{\partial x^1} = 0$.

又对于 $x^2 = \text{const.}$，同样可得

$$\underset{2}{\Omega}_{ij} = 0 \quad (i \neq j,\ i,\ j = 1,\ 3,\ \cdots,\ m),$$

$$\underset{2}{\Omega}_{ii}\underset{2}{\Omega}_{jj} = 0 \quad (i,\ j = 1,\ 3,\ \cdots,\ m).$$

这时可分为二种情形：

i) 当 $\underset{2}{\Omega}_{11} \neq 0$ 且其余都是 0 时，$\frac{\partial g_{33}}{\partial x^2} = 0,\ \cdots,\ \frac{\partial g_{mm}}{\partial x^2} = 0$.

ii) 当 $\underset{2}{\Omega}_{33} \neq 0$ 且其余都是 0 时，$\frac{\partial g_{11}}{\partial x^2} = 0,\ \frac{\partial g_{44}}{\partial x^2} = 0,\ \cdots,\ \frac{\partial g_{mm}}{\partial x^2} = 0$.

对于超曲面 $x^3 = \text{const.}$，可分为三种情形 i) $\underset{3}{\Omega}_{11} \neq 0$ 其余是 0；ii) $\underset{3}{\Omega}_{22} \neq 0$ 其余是 0；iii) $\underset{3}{\Omega}_{44} \neq 0$ 其余是 0.依此类推我们就可得出

定理 6. 设黎曼空间 V_m 容有 m 重直交同曲率超曲面系统，那么线素一定可以化为 $ds^2 = g_{11}(dx^1)^2 + \cdots + g_{mm}(dx^m)^2$，其中 g_{ii} 满足下列二条件：i) 每个 g_{ii} 中可包含 x^i；ii) 每个 x^j 最多只能一次地被包含在另一个 $g_{ii}(i \neq j)$ 之内.

参 考 文 献

[1] 谷超豪，1957.全测地曲面的一种推广，复旦学报，1，111－115.

［2］ 胡和生，1957.关于黎曼空间的两种秩数，复旦学报，2，346 - 349.

［3］ Vranceanu，M. G.，1935. Les espaces non holonomes，Memorial des sciences math. 76.

［4］ 谷超豪，1955.论平面素的平行移动与非完整流形，数学学报，5.383 - 392.

［5］ Яненко，H，H. 1954. К Теории вложния поверхностей многомерного евклидова пространства труды Моск. Матем 0 - Ва，Ⅲ，89 - 180.

［6］ Eisenhart，L. P.，1949，Riemannian geometry. 183 - 184.

（本文曾发表于《复旦大学（自然科学学报）》1960 年第 1 期，38 - 43）

论李-嘉当变换拟群的可约性及其在微分几何中的应用

胡和生
复旦大学

【摘要】 由解析函数领域中的偏微分方程组所定义的变换拟群称为李-嘉当拟群.在这变换拟群中使一点不变的变换全体在该点相切空间中诱导一个线性变换群,称为迷向群.本文研究迷向群的性质与变换拟群性质的联系,证明了如下的事实:1)设 G 为可迁的李-嘉当变换拟群,它的迷向群 G_1 为可约的(即容有不变平面)且包含放大,那么 G 本身也为非素性的,即它容有一系非素性流形.2)设 G_1 为完全可约的,且包含放大,那么群 G 有二系互补的非素性流形.特别当 G_1 只含有放大时,群 G 必为 $\bar{x}^i = e x^i + b^i$ 的形状.把这些性质应用到微分几何学中得到如下的结果:1)设黎曼空间 V_n 容许可迁共形变换群,其迷向群为可约且包含放大,那么空间为共形可分的,即它与一乘积空间相共形对应.特别迷向群只包含放大时,V_n 为共形平坦.2)设辛尺度空间 S_{2n} 容许共形变换群,其迷向群为可约,不变平面为非迷向的又包含放大,那么空间为共形可分的,即它与“乘积辛空间”相共形对应.特别当迷向群只包含放大时,S_{2n} 共形于辛空间.

1. 李-嘉当变换拟群是指作用于空间 $M_n(x^1, \cdots, x^n)$,由解析领域内的微分方程组

$$F_\sigma(x, \bar{x}, \partial \bar{x}, \cdots, \partial^p \bar{x}) = 0$$

所定义起来的变换拟群,其中元素为 M_n 中的解析变换

$$\bar{x}^i = \varphi^i(x^j).$$

李-嘉当变换拟群包含了有限连续变换拟群与无限连续变换拟群.在最近期内,这项

理论有了相当的发展.谷超豪对可递的连续变换拟群的可分性与可约性作了一系列的研究[1][2][3],大部分的结果是假定一阶迷向群是分解为两个线性群直积的情形,且这两个线性群作用于切空间的两个互补的子空间之中.

本文是讨论非素性的问题,但我们所考虑的迷向群只要求是具有不变平面及包含放大的.

如所知,群 G 称为非素性的,如果在空间 M_n 中存在坐标系统 $(x^1, x^2, \cdots, x^n)$,使群中一般变换具形状

$$\bar{x}^{i_1}=f^{i_1}(x^{j_1}),$$
$$\bar{x}^{i_2}=f^{i_2}(x^{j_1}, x^{j_2}) \quad (i_1, j_1=1, \cdots, q; i_2, j_2=q+1, \cdots, n).$$

这时 $x^{i_1}=\text{const}$ 为群 G 的非素性集.

我们要证:

定理 1 设 G 为可递的李-嘉当变换拟群,其迷向群具有不变平面且包含放大,则群为非素性的,且其非素性集切于不变平面.

如在空间 M_n 中存在坐标系统,使群中一般变换具形状

$$\bar{x}^{i_1}=f^{i_1}(x^{j_1}) \quad (i_1=1, \cdots, q),$$
$$\bar{x}^{i_2}=f^{i_2}(x^{j_2}) \quad (i_2=q+1, \cdots, n),$$

则群称为全非素性的.这时空间 M_n 有群 G 的二系非素性集 $x^{i_1}=\text{const}$, $x^{i_2}=\text{const}$.

定理 2 设可递的李-嘉当拟群,其迷向群有二个互补的不变平面 E_q 及 E_{n-q},且包含放大,则群为全非素性,相应的二系非素性集切于每点的不变平面 E_q 及 E_{n-q}.

本文中还证明了如下事实:

定理 3 设 G 为 n 维空间的可递的李-嘉当变换拟群,且其迷向群只包含放大,则 G 为 $n+1$ 参数有限连续变换拟群,变换方程为

$$\bar{x}^i=e^a x^i+b^i.$$

这些定理可以应用到容许可递共形运动群的黎曼空间与辛尺度空间去,其结果

如下：

定理 4 设黎曼空间 V_n 容许可递共形运动群，其迷向群具有不变平面且包含放大，则共形运动群是全非素性的，并且空间线素可化为

$$ds^2 = A(x^i)(g_{i_1j_1}(x^{k_1})dx^{i_1}dx^{j_1} + g_{i_2j_2}(x^{k_2})dx^{i_2}dx^{j_2})$$
$$(i=1, \cdots, n;\ i_1, j_1, k_1 = 1, \cdots, q;\ i_2, j_2, k_2 = q+1, \cdots, n).$$

定理 5 设黎曼空间 V_n 容许可递共形运动群，其迷向群只包含放大，则 V_n 是共形平坦的.

定理 6 设辛尺度空间 S_{2n} 容许可递共形运动群，其迷向群具有不变的非迷向平面且包含放大，则共形运动群是全非素性的并且空间辛尺度为

$$\begin{aligned}\Omega(x, dx) &= g_{ij}(x^k)[dx^i dx^j] \\ &= \rho(x^i)(g_{i_1j_1}(x^{k_1})[dx^{i_1}dx^{j_1}] + g_{i_2j_2}(x^{k_2})[dx^{i_2}dx^{j_2}]) \\ &= \rho(x^i)(\Omega_1(x^{i_1}, dx^{i_1}) + \Omega_2(x^{i_2}, dx^{i_2}))\end{aligned}$$
$$(i, j, k = 1, \cdots, 2n;\ i_1, j_1, k_1 = 1, \cdots, 2q;\ i_2, j_2, k_2 = 2q+1, \cdots, 2n).$$

定理 7 设辛尺度空间 S_{2n} 容许可递共形运动群，其迷向群只包含放大，则 S_{2n} 共形于辛空间.

2. 根据嘉当的研究[4]，对于可递的李-嘉当拟群(不论其定义方程的阶数如何)均存在 n 个法甫形式

$$\omega^i = \omega^i(x^j, u^\alpha, dx^j),$$

使群 G 中的一般变换

$$\bar{x}^i = \varphi^i(x^j)$$

连同适当选取的

$$\bar{u}^\alpha = \varphi^\alpha(x^j, u^\beta)$$

满足

$$\omega^i(\bar{x},\ \bar{u},\ d\bar{x})=\omega^i(x,\ u,\ dx),$$

并且这些 ω^i 满足

$$D\omega^i=\frac{1}{2}C^i_{jk}[\omega^j\omega^k]+C^i_{j\alpha}[\omega^j\omega^\alpha] \tag{1}$$

$$(i,\ j,\ k=1,\ \cdots,\ n;\ \alpha=n+1,\ \cdots,\ n+r).$$

这里 r 是迷向群的参数，$C^i_{j\rho}$，C^i_{jk} 均为常数，它们关于下指标是反称的，还满足条件：存在常数 C^α_{lm}，$C^\alpha_{i\beta}$，$C^\alpha_{\beta\gamma}$ 使

$$C^i_{j\beta}C^j_{k\gamma}-C^i_{j\gamma}C^j_{k\beta}=C^\alpha_{\gamma\beta}C^i_{k\alpha}, \tag{2}$$

$$C^k_{l\alpha}C^\alpha_{m\beta}-C^k_{m\alpha}C^\alpha_{l\beta}=C^i_{lm}C^k_{i\beta}+C^k_{il}C^i_{m\beta}-C^k_{im}C^i_{l\beta}, \tag{3}$$

$$C^k_{l\alpha}C^\alpha_{mn}+C^k_{m\alpha}C^\alpha_{nl}+C^k_{n\alpha}C^\alpha_{lm}=C^i_{lm}C^k_{in}+C^i_{mn}C^k_{il}+C^i_{nl}C^k_{im}$$

$$(i,\ j,\ k,\ l,\ m,\ n=1,\ \cdots,\ n;\ \alpha,\ \beta,\ \gamma=n+1,\ \cdots,\ n+r)$$

成立.此外

$$X_\alpha f=C^i_{j\alpha}x^j\frac{\partial f}{\partial x^i}$$

为迷向群的微分算子.

现设迷向群有不变平面 E_{n-q} 且包含放大.我们选取某一定点相切空间的基 e_1，…，e_n，使 $e_{i_2}(i_2=q+1,\ \cdots,\ n)$ 在 E_{n-q} 中；对 ω^i 也作适当的常系数线性变换，使 $\omega^{i_2}=0$ 定义 E_{n-q}，这时利用线性群及其李代数的关系，可知

$$C^{i_1}_{j_2\alpha}=0\begin{pmatrix} i_1=1,\ \cdots,\ q \\ j_2=q+1,\ \cdots,\ n\end{pmatrix}. \tag{4}$$

选取 ω^α，使 $C^i_{j,\ n+r}$ 对应于放大，因而可有

$$C^i_{j,\ n+r}=\delta^i_j. \tag{5}$$

这时(1)式可改写为

$$D\omega^{i_1}=\frac{1}{2}C^{i_1}_{jk}[\omega^j\omega^k]+C^{i_1}_{j_1\alpha}[\omega^{j_1}\omega^\alpha],$$

$$D\omega^{i_2}=\frac{1}{2}C^{i_2}_{jk}[\omega^j\omega^k]+C^{i_2}_{j\alpha}[\omega^j\omega^\alpha].$$

在(3)中置 $\rho=n+r$, $k=k_1$, $l=l_2$, $m=m_2$, 则有

$$C^{k_1}_{l_2m_2}=0$$

成立,因而 $\omega^{i_1}=0$ 为完全可积;设其初积分为 x^{i_1},以 x^{i_1} 为坐标的一部分,可知

$$\omega^{i_1}(\bar{x}^j, d\bar{x}^j, \bar{u}^\alpha)=\omega^{i_1}(x^j, dx^{j_1}, u^\alpha),$$
$$\omega^{i_2}(\bar{x}^j, d\bar{x}^j, \bar{u}^\alpha)=\omega^{i_2}(x^j, dx^j, u^\alpha)$$

的解具形状

$$\bar{x}^{i_1}=f^{i_1}(x^{j_1}),$$
$$\bar{x}^{i_2}=f^{i_2}(x^{j_1}, x^{j_2}),$$

因而群中的变换也具此形状,所以群是非素性的, $x^{i_1}=\text{const}$ 为切于 E_{n-q} 的非素性集,这样就证明了定理 1.

现设迷向群有二个互补不变平面 E_q 及 E_{n-q},且包含放大,则利用定理 1 可知有一系 q 维非素性集切于 E_q,又有一系 $n-q$ 维非素性集切于 E_{n-q}.我们选取坐标设它们的方程各为 $x^{i_2}=\text{const}$ 与 $x^{i_1}=\text{const}$($i_1=1, \cdots, q$; $i_2=q+1, \cdots, n$), 就可得群的变换为

$$\bar{x}^{i_1}=f^{i_1}(x^{j_1}),$$
$$\bar{x}^{i_2}=f^{i_2}(x^{j_2}),$$

因而定理 2 得到证明.由此即得

推论 设 G 为 n 维空间的可递的李-嘉当变换拟群,其迷向群包含放大且切空间分解为迷向群下不变平面 E_{q_1}, E_{q_2}, $\cdots$, E_{q_s} 的直和 ($q_1+q_2+\cdots+q_s=n$), 则群中一般变换形状为

$$\bar{x}^{i_1}=f^{i_1}(x^{j_1})\quad(i_1,\ j_1=1,\ \cdots,\ q_1),$$

$$\bar{x}^{i_2}=f^{i_2}(x^{j_2})\quad(i_2,\ j_2=q_{1+1},\ \cdots,\ q_2),$$

……

$$x^{i_s}=f^{i_s}(x^{j_s})\quad(i_s,\ j_s=q_{s+1},\ \cdots,\ n).$$

3. 我们在本节中证明定理 3.设 G 为可递的李-嘉当变换拟群,设其迷向群只包含放大,这时

$$D\omega^i=\frac{1}{2}C^i_{jk}[\omega^j\omega^k]+C^i_{j0}[\omega^j\omega^0],\tag{6}$$

$$D\omega^0=\frac{1}{2}C^0_{jk}[\omega^j\omega^k]+C^0_{j0}[\omega^j\omega^0]+C^0_{j\tau}[\omega^j\omega^\tau].\tag{7}$$

这些结构常数满足

$$C^i_{j0}=\delta^i_j,\tag{8}$$

$$C^i_{AB}C^A_{DE}+C^i_{AD}C^A_{EB}+C^i_{AE}C^A_{BD}=0,\tag{9}$$

$$C^0_{AB}C^A_{DE}+C^0_{AD}C^A_{EB}+C^0_{AD}C^A_{EB}=0,\tag{10}$$

于此 A, B 等表示指标 i, α, τ;又在(6),(7)中不出现的 C^i_{AB}, C^0_{AB} 均为 0.

从(9)利用(8)可得

$$C^k_{lm}=0\quad(k\neq m,\ k\neq l),\tag{11}$$

$$C^0_{m0}=C^l_{ml}\quad(l\ \text{固定,非作和}).\tag{12}$$

将(8),(11),(12)代入(6)式,令 $\widetilde{\omega}^0=\omega^0+\frac{1}{2}C^0_{0k}\omega^k$,并将$\widetilde{\omega}^0$仍记为 ω^0,则(6)式化为

$$D\omega^i=[\omega^i\omega^0].\tag{13}$$

又由(9),(10)经计算后得

$$C^0_{jk}=0,\ C^0_{j0}=0,\ C^0_{j\tau}=0.$$

从而 $D\omega^0=0$,群为有限,且 $\omega^0=dv$.

由(13)可求得

$$\omega^i=e^{-v}dx^i. \tag{14}$$

作下式的外微分

$$\omega^i(\bar{v},\ d\bar{x})=\omega^i(v,\ dx), \tag{15}$$

依嘉当引理利用(13)可得 $\bar{v}=v+a$,再从上式就可知群中的一般变换为

$$\bar{x}^i=e^a x^i+b^i.$$

这样就证明了定理 3.

4. 现讨论对黎曼空间共形运动群的应用.

设黎曼空间 V_n,其线素为

$$ds^2=g_{ij}(x)dx^i dx^j=\sum_{i=1}^{n}(\underset{0}{\omega}^i(x,\ dx))^2. \tag{16}$$

如果 V_n 容许可递共形运动群,则根据嘉当的一般等价理论[4],存在 n 个法甫形式

$$\omega^i(x,\ u,\ dx)=a^i_j(x,\ u)\underset{0}{\omega}^j(x,\ dx) \tag{17}$$

使具下列性质:

(i) 成立

$$\sum_{i=1}^{n}(\omega^i)^2=\sigma^2 ds^2. \tag{18}$$

(ii) $a^i_j(x,\ u)$满足

$$\sum_i a^i_j(x,\ u)a^i_k(x,\ u)=\sigma^2(x,\ u)\delta_{jk}. \tag{19}$$

(iii) 共形变换群决定于方程

$$\omega^i(X,\ U,\ dX)=\omega^i(x,\ u,\ dx). \tag{20}$$

(iv)

$$D\omega^i=\frac{1}{2}C^i_{jk}[\omega^j\omega^k]+C^i_{j\alpha}[\omega^j\omega^\alpha], \tag{21}$$

$$D\omega^{a}=\frac{1}{2}C_{jk}^{\alpha}[\omega^{j}\omega^{k}]+C_{j\beta}^{\alpha}[\omega^{j}\omega^{\beta}]+\frac{1}{2}C_{\beta\gamma}^{\alpha}[\omega^{\beta}\omega^{\gamma}]+C_{j\tau}^{\alpha}[\omega^{j}\omega^{\tau}] \tag{22}$$

$(i,j,k=1,\cdots,n;\ \alpha,\beta=n+1,\cdots,n+r;\ \tau=n+r+1,\cdots,n+r+s)$.

这里 C 是常数,关于下指标为反称,且存在常数 C_{ij}^{τ}, $C_{i\alpha}^{\tau}$, $C_{\alpha\beta}^{\tau}$, $C_{\alpha\lambda}^{\tau}$, $C_{i\lambda}^{\tau}$, $C_{\lambda\mu}^{\tau}$,使

$$C_{AB}^{i}C_{CD}^{A}+C_{AC}^{i}C_{DB}^{A}+C_{AD}^{i}C_{BC}^{A}=0, \tag{23}$$

$$C_{AB}^{\alpha}C_{CD}^{A}+C_{AC}^{\alpha}C_{DB}^{A}+C_{AD}^{\alpha}C_{BC}^{A}=0 \tag{24}$$

成立.这里 A, B, C 等表指标(i, j, $\cdots$, α, β, $\cdots$, τ, $\cdots$)的总集,且在(22)式中不出现的那些 C_{AB}^{i}, C_{AB}^{α}应置为 0.

同时 $A_{\alpha}=\| C_{j\alpha}^{i} \|$ 是迷向群所对应的李代数的基.

现在来证明定理 4.设 V_n 的共形运动的迷向群具有不变平面 E_q 且包含放大,则知迷向群必存在与 E_q 互补的不变平面 E_{n-q},这时我们利用定理 2,可知群为全非素性,且

$$D\omega^{i_1}=\frac{1}{2}C_{j_1k_1}^{i_1}[\omega^{j_1}\omega^{k_1}]+C_{j_1k_2}^{i_1}[\omega^{j_1}\omega^{k_2}]+C_{j_1\alpha}^{i_1}[\omega^{j_1}\omega^{\alpha}],$$

$$D\omega^{i_2}=\frac{1}{2}C_{j_2k_2}^{i_2}[\omega^{j_2}\omega^{k_2}]+C_{j_2k_1}^{i_2}[\omega^{j_2}\omega^{k_1}]+C_{j_2\alpha}^{i_2}[\omega^{j_2}\omega^{\alpha}]$$

$$(i_1,j_1,k_1=1,\cdots,q;\ i_2,j_2,k_2=q+1,\cdots,n).$$

这里并有

$$C_{j_1\rho}^{i_1}+C_{i_1\rho}^{j_1}=0,\ C_{j_2\rho}^{i_2}+C_{i_2\rho}^{j_2}=0 \quad (\rho=n+1,\cdots,n+r-1), \tag{25}$$

记 C_{jn+r}^{i} 为 C_{j0}^{i},它对应于放大.为方便计,C 的指标中所遇到的 $n+r$ 总用 0 代替.

令 $i=i_1$, $B=l_1$, $C=m_2$, $D=0$, 从(23)得到

$$C_{m_2l_1}^{i_1}=\delta_{l_2}^{i_1}C_{m_20}^{0}+C_{l_1\rho}^{i_1}C_{m_20}^{\rho},\ C_{j0}^{i}=\delta_{j}^{i}, \tag{26}$$

又令 $i=i_2$, $B=l_2$, $C=m_1$, $D=0$, 得到

$$C_{m_1l_2}^{i_2}=\delta_{l_2}^{i_2}C_{m_10}^{0}+C_{l_2\rho}^{i_2}C_{m_10}^{\rho}.$$

从而令

$$\widetilde{\omega}^{\alpha}=\omega^{\alpha}+C_{0k_2}^{\alpha}\omega^{k_2}+C_{0k_1}^{\alpha}\omega^{k_1},$$

则(21)式化为

$$D\omega^{i_1}=\left(\frac{1}{2}C_{j_1k_1}^{i_1}-C_{j_1\alpha}^{i_1}C_{0k_1}^{\alpha}\right)[\omega^{j_1}\omega^{k_1}]+C_{j_1\rho}^{i_1}[\omega^{j_1}\ \widetilde{\omega}^{\rho}]+\delta_{j_1}^{i_1}[\omega^{j_1}\ \widetilde{\omega}^{0}].$$

将$[\omega^{j_1}\omega^{k_1}]$的系数仍记为$\frac{1}{2}C_{j_1k_1}^{i_1}$，而$\widetilde{\omega}^{\alpha}$仍记为$\omega^{\alpha}$，则得

$$D\omega^{i_1}=\frac{1}{2}C_{j_1k_1}^{i_1}[\omega^{j_1}\omega^{k_1}]+C_{j_1\rho}^{i_1}[\omega^{j_1}\omega^{\rho}]+\delta_{j_1}^{i_1}[\omega^{j_1}\omega^{0}], \tag{27}$$

$$D\omega^{i_2}=\frac{1}{2}C_{j_2k_2}^{i_2}[\omega^{j_2}\omega^{k_2}]+C_{j_2\rho}^{i_2}[\omega^{j_2}\omega^{\rho}]+\delta_{j_2}^{i_2}[\omega^{j_2}\omega^{0}].$$

记

$$f_1=\sum_{i_1}(\omega^{i_1})^2,\ f_2=\sum_{i_2}(\omega^{i_2})^2, \tag{28}$$

则由(18)得

$$ds^2=\frac{1}{\sigma^2}\left(\sum_{i_1}(\omega^{i_1})^2+\sum_{i_2}(\omega^{i_2})^2\right)=\frac{1}{\sigma^2}(f_1+f_2). \tag{29}$$

为证明V_n的线素可化为定理4中所写的形态，我们需要证明：除一个公共因子外，f_1只依赖于x^{i_1}及dx^{i_1}，f_2只依赖于x^{i_2}，dx^{i_2}，这里x^{i_1}及x^{i_2}是$\omega^{i_1}=0$与$\omega^{i_2}=0$的初积分.

选取微分δ_1，使

$$\delta_1 x^{i_1}=0\quad 或\quad \omega^{i_1}(\delta_1)=0. \tag{30}$$

注意到

$$D\omega^{i_1}(\delta_1,d)=\delta_1\omega^{i_1}(d)-d\omega^{i_1}(\delta)=\delta_1\omega^{i_1}(d),$$

利用(27)，(25)及(26)，

$$\delta_1 f_1(d)=2\sum_{i_1}\omega^{i_1}(d)\delta_1\omega^{i_1}(d)=2\sum_{i_1}\omega^{i_1}(d)D\omega^{i_1}(\delta_1,d)$$

$$=-2\sum_{i_1}\omega^{i_1}(d)C^{i_1}_{j_1\alpha}\omega^{j_1}(d)\omega^{\alpha}(\delta_1)=-2\sum_{i_1}(\omega^{i_1}(d))^2\omega^0(\delta_1)$$
$$=-2\omega^0(\delta_1)f_1(d),$$

因此

$$\delta_1 \lg f_1(d)=-2\omega^0(\delta_1). \tag{31}$$

又选取 δ_2 使 $\omega^{i_2}(\delta)=0$，同样地得到

$$\delta_2 \lg f_2(d)=-2\omega^0(\delta_2). \tag{32}$$

我们现在来证明 ω^0 是全微分，因而要证明

$$D\omega^0=\frac{1}{2}C^0_{jk}[\omega^j\omega^k]+C^0_{j\beta}[\omega^j\omega^\beta]+C^0_{\beta\gamma}[\omega^\beta\omega^\gamma]+C^0_{j\tau}[\omega^j\omega^\tau]$$

的系数均为 0，为此要利用(25)，(26)及

$$C^{i_1}_{jk_2}=0,\ C^{i_2}_{jk_1}=0,\ C^{i_1}_{j_2\alpha}=0,\ C^{j_2}_{i_1\alpha}=0. \tag{33}$$

在(23)式中令 $B=\alpha$，$C=k$，$D=\beta$，且关于 i，k 缩短，得

$$C^i_{i\gamma}C^\gamma_{\beta\alpha}=0.$$

利用(25)，(26)得

$$C^0_{\alpha\beta}=0. \tag{34}$$

又令 $i=i_1$，$B=l_1$，$C=m_2$，$D=d$，且利用(33)，

$$C^0_{m_2\alpha}=0$$

成立，类似地 $C^0_{m\alpha}=0$，因此

$$C^0_{\alpha j}=0. \tag{35}$$

再令 $B=j$，$C=k$，$D=\tau$，得出

$$C^i_{\alpha j}C^\alpha_{k\tau}+C^i_{\alpha k}C^\alpha_{\tau j}=0. \tag{36}$$

令 $i=j=i_1$, $k=k_2$, 从上式可得出 $C^0_{k_2\tau}=0$, 类似地从上式可得 $C^0_{k_1\tau}=0$, 因而

$$C^0_{k\tau}=0. \tag{37}$$

在(24)式中令 $\alpha=0$, $B=i$, $C=j$, $D=0$, 利用(34),(35),(37)可求得

$$C^0_{ij}=0, \tag{38}$$

因而 $D\omega^0=0$, ω^0为全微分,令

$$\omega^0(d)=-\frac{1}{2}d\lg h(x,u), \tag{39}$$

与(31),(32)相比,得到

$$\frac{\partial}{\partial x^{i_1}}\lg\left(\frac{f_2(d)}{h(x,u)}\right)=0, \tag{40}$$

$$\frac{\partial}{\partial x^{i_2}}\lg\left(\frac{f_1(d)}{h(x,u)}\right)=0, \tag{41}$$

$$\frac{\partial}{\partial u^{\alpha}}\lg\left(\frac{f_1(d)}{f_2(d)}\right)=0. \tag{42}$$

由(40),(41)可知

$$f_1(d)=h(x,u)\varphi_1(d), \tag{43}$$

$$f_2(d)=h(x,u)\varphi_2(d). \tag{44}$$

这里 $\varphi_1(d)$与 x^{i_2}, dx^{i_2}, du^{α}无关,$\varphi_2(d)$与 x^{i_1}, dx^{i_1}, du^{α}无关.

再由(42)式可知

$$\frac{f_1(d)}{f_2(d)}=\frac{\varphi_1(d)}{\varphi_2(d)},$$

左方与 u^{α}无关.令 $u^{\alpha}=u^{\alpha}_0$, 于上式可得

$$\frac{f_1(d)}{f_2(d)}=\frac{\psi_1(x^{i_1},dx^{i_1})}{\psi_2(x^{i_2},dx^{i_2})},$$

因而

$$f_1(d)=k(x, u)\psi_1(x^{i_1}, dx^{i_1}), \tag{45}$$

$$f_2(d)=k(x, u)\psi_2(x^{i_2}, dx^{i_2}). \tag{46}$$

这里 ψ_1 为 dx^{i_1} 的二次形式，ψ_2 为 dx^{i_2} 的二次形式.将上二式代入(29)得

$$ds^2=\frac{1}{\sigma^2}\sum_i(\omega^i)^2=\frac{1}{\sigma^2}(f_1+f_2)=\frac{1}{\sigma^2}k(x, u)(\psi_1+\psi_2),$$

从而 $k(x, u)$与 u 无关，因此空间线素为

$$ds^2=\rho(x)(g_{i_1j_1}(x^{k_1})dx^{i_1}dx^{j_1}+g_{i_2j_2}(x^{k_2})dx^{i_2}dx^{j_2}),$$

V_n 共形对应于乘积空间，这样定理 4 得证.

现设黎曼空间 V_n 容许可递共形运动，其迷向群只包含放大，则由定理 3 的证明，可知这时群的变换方程为

$$\bar{x}^i=e^a x^i+b^i,$$

$$\omega^i=e^{-v}dx^i,$$

于是

$$ds^2=\frac{1}{\sigma^2}\sum_i(\omega^i)^2=\frac{1}{\sigma^2}(e^{-v})^2\sum_i(dx^i)^2.$$

因此空间是共形平坦的，定理 5 得到证明.

5. 现讨论辛尺度空间的共形运动.设辛尺度空间 S_{2n} 容许可迁共形运动群，迷向群包含放大且有非迷向的不变平面 E_{2q}，我们来确定空间尺度的形式.这时我们可选取辛标形使 $e_{i_1}(i_1=1, 2, \cdots, 2q)$在 E_{2q} 上，且为辛直交，$e_{i_2}(i_2=2q+1, \cdots, n)$在 E_{2n-2q} 上，也为辛直交，这时空间尺度可书为

$$\Omega(x, dx)=g_{ij}[dx^i dx^j]=[\underset{0}{\omega}^1\ \underset{0}{\omega}^{q+1}]+[\underset{0}{\omega}^2\ \underset{0}{\omega}^{q+2}]+\cdots+$$
$$[\underset{0}{\omega}^q\ \underset{0}{\omega}^{2q}]+\cdots+[\underset{0}{\omega}^{2q+1}\ \underset{0}{\omega}^{n+q+1}]+\cdots+[\underset{0}{\omega}^{n+q}\ \underset{0}{\omega}^{2n}],$$

同样利用嘉当一般等价理论，存在 n 个决定变换群的不变形式

$$\omega^i(x, u, dx) = a^i_j(x, u) \underset{0}{\omega}^j(x, dx), \tag{47}$$

并能使

$$\begin{aligned}\Omega &= [\underset{0}{\omega}^1 \underset{0}{\omega}^{q+1}] + \cdots + [\underset{0}{\omega}^q \underset{0}{\omega}^{2q}] + [\underset{0}{\omega}^{2q+1} \underset{0}{\omega}^{n+q+1}] + \cdots + [\underset{0}{\omega}^{n+q} \underset{0}{\omega}^{2n}] \\ &= \sigma^2\{[\omega^1\omega^{q+1}] + \cdots + [\omega^q\omega^{2q}] + [\omega^{2q+1}\omega^{n+q+1}] + \cdots + [\omega^{n+q}\omega^{2n}]\},\end{aligned} \tag{48}$$

这时(27)式仍然成立，但

$$C^{t'_1}_{s_1\rho} = C^{s'_1}_{t_1\rho},\ C^{t_1}_{s'_1\rho} = C^{s_1}_{t'_1\rho},\ C^{s'_1}_{t'_1\rho} + C^{t_1}_{s_1\rho} = 0$$

$$(\rho = n+1, \cdots, n+r-1;\ s_1, t_1 = 1, \cdots, q;\ s'_1, t'_1 = q+1, \cdots, 2q), \tag{49}$$

$$C^{t'_2}_{s_2\rho} = C^{s'_2}_{t_2\rho},\ C^{t_2}_{s'_2} = C^{s_2}_{t'_2\rho},\ C^{s'_2}_{t'_2\rho} + C^{t_2}_{s_2\rho} = 0$$

$$(s'_1 = q + s_1;\ s_2 = 2q+1, \cdots, n+q;\ s'_2 = n - q + s_2), \tag{50}$$

$$C^i_{j0} = \delta^i_j. \tag{51}$$

记

$$f_1 = \sum_{s_1}[\omega^{s_1}\omega^{s'_1}],\ f_2 = \sum_{s_2}[\omega^{s_2}\omega^{s'_2}]. \tag{52}$$

选取 δ_1 与 δ_2，

$$\delta_1 x^{i_1} = 0 \text{ 或 } \omega^{i_1}(\delta_1) = 0,\ \delta_2 x^{i_2} = 0 \text{ 或 } \omega^{i_2}(\delta_2) = 0, \tag{53}$$

由于

$$Df_1(d_1, d_2, \delta_1) = d_1 f_1(d_2, \delta_1) + d_2 f_1(\delta_1, d_1) + \delta_1 f_1(d_1, d_2) = \delta_1 f_1(d_1, d_2), \tag{54}$$

而另一方面

$$Df_1 = \sum_{s_1}[D\omega^{s_1}\omega^{s'_1}] - \sum_{s_1}[\omega^{s_1}D\omega^{s'_1}],$$

从此式利用(27)，(53)，(49)，(51)，(54)可得

$$\delta_1 f_1(d_1, d_2) = 2\omega^0(\delta_1) f_1(d_1, d_2),$$

即

$$\delta_1 \lg f_1(d_1, d_2) = 2\omega^0(\delta_1).$$

类似地

$$\delta_2 \lg f_2(d_1, d_2) = 2\omega^2(\delta_2).$$

利用 Jacobi 等式及(49),(50),(51),经计算可证得 $D\omega^0=0$. 所以 ω^0 是全微分,从而与黎曼空间时类似地可推出

$$f_1(d_1, d_2) = k(x, u)\psi_1(x^{i_1}, dx^{i_1}),$$

$$f_2(d_1, d_2) = k(x, u)\psi_2(x^{i_1}, dx^{i_1}).$$

这里 ψ_1 为 x^{i_1}, dx^{i_1} 的二次外微分形式,ψ_2 为 x^{i_2}, dx^{i_2} 的二次外微分形式,代入(48)得到辛尺度空间线素化为

$$\Omega(x, dx) = \rho^2(x)(\Omega_1(x^{i_1}, dx^{i_1}) + \Omega_2(x^{i_2}, dx^{i_2})).$$

这里 $\Omega_1(x^{i_1}, dx^{i_1})$ 与 $\Omega_2(x^{i_2}, dx^{i_2})$ 均为辛尺度空间,因而定理 6 证毕.

明显地如果辛尺度空间 S_{2n} 容许可迁共形运动,其迷向群只包含放大,则

$$\Omega(x, dx) = \frac{1}{\sigma^2}(e^{-v})([dx^1 dx^{n+1}] + [dx^2 dx^{n+2}] + \cdots + [dx^n dx^{2n}]),$$

故共形于辛空间,此即定理 7 的结论.

参 考 文 献

[1] Гу чао-хао(谷超豪), О приводимости бесконечных групп преобразований Картана, Нзв. ВУЗ. Математика (1958), №4(5), 60 - 66.

[2] Гу чао-хао(谷超豪), Некоторые общие свойства псевдогрупп преобразованнй и их применения к дифференциальной геометрии, Докт. Дисс. МГУ (1959).

［3］ 谷超豪，О группах движений простронств с симпдектической метрикой，Научные Доклацы Высш. Школь（1959）.

［4］ Cartan，E.，La structure des groupes infinis，Seminaire de Math. exposés G. et H.（1937）.（见 Oeuvres complètes，partie Ⅱ，vol.2，pp.1335－1384.）

（本文曾发表于《1960上海市科学技术论文选集数学·化学》，上海科学技术出版社出版，1962年，100－109）

论容有不可迁共形变换群的黎曼空间*

胡和生

复旦大学数学研究所

§1. 引　　言

近年来，关于黎曼空间共形变换群方面有一系列的研究，T. Nagano[1]证明了如下的事实：非共形平坦的正定黎曼空间 V_n 如容有共形变换群 G_r，则必有另一黎曼空间 $\overline{V}_n$，它共形于 V_n 而以 G_r 为其运动群.由此可知，容许共形变换群的黎曼空间可分为二类，一类是共形平坦空间，另一类是和容有运动群的非共形平坦空间互相共形对应的空间.能作为黎曼空间的共形变换群也有二类，一类是欧氏空间的共形变换群及其子群，另一类是可作为黎曼空间运动群的变换群，但是这二类有公共部分，因为欧氏空间共形变换群的某些子群也可以作为某些黎曼空间的运动群.

本文的目的是对容有不可迁共形变换群的黎曼空间作一些深入的研究，我们得到：一个变换群 G_r 能作为黎曼空间的不可迁共形变换群的充要条件是，它能作为黎曼空间的不可迁运动群，并且黎曼空间 V_n 能容有不可迁共形变换群 G_r 的充要条件是，V_n 共形于一个以 G_r 为运动群的黎曼空间 $\overline{V}_n$.此外，我们还确定了容有不可迁共形变换群的所有黎曼线素.

---

* 本文 1963 年 1 月 29 日收到.

§2. 具有不可迁共形变换群的黎曼空间

如所知,共形变换群的迷向群是相似群及其子群.如果 P 点的迷向群是直交群或其子群,则称 P 为等长点;如果 P 点的迷向群是相似群或其相似子群,则称 P 为相似点.凡具有可迁共形变换群的黎曼空间,或者每点都是等长点,或者每点都是相似点;在后面的情况下空间本身是共形平坦的[2].

假设线素是 $g_{ij}(x)dx^i dx^j (i, j=1, 2, \cdots, n)$ 的 n 维黎曼空间 V_n 容有一个不可迁共形变换群 G_r:

$$\bar{x}^i = f^i(x^j, a^\sigma) \quad (\sigma=1, \cdots, r),$$

因而成立

$$g_{ij}(\bar{x})d\bar{x}^i d\bar{x}^j = \rho(x^1, \cdots, x^n, a^1, \cdots, a^r) g_{ij}(x)dx^i dx^j, \tag{1}$$

这里 $\rho(x, a)$ 是适当的函数.又设 $\xi^i_{\sigma 1}$ 是 G_r 的常系数线性无关的无穷小向量的完全系统,而且 $\| \xi^i_{\sigma 1} \|$ 的秩为 $n-k$,则可选取适当的坐标系,使得群的有限方程化为[3]

$$\bar{x}^1 = x^1, \ \bar{x}^2 = x^2, \ \cdots, \ \bar{x}^k = x^k, \ \bar{x}^\alpha = \varphi^\alpha(x^{k+1}, \cdots, x^n, a', \cdots, a^r)$$
$$(a=k+1, \cdots, n) \tag{2}$$

容易证明这时 $\xi^a_{\sigma 1}=0$,而 $x^a = c^a (a=1, 2, \cdots, k)$ 是一系 $n-k$ 维的最小不变流形;并且这时(1)式包括下面三组方程:

$$g_{ab}(\bar{x}^\gamma, x^c) = \rho(x, a) g_{ab}(x^\beta, x^c), \tag{3}$$

$$g_{a\alpha}(\bar{x}^\gamma, x^c) = \rho(x, a) g_{a\beta}(x^\delta, x^c) \frac{\partial x^\beta}{\partial \bar{x}^\alpha}, \tag{4}$$

$$g_{\alpha\beta}(\bar{x}^\gamma, x^c) = \rho(x, a) g_{\gamma\delta}(x^\varepsilon, x^c) \frac{\partial x^\gamma}{\partial \bar{x}^\alpha} \frac{\partial x^\delta}{\partial \bar{x}^\beta} \tag{5}$$

$$(a, b, c=1, \cdots, k; \alpha, \beta, \gamma, \delta, \varepsilon = k+1, \cdots, n).$$

现在，把变换 $\bar{x}^\alpha=\varphi^\alpha(x^\beta, a^\sigma)$ 的集合记为$\widetilde{G}_r$.很显然地(3)式表明了，在每一个最小不变流形上，g_{ab} 是群$\widetilde{G}_r$ 的变换因子为 $\rho(x, a)$ 的数量场*，(4)式表明了，当 $x^a=c^a$ 时，$g_{a\alpha}$(a 固定)是$\widetilde{G}_r$的不变方向场，而且这个方向的向量在变换后长度是原来的 ρ 倍；(5)式表明了，$\widetilde{G}_r$是最小不变流形上的可迁共形变换群.

我们首先考察方程

$$f(\bar{x}^\alpha, x^c)=\rho(x, a)f(x^\alpha, x^c), \tag{6}$$

且用 $a_\sigma^\mu=a_\sigma^\mu(t)(\sigma, \mu=1, \cdots, r)$ 表示 $\xi_{\sigma 1}^i(x)$所生成的单参数变换群.把它代入(6)，关于 t 微分后再置 $t=0$，我们看出，以 $\rho(x, a)$为变换因子的数量场必须满足微分方程：

$$X_\sigma f\equiv\xi_{\sigma 1}^i\frac{\partial f}{\partial x^i}=\psi_\sigma(x)f(x),\ \psi_\sigma(x)=\left.\frac{d\rho}{dt}\right|_{t=0} \tag{7}$$

或

$$X_\sigma \lg f=\psi_\sigma. \tag{7'}$$

反过来，容易证明：(7)的解必满足(6)式.

其次，考察最小不变流形的性质.由于 $\bar{x}^\alpha=\varphi^\alpha(x^\beta, a^\sigma)$ 是最小不变流形上的可迁共形变换群，最小不变流形上的每一点或者都是相似点，或者都是等长点.我们分别称它为相似点流形或等长点流形.又因为 $\bar{x}^\alpha=\varphi^\alpha(x^\beta, a^\sigma)$ 是这些最小不变流形共同的共形变换群，任意两个最小不变流形 $x^a=c_1^a$ 与 $x^a=c_2^a$ 上的对应点 $P(c_1^a, x_0^\alpha)$与 $P(c_2^a, x_0^\alpha)$的安定群相同，从而迷向群也相同，因此我们只须考察二种情况：i) 所有最小不变流形是等长点流形；ii) 所有最小不变流形都是相似点流形.

§ 3. 对情况 i)的讨论

这时根据 E. Cartan[4] 的结果得知最小不变流形上必有一个在$\widetilde{G}_r$下不变的黎曼线素

---

* 当一数量场 $f(x^\alpha)$满足 $f(\bar{x}^\alpha)=\rho(x, a)f(x^\beta)$ 时，就称为在$\widetilde{G}_r$上以 $\rho(x, a)$为变换因子的数量场.

$a_{\alpha\beta}(x)dx^{\alpha}dx^{\beta}$存在;在这里我们将采取如下的途径来作出 $a_{\alpha\beta}(x)$.设 P 是在最小不变流形 $x^{a}=c^{a}$ 上的一个固定点,选取坐标系使在这点 $g_{\alpha\beta}=\delta_{\alpha\beta}$,而 P 点的坐标为 x_{0}^{α},对这流形上的任一点 x^{α} 必有 $\widetilde{G}_{r}$ 中的变换 T_{b} 使 $x_{0}^{\alpha}=\varphi^{\alpha}(x^{\beta}, b^{\sigma})$,并且

$$a_{\alpha\beta}(x)=\delta_{rs}\frac{\partial\varphi^{r}}{\partial x^{\alpha}}\frac{\partial\varphi^{s}}{\partial x^{\beta}}$$

因此,很明显地黎曼线素 $d\tau^{2}=a_{\alpha\beta}(x)dx^{\alpha}dx^{\beta}$ 是以 $\widetilde{G}_{r}$ 为运动群的.另一方面,$\widetilde{G}_{r}$ 又是 $x^{a}=c^{a}$ 上的共形变换群,所以依据共形变换的定义得知 $d\tau^{2}$ 是和流形 $x^{a}=c^{a}$ 的线素共形的.因此,在情况 i)下,对每个最小不变流形 V_{n-k},必存在和它共形的而以 $\widetilde{G}_{r}$ 为其运动群的流形 $\bar{V}_{n-k}$.这样,就可写下

$$g_{\alpha\beta}(c^{a}, x^{\gamma})dx^{\alpha}dx^{\beta}=A(c^{a}, x^{\gamma})a_{\alpha\beta}(c^{a}, x^{\gamma})dx^{\alpha}dx^{\beta}, \tag{8}$$

并且 $\xi_{\sigma 1}^{\alpha}(\sigma=1, 2, \cdots, r)$ 一方面是 $g_{\alpha\beta}(c^{a}, x^{\gamma})dx^{\alpha}dx^{\beta}$ 的共形变换群的无穷小向量,而另一方面又是 $a_{\alpha\beta}(c^{a}, x^{\gamma})dx^{\alpha}dx^{\beta}$ 中的运动群的无穷小向量,所以

$$\xi_{\sigma 1}^{\gamma}\frac{\partial g_{\alpha\beta}}{\partial x^{\gamma}}+g_{\alpha\gamma}\frac{\partial\xi_{\sigma 1}^{\gamma}}{\partial x^{\beta}}+g_{\beta\gamma}\frac{\partial\xi_{\sigma 1}^{\gamma}}{\partial x^{\alpha}}=\psi_{\sigma}g_{\alpha\beta}, \tag{9}$$

$$\xi_{\sigma 1}^{\gamma}\frac{\partial a_{\alpha\beta}}{\partial x^{\gamma}}+a_{\alpha\gamma}\frac{\partial\xi_{\sigma 1}^{\gamma}}{\partial x^{\beta}}+a_{\beta\gamma}\frac{\partial\xi_{\sigma 1}^{\gamma}}{\partial x^{\alpha}}=0 \quad (\sigma=1, 2, \cdots, r). \tag{10}$$

将 $g_{\alpha\beta}(c^{a}, x^{\gamma})=A(c^{a}, x^{\gamma})a_{\alpha\beta}(c^{a}, x^{\gamma})$ 代入(9),并利用(10)来改写它,就得到

$$X_{\sigma}\lg A=\psi_{\sigma}. \tag{11}$$

从上式和(7′)式的比较看出:以 $\rho(x, a)$ 为变换因子的数量场必是形式

$$f(x)=c(x^{a})A(x), \tag{12}$$

这就是说,

$$s_{ab}(x^{c}, x^{\gamma})=c_{ab}(x^{c})A(x^{c}, x^{\gamma}). \tag{13}$$

这里 c_{ab} 是 x^{c} 的任意函数.当然为保证 ds^{2} 的正定性,对 c_{ab} 还要附加其他要求.

为研究系数 $g_{\alpha\beta}$ 起见，我们先注意到：设 $\lambda_\alpha(x)$ 是在群 $\widetilde{G}_r$ 下 $\overline{V}_{n-k}$ 的不变向量场，即

$$\lambda_\alpha(\bar{x})=\lambda_\beta(x)\frac{\partial x^\beta}{\partial \bar{x}^\alpha}, \tag{14}$$

则 $\mu_\gamma(\bar{x})=A(x)\lambda_\gamma(x)$ 满足

$$\mu_\gamma(\bar{x})=\rho\mu_\beta(x)\frac{\partial x^\beta}{\partial \bar{x}^\gamma}. \tag{15}$$

反过来，如果对 $\mu_\gamma(x)$ 成立(15)，那么 $\lambda_\alpha=\frac{1}{A}\mu_\alpha$ 就满足(14). 由(4)可见 $g_{a\beta}$ 恰满足(15)，因而系数 $g_{a\alpha}$ 由空间 $\overline{V}_{n-k}$ 在运动群 $\widetilde{G}_r$ 下的不变向量场所决定.

这样一来，如果 $g_{ij}dx^idx^j$ 容有不可迁共形变换群 G_r，且其最小不变流形是等长点流形，则在适当坐标系下 G_r 的方程化为(12)，而且线素为

$$ds^2=A(x)(c_{ab}(x^c)dx^adx^b+2c_{as}(x^c)\mu_\alpha^s(x^\gamma)dx^adx^\alpha+a_{\alpha\beta}(x)dx^\alpha dx^\beta). \tag{16}$$

这里 $x^a=c^a$ 是最小不变流形，$\widetilde{G}_r$ 是 $a_{\alpha\beta}(x)dx^\alpha dx^\beta$ 的可迁运动群，$\mu_\alpha^s(s=1, \cdots, p)$ 是 $\widetilde{G}_r$ 的一组全系独立的不变的共变向量场，而 $c_{ab}(x^c)$，$c_{as}(x^c)$ 是 x^c 的任意函数.

§4. 对情况 ii)的讨论和结论

这时，最小不变流形 V_{n-k} 上的每点都是相似点，所以 V_{n-k} 都是共形平坦的，并且 V_n 的线素可写为

$$ds^2=g_{ab}(x^c, x^\gamma)dx^adx^b+2g_{a\alpha}(x^ax^\gamma)dx^adx^\gamma+$$
$$B(x^\gamma, x^a)[(dx^{k+1})^2+\cdots+(dx^n)^2],$$

而 $B(x^\gamma, x^a)[(dx^{k+1})^2+\cdots+(dx^n)^2]$ 以 $\widetilde{G}_r: \bar{x}^\alpha=\varphi^\alpha(x^\beta, a^\sigma)$ 为可迁共形变换群. 如果 $\widetilde{G}_r$ 容有以 $\rho(x, a)$ 为变换因子的数量场 $f(f\neq0)$，则具有线素

$$d\tau^2=\frac{B(x^\gamma, x^a)}{f(x^\gamma, x^a)}[(dx^{k+1})^2+\cdots+(dx^n)^2]$$

的共形平坦空间是以$\widetilde{G}_r$为其可迁运动群的.因此,$\widetilde{G}_r$的迷向群也不会是欧氏空间E_{n-k}的相似子群.所以V_{n-k}上的点都是等长点,这就和情况 ii)的假设相违背.

如果群$\widetilde{G}_r$除$f=0$外,不再容有以$\rho(x, a)$为变换因子的数量场,g_{ab}必须为0,从而就不能保证线素的正定性,所以这种情况也不可能发生.总之情况 ii)就是不会存在的.

综合上述结果,我们得到

定理 如黎曼空间V_n容有不可迁共形变换群G_r,则V_n必共形于一个以G_r为运动群的黎曼空间$\overline{V}_n$,而且V_n的线素必具有型如(16)的表达式.

由此得到

推论 共形平坦空间中共形变换群的不可迁子群一定可作为某一黎曼空间的运动群.

根据这个定理,如果我们要作出一切的容有不可迁的共形变换群的黎曼空间,只要先作出容有不可迁运动群的黎曼空间,然后再作出和它共形的黎曼空间就可以了.

值得注意的是,当

$$\bar{x}^\alpha=\varphi^\alpha(x^\beta, a^\sigma)\quad(\alpha, \beta=k+1, \cdots, n; \sigma=1, \cdots, r)\tag{17}$$

为$n-k$维黎曼空间的可迁共形变换群时,不可迁群

$$\bar{x}^a=x^a, \bar{x}^\alpha=\varphi^\alpha(x^\beta, a^\sigma)\quad(a=1, \cdots, k)$$

未必能作为n维黎曼空间的共形变换群.这是因为,如果取(17)为$n-k$维欧氏空间的完全共形群,那么在群的无穷小变换的微分算子中包含$X_\alpha=\frac{\partial}{\partial x^\alpha}$, $X_0=x^\alpha\frac{\partial}{\partial x^\alpha}$,它们所对应的$\psi$分别为

$$\psi_\alpha=0, \psi_0=1$$

而由此就容易推出$f=0$.这是和运动群不相同的一个性质.

参考文献

[1] T. Nagano, On conformal transformations of Riemannian spaces, Jour. Math. Soc. Japan, 10(1958), 79-93.

[2] S. Ishihara and M. Obata, On the group of conformal transformations of a Riemannian manifold, Proc. Japan. Acad., 31(1955), 426-429.

[3] L. P. Eisenhart, Riemannian geometry (1949).

[4] E. Cartan, Leçons sur la géométrie des espaces de Riemann (1951).

On Riemannian Spaces Admitting Intransitive Group of Conformal Transformations

Hu He-sheng

The object of this paper is to investigate the Riemannian spaces which admit intransitive group of conformal transformations, we obtain the following

Theorem: The necessary and sufficient condition for a group of transformations Gr be the intransitive group of conformal transformations of an n-dimensional Riemannian space is that G_r be the intransitive group of motions of an n-dimensional Riemannian space. Furthermore, the necessary and sufficient condition for a Riemannian V_n admitting intransitive group of conformal transformations is that V_n be conformal to a Riemannian space $\overline{V}_n$ which admits G_r as its intransitive group of motions.

From these results we see that every intransitive subgroup of the conformal group of n-dimensional Enclidean space must be the group of motions of certain Riemannian space.

(本文曾发表于《复旦大学学报》1963 年第 8 卷第 2 期,227-230)

关于齐性黎曼空间的运动群与迷向群

胡和生

§1. 引　　言

黎曼空间运动群的研究已有久远历史，其根本目的是要将黎曼空间按照运动群来进行分类，即求出所有容许运动群的黎曼空间及其相应的运动群，并研究容许运动群的各种黎曼空间的几何性质.由于容许不可迁运动群的黎曼空间的研究要归结到容许可迁运动群的研究去[1]，因此重要的是研究齐性黎曼空间(即容许可迁运动群的黎曼空间).但是因为问题本身的复杂性，至今有许多问题还未解决.E. Cartan[1]利用李群的一般理论来进行这项工作，从而为研究齐性黎曼空间建立一个有力的工具——即根据迷向群的性质来研究齐性黎曼空间，这促进了齐性黎曼空间的发展.但是他所给出的一般方法对解决具体问题还有很多困难，所以在一段时间内这方面的研究有所停顿.但近十几年来，以E. Cartan的方法为基础的工作又活跃起来，有了很多的发展，当迷向群不可约时，O. B. Мантуров[2][3]利用李代数的表示理论得到迷向群不可约的不是对称空间的齐性黎曼空间，从而在这个方面将Cartan的研究推进了一步.当迷向群可约且分为互补不变平面上旋转群的直积时，Wakakuwa[4]在迷向群无不变向量场的条件下得到空间是乘积空间，因而就把决定线素的问题归结到维数较低的齐性黎曼空间去.但是在迷向群具不变向量时，情况就要困难得多，并且在研究容许不可迁运动群的黎曼空间时就经常要涉及到具不变向量场的齐性黎曼空间.谷超豪在[5,6,7]文中研究了具不变向量场的齐性黎曼空间，他在迷向群的不变向量全体所成子空间 E_q 的正交补 E_{n-q} 为不可约的情况下完全解决了这个问题.他在这项工作中尽量地利用群的结构常数的关系来得出空间的几何性质，从而对 Cartan 的方法有了一定的推进，文中所建立的工具对进一步开展这方面的研究很有影响.

作者在此基础上，在[8]文中进一步研究了具不变向量场的齐性黎曼空间，在迷向群分为两个互补平面上旋转群的直积且其中至少有一个没有交换旋转时，决定了空间的线素(包括把问题归结为在低维齐性黎曼空间决定线素的问题).这里的互补不变平面并不限于不可约的，我们将在§2 中叙述这一结果，它包括了[7]，[11]中的一些结果.

对于迷向群可约的齐性黎曼空间的研究过去都是在迷向群可约且分为直积的情况进行的，对于迷向群可约但不分为互补不变平面上旋转群的直积时，除了对于对称空间之外至今还未看到显著结果.然而这种情况应该是复杂而重要的，例如从[5]—[8]的研究中就可以看到在许多情况下迷向群分为直积的齐性黎曼空间 V_n 的研究要归结到迷向群可约而不分为直积的维数低于 n 的齐性黎曼空间的研究去.因而迷向群可约而不分为直积的齐性黎曼空间的研究将是今后运动群研究中的一个十分重要的问题.在文[9]中作者开始进行这方面的研究，在本文§2 中，我们介绍文[9]中的一个定理.

由于黎曼空间按运动群分类的复杂性，人们从另一途径来研究齐性黎曼空间，也就是按照黎曼空间完全运动群阶数大小依次得出完全运动群取得较大阶数的黎曼空间，并研究其性质.Fubini 首先发现运动群阶数会出现空隙，王宪钟[12]，Егоров[13]，Yano[14]等人得到第一，二空隙，并研究了相应的黎曼空间的性质.作者在[10]文中利用了不可约旋转群的 Teleman 定理[11]及本文§2 结果及一系列已知定理[4—7]，给出了依次得到齐性黎曼空间完全运动群的空隙的方法，并在 n 有适当限制时，确定了最初八个空隙并定出完全运动群取得最大的十五个阶数所对应的黎曼空间的线素.依据其中的方法，空隙还可逐个的确定下去，本文§3 中将叙述这方面的结果.

在本文§4 中，我们要介绍作者在[16]文中关于黎曼空间相似变换群的研究结果，这项工作是在齐性黎曼空间理论的基础上进行的，我们把相似群的研究归结到较低维数黎曼空间的运动群的问题上去，从我们的结果中可推出：一个变换群如能作为黎曼空间(除欧氏空间外)的可迁与不可迁相似群，则必可作为一黎曼空间的运动群.

本文中所讨论的黎曼空间线素都是正定的.

作者对苏步青教授的鼓励与帮助表示感谢.

§2. 迷向群可约的齐性黎曼空间

设线素正定的 n 维黎曼空间 V_n 容许可迁运动群 G_r，则 G_r 的直交可容许标形族的两个邻近标形之间的推移关系为

$$dM=\omega^i e_i,\ de_i=\omega_i^j e_j.\quad (i,\ j=1,\ 2,\ \cdots,\ n)$$

如所知，ω^i，ω_j^i 都是群的不变形式，ω^i 是 n 个独立的不变形式，它与 ω_j^i 中的 $r-n$ 个独立的 $\omega^\alpha(\alpha=n+1,\ \cdots,\ r)$ 一起作成群 G_r 的全系不变形式系统. G_r 的 Maurer-Cartan 方程为

$$\left.\begin{aligned}&D\omega^i=\frac{1}{2}C_{jk}^i[\omega^i\omega^k]+C_{j\rho}^i[\omega^j\omega^\rho],\\&D\omega^\alpha=\frac{1}{2}C_{ij}^\alpha[\omega^i\omega^j]+C_{j\beta}^\alpha[\omega^j\omega^\beta]+\frac{1}{2}C_{\beta\gamma}^\alpha[\omega^\beta\omega^\gamma],\\&(i,\ j,\ k=1,\ \cdots,\ n;\ \alpha,\ \beta,\ \gamma,\ \rho=n+1,\ \cdots,\ r)\end{aligned}\right\}\tag{2.1}$$

于此 C_{jk}^i，$\cdots$，$C_{\beta\gamma}^\alpha$ 为 G_r 的结构常数，由于齐性黎曼空间是化约的[18]，因此适当选择 ω^i，ω^α 可使 $C_{\sigma j}^\rho=0$，结构常数除了关于下指标的反称性外，还满足下列 Jacobi 恒等式

$$C_{l\rho}^i C_{j\sigma}^\rho - C_{l\sigma}^i C_{j\rho}^l = C_{\sigma\rho}^\tau C_{j\tau}^i,\tag{2.2}$$

$$O=C_{lm}^i C_{i\rho}^k + C_{il}^k C_{m\rho}^i - C_{im}^k C_{l\rho}^i,\tag{2.3}$$

$$C_{l\tau}^k C_{mh}^\tau + C_{m\tau}^k C_{hl}^\tau + C_{h\tau}^k C_{lm}^\tau = C_{lm}^i C_{ih}^k + C_{mh}^i C_{il}^k + C_{hl}^i C_{im}^k,\tag{2.4}$$

并且 $C_{j\rho}^i=-C_{i\rho}^j(\rho=n+1,\ \cdots,\ r)$ 构成迷向群李代数的基.

1° 我们研究了这样的齐性黎曼空间 V_n，它的运动群 G_r 的迷向群分为直积，分别作用在互补的不变平面 E_q 与 E_{n-q} 上，且迷向群在 E_{n-q} 上没有交换旋转(这时 E_{n-q} 可以是可约的，也可以是不可约的). 不失一般性，可假定迷向群在 E_{n-q} 上无不变向量，这是因

为,假如它有 a 个不变向量我们可将它们放入 E_q 中使成 $q'=q+a$ 维平面 E_{a+q},而其互补平面 $E_{n-(q+a)}$ 无不变向量,这样便可对 $E_{q'}$, $E_{n-q'}$ 进行讨论.

我们选好直交可容许标形使 $e_{i'}$ ($i'=1, 2, \cdots, q$) 和 $e_{i''}$ ($i''=q+1, \cdots, n$) 分别成为 E_q 和 E_{n-q} 上的基向量.设 E_{n-q} 在迷向群下有 s 个不变子空间 $\left(显然 1\leqslant s\leqslant\left[\frac{n-q}{2}\right]\right)$,则利用结构常数与若干已知结果我们首先证明了 Maurer - Cartan 方程(2.1)可化简为

$$\left.\begin{aligned}
&D\omega^{i'}=\frac{1}{2}C^{i'}_{j'k'}[\omega^{j'}\omega^{k'}]+C^{i'}_{j'\rho'}[\omega^{j'}\omega^{\rho'}],\\
&D\omega^{i''_1}=\frac{1}{2}C^{i''_1}_{j''k''}[\omega^{j''}\omega^{k''}]+C^{(1)}_{k'}[\omega^{i''_1}\omega^{k'}]+C^{i''_1}_{j''_1\rho''}[\omega^{j''_1}\omega^{\rho''}],\\
&D\omega^{i''_2}=\frac{1}{2}C^{i''_2}_{j''k''}[\omega^{j''}\omega^{k''}]+C^{(2)}_{k'}[\omega^{i''_2}\omega^{k'}]+C^{i''_2}_{j''_2\rho''}[\omega^{j''_1}\omega^{\rho''}],\\
&\cdots\cdots\\
&D\omega^{i''_s}=\frac{1}{2}C^{i''_s}_{j''k''}[\omega^{j''}\omega^{k''}]+C^{(s)}_{k'}[\omega^{i''_s}\omega^{k''}]+C^{i''_s}_{j''_s\rho''}[\omega^{j''_s}\omega^{\rho''}].
\end{aligned}\right\}\tag{2.5}$$

设 $C^{(1)}_{i'}$, $C^{(2)}_{i'}$, $\cdots$, $C^{(s)}_{i'}$ 中有 l 组线性独立的,记为 $C^{(a)}_{i'}$ ($a=1, \cdots, l$),有 t 组为 0,即 $C^{(\lambda)}_{i'}=0$ ($\lambda=l+1, \cdots, l+t$),其余 $s-l-t$ 组为 $C^{(a)}_{i'}$ 的线性组合,即 $C^{(u)}_{k}=A^u_a C^{(a)}_{k}$,($u=l+t+1, \cdots, s$),$A^u_a$ 为常数,我们证明了

定理 2.1 如果齐性黎曼空间 V_n 的迷向群分解为作用于互补子空间 E_q 和 E_{n-q} 的旋转群的直积,而且在 E_{n-q} 中旋转群无不变向量,也无交换旋转,那么 V_n 的线素可写成形状

$$ds^2=ds_1^2+\sum_a e^{-2x^a}d\sigma_a^2+\sum_u e^{-2A^u_e x^e}d\sigma_u^2+\sum_\lambda d\sigma_\lambda^2.\tag{2.6}$$

这里 ds_1^2 为迷向群具不变向量场的 q 维齐性黎曼空间,其中还存在 l 个独立的数量函数 x^1, $\cdots$, x^l,使在运动群下成立 $\bar{x}^1=x^1+c^1$, $\cdots$, $\bar{x}^l=x^l+c^l$;$d\sigma_a^2$, $d\sigma_u^2$, $d\sigma_\lambda^2$ 是依赖于 $x^{i''}$, $dx^{i''}$ 的二次微分形式,并且

$$\sum_a d\sigma_a^2+\sum_u d\sigma_u^2+\sum_\lambda d\sigma_\lambda^2\tag{2.7}$$

为一 $n-q$ 维齐性黎曼空间的线素,且在空间还可选取适当的不变发甫形式 $\Omega^{i''}$,使

$$d\sigma_a^2=\sum_{i''_a}(\Omega^{i''_a})^2,\ d\sigma_u^2=\sum_{i''_u}(\Omega^{i''_u})^2,\ d\sigma_\lambda^2=\sum_{i''_\lambda}(\Omega^{i''_\lambda})^2. \tag{2.8}$$

根据这定理便可明了所论的齐性黎曼空间的决定可归结到低维齐性黎曼空间的决定去.并且我们也证明了线素为 ds_1^2 的 q 维齐性黎曼空间的运动群 G_{r_1} 含有 r_1-l 维的正常子群 H,而且 G_{r_1} 关于它的商群是阿贝尔群.这里的 H 包含一个定点的安定群.这样的空间我们记为 $V_{q,l}$.

当 E_{n-q} 为不可约时从定理 2.1 就推出谷超豪与 Г. И. Кручкович 的一个定理[7].

对线素(2.7),我们区分了各种情况,并进一步加以分析之后,还能把 V_n 的线素决定得更具体些,我们得到

定理 2.2 定理(2.1)中的线素(2.7)(除若干例外情况外)必具下列形状之一.

a) $$ds^2=ds_1^2+ds_2^2, \tag{2.9}$$

即 V_n 为乘积空间,并且 ds_1^2 所对应的黎曼空间容许以 $\omega^{i'}$, $\omega^{\rho'}$ 为不变形式的运动群 $G^{(1)}$, ds_2^2 所对应的黎曼空间容许以 $\omega^{i''}\omega^{\rho''}$ 为不变形式的运动群 $G^{(2)}$.

b) $$ds^2=g_{i'j'}(x^{k'})dx^{i'}dx^{j'}+e^{-2x^1}\sum_{i''_1}(dx^{i''_1})^2+\cdots+e^{-2x^{i''_s}}\sum_{i''_s}(dx^{i''_s})^2, \tag{2.10}$$

这里 $g_{i'j'}(x^{k'})dx^{i'}dx^{j'}$ 为上面所描述的空间 $V_{q,s}$ 的线素

c) $$ds^2=ds_1^2+\sum_{a=1}^{l}e^{-2x^a}d\sigma_a^2+ds_2^2, \tag{2.11}$$

这里 $d\sigma_a^2$ 是欧氏线素 $\sum_{i''_a}(dx^{i''_a})^2$, ds_1^2 是 $V_{q,l}$ 的线素, ds_2^2 为另一没有什么特别限制的齐性黎曼空间线素.

d) $$ds^2=ds_1^2+\sum_a e^{-2x^a}d\sigma_a^2+\sum_u e^{-A_e^u x^e}d\sigma_u^2+ds_2^2, \tag{2.12}$$

式中 $d\sigma_a^2$, $d\sigma_u^2$ 为欧氏线素 $\sum(dx^{i''_a})^2$, $\sum(dx^{i''_u})^2$, ds_1^2 是 $V_{q,l}$ 的线素, ds_2^2 为另一齐性黎曼空间的线素.

运动群相应地采取形状

$$\text{a}')\qquad \left.\begin{aligned}\bar{x}^{i'}&=\varphi^{i'}(x^{j'},\ C^{j'},\ C^{\rho'}),\\ \bar{x}^{i''}&=\varphi^{i''}(x^{j''},\ C^{j''},\ C^{\rho''}).\end{aligned}\right\}\ G_r=G^{(1)}\times G^{(2)}. \tag{2.9$'$}$$

$$\text{b}')\qquad \left.\begin{aligned}&\bar{x}^{a}=x^{a}+C^{a}, && (a=1,\ \cdots,\ s)\\ &\bar{x}^{p}=\varphi^{p}(x^{j'},\ C^{j'},\ C^{\rho'}), && (p=s+1,\ \cdots,\ q)\\ &\bar{x}^{i''_1}=e^{c1}(a^{i''_1}_{j''_1}x^{j''_1}_1+C^{i''_1}),\\ &\cdots\cdots\\ &\bar{x}^{i''_s}=e^{cs}(a^{i''_s}_{j''_s}x^{i''_s}+C^{i''_s}).\end{aligned}\right\} \tag{2.10$'$}$$

$$\text{a}')\qquad \left.\begin{aligned}&\bar{x}^{a}=x^{a}+C^{a},\\ &\bar{x}^{p}=\varphi^{p}(x^{j'},\ C^{j'},\ C^{\rho'}),\\ &\bar{x}^{i''_a}=e^{c^a}(a^{i''_a}_{j''_a}x^{j''_a}+C^{i''_a}),\\ &\bar{x}^{i''_\lambda}=f^{i''_\lambda}(x^{j''_\lambda},\ C^{a}).\end{aligned}\right\} \tag{2.11$'$}$$

$$\text{d}')\qquad \left.\begin{aligned}&\bar{x}^{a}=x^{a}+C^{a},\\ &\bar{x}^{p}=\varphi^{p}(x^{j'},\ C^{j'}),\\ &\bar{x}^{i''_a}=e^{c_a x^a}(a^{i''_a}_{j''_a}x^{j''_a}+C^{i''_a}),\\ &\bar{x}^{i''_u}=e^{A^y_e x^e}(a^{i''_u}_{j''_u}x^{j''_u}+C^{j''_u}),\\ &\bar{x}^{i''_\lambda}=f^{i''_\lambda}(x^{j''_\lambda},\ C^{a}).\end{aligned}\right\} \tag{2.12$'$}$$

由于容许不可迁运动群的黎曼空间的线素可由其最小不变流形上的线素和这个流形上的不变发甫形式(即不变的共变向量场)所决定,因此,上述的结果立即可以用来建造一类容许不可迁运动群的黎曼空间,并写出它的线素,特别是,我们得到

定理 2.3 如果容许不可迁运动群的黎曼空间 V_n 的迷向群有两个互补不变平面 E_q 及 E_{n-q},迷向群分解为作用在这两平面上的旋转群的直积,而且在 E_{n-q} 上是不可约的,又无交换旋转,则空间 V_n 是半可约的,并且其线素具形状

$$ds^2=g_{\alpha\beta}(x^\gamma)dx^\alpha dx^\beta+C(x^a)g_{i''j''}(x^{k''})dx^{i''}dx^{j''},$$

或

$$ds^2 = g_{\alpha\beta}(x^\gamma)dx^\alpha dx^\beta + C(x^a)e^{-2x^{k+1}} \sum_{i''=q+1}^{n}(dx^{i''})^2.$$

式中 $a=1, 2, \cdots, k$；$\alpha, \beta, \gamma=1, 2, \cdots, q$；$i'', j'', k''=q+1, \cdots, n$；$n-k$ 是最小不变流形的维数.

定理 2.1 与 2.3 包含了 Teleman 在[11]文中的定理 1.

2°. 我们在[9]文中开始进行迷向群可约但不分为直积的齐性黎曼空间的研究.所得的结果在逐个的确定黎曼空间完全运动群的空隙时是有用处的.

如所知[4]，齐性黎曼空间 V_n 如迷向群可约且分为互补不变平面 E_q 与 E_{n-q} 上诱导群 $O(q)$ 与 $O(n-q)$ 的直积时，V_n 必为 q 维常曲率空间与 $n-q$ 维常曲率空间的直积，即 $V_n=S_q\times S_{n-q}$. 我们要问，当上述关于迷向群的假定中去掉直积这一条件时，情况将会如何？在[9]文中我们回答了这一问题.首先证明了

引理 2.1 直交群没有非平凡的三阶不变张量 $T^i_{jk}=-T^i_{kj}$（即一阶反变，二阶共变且关于共变指标是反称的三阶张量）.

引理 2.2 n 维欧氏空间可约旋转群 $H_{r'}$ 如不分解其互补不变平面 E_q 与 E_{n-q} 上旋转群的直积时，则可适当选取李代数的基使具形状

$$C_{\rho_1}=\begin{pmatrix} C^{i'}_{j'\rho_1} & 0 \\ 0 & 0 \end{pmatrix},\ C_{\rho_2}=\begin{pmatrix} 0 & 0 \\ 0 & C^{i''}_{j''\rho_2} \end{pmatrix},\ C_{\rho_3}=\begin{pmatrix} C^{i'}_{j'\rho_3} & 0 \\ 0 & C^{i''}_{j''\rho_3} \end{pmatrix}. \tag{2.13}$$

$$(i', j'=1, \cdots, q;\ i'', j''=q+1, \cdots, n;\ \rho_1=1, \cdots, r_1;$$
$$\rho_2=r_1+1, \cdots, r_1+r_2;\ \rho_3=r_1+r_2+1, \cdots, r')$$

并且$\{C_{\rho_1}\}$，$\{C_{\rho_2}\}$，$\{C_{\rho_3}\}$各自构成理想子代数.又$(C^{i'}_{j'\rho_3})$和$(C^{i''}_{j''\rho_3})$为同构的线性李代数的基.

我们也充分应用了 Jacobi 恒等式(2.2)～(2.4)以及谷超豪在“齐性空间几何学”[23]中的下述结果来化简运算：设 ω^i，ω^ρ 为一组独立的发甫形式，成立

$$D\omega^i=\frac{1}{2}C^i_{jk}[\omega^j\omega^k]+C^i_{j\rho}[\omega^j\omega^\rho],$$

式中 C^i_{jk}，$C^i_{j\rho}$ 为常数，$C_\rho=(C^i_{j\rho})$ 为一组独立的反称阵，则 $D\omega^\rho$ 可用 ω^i，ω^ρ 来表达.从而证明了

定理 2.4 设齐性黎曼空间 $V_n(n>4)$ 的运动群的迷向群 $G(P)$ 有互补不变平面 E_q 与 E_{n-q}，又 $G(P)$ 在 E_q 与 E_{n-q} 上的诱导群分别是直交群 $O(q)$ 与 $O(n-q)$，但 $G(P)\neq O(q)\times O(n-q)$，则当 $n\neq 7$ 时只可能有 $q=n-q$ 且 V_n 是欧氏空间，而当 $n=7$ 时，则 $q=3,4$，V_7 是 E_7 或是 $E_4\times S_3(k>0)$，相反地，偶数维欧氏空间，E_7 及 $E_4\times S_3(k>0)$ 均容有迷向群满足上述要求的可迁运动群.

这里 $S_3(k>0)$ 表示正常曲率空间.

我们指出，在研究迷向群可约但不分为直积的齐性黎曼空间时，引理 2.2 很有用处.

§3. 关于黎曼空间完全运动群的空隙[15]

如所知，常曲率空间 S_n(包括欧氏空间)容许最多参数的完全运动群，参数个数等于 $\frac{n(n+1)}{2}$. 记 r 为正定黎曼空间 V_n 的完全运动群的参数个数(或称阶数)，王宪钟[12]首先得出：当 $n\neq 4$ 时，r 不能满足不等式

$$\frac{n(n-1)}{2}+1<r<\frac{n(n+1)}{2}, \tag{3.1}$$

这称为运动群的第一空隙，矢野[14]继续研究这个问题得出完全运动群阶数为 $\frac{n(n+1)}{2}+1$ 时的黎曼空间的线素形式.Егоров[13]得出了完全运动群的第二个空隙是

$$\frac{1}{2}(n-1)(n-2)+3<r<\frac{1}{3}n(n-1)+1. \tag{3.2}$$

如所知，n 维齐性黎曼空间的运动群阶数为 $r=n+r_1$，于此 r_1 是迷向群的阶数，且当 r_1 是最大的迷向群的阶数时，r 就是完全运动群的阶数，因此迷向群的阶数在这里起了很大作用，黎曼空间的迷向群是直交群或其子群(总称旋转群).王宪钟的上述结果就

是依据 n 维空间直交群的有关定理导出的.C. Teleman[11]证明了下述的定理：n 维直交群的不可约（irreducible）的可析（Separable）子群的参数个数（直交群本身除外）$\leqslant \left[\frac{n}{2}\right]^2$. 这个定理可以为研究黎曼空间的一般的空隙提供一个基础.但为了真正解决问题，需要依赖于利用黎曼空间一系列已知性质以及本文定理 2.1 及 2.4 等等.谷超豪在 1959 年专门组课程中对求一般的空隙大略地叙述了一些想法.作者在历次讲义中把这个问题进行了具体的论证和计算，给出了依次得到齐性黎曼空间完全运动群的空隙的方法，并在 n 有适当的限制时，确定了最初八个空隙并定出完全运动群取得最大的 15 个阶数时对应的黎曼空间的线素，依据其中的方法空隙可逐个确定下去，我们把结果整理成[10]文，现在简述文中的思想和主要结果.

要求出 V_n 的运动群的空隙，就需要依据阶数大小依次列出直交群的子群的阶数，设 $O(n)$ 的最大不可约子群记为 $N(n)$，则由 Teleman 上述定理可知：当 $N(n)$ 是可析时，其阶数 $\leqslant \left[\frac{n}{2}\right]^2$，又由于迷向群为不可析的齐性黎曼空间一定是常曲率的，因此在讨论完全运动群的空隙时可不加考虑.这样，我们可看到当 n 有适当限制时，我们只要考察直交群的可约子群，并将可约子群的阶数依大小次序排出，显然当可约子群的不可约不变平面的维数大时，迷向群阶数才可能较大，而且还可以证明，当群的阶数相当大时，迷向群一定分为直积.我们得到：

引理 3.1 设 H 为 n 维全直交群 $O(n)$ 的一个子群，E_q、E_{n-q} 是两个互补的不变平面，设 H 在 E_{n-q} 上的诱导 K 为不可约的，又 H 的阶数 $> \left[\frac{n-q}{2}\right]^2+\frac{q(q-1)}{2}$，那么 H 必分解为 K 和 L 的直积，于此 L 为 H 在 E_q 上的诱导.

引理 3.2 设 $V_n(n>2)$ 为非常曲率的齐性黎曼空间，其完全运动群的阶数 $> \left[\frac{n}{2}\right]^2+n$，那么迷向群 H 必具形状

$$H=G(q)\times O(n-q).$$

于此 $G(q)$ 为 $O(q)$ 的一个子群.又 $q \leqslant \frac{n}{2}$，并且 V_n 的线素为

i) $$ds^2 = ds_1^2 + ds_2^2,$$

ds_2^2是 $n-q$ 维常曲率空间的线素，ds_1^2是迷向群为 $G(q)$的 q 维齐性黎曼空间的线素；

ii) $$ds^2 = ds_1^2 + e^{-2kx^1}\sum_{\alpha=q+1}^{n}(dx^\alpha)^2,\ (k\ \text{常数})$$

这里 ds_1^2是以 $G(q)$为迷向群的 q 维齐性黎曼空间的线素，且运动群方程中包含 $\bar{x}^1 = x^1 + c^1$.

这个引理与 Teleman 文[11]中定理 10 相类似，但比他的结果具体，并改正了其中不确切的地方.

在可约迷向群依阶数的大小次序排出后，我们利用齐性黎曼空间中一系列已知结果得出它所对应的黎曼空间的线素，但是这样确定出来的黎曼空间中有一部分空间并不以给定的迷向群所对应的运动群为完全运动群，我们就必须去掉这部分线素.在这里我们应用乘积空间及混合运动与非混合运动方面的结果来处理这一问题.

在确定空隙的过程中也需要经常用到引理 2.2 及

引理 3.3 设半可约的黎曼空间 V_n的线素为

$$ds^2 = ds_1^2 + e^{-x^1} ds_2^2.$$

式中

$$ds_1^2 = (dx^1)^2 + e^{-kx^1}\sum_{a=2}^{q}(dx^a)^2,\ ds_2^2 = \sum_{q+1}^{n}(dx^\alpha)^2.$$

如果 $k \neq 0,\ 1$，则空间无单参数混合运动，$k=0,\ 1$ 时有混合运动.

在本文中也得出如下结论：非共形平坦的黎曼空间 $V_n(n>8)$ 的完全运动群的最大阶数是 $\dfrac{(n-1)(n-2)}{2}+3$，并且这时空间

$$V_n = S_2 \times S_{n-2},$$

且 S_2的常曲率 k_1与 S_{n-2}的常曲率 k_2满足

$$k_1 + k_2 \neq 0.$$

现把这八个空隙及有关的线素列表于后.为方便计，我们引入下列记号

$S_n(k<0)$，n 维负常曲率空间，S_n 为 n 维常曲率空间.

P_n，n 维特殊次射影空间，即

$$P_n=E_1\times S_{n-1}.$$

g_n，完全运动群为 n 参数的 n 维群空间.

$g_n(n-1)$，具有 $n-1$ 参数正常子群且完全运动群为 n 参数的 n 维群空间.

C 常数，K 常数.

序号	齐性黎曼空间或空间的线素	完全运动群的阶数	空　隙
1	S_n	$\dfrac{n(n+1)}{2}$	第一空隙
2	$P_n=S_{n-1}\times E_1$（S_{n-1} 不是欧氏的）	$\dfrac{n(n-1)}{2}+1\ (n>4)$	
3	$S_{n-2}\times S_2$（S_{n-2}，S_2 中至少有一个不是欧氏的）	$\dfrac{(n-1)(n-2)}{2}+3\ (n>8)$	第二空隙
4	$S_2(k<0)$ 与 E_{n-2} 的一类半可约空间，线素为 $ds^2=(dx^1)^2+e^{-2cx^1}(dx^2)^2+e^{-2x^1}\sum\limits_{\alpha=3}^{n}(dx^\alpha)^2$ （C 常数，$C\neq 0$，1）	$\dfrac{(n-1)(n-2)}{2}+2\ (n>8)$	第三空隙
5	$S_{n-3}\times S_3$（S_{n-3}，S_3 中至少有一个不是欧氏的）	$\dfrac{(n-2)(n-3)}{2}+6\ (n>10)$	
6	a. $S_{n-3}\times P_3=S_{n-3}\times S_2\times E_1$（$S_3$ 与 S_2 均不是欧氏的） b. $S_{n-2}(k<0)$ 与 E_2 的一类半可约空间，线素为 $ds^2=(dx^1)^2+e^{-2cx^1}\sum\limits_{a=2}^{s}(dx^a)^2+e^{-2x^1}\sum\limits_{\alpha=4}^{n}(dx^\alpha)^2\ (C\neq 0,\ 1)$ c. $S_{n-3}\times V_3$，V_3 的线素为 $ds_2^2=\left\{dx^1+\dfrac{x^2dx^3-x^3dx^2}{1+\frac{K}{4}[(x^2)^2+(x^3)^2]}\right\}^2+\dfrac{(dx^2)^2+(dx^3)^2}{\left\{1+\frac{K}{4}[(x^2)^2+(x^3)^2]\right\}^2}$	$\dfrac{(n-2)(n-3)}{2}+4\ (n>10)$	第四空隙

续 表

序号	齐性黎曼空间或空间的线素	完全运动群的阶数	空　隙
7	a. $S_{n-3}\times g_3$（没有单参数混合运动） b. $ds^2=g_{ab}dx^a dx^b+e^{-2x^1}\sum\limits_{\alpha=4}^{n}(dx^\alpha)^2$ $a, b, c=1, 2, 3$；$g_{ab}dx^a dx^b$ 为 $g_3(2)$的线素 （没有单参数混合运动）	$\dfrac{(n-2)(n-3)}{2}+3\ (n>10)$	第五空隙
8	$S_{n-4}\times S_3$（S_{n-4}，S_3 中至少有一个不是欧氏的）	$\dfrac{(n-3)(n-4)}{2}+10\ (n>12)$	
9	$S_{n-4}\times V_4$（$S_{n-4}\neq E_{n-4}$），V_4是容有 4 参数不可约运动群的 4 维齐性黎曼空间	$\dfrac{(n-3)(n-4)}{2}+8\ (n>14)$	第六空隙
10	a. $S_{n-4}\times S_3\times E_1$　（S_{n-4}，S_3都不是欧氏的） b. $ds^2=(dx^1)^2+e^{-2cx^1}\sum\limits_{\alpha=2}^{4}(dx^\alpha)^2+e^{-2x^1}\sum\limits_{\alpha=5}^{n}(dx^\alpha)^2(h\neq 0, 1)$	$\dfrac{(n-3)(n-4)}{2}+7\ (n>14)$	
11	$S_{n-4}\times S_2\times S_2$（$S_{n-4}$，$S_2$，$S_2$ 中至少有 2 个不是欧氏的）	$\dfrac{(n-3)(n-4)}{2}+6\ (n>14)$	
12	a. $S_{n-4}\times V_4$，V_4 是迷向群为 $O(2)\times O(1)\times O(1)$ 的 4 维齐性黎曼空间 b. $ds^2=ds_1^2+e^{-2x^1}\sum\limits_{\alpha=5}^{n}(dx^\alpha)^2$，$ds_1^2$ 为 S_{n-4} 的线素	$\dfrac{(n-3)(n-4)}{2}+5\ (n>14)$	
13	a. $S_{n-4}\times g_4$（$S_{n-4}\neq E_{n-4}$，没有单参数混合运动） b. $ds^2=g_{ab}(x^c)dx^a dx^b+e^{-2x^1}\sum\limits_{\alpha=5}^{n}(dx^\alpha)^2$（$a, b, c=1, \cdots, 4$）$g_{ab}(x^c)dx^a dx^b$ 是$g_4(3)$ 的线素	$\dfrac{(n-3)(n-4)}{2}+4\ (n>14)$	第七空隙
14	$S_{n-5}\times S_5$　（S_5，S_{n-5}中至少有一个不是欧氏的）	$\dfrac{(n-4)(n-5)}{2}+15\ (n>14)$	
15	a. $S_{n-4}\times S_4\times E_1$（$S_{n-4}$，$S_4$ 都不是欧氏的） b. $ds^2=(dx^1)^2+e^{-2cx^1}\sum\limits_{\alpha=2}^{5}(dx^\alpha)^2+e^{-2x^1}\sum\limits_{\alpha=6}^{n}(dx^\alpha)^2(C\neq 0, 1)$	$\dfrac{(n-4)(n-5)}{2}+11\ (n>16)$	第八空隙

§4. 黎曼空间的相似变换群

对于黎曼空间的相似变换群，曾有不少作者进行过研究，已有一定的结果[19,20].研究黎曼空间相似变换群的根本目的是要确定出所有容有相似变换群的黎曼空间，并求出可作为黎曼空间相似群的所有变换群，已有结果并未就这个问题作出完全的回答.

我们在[16]文中证明了正定黎曼空间的相似变换群可以从另一个较低维数的黎曼空间的运动群中构造出来，从而将研究相似群的问题归结到运动群的研究去.在这个意义下，我们给予上述问题一个完全的回答.我们要利用谷超豪在研究齐性黎曼空间时所得到的下述引理：正定齐性黎曼空间如容许真正的相似变换则必为欧氏空间.此外由于黎曼空间相似变换群与运动群的密切联系，齐性黎曼空间中的思想及§3 中有关空隙性的结果在确定容许相似变换群的黎曼空间时很有用处.

在可迁相似变换群方面，得到如下结果：

定理 4.1 n 维黎曼空间 V_n（不是欧氏空间）的可迁相似变换群 G_r 有正常子群 G_{r-1}，它是 V_n 的一个不可迁运动群，最小不变流形为 $n-1$ 维的.相反地，一变换群 G_r 如有可作为 n 维黎曼空间的不可迁运动群（最小不变流形 $n-1$ 维）的正常子群 G_{r-1}，则 G_r 必可作为一 n 维黎曼空间的可迁相似变换群.

推论 1 单纯可迁群 G_r 能作为黎曼空间 V_n 的相似群的充要条件是它含有 $r-1$ 参数的正常子群.

这是 K. Yano 的一个结果[19].又从定理 4.1 的证明中可见

推论 2 一变换群如能作为黎曼空间（欧氏空间除外）的可迁相似变换群，则必可作为一黎曼空间的运动群.

定理 4.2 设 $X_1, \cdots, X_{r-1}$ 是坐标为$(x^2, \cdots, x^n)$的 $n-1$ 维黎曼空间的可迁运动群 G_{r-1} 的独立微分算子，则存在 $X_r=\xi^i(x^1, \cdots, x^n)\dfrac{\partial}{\partial x^i}$ $(i=1, 2, \cdots, n)$，使得由 $X_1, \cdots, X_r$ 生成的变换群是 n 维黎曼空间的一个相似变换群，并且黎曼空间的所有相似

变换群都可这样地得到.

这个定理给出从运动群得出相似群的具体做法.

定理 4.3 容许可迁相似变换群的黎曼空间只有

i) 欧氏空间,

ii) 线素具 $ds^2=\sigma(x)h_{ij}(x)dx^idx^j$ 的空间,其中 $h_{ij}(x)dx^idx^j$ 为容许不可迁运动群的黎曼空间,最小不变流形为 $n-1$ 维的 $\sigma(x)=$ 常数,函数 σ 满足 $X_{\alpha\sigma}=C_\alpha\sigma$ (C_α: 常数).这里 X_α 是 G_r 的微分算子.

利用定理 4.3 及齐性黎曼空间理论可得

定理 4.4 除欧氏空间外,容许最大参数可迁相似变换群的黎曼空间 V_n 的线素必可化为

$$ds^2=(dx^1)^2+(x^1)^2g_{ab}(x^c)dx^adx^b \quad (a, b, c=1, \cdots, n). \tag{4.1}$$

式中 $g_{ab}(x^c)dx^adx^b$ 是常曲率线素,且这样的空间 V_n 所容许的完全相似群的参数为 $r=\frac{n(n-1)}{2}+1$, 其逆亦真.

利用运动群的第一空隙性定理,我们得到除欧氏空间及次射影空间(4.1)之外,完全相似群的阶数以 $r=\frac{(n-1)(n-2)}{2}+2$ 为最大,并且我们确定了相应的黎曼空间的线素.

我们对黎曼空间的不可迁相似变换群进行了研究,得到

定理 4.5 正定黎曼空间 V_n 如果容许不可迁相似变换群 G_r,则必存在另一黎曼空间 $\overline{V}_n$ 共形于 V_n,而以 G_r 为其不可迁运动群.

由这个定理可知:一变换群如可作为黎曼空间的不可迁相似群必可作为另一黎曼空间的不可迁运动群,反过来显然不成立.但我们有

定理 4.6 n 维空间中具 s 维最小不变流形的 r 参数不可迁变换群 G_r 能作为黎曼空间的不可迁相似变换群的充要条件是 G_r 有 $r-1$ 参数的正常运动子群,其最小不变流形为 $s-1$ 维.

定理 4.7 容许最大阶数不可迁相似群的黎曼空间 V_n(除欧氏空间外)的线素必可

化为

$$ds^2 = e^{2x^1}(h_{ab}(x^1)dx^a dx^b + \sigma(x^1) g_{\alpha\beta}(x^\gamma) dx^\alpha dx^\beta) \tag{4.2}$$

$$(a, b=1, 2; \alpha, \beta, \gamma=3, \cdots, n),$$

式中 $g_{\alpha\beta}(x^\gamma)dx^\alpha dx^\beta$ 是 $n-2$ 维常曲率空间的线素,这样的空间 V_n 所容许的相似群参数个数为 $r=\frac{(n-1)(n-2)}{2}+1$.

定理 4.5 可推广到黎曼空间的不可迁共形变换群去[17],而得到

定理 4.8 任何变换群如果可作为黎曼空间的不可迁共形变换群就必可作为黎曼空间的运动群.

把这个定理与 Ishihara 与 Obata 定理联系起来,就可得出:黎曼空间的共形变换群的研究归结为运动群及欧氏空间可迁共形变换群去,并且由于仿射变换群的研究也归结为相似群的研究,因而从本节的论述看来可以得出结论:正定黎曼空间的相似,共形,仿射变换群和不可迁运动群的研究几乎一概可以归结为齐性黎曼空间的研究去.

参考文献

[1] Cartan, E., Leçons sur la geometrie des espaces de Riemann, 2 edition, 1951, ch. Ⅻ.

[2] Manturov, O. V., Riemannian spaces with orthogonal and symplectic motion groups and an irreducible rotation group, Dokl, Akad, Nauk SSSR 141, 1961, 1034 - 1037.

[3] Manturov O, V., Homogeneous non-symmetric Riemannian spaces with an irreducible rotation group, Dokl. Akad Nauk SSSR, 141 1961, 792 - 795.

[4] Wakakuwa, H., On n-dimensional Riemannian spaces admitting some groups of motions of order less than $\frac{1}{2}n(n-1)$, Tôhoku Math. Journ., 6, 1954, 121 - 134.

[5] 谷超豪:具不变向量场的齐性黎曼空间,复旦学报,1959,第二期,12 - 25;1960,第一期,19 - 37.

[6] Гу Чао-хао(谷超豪), О Некоторых типах однородных Римановых пространств, ДАН СССР, 122, 1958, 171 - 174.

[7] Кручкович, Г. И., Гу Чао-хао (谷超豪), Признак полуприводимости однородных Римановых пространств, ДАН, 120, 1958, 1183 - 1186.

[8] 胡和生：黎曼空间的运动群与迷向群(Ⅰ),数学学报,14,1964,第6期,91—100.

[9] 胡和生：关于齐性黎曼空间的运动群(Ⅰ).迷向群可约的齐性黎曼空间(未发表).

[10] 胡和生：关于齐性黎曼空间的运动群(Ⅱ).完全运动群的空隙性问题(未发表).

[11] Teleman C., Sur les groupes de mouvement maximums des espaces riemanniens V_n. Revue de math. pures et appliquies, 5, 1960, 27 - 57.

[12] Wang, H. C., (王宪钟), On Finsler spaces with complete integrable equations of Killing Journ. London Math. Soc. 22, 1947, 5 - 7.

[13] Егоров, И. П., Римановые пространства второй лакднарности, ДАН 111, 1956, 276 - 279.

[14] Yano, K., On n-dimensional Riemannian space admitting a group of motions of order $\frac{1}{2}n(n-1)+1$. Trans. Amer. Math. Soc. 74, 1953, 260 - 279.

[15] 苏步青：现代微分几何概论,第二章.

[16] 胡和生：黎曼空间的相似变换群,未发表.

[17] 胡和生：论容有不可迁共形变换群的黎曼空间,复旦学报,8,1963,227 - 230.

[18] Рашевский, П. К., О геометрии однородных пространств, Тр. семин. по вект. и тенз. анализу, 9, 1950, 47 - 74.

[19] Yano, K., On groups of homothetic transformations in Riemannian spaces, Journ. of Indian Math. Soc. New series 15 - 17, 1951 - 1953, 105 - 117.

[20] Knebelman, M. S. and Yano, K., On homothetic mappings of Riemann spaces, Proc. of Amer. Math. Soc. 12, 1961, 300 - 303.

[21] Кручкович, Г. И., О движениях в полуприводимых Римановых пространствах, У. М. Н., Ⅶ 1952, 149 - 156.

[22] Ishihara, S. and Obata, M., On the group of conformal transformations of a Riemannian manifold, Proc. Japan. Acad., 31, 1955, 426 - 429.

[23] 谷超豪：齐性空间微分几何学,1964.

(本文曾发表于《数学论文集》,1964年,9 - 19)

关于黎曼空间的运动群与迷向群(I)*

胡和生

复旦大学数学研究所

§1. 引　言

如所知，容许运动群的黎曼空间的决定是十分复杂的一个问题，至今还未完全解决.近年来这问题的研究有了较大的发展，在利用迷向群的性质来决定空间的线素方面取得了不少的成就.例如，在迷向群可约时，Wakakuwa[1]讨论了迷向群分解为作用在两个互补子空间的旋转群的直积且迷向群无不变向量的情形，这时空间必为乘积空间，由此可以把决定线素的问题归结为在维数较低的齐次黎曼空间决定线素的问题.谷超豪[2]研究了迷向群具有不变向量的齐次黎曼空间，并阐明了这种空间的线素具有多种多样的形式.当迷向群在不变向量所成子空间的正交补上为不可约时，他完全解决了这个问题.

在容许可迁运动群的黎曼空间的构成中，迷向群容有不变向量的空间占了一大类，而且在构成容许不可迁运动群的黎曼空间时主要问题也是要作出迷向群具有不变向量场的齐次黎曼空间[3]，所以在研究容许运动群的黎曼空间的问题中，迷向群具不变向量的空间显得重要.为了弄清这样的空间，本文要研究这样的情形，就是迷向群具有不变向量，而且在它所实质作用的子空间中(即不变向量的正交补上)分解为作用在两个互补的子空间的旋转群的直积.我们假定了这两个旋转群中至少有一个是没有交换旋转的，分析了运动群的结构常数，从而完全决定了空间的线素(包括把问题归结为在低维齐次黎

* 1963年9月16日收到，1964年7月23日收到修改稿.

曼空间的决定).对于两个群都有交换旋转的情形将在下文中进行讨论.

§2. Maurer-Cartan 方程

设一个线素正定的 n 维黎曼空间 V_n 容许可迁运动群 G_r，则群 G_r 的可容许直交标形族的两个邻近标形之间的推移关系为

$$dM=\omega^i e_i,\ de_i=\omega_i^j e_j \quad (i,\ j=1,\ \cdots,\ n).$$

已知 ω^i，ω_j^i 都是群的不变形式；ω^i 是 n 个独立的不变形式，与 ω_i^j 中 $r-n$ 个独立的 ω^α $(\alpha=n+1,\ \cdots,\ r)$ 一起作成群 G_r 的全系不变形式系统[2].G_r 的 Maurer-Cartan 方程为

$$\begin{aligned} D\omega^i &= \frac{1}{2}C_{jk}^i[\omega^j\omega^k]+C_{j\rho}^i[\omega^i\omega^\rho], \\ D\omega^\alpha &= \frac{1}{2}C_{jk}^\alpha[\omega^j\omega^k]+C_{j\beta}^\alpha[\omega^j\omega^\beta]+\frac{1}{2}C_{\beta\gamma}^\alpha[\omega^\beta\omega^\gamma] \end{aligned} \tag{1}$$

$$(i,\ j,\ k=1,\ \cdots,\ n;\ \alpha,\ \beta,\ \gamma,\ \rho=n+1,\ \cdots,\ r).$$

于此 C_{jk}^i，$\cdots$，$C_{\beta\gamma}^\alpha$ 为 G_r 的结构常数，它们除了关于下指标的反称性外，还满足下列 Jacobi 恒等式

$$C_{l\rho}^i C_{j\sigma}^l - C_{l\sigma}^i C_{j\rho}^l = C_{\sigma\rho}^\tau C_{j\tau}^i, \tag{2}$$

$$C_{l\tau}^k C_{m\rho}^\tau - C_{m\tau}^k C_{l\rho}^\tau = C_{lm}^i C_{i\rho}^k + C_{il}^k C_{m\rho}^i - C_{im}^k C_{l\rho}^i, \tag{3}$$

$$C_{l\tau}^k C_{mh}^\tau + C_{m\tau}^k C_{hl}^\tau + C_{h\tau}^k C_{lm}^\tau = C_{lm}^i C_{ih}^k + C_{mh}^i C_{il}^k + C_{hl}^i C_{im}^k, \tag{4}$$

并且 $C_{j\rho}^i=-C_{i\rho}^j$ $(\rho=n+1,\ \cdots,\ r)$ 构成迷向群李代数的基[3].

现在我们来研究这样的齐次黎曼空间 V_n，它的运动群 G_r 的迷向群分为直积，分别作用在互补的不变平面 E_q 与 E_{n-q} 上，且迷向群在 E_{n-q} 中没有交换旋转(这时 E_{n-q} 可以是可约的，也可以是不可约的).

不失一般性，可假定迷向群在 E_{n-q} 上没有不变向量.这是因为，假如它有 a 个不变向量*，可将它们放入 E_q 中使成 $q'=q+a$ 维平面 E_{q+a}，而它的互补平面 $E_{n-(q+a)}$ 无不变

* 如 $a>1$，则 E_{n-q} 必有交换旋转.

向量，这样便可对 $E_{q'}$，$E_{n-q'}$ 进行讨论，它们满足假定的条件，且 $E_{n-q'}$ 上无不变向量.

我们选好直交标形 $e_{i'}$（$i'=1, \cdots, q$）和 $e_{i''}$（$i''=q+1, \cdots, n$）分别成为 E_q 和 E_{n-q} 上的基向量.由于迷向群分解为直积，因而它的李代数的基可选为

$$A_{\rho'}=\left\|\begin{matrix} C^{i'}_{j'\rho'} & O \\ O & O \end{matrix}\right\|,\ A_{\rho''}=\left\|\begin{matrix} O & O \\ O & C^{i''}_{j''\rho''} \end{matrix}\right\| \left(\begin{matrix} i', j'=1, \cdots, q \\ i'', j''=q+1, \cdots, n \end{matrix}\right),$$

这里 $A_{\rho'}$，$A_{\rho''}$ 分别为作用在 E_q 及 E_{n-q} 上的迷向群的诱导群的李代数的基，因而

$$C^{i'}_{j'\rho''}=C^{i''}_{j''\rho'}=C^{i'}_{j''\rho}=C^{i''}_{j'\rho}=0, \tag{5}$$

在(3)式中令 $k=k'$, $l=l'$, $m=m''$, $\rho=\rho''$, 得到

$$C^{k'}_{i''l'}C^{i''}_{m''\rho''}=0,$$

由于迷向群在 E_{n-q} 中没有不变的共变向量，

$$C^{k'}_{i''l'}=0. \tag{6}$$

其次，在(3)式中令 $k=k'$, $l=l''$, $m=m''$, $\rho=\rho''$, 得到

$$C^{k'}_{i''l''}C^{i''}_{m''\rho''}+C^{k'}_{m''i''}C^{i''}_{l''\rho''}=0,$$

对固定 k'，可见 $C^{k'}_{m''l''}$ 是 E_{n-q} 中迷向群的不变二阶反称张量.利用[2]中引理 2，可知 $C^{k'}_{m''l''}$ 所生成的单参数旋转与 E_{n-q} 中迷向群中任一旋转可交换，根据假定就必须成立

$$C^{k'}_{m''l''}=0. \tag{7}$$

再在(3)式中令 $k=k''$, $l=l'$, $m=m'$, $\rho=\rho''$, 并利用 E_{n-q} 中没有不变向量就可推出

$$C^{i''}_{l'm'}=0. \tag{8}$$

利用(5)～(8)式，我们可将(1)中第一式改写为

$$\begin{aligned} D\omega^{i'} &=\frac{1}{2}C^{i'}_{j'k'}[\omega^{j'}\omega^{k'}]+C^{i'}_{j'\rho'}[\omega^{j'}\omega^{\rho'}], \\ D\omega^{i''} &=\frac{1}{2}C^{i''}_{j''k'}[\omega^{j''}\omega^{k'}]+\frac{1}{2}C^{i''}_{j''k''}[\omega^{j''}\omega^{k''}]+C^{i''}_{j''\rho''}[\omega^{j''}\omega^{\rho''}]. \end{aligned} \tag{9}$$

为进一步化简上述方程,在(3)中令 $k=k''$, $l=l''$, $m=m'$, $\rho=\rho''$,并利用齐性黎曼空间是化约(Reductive)的这一性质[7],因而可选适当的 ω^i, ω^ρ 使 $C^{\tau}_{i\rho}=0$,因此

$$C^{i''}_{l''m'}C^{k''}_{i''\rho''}-C^{k''}_{i''m'}C^{i''}_{l''\rho''}=0. \tag{10}$$

又易见 $C^{i''}_{l''m'}$ 的关于指标 i'', l'' 的反称部分及对称部分均满足(10),但由假定迷向群在 E_{n-q} 上没有交换旋转,反称部分为 0,所以

$$C^{i''}_{l''m'}=C^{l''}_{i''m'}, \tag{11}$$

依(10)式可知:$C^{i''}_{l''m'}$ 的对应于同一特征值的特征向量集合必为迷向群的不变子空间.

设 E_{n-q} 在迷向群下有 s 个不变子空间 $\left(\text{显然 } 1\leqslant s\leqslant\left[\dfrac{n-q}{2}\right]\right)$,则可选取适当的直交标形使 $C^{i''}_{l''m'}$ 采取形式

$$\| C^{i''}_{l''m'} \| = \begin{Vmatrix} C^{(1)}_{m'}\delta^{i''_1}_{j''_1} & & & O \\ & C^{(2)}_{m'}\delta^{i''_2}_{j''_2} & & \\ & & \ddots & \\ O & & & C^{(s)}_{m'}\delta^{i''_s}_{j''_s} \end{Vmatrix}, \tag{12}$$

这样一来,Maurer-Cartan 方程(9)化简为

$$\begin{aligned}
D\omega^{i'} &= \frac{1}{2}C^{i'}_{j'k'}[\omega^{j'}\omega^{k'}]+C^{i'}_{j'\rho'}[\omega^{j'}\omega^{\rho'}],\\
D\omega^{i''_1} &= \frac{1}{2}C^{i''_1}_{j''k''}[\omega^{j''}\omega^{k''}]+C^{(1)}_{k'}[\omega^{i''_1}\omega^{k'}]+C^{i''_1}_{j''_1\rho''}[\omega^{j''_1}\omega^{\rho''}],\\
D\omega^{i''_2} &= \frac{1}{2}C^{i''_2}_{j''k''}[\omega^{j''}\omega^{k''}]+C^{(2)}_{k'}[\omega^{i''_2}\omega^{k'}]+C^{i''_2}_{j''_2\rho''}[\omega^{j''_2}\omega^{\rho''}],\\
&\cdots\cdots\\
D\omega^{i''_s} &= \frac{1}{2}C^{i''_s}_{j''k''}[\omega^{j''}\omega^{k''}]+C^{(s)}_{k'}[\omega^{i''_s}\omega^{k'}]+C^{i''_s}_{j''_s\rho''}[\omega^{j''_s}\omega^{\rho''}].
\end{aligned} \tag{13}$$

这里 i''_1, i''_2, …, i''_s 是 E_{n-q} 中各个不可约不变子空间对应的指标.

§3. 空间线素的决定

现在决定这类空间的线素，在(4)式中令 $k=k''_1$，$h=h''_1$，$l=l'$，$m=m'$，我们得到

$$C^{k''_1}_{h''_1\tau''}C^{\tau''}_{l'm'}=-C^{i'}_{l'm'}C^{(1)}_{i'}\delta^{k''_1}_{h''_1}.$$

特别取 $k''_1=h''_1$，就成立

$$C^{(1)}_{i'}C^{i}_{l'm'}=0. \tag{14}$$

在(3)式中令 $k=k''_1$，$l=l''_1$，$m=m'$，$\rho=\rho'$ 又得到

$$C^{k''_1}_{i'l''_1}C^{i'}_{m'\rho'}=0,$$

由此

$$C^{(1)}_{i'}C^{i'}_{m'\rho'}=0, \tag{15}$$

因而

$$D(C^{(1)}_{i'}\omega^{i'})=0, \tag{16}$$

我们类似地得到

$$D(C^{(2)}_{i'}\omega^{i'})=0,\ \cdots,\ D(C^{(s)}_{i'}\omega^{i'})=0. \tag{17}$$

设 $C^{(1)}_{i'}$，$C^{(2)}_{i'}$，…，$C^{(s)}_{i'}$ 中有 l 组线性独立的，记为 $C^{(a)}_{i'}(a=1,\cdots,l)$，有 t 组为 0，即 $C^{(\lambda)}_{i'}=0(\lambda=l+1,\cdots,l+t)$，其余 $s-l-t$ 组为 $C^{(a)}_{i'}$ 的线性组合，即 $C^{(u)}_{k}=A^{u}_{a}C^{(a)}_{k}(u=l+t+1,\cdots,s)$，$A^{u}_{a}$ 为常数.

根据(16)，(17)式，这时我们可令 $C^{(a)}_{i'}\omega^{i'}=dx^{a}(a=1,2,\cdots,l)$，因而从(13)得出

$$D\omega^{i''_a}=\frac{1}{2}C^{i''_a}_{j''k''}[\omega^{j''}\omega^{k''}]+[\omega^{i''_a}dx^{a}]+C^{i''_a}_{j''_a\rho''}[\omega^{i''_a}\omega^{\rho''}], \tag{18}$$

$$D\omega^{i''_u}=\frac{1}{2}C^{i''_u}_{j''k''}[\omega^{j''}\omega^{k''}]+A^{u}_{a}[\omega^{i''_u}dx^{a}]+C^{i''_u}_{j''_u\rho''}[\omega^{i''_u}\omega^{\rho''}], \tag{19}$$

$$D\omega^{i''_\lambda}=\frac{1}{2}C^{i''_\lambda}_{j''k''}[\omega^{j''}\omega^{k''}]+C^{i''_\lambda}_{j''_\lambda\rho''}[\omega^{j''_\lambda}\omega^{\rho''}]. \tag{20}$$

由(18),(19),(20)可见,$\omega^{i''}=0$ 为完全可积,设 $x^{i''}$ 是它的一系独立的初积分.又 $\omega^{i''}=0$, $\omega^{\rho''}=0$, $dx^a=0$ 为发甫形式 $\omega^{i''}$ 的特征系统,因此为完全可积.不但如此,$\omega^{i''}=0$, $\omega^{\rho''}=0$ 也是完全可积.事实上,外微分(18),(19),(20)式,左边应为 0,而右边的$[\omega^{j''}, dx^b, dx^c]$的系数为 $C^{i''}_{j''\rho''}C^{\rho''}_{bc}$,因此得

$$C^{i''}_{j''\rho''}C^{\rho''}_{bc}=0,$$

由此得 $C^{\rho''}_{bc}=0$,所以当 $\omega^{j''}=0$, $\omega^{\rho''}=0$ 成立时就有 $D\omega^{\rho''}=0$, $D\omega^{j''}=0$ *.这表明 $\omega^{i''}=0$, $\omega^{\rho''}=0$ 为完全可积,设 $x^{i''}$, $u^{\rho''}$ 为一全系的初积分,由此可见,$\omega^{i''}$ 为 $dx^{i''}$ 的线性组合,$\omega^{\rho''}$ 为$dx^{i''}$和$du^{\rho''}$的线性组合,而且从(18)还可见到,$\omega^{i''_a}$ 中不含 x^b($b=1, 2, \cdots, l$; $b\neq a$).

选 δ_1 使 $\delta_1x^1=1$, $\delta_1x^2=\cdots=\delta_1x^s=0$, $\delta_1x^{i''}=\delta_1u^a=0$,可见 $\omega^{i''_1}(\delta_1)=0$. 又由

$$D\omega^{i''_1}(\delta_1, d)=\delta_1\omega^{i''_1}(d)-d\omega^{i''_1}(\delta_1)=\delta_1\omega^{i''_1}(d)=\frac{\partial}{\partial x'}\omega^{i''_1}(d)$$

及(18)式,就可得

$$\frac{\partial\omega^{i''_1}(d)}{\partial x^1}=-\omega^{i''_1}(d),$$

因而

$$\omega^{i''_1}(d)=e^{-x^1}\omega_0^{i''_1}(d) \text{ 或 } \omega_0^{i''_1}(d)=e^{x'}\omega^{i''_1}(d),$$

这里 $\omega_0^{i''_1}$ 不依赖于 $x^1, \cdots, x^l$.类似地,我们得到

$$\omega_0^{i''_a}(d)=e^{x^a}\omega^{i''_a}(d)\ (a=1, \cdots, l), \tag{21}$$

---

* 这里应用了[6]中一个结论:设 ω^i, ω^ρ 为一组独立的发甫形式,成立

$$D\omega^i=\frac{1}{2}C^i_{jk}[\omega^j\omega^k]+C^i_{j\rho}[\omega^j\omega^\rho],$$

式中 C^i_{jk}, $C^i_{j\rho}$为常数,$C_\rho=(C^i_{j\rho})$ 为一组独立的反称阵,则 $D\omega^\rho$可用ω^i, ω^ρ来表达,即

$$D\omega^\rho=\frac{1}{2}C^\rho_{jk}[\omega^j\omega^k]+C^\rho_{i\sigma}[\omega^i\omega^\sigma]+\frac{1}{2}C^\rho_{\sigma\tau}[\omega^\sigma\omega^\tau].$$

$$\omega_0^{i''_u}(d)=e^{A_a^u x^a}\omega^{i''_u}(d),\tag{22}$$

$\omega_0^{i''_a}$，$\omega_0^{i''_u}$ 只依赖于 $x^{i''}$，$u^{\rho''}$.利用(21)，(22)式及(18)，(19)式计算 $D\omega_0^{i''_a}$，$D\omega_0^{i''_u}$，得到

$$\begin{aligned}D\omega_0^{i''_a}=&\frac{1}{2}\sum_{b,c}e^{(\delta_e^a-\delta_e^b-\delta_e^c)x^e}C^{i''_a}_{j''_b k''_c}[\omega_0^{j''_b}\omega_0^{k''_c}]+\\&\frac{1}{2}\sum_{u,v}e^{(\delta_e^a-A_e^u-A_e^v)x^e}C^{i''_a}_{j''_u j''_v}[\omega_0^{j''_u}\omega_0^{k''_v}]+\\&\sum_{b,u}e^{(\delta_e^a-\delta_e^b-A_e^u)x^e}C^{i''_a}_{j''_b k''_u}[\omega_0^{j''_b}\omega_0^{k''_u}]+C^{i''_a}_{j''_a\rho''}[\omega_0^{j''_a}\omega^{\rho''}],\end{aligned}\tag{23}$$

$$\begin{aligned}D\omega_0^{i''_u}=&\frac{1}{2}\sum_{b,c}e^{(A_a^u-\delta_a^b-\delta_a^c)x^a}C^{i''_u}_{j''_b k''_c}[\omega_0^{j''_b}\omega_0^{k''_c}]+\\&\frac{1}{2}\sum_{v,w}e^{(A_a^u-A_a^v-A_a^w)x^a}C^{i''_u}_{j''_v k''_w}[\omega_0^{j''_v}\omega_0^{k''_w}]+\\&\sum_{b,w}e^{(A_a^u-\delta_a^b-A_a^w)x^a}C^{i''_u}_{j''_b k''_w}[\omega_0^{j''_b}\omega_0^{k''_w}]+C^{i''_u}_{j''_u\rho''}[\omega_0^{j''_u}\omega^{\rho''}],\end{aligned}\tag{24}$$

$$\begin{aligned}D\omega^{i''_\lambda}=&\frac{1}{2}\sum_{v,w}e^{(-A_e^u-A_e^v)x^e}C^{i''_\lambda}_{j''_v k''_w}[\omega_0^{j''_v}\omega_0^{k''_w}]+\\&\sum_{b,u}e^{(-\delta_e^u-A_e^u)x^e}C^{i''_\lambda}_{j''_b k''_u}[\omega_0^{j''_b}\omega_0^{k''_u}]+\\&\frac{1}{2}C^{i''_\lambda}_{j''_\mu k''_\nu}[\omega^{j''_\mu}\omega^{k''_\nu}]+C^{i''_\lambda}_{j''_\lambda\rho''}[\omega^{j''_\lambda}\omega^{\rho''}]\end{aligned}\tag{25}$$

$(a, b, c, e=1, \cdots, l; u, v, w=l+t+1, \cdots, s; \lambda, \mu, \nu=l+1, \cdots, l+t)$. 因为 $\omega_0^{i''_a}$，$\omega_0^{i''_u}$，$\omega^{i''_\lambda}$ 和 $x^1, \cdots, x^l$ 均无关，上列各式的右边不应包含 $x^1, \cdots, x^l$（在计算中我们已经利用这一事实）.所以在(23)—(25)式中形式上虽似含有 $x^1, \cdots, x^l$ 的项，而实际上应该和这些变量无关.考察 $e^{(A_e^u-\delta_e^b-\delta_e^c)x^e}C^{i''_u}_{j''_b k''_c}[\omega_0^{j''_b}\omega_0^{k''_c}]$ 项作为例子，要使它们与 $x^1, \cdots, x^l$ 无关就必须有 $C^{i''_u}_{j''_b k''_c}=0$ 或者 $(A_e^u-\delta_e^b-\delta_e^c)x^e=0$ 从此就看到：$\omega_0^{i''_a}$，$\omega_0^{i''_u}$，$\omega^{i''_\lambda}$ 为完全可积的独立的发甫系统，其微分满足形为

$$D\Omega^{i''}=\frac{1}{2}C^{i''}_{j''k''}[\Omega^{j''}\Omega^{k''}]+C^{i''}_{j''\rho''}[\Omega^{j''}\omega^{\rho''}]\tag{26}$$

的方程，这里已采用记号

$$\Omega^{i''_a}=\omega_0^{i''_a},\ \Omega^{i''_u}=\omega_0^{i''_u},\ \Omega^{i''_\lambda}=\omega^{i''_\lambda}. \tag{27}$$

因此,$\Omega^{i''}$构成一个 $n-q$ 维齐次黎曼空间的不变形式系统.由此可得

定理 1. 如果齐次黎曼空间 V_n的迷向群分解为作用于互补子空间 E_q和 E_{n-q}的旋转群的直积,而且在 E_{n-q}中诱导群没有不变向量,也没有交换旋转,那么 V_n的线素可写成形状

$$ds^2=ds_1^2+\sum_a e^{-2x^a}d\sigma_a^2+\sum_u e^{-A_e^u x^e}d\sigma_u^2+\sum_\lambda d\sigma_\lambda^2. \tag{28}$$

这里 ds_1^2为迷向群具不变向量场的 q 维齐次黎曼空间;其中还存在 l 个独立的数量函数 $x^1,\cdots,x^l$,使在运动群下成立,$\bar{x}^1=x^1+c^1,\cdots,\bar{x}^l=x^l+c^l$;$d\sigma_a^2$, $d\sigma_u^2$, $d\sigma_\lambda^2$是依赖于 $x^{i''}$, $dx^{i''}$的二次微分形式,并且

$$\sum_a d\sigma_a^2+\sum_u d\sigma_u^2+\sum_\lambda d\sigma_\lambda^2 \tag{29}$$

为一 $n-q$ 维齐次黎曼空间的线素,在这空间还可选取适当的不变发甫形式 $\Omega^{i''}$使

$$d\sigma_a^2=\sum_{i''_a}(\Omega^{i''_a})^2,\ d\sigma_u^2=\sum_{i''_u}(\Omega^{i''_u})^2,\ d\sigma_\lambda^2=\sum_{i''_\lambda}(\Omega^{i''_\lambda})^2. \tag{30}$$

根据这定理,便可明了所论的齐次黎曼空间的决定可归结为低维齐次黎曼空间的决定.

附带指出,从 Gartan-Maurer 方程看出:ds_1^2的运动群 G_{r_1}含有 r_1-l 维的正常子群 H,而且 G_{r_1}关于它的商群是阿贝尔群.这里的 H 包含一个定点的安定群.这样的空间以后记为 $V_{q,l}$.

我们现在考察一种特殊情形,即迷向群在 E_{n-q}上为不可约(并且没有交换旋转)的情形.这时,由(13)式得知齐次黎曼空间中的不变形式要满足

$$D\omega^{i'}=\frac{1}{2}C^{i'}_{j'k'}[\omega^{j'}\omega^{k'}]+C^{i'}_{j'\rho'}[\omega^{j'}\omega^{\rho'}],$$

$$D\omega^{i''}=\frac{1}{2}C^{i''}_{j''k''}[\omega^{j''}\omega^{k''}]+C_{k'}[\omega^{i''}\omega^{k'}]+C^{i''}_{j''\rho''}[\omega^{j''}\omega^{\rho''}].$$

如果一切 $C_{k'}=0$，则 V_n 的线素具有形状

$$ds^2=\sum_{i'}(w^{i'})^2+\sum_{i''}(w^{i''})^2=g_{i'j'}(x^{k'})dx^{i'}dx^{j'}+g_{i''j''}(x^{k''})dx^{i''}dx^{j''} \tag{31}$$

$$(i'',\ j'',\ k''=q+1,\ \cdots,\ n;\ i',\ j',\ k'=1,\ \cdots,\ q).$$

如果至少有一个 $C_{k'}\neq 0$，则线素具有形状

$$ds^2=g_{i'j'}(x^{k'})dx^{i'}dx^{j'}+e^{-2x^1}\sum_{i''}(dx^{i''})^2. \tag{32}$$

因此得到

推论. 如果齐次黎曼空间 V_n 的迷向群有两个互补不变平面 E_q 与 E_{n-q}，迷向群分为作用在这两平面上的旋转群的直积，且在 E_{n-q} 上的诱导群为不可约而又无交换旋转，则 V_n 是乘积空间或者是半可约空间，而且它们的线素具有形状(31)或(32).

这就是谷超豪与 Кручкович 的结果[4].

§4. V_n 的线素的具体分析

为将空间决定得更具体些，我们对线素(28)再区分为下列几种情况进一步加以分析.

(i) $l=0$ 的情况，因此 $p=s-l-t=0$，$t\neq 0$. 这时空间显然是乘积空间，线素为

$$ds^2=ds_1^2+ds_2^2, \tag{33}$$

而 ds_1^2 所对应的黎曼空间容许以 $\omega^{i'}$，$\omega^{\rho'}$ 为不变形式的运动群 $G^{(1)}$，ds_2^2 所对应的黎曼空间容许以 $\omega^{i''}$，$\omega^{\rho''}$ 为不变形式的运动群 $G^{(2)}$，群 G_r 分为直积 $G_r=G^{(1)}xG^{(2)}$，其方程具有形状

$$\begin{aligned}\bar{x}^{i'}&=\varphi^{i'}(x^{j'},\ c^{j'},\ c^{\rho'}),\\ \bar{x}^{i''}&=\varphi^{i''}(x^{j''},\ c^{j''},\ c^{\rho''}).\end{aligned} \tag{34}$$

(ii) $l\neq 0$，$t=0$，$p=0$ 的情形.这时(23)式化为

$$D\omega_0^{i''_a}=C^{i''_a}_{j''_a\rho''}[\omega_0^{j''_a}\omega^{\rho''}]\quad(a=1,\cdots,s).$$

因此 $d\sigma_a^2=\sum\limits_{i''_a}(\omega_0^{i''_a})^2$ 是相互无关的 s 个齐次空间的线素.由于它们都容许相似群,$d\sigma_a^2$ 必为欧氏线素,所以 V_n 的线素为

$$ds^2=g_{i'j'}(x^{k'})dx^{i'}dx^{j'}+e^{-2x^1}\sum_{i''_1}(dx^{i''_1})^2+\cdots+e^{-2x^s}\sum_{i''_s}(dx^{i''_s})^2,\tag{35}$$

并且 $g_{i'j'}(x^{k'})dx^{i'}dx^{j'}$ 为上节所述空间 $V_{q,s}$ 的线素.群 G_r 的方程为

$$\begin{aligned}
&\bar{x}^a=x^a+c^a\quad(a=1,\cdots,s),\\
&\bar{x}^\rho=\varphi^\rho(x^{j'},c^{j'},c^{\rho'})\quad(p=s+1,\cdots,q),\\
&\bar{x}^{i''_1}=e^{c^1}(a^{i''_1}_{j''_1}x^{j''_1}+c^{i''_1}),\\
&\cdots\cdots\\
&\bar{x}^{i''_s}=e^{c^s}(a^{i''_s}_{j''_s}x^{j''_s}+c^{i''_s}).
\end{aligned}\tag{36}$$

相反地,如果给定一个 q 维齐次黎曼空间 $V_{q,s}$,那么必有不变形式 ω^a, ω^p, $\omega^{\rho'}$,使

$$D\omega^a=0,\ D\omega^p=c^p_{i'j'}[\omega^{i'}\omega^{j'}]+c^p_{h\rho'}[\omega^h\omega^{\rho'}]\ (p,\ h=s+1,\cdots,q).$$

令 $\omega^a=dx^a$,作起(35)式,其中 $g_{i'j'}(x^{k'})dx^{i'}dx^{j'}$ 为 $V_{q,s}$ 的线素,那么就得出这一类型的齐次黎曼空间,在这一情形中至少有 s 个独立的不变向量场.

(iii) $l\neq 0$, $p=0$, $t\neq 0$ 的情况.这时

$$D\omega_0^{i''_a}=C^{i''_a}_{j''_a\rho''}[\omega_0^{j''_a}\omega^{\rho''}],$$

$$D\omega^{i''_\lambda}=\frac{1}{2}C^{i''_\lambda}_{j''_\mu k''_\nu}[\omega^{j''_\mu}\omega^{k''_\nu}]+C^{i''_\lambda}_{j''_\lambda\rho''}[\omega^{j''_\lambda}\omega^{\rho''}].$$

从而

$$ds^2=ds_1^2+\sum_{a=1}^{l}e^{-2x^a}d\sigma_a^2+ds_2^2.\tag{37}$$

这里 $d\sigma_a^2$ 是欧氏线素 $\sum\limits_{i''_a}(dx^{i''_a})^2$, ds_1^2 是 $V_{q,l}$ 的线素,ds_2^2 为另一没有什么特别限制的齐

次黎曼空间的线素.群 G_r 的方程为

$$\begin{aligned}
&\bar{x}^{a}=x^{a}+c^{a},\\
&\bar{x}^{p}=\varphi^{p}(x^{j'},\ c^{j'},\ c^{\rho'}),\\
&\bar{x}^{i''_a}=e^{c^a}(a^{i''_a}_{j''_a}x^{j''_a}+c^{i''_a}),\\
&\bar{x}^{i''_\lambda}=f^{i''_\lambda}(x^{j''_\lambda},\ c^{i''},\ c^{\rho''}).
\end{aligned}\tag{38}$$

相反地,给定一个 q 维齐次黎曼空间 $V_{q,l}$ 和齐次黎曼空间 ds_2^2 时,就可作出这一类型的齐次黎曼空间 V_n.

(iv) $l\neq 0$, $p\neq 0$,但(23),(24),(25)出现的关于 x^l 的一次式

$$\begin{gathered}
(A^u_e-\delta^b_e-\delta^c_e)x^e,\ (A^u_e-A^v_e-A^w_e)x^e,\ (A^u_e-\delta^b_e-A^w_e)x^e,\ (A^u_e+A^v_e)x^e,\\
(\delta^u_e+A^u_e)x^e,\ (\delta^a_e-A^u_e-A^v_e)x^e,\ (\delta^a_e-\delta^b_e-A^u_e)x^e
\end{gathered}\tag{39}$$

均不恒等于 0 的情况.我们有

$$\begin{aligned}
D\omega_0^{i''_a}&=C^{i''_a}_{j''_a\rho''}[\omega_0^{j''_a}\omega^{\rho''}],\\
D\omega_0^{i''_u}&=C^{i''_u}_{j''_u\rho''}[\omega_0^{j''_u}\omega^{\rho''}],\\
D\omega^{i''_\lambda}&=\frac{1}{2}C^{i''_\lambda}_{j''_\mu k''_\nu}[\omega^{j''_\mu}\omega^{k''_\nu}]+C^{i''_\lambda}_{j''_\lambda\rho''}[\omega^{j''_\lambda}\omega^{\rho''}].
\end{aligned}$$

V_n 的线素是

$$ds^2=ds_1^2+\sum_a e^{-2x^a}d\sigma_a^2+\sum_u e^{-A^u_e x^e}d\sigma_u^2+ds_2^2,\tag{40}$$

式中 $d\sigma_a^2$, $d\sigma_u^2$ 均为欧氏线素,即 $d\sigma_a^2=\sum_{i''_a}(dx^{i''_a})^2$, $d\sigma_u^2=\sum_{i''_u}(dx^{i''_u})^2$. 这是因为它们都是容有相似变换的齐次黎曼空间的线素的缘故.ds_2^2 为另一黎曼空间的线素,ds_1^2 是 $V_{q,l}$ 的线素,这时运动群的形状为

$$\begin{aligned}
&\bar{x}^{a}=x^{a}+c^{a},\\
&\bar{x}^{p}=\varphi^{p}(x^{j'},\ c^{j'}),\\
&\bar{x}^{i''_a}=e^{c^a x^a}(a^{i''_a}_{j''_a}x^{j''_a}+c^{i''_a}),
\end{aligned}\tag{41}$$

$$\bar{x}^{i''_u} = e^{A^u_e x^e} (a^{i''_u}_{j''_u} x^{j''_u} + c^{i''_u}),$$

$$\bar{x}^{i''_\lambda} = f^{i''_\lambda}(x^{j''_\lambda},\ c^{i''},\ c^{\rho''}).$$

相反地,给定 $V_{q,\,l}$ 的线素和任一齐次黎曼空间的线素 ds_2^2 时,就可作出这样的空间 V_n.特别,当 $t=0$ 时,ds_2^2 并不出现.

定理 2. 除了在(39)中各一次式有恒等于 0 的情形以外,定理 1 中的线素(28)必采取形状(33),(35),(37)或(40)的形状,而且运动群相应地采取(34),(36),(38)或(41).

这样,除了若干例外的情形外,我们已经把空间的线素很清楚地确定出来.

§5. 不可迁的情形

由于容许不可迁运动群的黎曼空间 V_n 的线素可由其最小不变流形上的线素(它容许可迁运动群)和这个流形上的不变的发甫形式的系数(即不变的共变向量场)所决定,因此,上述的结果立即可以用来建造一类容许不可迁运动群的黎曼空间,并写出它的线素,这里我们不详细来叙述.而特别把 §3 的推论扩充到不可迁群的情形.

定理 3. 如果容许不可迁运动群的黎曼空间 V_n 的迷向群有两个互补的不变平面 E_q 及 E_{n-q},迷向群分解为作用在这两平面上的旋转群的直积,而且在 E_{n-q} 上是不可约的,又无交换旋转,则空间 V_n 是半可约的,并且其线素具有形状

$$ds^2 = g_{\alpha\beta}(x^\gamma)dx^\alpha dx^\beta + c(x^a)g_{i''j''}(x^{k''})dx^{i''}dx^{j''}, \tag{42}$$

或

$$ds^2 = g_{\alpha\beta}(x^\gamma)dx^\alpha dx^\beta + c(x^a)e^{-2x^{k+1}} \sum_{i''=q+1}^{n} (dx^{i''})^2, \tag{43}$$

式中 $a=1,\ 2,\ \cdots,\ k$;$\alpha,\ \beta,\ \gamma=1,\ 2,\ \cdots,\ q$;$i'',\ j'',\ k''=q+1,\ \cdots,\ n$,$n-k$ 为最小不变流形的维数.

证. 由 §3 的推论可知,最小不变流形 V_{n-k}($x^a=c^a$,c^a 为常数)的线素可写成

$$ds_1^2 = g_{i'j'}(x^{k'},\ c^a)dx^{i'}dx^{j'} + c(x^a)g_{i''j''}(x^{k''})dx^{i''}dx^{j''} \tag{44}$$

$$(i',\ j',\ k'=k+1,\ \cdots,\ q;\ i'',\ j'',\ k''=q+1,\ \cdots,\ n;\ a=1,\ 2,\ \cdots,\ q)$$

或

$$ds_1^2=g_{i'j'}(x^{k'},\ c^a)dx^{i'}dx^{j'}+c(x^a)e^{-2x^{k+1}}\sum_{i''=q+1}^{n}(dx^{i''})^2, \tag{45}$$

依假定，迷向群有不变平面 E_q 及 E_{n-q}，且在 E_{n-q} 上为不可约的，因而迷向群的不变向量必在 E_q 上，设 G_r 在最小不变流形上独立的共变向量场共有 l 个：$\lambda_i^p(x^j)(p=k+1,\ \cdots,\ k+l;\ i,\ j=k+1,\ \cdots,\ n)$，则 $\lambda_i^p dx^i$ 为 G_r 的 l 个独立的不变形式，可将它取为 $\omega^{k+1},\ \cdots,\ \omega^{k+l}$ 这时成立

$$D\omega^p=C_{j'k'}^p[\omega^{j'}\omega^{k'}](p=k+1,\ \cdots,\ k+l).$$

从而得到 $\omega^p=\lambda_{i'}^p(x^{j'})dx^{i'}$，即 λ_i^p 中只有 $\lambda_{i'}^p$ 不等于 0，且只依赖于 $x^{j'}$.利用这个事实及(44)，(45)式，并根据[3]可知，空间 V_n 的线素为

$$ds^2=g_{ab}(x^c)dx^a dx^b+c_{ap}(x^b)\lambda_{i'}^p(x^{j'})dx^a dx^{i'}+$$
$$g_{i'j'}(x^a)dx^{i'}dx^{j'}+c(x^a)g_{i''j''}(x^{k''})dx^{i''}dx^{j''}$$

或

$$ds^2=g_{ab}(x^c)dx^a dx^b+c_{ap}(x^b)\lambda_{i'}^p(x^{j'})dx^a dx^{i'}+$$
$$g_{i'j'}(x^a)dx^{i'}dx^{j'}+c(x^a)e^{-2x^{k+1}}\sum_{i''}(dx^{i''})^2$$
$$(p=k+1,\ \cdots,\ k+l),$$

这样，我们就得到型如(42)或(43)的线素.

最后，我们指出这个定理与 Teleman 在[5]文中定理 1 的联系. Teleman 是在 V_n 的运动群具有不变平面 E_q，$E_{n-q}\left(n-q>\frac{n}{2}\right)$ 而且迷向群在 E_{n-q} 上的诱导群的全直交群的假定下证明了 V_n 是乘积空间或半可约空间.事实上容易看出：这时迷向群分解为直积，并且由于迷向群在 E_{n-q} 上诱导群是全直交群，它无交换旋转，因而 Teleman 的定理实际上是定理 3 和上节推论的特殊情形.

参 考 文 献

[1] Wakakuwa, H., On n-dimensional Riemannian spaces admitting some gronps of motions of order less than $\frac{\lambda}{2}$n(n—1). *Tôhoku Math. Journ.* (2), **6**(1954), 121-134.

[2] 谷超豪,具不变向量场的齐次黎曼空间,复旦学报(1959)第二期,12-25;(1960)第一期,19-37.

[3] Cartan, E., Leçons sur la geometrie des espaces de Riemann 2 edition, (1951), ch Ⅻ.

[4] Кручкович Г.И., Гу Чао-хао, Признак полуприводимости однородных римановых пространств, *ДАН*, **120**(1958), 1183-1186.

[5] Teleman C., Sur les groups de mouvement maximums des espaces riemanniens Vn, *Revue de math. pures et appliquies*, **5**(1960), 27-57.

[6] 谷超豪,齐性空间几何学.

[7] Рашевский П.К., О геометрии однородных пространств, *Труды Сет. по Век. п Тенз. анализу*, **9** (1952), 49-74.

(本文发表于《数学学报》1964 年第 14 卷第 6 期,828-836)

On the Lacunae of Complete Groups of Motions of Homogeneous Riemannian Spaces

Hu Hou-sung (胡和生)

Department of Mathematics, Futan University, Shanghai,

It is well known that a group of motions in an n-dimensional Riemannian space V_n attains the maximum order $n(n+1)/2$ when and only when V_n is of constant curvature. Wang Hsien-chung[1] first demonstrated that in a V_n for $n \neq 4$, there does not exist a group of motions of order r such that

$$\frac{1}{2}n(n+1) > r > \frac{1}{2}n(n-1)+1.$$

This gives rise to the first lacuna of groups of motions. Afterwards K. Yano[2] determined V_n's admitting a group of motions of order $\frac{1}{2}n(n-1)+1$. As the second lacuna Egorov[3] obtained

$$\frac{1}{2}n(n-1)+1 > r > \frac{1}{2}(n-1)(n-2)+3$$

for non-Einstein spaces.

In attempting to find more lacunae of complete groups of motions in homogeneous Riemannian spaces, we have established the following

Theorem. *Suppose that in a homogeneous Riemannian space* $V_n (n > 4,\ n \neq 6)$

the linear group of isotropy $G(p)$ *of a group of motion* G_r *has two complementary invariant planes of dimensions a and* $n-q$ *respectively such that the induced groups of* $G(p)$ *in these planes are orthogonal groups* $O(q)$ *and* $O(n-q)$ *and* $G(p) \neq O(q) \times O(n-q)$; *then* i) $q=n-q$ *for* $n \neq 7$ *and* V_n *becomes a Euclidean space* E_n, *and* ii) $q=3, 4$ for $n=7$ *and* V_7 *is* E_7 *or the direct product of* E_4 *and* S_3 *of positive constant curvature, and vice versa.*

In virtue of some theorems due to Teleman[4], Gu Chao-hao[5,6], and Wakakuwa[7], we have reached an available method of obtaining lacunae of complete groups of motions in homogeneous Riemannian spaces. In fact, eight lacunae have been found with fourteen associate matrices of spaces admitting complete groups of motions of a greater order. In a similar way we can further give some more lacunae. The results are summarized in the following list, in which we adopt the notations:

E_n: n-dimensional Euclidean space.

S_n: n-dimensional space of constant curvature.

$S_n(k<0)$: n-dimensional space of negative constant curvature.

P_n: n-dimensional particular subprojective space with the metric $ds^2 = (dx^1)^2 + g_{\alpha\beta}(x^\gamma)dx^\alpha dx^\beta (\alpha, \beta, \gamma = 2, \cdots, n)$ where $g_{\alpha\beta}(x^\gamma)dx^\alpha dx^\beta$ is a metric of S_{n-1}.

g_n: n-dimensional Riemannian space whose complete group of motions is a simply transitive group. In particular, when the group of motions has an $(n-1)$-dimensional invariant subgroup the space g_n is denoted by g'_n.

c: constant.

Every metric in the list may be multiplied by a constant factor.

Homogeneous Riemannian Spaces or Metric Forms of the Spaces	Order of Complete Group of Motion	Lacuna
1. S_n	$\frac{n(n+1)}{2}$	1st lacuna
2. $P_n = S_{n-1} \times E_1$ $(S_{n-1} \neq E_{n-1})$.	$\frac{n(n-1)}{2} + 1\ (n > 4)$	2nd lacuna
3. $S_{n-2} \times S_2$ $(S_{n-2} \neq E_{n-2}$ or $S_2 \neq E_2)$.	$\frac{(n-1)(n-2)}{2} + 3\ (n > 8)$	
4. A kind of semiproduct space of $S_2\ (k < 0)$ and E_{n-2} whose metric is $ds^2 = (dx^1)^2 + e^{-2cx^1}(dx^2)^2 + e^{-2x^1}\sum_{a=3}^{n}(dx^a)^2\ (c \neq 0,\ 1)$.	$\frac{(n-1)(n-2)}{2} + 2\ (n > 8)$	3rd lacuna
5. $S_{n-3} \times S_3$ $(S_{n-3} \neq E_{n-3}$ or $S_3 \neq E_3)$.	$\frac{(n-2)(n-3)}{2} + 6\ (n > 10)$	
6. a. $S_{n-3} \times P_3 = S_{n-3} \times S_2 \times E_1$ $(S_{n-3} \neq E_{n-3},\ S_2 \neq E_2)$. b. $ds^2 = (dx^1)^2 + e^{-2cx^1}\sum_{a=2}^{3}(dx^a)^2 + e^{-2x^1}\sum_{a=4}^{n}(dx^a)^2\ (c \neq 0,\ 1)$. c. $V_3 \times S_{n-3}$ the metric of V_3 is $ds_0^2 = \left\{dx^1 + A\frac{x^2dx^3 - x^3dx^2}{1+\frac{K}{4}[(x^2)^2+(x^3)^2]}\right\}^2 + \frac{(dx^2)^2+(dx^3)^2}{\left\{1+\frac{K}{4}[(x^2)^2+(x^3)^2]\right\}^2}$. A, K are constants $(A \neq 0)$.	$\frac{(n-2)(n-3)}{2} + 4\ (n > 10)$	4th lacuna

(continued)

Homogeneous Riemannian Spaces or Metric Forms of the Spaces	Order of Complete Group of Motion	Lacuna
7. a. $S_{n-3} \times g_3$ (does not admit mixed motion). b. $ds^2 = g_{ab}(x^c)dx^a dx^b + e^{-2x1}\sum_{a=4}^{n}(dx^a)^2$; $a, b, c = 1, 2, 3$ $g_{ab}(x^c)dx^a dx^b$ is the metric of g'_3 (does not admit mixed motion).	$\frac{(n-2)(n-3)}{2}+3\ (n>12)$	5th lacuna
8. $S_{n-4} \times S_4$ ($S_{n-4} \neq E_{n-4}$ or $S_4 \neq E_4$).	$\frac{(n-3)(n-4)}{2}+10(n>12)$	
9. a. $S_{n-4} \times S_3 \times E_1$ ($S_{n-4} \neq E_{n-4}$, $S_3 \neq E_3$). b. $ds^2 = (dx^1)^2 + e^{-2cx1}\sum_{a=2}^{4}(dx^a)^2 +$ $e^{-2x1}\sum_{a=5}^{n}(dx^a)^2$ ($c \neq 0, 1$).	$\frac{(n-3)(n-4)}{2}+7\ (n>14)$	6th lacuna
10. $S_{n-4} \times S_2 \times S_2$ at least two of S_{n-4}, S_2, S_2 are not Euclidean.	$\frac{(n-3)(n-4)}{2}+6\ (n>14)$	
11. a. $S_{n-4} \times V_4$. b. $ds^2 = ds_0^2 + e^{-2x1}\sum_{a=5}^{n}(dx^a)^2$, V_4 are some of 4-dimensional Riemannian spaces whose linear group of isotropy is $O(2) \times O(1) \times O(1)$, ds_0^2 is the metric of V_4. c. $ds^2 = 4e^{-2x3}(dx^1 + x^2dx^4 - x^4dx^2)^2 +$ $e^{2x3}(dx^2)^2 + (dx^3)^2 +$ $4e^{-4x3}(dx^4)^2$.	$\frac{(n-3)(n-4)}{2}+5\ (n>14)$	

(continued)

Homogeneous Riemannian Spaces or Metric Forms of the Spaces	Order of Complete Group of Motion	Lacuna
12. a. $S_{n-4} \times g_4$ ($S_{n-4} \neq E_{n-4}$ does not admit mixed motion). b. $ds^2 = g_{ab}(x^c)dx^a dx^b + e^{-2x^1} \sum_{a=5}^{n} (dx^a)^2$, $g_{ab}(x^c)dx^a dx^b$ is the metric of g'_4 (does not admit mixed motion) $a, b, c = 1, \cdots, 4$.	$\frac{(n-3)(n-4)}{2} + 4 \ (n > 14)$	7th lacuna ($n > 15$)
13. $S_{n-5} \times S_5$ ($S_{n-5} \neq E_{n-5}$ or $S_5 \neq E_5$).	$\frac{(n-4)(n-5)}{2} + 15 \ (n > 14)$	
14. a. $S_{n-5} \times S_4 \times E_1$ ($S_{n-5} \neq E_{n-5}$, $S_4 \neq E_4$). b. $ds^2 = (dx^1)^2 + e^{-2cx^1} \sum_{a=2}^{5} (dx^a)^2 +$ $e^{-2x^1} \sum_{a=6}^{n} (dx^a)^2 \ (c \neq 0, 1)$.	$\frac{(n-4)(n-5)}{2} + 11 \ (n > 16)$	8th lacuna

July 16, 1964

References

[1] Wang, H. C. 1947 *J. London Math. Soc.*, **22**, 5-9.

[2] Yano, K. 1953 *Trans. Amer. Math. Soc.*, **74**, 260-279.

[3] Егоров И. П. 1956 *ДАН*, **111**, 276-279.

[4] Teleman, C. 1960 *Revue de math. pures et appliquies*, **5**, 27-57.

[5] Кручкович Г. И. и Гу Чао-хао 1958 *ДАН*, **120**, 1183-1186.

[6] Гу Чао-хао 1958 *ДАН*, **122**, 171-174.

[7] Wakakuwa, H. 1954 *Tohoku Math. Journ.*, **6**, 121-134.

(本文曾发表于 *Scientia Sinica* (*Notes*), 1965, 14(1), 134-136)

球对称引力场方程的严格解

胡和生

提　　要

压力忽略不计时(即 $p=0$ 时)球对称引力场方程的一般解,过去虽先后有不少人研究过,但都没有做完全,有各自的遗漏,更主要的缺点是他们得出的都只是度规的系数所满足的隐函数方程,并没有得出度规的明显表达式,因而不利于应用它们去讨论物理学上的问题.本文改善了这种情况:① 给出了 $p=0$ 球对称引力场方程的全部严格解的明显表达式,这些解具有三种类型及一种例外情况.② 确定了这些解都有奇点,对奇点类型进行分类,分析了奇点的性质,给出初步的物理解释.

一、引　　言

引力场方程

$$R_{ij}-\frac{R}{2}g_{ij}=\frac{8\pi G}{c^4}T_{ij}\quad (i,\ j=0,\ 1,\ 2,\ 3)$$

已知具有物理意义的解不多,因而需要多作出一些解,以适应分析问题的需要.

在相对论天体物理中,常用到压力忽略不计时(即 $p=0$ 时)有物质分布的球对称引力场方程的严格解,例如 Oppenheimer[1] 等研究引力坍缩就利用 Tolman 所获得的解.虽然 $p=0$ 这个模型是过于简单化的,但一般认为,从今天来看在某些天体问题中也还是一个可利用的近似.在一定条件下,有人通过数值计算发现,相应于 $p=0$ 与 $p=p(\rho)$ (ρ 为密度)的两组解之间定性上有相似之处[2].因而,求出 $p=0$ 时的严格解对分析问题是有利的.

如所知,常用的 Tolman 解只是 $p=0$ 的球对称解中特殊的一类. $p=0$ 时球对称引力场方程的一般解虽然先后有人进行过讨论[3-5],但是,都没有做完全,都有各自的遗漏.更主要的缺点是,他们所得出的解都是用隐函数表达的,实际上只列举了度规的系数所满足的一套隐函数方程,没有能够写出度规的明显表达式,这就不利于讨论物理学上的问题,所以在最近的文献(如[6],[7])中,虽然有的也提到 Bondi 的结果[4],但实际上加以利用的,还只能是 Tolmam 解.

本文的目的就是想改善这些情况,主要结果是:

给出 $p=0$ 时有物质分布的球对称引力场方程的全部严格解的明显表达式.这些解具有三种类型与一种例外情况.从这些解中可见到每种解均有奇点,并对奇点的类型进行分类,对奇点的性质进行一些分析,给出物理解释.

二、$p=0$ 球对称引力场方程严格解的三种类型

我们采用随动坐标系,且选取适当单位,使 $c=G=1$,这时度规为[1]

$$ds^2=d\tau^2-e^{\bar{\omega}}dR^2-e^{\omega}(d\theta^2+\sin^2\theta d\varphi^2), \tag{1.1}$$

其中 ω, $\bar{\omega}$ 是 τ, R 的函数,又能量动量张量为

$$T_0^0=\rho,\text{其余 } T_j^i=0 \quad (i,\ j=0,\ 1,\ 2,\ 3). \tag{1.2}$$

Einstein 场方程为

$$8\pi T_1^1=0=e^{-\omega}-e^{-\bar{\omega}}\frac{\omega'^2}{4}+\ddot{\omega}+\frac{3}{4}\dot{\omega}^2, \tag{1.3}$$

$$8\pi T_2^2=8\pi T_3^3=0=-e^{-\bar{\omega}}\left(\frac{\omega''}{2}+\frac{\omega'^2}{4}-\frac{\bar{\omega}'\omega'}{4}\right)+\frac{\ddot{\bar{\omega}}}{2}+\frac{\dot{\bar{\omega}}^2}{4}+\frac{\ddot{\omega}}{2}+\frac{\dot{\omega}^2}{4}+\frac{\dot{\bar{\omega}}\dot{\omega}}{4}, \tag{1.4}$$

$$8\pi T_4^4 = 8\pi\rho = e^{-\omega} - e^{-\bar{\omega}}\left(\frac{3}{4}\omega'^2 + \omega'' - \frac{\omega'\bar{\omega}'}{2}\right) + \frac{\dot{\omega}^2}{4} + \frac{\dot{\bar{\omega}}\dot{\omega}}{2}, \tag{1.5}$$

$$8\pi e^{\bar{\omega}} T_4^1 = -8\pi T_1^4 = 0 = \frac{\omega'\dot{\omega}}{2} - \frac{\dot{\bar{\omega}}\omega'}{2} + \dot{\omega}', \tag{1.6}$$

这里 $\omega' = \frac{\partial\omega}{\partial R}$, $\dot{\omega} = \frac{\partial\omega}{\partial\tau}$,在 $\omega' \neq 0$ 的假定下,Tolman 已作出最后一式的积分

$$e^{\bar{\omega}} = e^{\omega}\frac{\omega'^2}{4f^2(R)}, \tag{1.7}$$

其中 $f^2(R)$为 R 的任意函数,且 $f(R) > 0$. 所谓 Tolman 解就是在 $f=1$ 时所得出的度规.将(1.7)式代入(1.3)式得到

$$e^{\omega}\left(\ddot{\omega} + \frac{3}{4}\dot{\omega}^2\right) + (1 - f^2) = 0. \tag{1.8}$$

记 $1 - f^2 = \lambda$, 又令

$$e^{-\frac{\omega}{2}} d\tau = d\sigma \tag{1.9}$$

以引入新的自变量 σ.经过这样一个依赖于解的变换以后,方程(1.8)化为

$$\frac{d^2\omega}{d\sigma^2} + \frac{1}{4}\left(\frac{d\omega}{d\sigma}\right)^2 + \lambda = 0, \tag{1.10}$$

这时 R 是作为参数进入方程的,这个方程可以通过求积分解出,但解的形式和 λ 的正或负或零有关.

(1) $\lambda > 0$ 的解

记 $4\lambda = \alpha^2$, 由(1.10)通过两次积分可得出

$$e^{\omega} = G^2\cos^4\frac{\alpha}{4}(\sigma - F), \tag{1.11}$$

这里 G, F 为 R 的任意函数,再解方程(1.9)可得

$$\frac{\alpha}{2}(\sigma - F) + \sin\frac{\alpha}{2}(\sigma - F) = \frac{\alpha}{G}\tau + H,$$

H 为 R 的任意函数,为方便计,记

$$x+\sin x=y$$

的逆函数为 $x=\varphi(y)$,由此 $\varphi'(y)=\frac{1}{2}\sec^2\left[\frac{1}{2}\varphi(y)\right]$,$\varphi(y)$为 y 的单调增加函数.利用这个记号得出

$$e^{\omega}=G^2\cos^4\left[\frac{1}{2}\varphi\left(\frac{\alpha}{G}\tau+H\right)\right], \tag{1.12}$$

再计算 $e^{\bar{\omega}}$ 得

$$e^{\bar{\omega}}=\frac{G^2\cos^4\left[\frac{1}{2}\varphi\left(\frac{\alpha}{G}\tau+H\right)\right]\left\{\frac{2G'}{G}-\mathrm{tg}\left[\frac{1}{2}\varphi\left(\frac{\alpha}{G}\tau+H\right)\right]\sec^2\left[\frac{1}{2}\varphi\left(\frac{\alpha}{G}\tau+H\right)\right]\left[\left(\frac{\alpha}{G}\right)'\tau+H'\right]\right\}^2}{4-\alpha^2}, \tag{1.13}$$

这样我们就得出了空间度规的明显表达式.

现在来求 ρ 的表达式,把(1.8)式乘以 $e^{\frac{1}{2}\omega}\dot{\omega}$,积分后得到

$$\frac{1}{2}e^{\frac{3}{2}\omega}\dot{\omega}^2=2(f^2-1)e^{\frac{\omega}{2}}+2f_1(R),$$

为计算 $f_1(R)$的表达式,利用 e^{ω}的表达式(1.12)求出

$$\dot{\omega}=-\mathrm{tg}\left[\frac{1}{2}\varphi\left(\frac{\alpha}{G}\tau+H\right)\right]\sec^2\left[\frac{1}{2}\varphi\left(\frac{\alpha}{G}\tau+H\right)\right]\frac{\alpha}{G},$$

代入前面式内可得出 $f_1(R)=\frac{\alpha^2}{4}G$.

再将这些结果代入(1.5),又利用(1.7)就可得到 ρ 的表达式为

$$\rho=\frac{1}{16\pi}\frac{(G\alpha^2)'}{\omega' e^{\frac{3}{2}\omega}}\quad(\alpha^2=4\lambda). \tag{1.14}$$

(2) $\lambda<0$ 的解

我们记 $4\lambda^2=-\alpha^2$，用类似的计算可得到场方程的解

$$e^{\omega}=G^2 sh^4\left[\frac{1}{2}\psi\left(\frac{\alpha}{G}\tau+H\right)\right], \tag{1.15}$$

$$e^{\bar{\omega}}=\frac{G^2 sh^4\left[\frac{1}{2}\psi\left(\frac{\alpha}{G}\tau+H\right)\right]\left\{\frac{2G'}{G}+cth\left[\frac{1}{2}\psi\left(\frac{\alpha}{G}\tau+H\right)\right]sh^{-2}\left[\frac{1}{2}\psi\left(\frac{\alpha}{G}\tau+H\right)\right]\left[\left(\frac{\alpha}{G}\right)'\tau+H'\right]\right\}^2}{4+\alpha^2}, \tag{1.16}$$

这里 $\psi(y)$是 $y=-x+sh\,x$ 的逆函数，$\psi(y)$是 y 的单调函数，G，H 均为R 的任意函数，并且 ρ 的表达式仍为

$$\rho=\frac{1}{16\pi}\frac{(G\alpha^2)'}{\omega' e^{\frac{3}{2}\omega}} \quad (\alpha^2=-4\lambda). \tag{1.17}$$

(3) $\lambda=0$ 即相应于 $f=1$，就是 Tolman 所解出的：

$$e^{\omega}=(A\tau+B)^{\frac{4}{3}}, \tag{1.18}$$

$$e^{\bar{\omega}}=\frac{4}{9}(A\tau+B)^{-\frac{2}{3}}(A'\tau+B')^2, \tag{1.19}$$

A，B 为R 的任意函数.可以直接验证这三组解都满足(1.4)式.

三、例外情形：$\omega'=\dfrac{\partial\omega}{\partial R}=0$ 时的解

前面三种类型都是在 $\omega'\neq 0$ 时得出的，而当 $\omega'=0$ 时的解具有它的特殊形式.这时 $\omega=\omega(t)$，(1.6)式自然成立，而(1.3)、(1.4)、(1.5)分别化为

$$\ddot{\omega}+\frac{3}{4}\dot{\omega}^2+e^{-\omega}=0, \tag{2.1}$$

$$\ddot{\bar{\omega}}+\ddot{\omega}+\frac{1}{2}(\dot{\bar{\omega}}^2+\dot{\omega}\,\dot{\bar{\omega}}+\dot{\omega}^2)=0, \tag{2.2}$$

$$8\pi\rho=\frac{1}{4}(2\dot{\bar{\omega}}+\dot{\omega})\dot{\omega}+e^{-\omega}. \tag{2.3}$$

由(2.1)式解 e^{ω} 的方法与前同,即通过(1.9)式引入 σ,把(2.1)式化为(1.10)式,从而求出

$$e^{\omega}=G^2\cos^2\frac{\alpha}{4}(\sigma-F),$$

这儿 F 为常数.再消去 $\sigma-F$,得

$$e^{\omega}=G^2\cos^4\left[\frac{1}{2}\varphi\left(\frac{2}{G}\tau+H\right)\right]\quad (G,\ H\ 为常数), \tag{2.4}$$

φ 的意义同前,但注意这里 $G,\ H$ 为常数.

解 $e^{\bar{\omega}}$ 的方法比较复杂一些,仍然利用 σ 作为自变量,用(2.1)及(1.9)可把(2.2)化为

$$\frac{d^2\bar{\omega}}{d\sigma^2}+\frac{1}{2}\left(\frac{d\bar{\omega}}{d\sigma}\right)^2-\frac{1}{4}\left(\frac{d\omega}{d\sigma}\right)^2-1=0. \tag{2.5}$$

由(2.4)计算得出 $\frac{d\omega}{d\tau}$,再令

$$\sigma-F=W, \tag{2.6}$$

$$e^{\frac{\bar{\omega}}{2}}=z, \tag{2.7}$$

则(2.5)可化为

$$\frac{d^2z}{dW^2}-\left(\frac{1}{2}\sec^2\frac{1}{2}W\right)z=0.$$

此方程有一特解 $z=\operatorname{tg}\frac{W}{2}$,利用常数变易法可求得通解为

$$z=K\left(1+\frac{W}{2}\operatorname{tg}\frac{W}{2}\right)+L\operatorname{tg}\frac{W}{2},$$

这里 K,L 为 R 的任意函数.由此得到

$$e^{\frac{\bar{\omega}}{2}}=K(R)\left\{1+\frac{\varphi\left(\frac{2}{G}\tau+H\right)}{2}\operatorname{tg}\left[\frac{1}{2}\varphi\left(\frac{2}{G}\tau+H\right)\right]+L(R)\operatorname{tg}\left[\frac{1}{2}\varphi\left(\frac{2}{G}\tau+H\right)\right]\right\}, \tag{2.8}$$

这样，我们就得到 $p=0$ 时球对称引力场方程在 $\omega'=0$ 时的解(2.4)、(2.8)，这里 $\varphi(y)$ 仍然是 $y=x+\sin x$ 的逆函数.将(2.4)、(2.8)代入(2.3)就可得出 ρ 的表达式.

至此，我们得出了一切可能的解.现在我们把二、三结果归结为下述形式：

定理 压力忽略不计时(即 $p=0$ 时)具有物质分布的球对称引力场方程的全部严格解可划分为三种类型及一种例外情况.具体来说，在 $\omega'\neq 0$ 时解具有三种类型.

(1) 当 $\lambda>0$，解为(1.12)—(1.13)，

(2) 当 $\lambda<0$，解为(1.15)—(1.16)，

(3) 当 $\lambda=0$，解为(1.18)—(1.19)，

而当 $\omega'=0$ 时，解具有形式(2.4)、(2.8).

下面我们指出两个注意事项：

1. 如果一区间中同时有 $\lambda>0$，$\lambda<0$，$\lambda=0$ 的情况.例如在 $R\in(a,b)$ 时，$\lambda<0$；在 $R\in(b,c)$ 时，$\lambda>0$；又 $\lambda(b)=0$，则我们可以在 (a,b) 依(1)求解，在 (b,c) 依(2)求解，但 G，H 要选好，使在 $R=b$ 时两边能联结好.

2. 熟知的 Friedman 解中 $k>0$ 情况属于类型(1)；$k<0$ 情况属于类型(2)；$k=0$ 情况属于类型(3).

四、奇点的分析

(1) $\lambda>0$ 时奇点的分析

a. 中心奇点.注意到 $\varphi(y)$ 是 y 的单调增加函数，并且从 $-\infty$ 增加到 $+\infty$，因此对每一固定的 R，我们总能找到 τ 使 $\frac{1}{2}\varphi\left(\frac{\alpha}{G}\tau+H\right)$ 具有 $\left(n+\frac{1}{2}\right)\pi$ 形式；从(1.12)式可见，

对于这种 τ 值 $e^{\omega}=0$，这时度规发生奇性.

由于向径 R 的球面面积是 $4\pi e^{\omega(R,\tau)}$，因而 $e^{\omega}=0$ 表示物质所构成的球面面积为 0，即物质趋向中心.所以 $e^{\omega}=0$ 是中心奇点，且由 ρ 的表达式可知，在中心奇点 $\rho\to\infty$.

对于每一固定的 R 总有 τ 使 $e^{\omega(R,\tau)}=0$. 我们可以认为这奇点是由物质在中心碰撞所形成. 取 $r^2=e^{\omega}$，画出 (r,τ) 平面上物质运动的世界线（如图 1），图中表示了每个球面有各自到达中心的时间.

b. 非中心碰撞型奇点. 当 $e^{\bar{\omega}}=0$ 但 $e^{\omega}\neq 0$ 时度规也退化，这是另一种类型的奇点，我们现在来考察这种奇点.

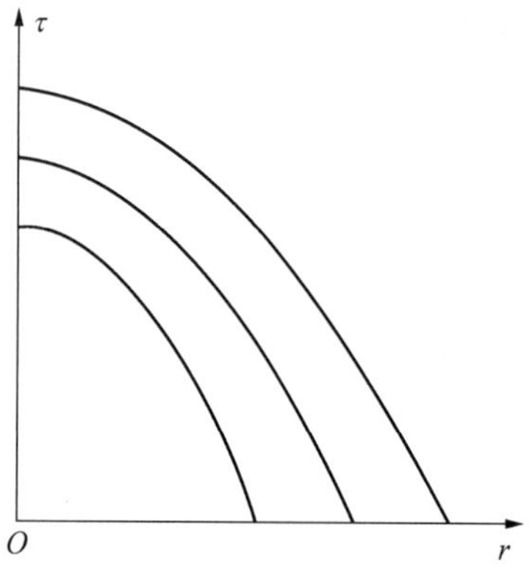

图 1

注意到：取定 R_0 后，R_0 所表示的物质和 R 所表示的物质在 τ 时的距离为

$$D(R_0, R, \tau)=\int_{R_0}^{R} e^{\frac{\bar{\omega}}{2}}dR,$$

在 (D,τ) 平面上，当 R 等于固定常数时，

$$D=D(R_0, R, \tau)$$

表示物质运动的一条世界线，当 R 变动时，它表示 (D,τ) 平面上单参数世界线族.注意到 $\dfrac{\partial D}{\partial R}=e^{\frac{\bar{\omega}}{2}}$，可知这一族世界线的包络由

$$\begin{cases} D=D(R_0, R, \tau), \\ \dfrac{\partial D}{\partial R}=e^{\frac{\bar{\omega}}{2}}=0 \end{cases}$$

所决定.所以满足 $e^{\bar{\omega}}=0$ $(e^{\omega}\neq 0)$ 的点，正是世界线与邻近世界线相切的点，这表示物质在这种奇点开始碰撞，所以使 $e^{\bar{\omega}}=0$ 的点为碰撞型的奇点.这时密度 $\rho\to\infty$.

c. $\alpha^2(R)\to 4$ 时奇点的分析.

设 α 为在 $[R_0, R_1]$ 定义的光滑函数，且 $R<R_1$ 时 $\alpha^2<4$，$\alpha^2(R_1)=4$，那么当 $R\to R_1$ 时 $e^{\bar{\omega}}\to\infty$，因此这也是一奇点.这时

$$\lim_{R\to R_1} D(R_0, R, \tau)=\lim_{R\to R_1}\int_{R_0}^{R}\frac{A(R,\tau)}{\sqrt{4-\alpha^2}}dR,$$

式中 $A(R, \tau)$表示(1.13)式中分子的开方，这时可举出两种情形：

c_1. 这个极限趋向无穷，这表示 $R \to R_1$ 时物质趋向无限远处.

c_2. 设右边的极限为有限，则令

$$D(R_0, R, \tau)=\int_{R_0}^{R} \frac{A(R, \tau)}{\sqrt{4-\alpha^2}} dR$$

作为新的自变量代替 R，又设 $\int_{R_0}^{R} \frac{\partial A(R, \tau)/\partial \tau}{\sqrt{4-\alpha^2}} dR$ 当 $R \to R_1$ 也为一致收敛，那么

$$dD=\frac{A(R, \tau)}{\sqrt{4-\alpha^2}} dR+\left(\int_{R_0}^{R} \frac{\partial A(R, \tau)/\partial \tau}{\sqrt{4-\alpha^2}} dR\right) d\tau$$

或

$$e^{\bar{\omega}} dR^2=\left[dD-\left(\int_{R_0}^{R} \frac{\partial A(R, \tau)/\partial \tau}{\sqrt{4-\alpha^2}} dR\right) d\tau\right]^2,$$

度规的奇性被消去了，实际上这是坐标奇点.

(2) $\lambda<0$ 时奇点的分析

a. 中心奇点. 由 $\psi\left(\frac{\alpha}{G} \tau+H\right)$ 的性质，我们得到对每个 R 能找到唯一的 τ 使 $\psi\left(\frac{\alpha}{G} \tau+H\right)=0$，由此 $e^{\omega}=0$，$e^{\bar{\omega}}=0$ 所得到的奇点是中心奇点.

b. 非中心的碰撞奇点. 当 $e^{\omega} \neq 0$ 而 $e^{\bar{\omega}}=0$ 时表示碰撞型奇点.

(3) $\lambda=0$ 时奇点的分析

a. 中心奇点.对每一固定的 R，取 $\tau=-B(R)/A(R)$ 得到 $e^{\omega(R, \tau)}=0$，这是中心奇点.

b. 非中心的碰撞奇点.当 $A(R) \tau+B(R) \neq 0$ 而 $A'(R) \tau+B'(R)=0$ 时，为非中心的碰撞奇点.

(4) 例外情形即 $\omega'=\frac{\partial \omega}{\partial R}=0$ 时的奇点

a. 中心奇点. $e^{\omega}=0$ 为中心奇点，又由于 G, H

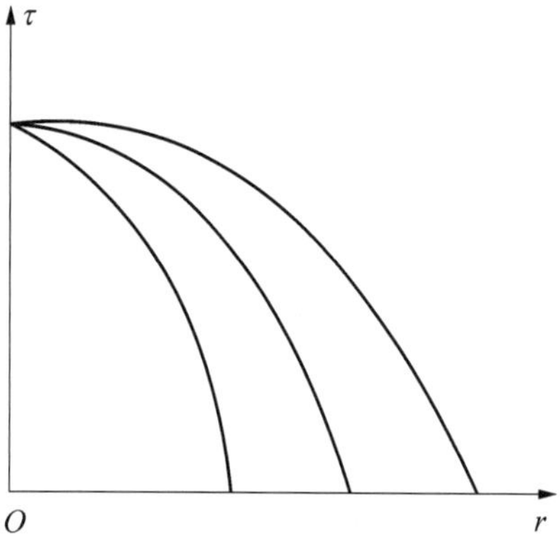

图 2

为常数,所以不论 R 等于多少,有相同的 τ 值使 $e^{\omega}=0$. 这表示分布在不同 R 的物质同时到达中心.

b. 非中心的碰撞奇点. $e^{\bar{\omega}}=0$ 也是由物质碰撞所产生的奇点,但此时有一特点:在不同时间 τ,分布在不同 R 的物质所构成的球面有相同的面积.

综合上述对奇点的讨论,可知 $p=0$ 时球对称引力场方程的任一严格解均具有奇点.奇点可在球的中心也可不在中心,这两种奇点处均有 $\rho\to\infty$. 由于 $\rho\to\infty$ 能推出数量曲率→∞,所以它们是真性的奇点.如果要使问题的解能越过奇点而继续制作下去,就必须对这种碰撞的物理过程进行研究,即要弄清物质在发生碰撞之后的运动规律,看来,单用广义相对论是不能解决这种问题的.

参考资料

[1] J. R. Oppenheimer and H. Snyder, *Phys. Rev.*, **56**, 455 (1939).

[2] M. M. May and R. H. White, *Phys. Rev.*, **141**, 1232 (1966).

[3] 朗道、栗弗席兹,《场论》中译本第 345 页(1959).

[4] H. Bondi, *Mon. Not. R. Astr. Soc.*, **107**, 410 (1947).

[5] 韩继昌.《南京大学学报》(1965).

[6] W. B. Bonnor, *Mon. Not. R. Astr. Soc.*, **159**, 261(1972).

[7] 谷超豪,《复旦学报(自然科学版)》1973 年第 3 期.

(本文曾发表于《复旦学报(自然科学版)》1974 年第 1 期,92 - 98;英文版曾发表于于 *Frontiers of Mathematics in China*, 2006, **2**, 169 - 177)

关于一类 SU_2 群的规范场

胡和生

摘　　要

本文讨论一种特殊类型的 SU_2 规范场，它的规范势在同位旋空间和普通空间作同一旋转时能保持不变，我们称这种势为同步球对称的．我们给出了这种规范势的表达式并得到如下结果：对于 SU_2 群的静态同步球对称规范场，规范势是由场的强度和源（除规范变换外）所唯一确定的．并且，两个与静态同步球对称规范场相等价的 SU_2 规范场如有相同的强度与源，则这两个规范场也等价．

§1．引　　言

Bohm-Aharonov 实验说明了在某种多连通区域内，电磁场强度 $f_{\mu\nu}$ 并不足以确定电磁势．文[1—2]从整体的角度考虑电磁场与规范场的关系，说明规范场的强度是否能确定电磁势要看时空区域的拓扑性质．对非阿贝尔群的规范场，杨振宁和吴大峻在文[3]中指出，即使在局部的情况，规范场强度也不能决定规范势，他们具体地作出了两个 SU_2 规范场，它们的强度 $f_{\mu\nu}^k$ 完全相等，但一个场是有源的，另一个场是无源的，因而它们总是不能用规范变换来相互变换的．从而就产生了这样的问题：两个 SU_2 群的规范场，如果它们的强度相同，源相同，何时会相互等价？我们对一种特殊类型的 SU_2 规范场来讨论这个问题，这类规范场的特点是：它的规范势在同位旋和普通空间作同一旋转时能保持不变，我们称这种势为同步球对称的．这种规范场在物理上较有意义，文[3—6]中所讨论的 SU_2 规范场都属于这种类型．

在本文中我们给出同步球对称规范势的表达式并得到如下结果：对于 SU_2 群的静态同步球对称规范场，规范势是由场的强度和源（除规范变换外）所唯一确定的.并且，两个与静态同步球对称规范场相等价的 SU_2 群规范场如有相同的强度与源，则这两个规范场也等价.

§2. SU_2 群的同步球对称规范场

我们考虑四维平坦时空中 SU_2 群的规范场，选取 SU_2 群的李代数 SU_2' 的基 $X_i\,(i=1,2,3)$ 使

$$[X_i, X_j]=\varepsilon_{ijk}X_k \quad (i, j, k=1, 2, 3), \tag{2.1}$$

规范势记为 $b_\lambda^i(x, t)\,(\lambda=0, 1, 2, 3)$，当李代数 SU_2' 受到一个旋转（伴随变换）

$$X_i \to \bar{X}_i=a_{ij}X_j \tag{2.2}$$

时，(2.1)不变，而 b_λ^i 受到变换

$$b_\lambda^i \to \bar{b}_\lambda^i(x, t)=a_{ij}b_\lambda^j \quad (\lambda=0, 1, 2, 3). \tag{2.3}$$

又当四维时空中的三维空间受到旋转

$$x^i \to \tilde{x}^i=a_{ij}x^j \tag{2.4}$$

时，

$$\begin{aligned} b_k^i &\to \tilde{b}_k^i(\tilde{x}, t)=b_j^i(x, t)a_{kj}, \\ b_0^i &\to \tilde{b}_0^i(\tilde{x}, t)=b_0^i(x, t). \end{aligned} \tag{2.5}$$

而当李代数和空间同时经受旋转(2.2)、(2.4)时，

$$b_k^i \to \bar{b}_k^i(\tilde{x}, t)=b_j^l(x, t)a_{il}a_{kj}, \tag{2.6A}$$

$$b_0^i \to \bar{b}_0^i(\tilde{x}, t)=b_0^j(x, t)a_{ij}. \tag{2.6B}$$

如果在变换(2.6A)、(2.6B)下规范势不变，即

$$\bar{b}_{\mu}^{i}(\tilde{x}(x),\ t)=b_{\mu}^{i}(\tilde{x},\ t), \tag{2.7}$$

则称相应的规范场为同步球对称的.

事实上，同步球对称的规范场是一种球对称的规范场.这是因为：SU_2 在 SU_2'中的伴随表示就是旋转群，从而变换(2.3)是一种平凡的规范变换.

首先我们证明

定理 1 同步球对称的规范势必具形式

$$\begin{aligned} b_i^l &= \varepsilon_{ilk}x^k V(r,\ t)+\delta_{il}S(r,\ t)+x^i x^l T(r,\ t), \\ b_0^l &= x^l U(r,\ t), \end{aligned} \tag{2.8}$$

式中 U, V, S, T 都是 r, t 的任意函数，而 $r=\sqrt{(x^1)^2+(x^2)^2+(x^3)^2}$.

证明.通过简单计算可知，规范势(2.8)满足(2.7)式，因而是同步球对称的.

现证明同步球对称的规范势必具形式(2.8).从(2.6B)及(2.7)可见 $b_0^l(x,\ t)$ 必须是旋转群 SO_3 下的不变向量场，取点 $P_0(r,\ 0,\ 0)$，记 $V^l=b_0^l(r,\ 0,\ 0,\ t)$，它必须在使点 $P_0(r,\ 0,\ 0)$ 不变的旋转群 G_1(即以 OP_0 为轴的旋转所成的群)下保持不变，因此得到

$$\begin{cases} V^2=V^2\cos\theta-V^3\sin\theta, \\ V^3=V^2\sin\theta+V^3\cos\theta, \end{cases}$$

即$(V^2,\ V^3)$为平面上旋转群的不变向量，从而得到 $V^2=V^3=0$，所以可记

$$b_0^l(r,\ 0,\ 0,\ t)=rU(r,\ t)\delta_1^l,$$

取 (a_{ij}) 为使 P_0 变为 $P(x^1,\ x^2,\ x^3)$ 的任一旋转，因而 $a_{j1}=\dfrac{x^j}{r}$，利用(2.7)式可见

$$b_0^l(x,\ t)=a_{lj}b_0^j(r,\ 0,\ 0,\ t)=U(r,\ t)x^l.$$

又从(2.6A)及(2.7)可见，$b_i^l(x,\ t)$ 是旋转群下的不变张量场，记 $V_{jl}=b_j^l(r,\ 0,\ 0,\ i)$，它必须是群 G_1 的二阶不变张量，从而易见，$V_{12}=V_{13}=0$，$V_{\alpha\beta}(\alpha,\ \beta=2,\ 3)$ 为平面上

旋转群的二阶不变张量,因此(V_{lj})必须取形式

$$\begin{pmatrix} V_{11} & 0 & 0 \\ 0 & V_{22} & V_{23} \\ 0 & -V_{23} & V_{22} \end{pmatrix}.$$

记

$$V=\frac{V_{23}}{r}, \quad T=\frac{V_{11}-S}{r^2}, \quad S=V_{22},$$

从此可见

$$b_i^l(r, 0, 0, t)=\varepsilon_{ilk}\delta_1^k rV+\delta_{il}S+r^2\delta_1^l\delta_1^i T.$$

仍取(a_{ij})为使P_0变为(x^1, x^2, x^3)的任一旋转,我们运用ε_{ijk}是SO_3的不变张量这一性质:

$$a_{lh}a_{ij}\varepsilon_{jhk}=\varepsilon_{ild}a_{dk} \tag{2.9}$$

由(2.7)就得出

$$b_i^l(x, t)=a_{lh}a_{ij}b_j^h(r, 0, 0, t)=\varepsilon_{ilk}x^kV(r, t)+\delta_{li}S(r, t)+x^lx^iT(r, t). \tag{2.8}$$

定理证毕.

在本文中我们研究静态的情况,即U, V, S, T均和t无关.

为了后文的需要,我们在这里先给出SU_2群同步球对称规范场的强度的表达式.由SU_2群规范场强度公式

$$f_{\mu\nu}^l=b_{\mu,\nu}^l-b_{\nu,\mu}^l-\varepsilon_{lmn}b_\mu^m b_\nu^n \tag{2.9}$$

及(2.8)式得到它的具体形式为

$$\begin{aligned} f_{ij}^l&=A\varepsilon_{ijk}x^kx^l+B(\delta_{li}x^j-\delta_{lj}x^i)+C\varepsilon_{lij}, \\ f_{i0}^l&=Hx^ix^l+E\delta_{li}+F\varepsilon_{ilk}x^k. \end{aligned} \tag{2.10}$$

这里

$$A=\frac{V'(r)}{r}+S(r)T(r)-V^2(r), \tag{2.11A}$$

$$B=\frac{S'(r)}{r}-T(r)-V(r)S(r)-r^2T(r)V(r), \tag{2.11B}$$

$$C=-(2V(r)+S^2(r)+V'(r)r+r^2S(r)T(r)), \tag{2.11C}$$

$$H=U(r)V(r)-\frac{U'(r)}{r}, \tag{2.11D}$$

$$E=-U(r)[1+r^2V(r)], \tag{2.11E}$$

$$F=S(r)U(r). \tag{2.11F}$$

§3. 主 要 结 果

在本节中，我们要得出本文的主要结果，即对同步球对称的规范场，研究强度和源是否能唯一确定规范势的问题.

设

$$\begin{aligned} b_i^l&=\varepsilon_{ilk}x^kV+\delta_{il}S+x^ix^lT,\\ b_0^l&=x^lU, \end{aligned} \tag{2.8}$$

为一个 SU_2 规范场的规范势，其规范场强度为(2.10)，又设

$$\begin{aligned} \underset{1}{b}{}_i^l&=\varepsilon_{ilk}x^kV_1+\delta_{il}S_1+x^ix^lT_1,\\ \underset{1}{b}{}_0^i&=x^iU_1, \end{aligned} \tag{3.1}$$

为另一个 SU_2 规范场的规范势，它的强度为

$$\begin{aligned} \underset{1}{f}{}_{ij}^l&=A_1\varepsilon_{ijk}x^kx^l+B_1(\delta_{li}x^j-\delta_{lj}x^i)+C_1\varepsilon_{lij},\\ \underset{1}{f}{}_{i0}^l&=H_1x^ix^l+E_1\delta_{li}+F_1\varepsilon_{ilk}x^k, \end{aligned} \tag{3.2}$$

其中 A_1, B_1, C_1, H_1, E_1, F_1 的表达式类似于(2.11),只需把相应的 S, T, U, V 改为 S_1, T_1, U_1, V_1 就成.

为使规范场 b_μ^i 与 $\underset{1}{b}{}_\mu^i$ 的强度相同,就有方程

$$\frac{V'}{r}+ST-V^2=\frac{V_1'}{r}+S_1T_1-V_1^2, \tag{3.3A}$$

$$\frac{S'}{r}-T-VS-r^2TV=\frac{S_1'}{r}-T_1-V_1S_1-r^2T_1V_1, \tag{3.3B}$$

$$2V+S^2+V'r+r^2ST=2V_1+S_1^2+V_1'r+r^2S_1T_1, \tag{3.3C}$$

$$UV-\frac{U'}{r}=U_1V_1-\frac{U_1'}{r}, \tag{3.3D}$$

$$U(1+r^2V)=U_1(1+r^2V_1), \tag{3.3E}$$

$$SU=S_1U_1, \tag{3.3F}$$

这里“′”表示对 r 的导数.

四维时空中规范场 b_μ^i 的源为[7]

$$J_\mu^k=g^{\nu\lambda}f_{\mu\nu|\lambda}^k=g^{\nu\lambda}(f_{\mu\nu,\lambda}^k+b_\lambda^i f_{\mu\nu}^j\varepsilon_{kij}). \tag{3.4}$$

式中 $f_{\mu\nu|\lambda}^k$ 是 $f_{\mu\nu}^k$ 的规范导数,为使规范场 b_μ^i 与规范场 $\underset{1}{b}{}_\mu^i$ 的源相同,则应成立

$$g^{\nu\lambda}b_\lambda^i f_{\mu\nu}^j\varepsilon_{kij}=g^{\nu\lambda}\underset{1}{b}{}_\lambda^i f_{\mu\nu}^j\varepsilon_{kij},$$

或

$$(b_l^i-\underset{1}{b}{}_l^i)f_{0l}^j\varepsilon_{kij}=0,$$

$$(b_l^i-\underset{1}{b}{}_l^i)f_{hl}^j\varepsilon_{kij}-(b_0^i-\underset{1}{b}{}_0^i)f_{h0}^j\varepsilon_{kij}=0.$$

再把(2.8),(2.10)代入,经化简后,就得到源相等的条件为

$$(Ar^2+C)(V-V_1)+Br^2(T-T_1)+B(S-S_1)-E(U-U_1)=0, \tag{3.5A}$$

$$2B(V-V_1)-C(T-T_1)-A(S-S_1)-F(U-U_1)=0, \tag{3.5B}$$

$$Cr^2(T-T_1)+(Ar^2+2C)(S-S_1)+Fr^2(U-U_1)=0, \tag{3.5C}$$

$$E(V-V_1)-F(S-S_1)=0, \tag{3.5D}$$

因而要求出有相同强度与相同源的规范势 b_μ^i 及 $\underset{1}{b}{}_\mu^i$，就需从二套方程(3.3A)～(3.3F)及(3.5A)～(3.5D)中得出除平凡解 $V=V_1$，$S=S_1$，$T=T_1$ 及 $U=U_1$ 之外的解 U, V, S, T, U_1, V_1, S_1, T_1. 通过对这二套方程的分析可以知道，它们的非平凡解只能属于下述两种情形之一(证明见附录)

(a) 成立 $$f_{\lambda\mu}^k=0.$$

(b) 成立

$$\begin{aligned} &V=V_1=-\frac{1}{r^2},\ S=S_1=0,\\ &T(r),\ T_1(r),\ U(r)\ \text{为}\ r\ \text{的任意函数}.\\ &U_1=U+\frac{k}{r},\ k\ \text{为任意常数}. \end{aligned} \tag{3.6}$$

这时

$$\begin{aligned} &f_{ij}^l=\varepsilon_{ijk}\frac{x^kx^l}{r^4}.\\ &f_{i0}^l=\left(-\frac{U}{r^2}-\frac{U'}{r}\right)x^ix^l, \end{aligned} \tag{3.7}$$

$U(r)$ 为 r 的任意函数.

再注意到对于一般规范场都成立的下述一些事项：

1. 两个规范势 b_μ^i 与 $\underset{1}{b}{}_\mu^i$ 等价的条件为

$$\underset{1}{b}{}_\mu^k=A_l^k(\alpha(x))\frac{\partial\alpha^l}{\partial x^\mu}+B_l^k(\alpha(x))b_\mu^l, \tag{3.8}$$

式中 α^l 为群 G(现为 SU_2)的元素 g 的参数，现为 x 的函数，B_l^k 是元素 g 在李代数 G' 中伴随表示的矩阵元，而 A_l^k 由无限小元素

$$g(\alpha)g^{-1}(\alpha+d\alpha)=I+A_l^k(\alpha)d\alpha^lX_k \tag{3.9}$$

所确定.

2. 如果把(3.8)看成 $\alpha(x)$ 的微分方程，则利用 B^i_j；A^i_j 之间所存在的一些恒等式可把它的可积条件写成

$$\underset{1}{f}{}^i_{\lambda\mu}=B^i_j(\alpha(x))f^j_{\lambda\mu}, \tag{3.10}$$

把它们再求导一次，又利用关系式(3.8)，(3.10)及函数 A^i_j，B^i_j 之间的一些关系，可得

$$\underset{1}{f}{}^i_{\lambda\mu|\nu}=B^i_j(\alpha(x))f^j_{\lambda\mu|\nu}. \tag{3.11}$$

现在看到，对于情况(a)，由于 $f^i_{\lambda\mu}=\underset{1}{f}{}^i_{\lambda\mu}=0$，所以(3.8)的可积条件是恒等成立的.因此必存在规范变换使 b^i_μ 变为$\underset{1}{b}{}^i_\mu$，实际上，任何强度等于0的规范场一定与势 $b^i_\mu=0$ 的规范场在局部意义下相等价.

对于情况(b)，我们从(2.20)式计算 $f^k_{\mu\nu}$ 及 $\underset{2}{f}{}^k_{\mu\nu}$ 的协变导数，首先注意到公式

$$\delta^k_l-\frac{\varepsilon_{kpd}\varepsilon_{lph}x^hx^d}{r^2}=\frac{1}{r^2}x^kx^l,$$

这是可以直接验证的.利用它可得

$$\begin{cases}\underset{1}{f}{}^k_{ij|l}=f^k_{ij|l}=\varepsilon_{ijl}\dfrac{x^k}{r^4}-3\varepsilon_{ijh}\dfrac{x^hx^kx^l}{r^6},\\[2ex] \underset{1}{f}{}^k_{i0|l}=f^k_{i0|l}=H\delta^i_lx^k+\left(\dfrac{H'}{r}+\dfrac{H}{r^2}\right)x^ix^kx^l,\\[2ex] \underset{1}{f}{}^k_{i0|0}=f^k_{i0|0}=0,\\[2ex] \underset{1}{f}{}^k_{ij|0}=f^k_{ij|0}=0.\end{cases} \tag{3.12}$$

这里

$$H=-\frac{U}{r^2}-\frac{U'}{r}. \tag{3.13}$$

容易看到，在这情形，(3.10)可归结为

$$x^i=B^i_j(\alpha(x))x^j, \tag{3.14}$$

而(3.11)也归结为同样的一组方程，所以(3.11)是(3.10)的推论.又(3.14)关于 $\alpha(x)$ 是显

然有解的，它的解就是使(x^1, x^2, x^3)点不动的转动，应用微分方程中一个已知定理[7]，就知道(3.8)有解，b^i_μ 与$\underset{1}{b}{}^i_\mu$ 也是相互等价的.

总结上述结果，我们得到

定理 2 对于 SU_2 群的静态同步球对称规范场，规范势是由场的强度和源(除规范变换外)所唯一确定的.

由于同步球对称的规范势经过规范变换之后未必具有(2.8)的形式，所以我们要对这个问题作进一步的讨论，即要考虑如下的问题：如两个规范势 b^i_μ 与$\underset{1}{b}{}^i_\mu$ 经规范变换后，有相同的强度和源，问它们是否规范等价？

在上述定理 2 的基础上，我们还可以证明定理 3.

定理 3 两个与静态同步球对称规范场相等价的 SU_2 规范场，如有相同的强度与源，则这两个规范场也等价.

换言之，如果两个静态球对称规范场 $\mathscr{F}$ 和 $\mathscr{F}_1$ 经规范变换后有相同的强度和源，那么 $\mathscr{F}$ 和 $\mathscr{F}_1$ 必为等价.

证明： 设 $\mathscr{F}$ 与 $\mathscr{F}_1$ 是二个静态同步球对称规范场，它们的强度分别是 $f^a_{\mu\nu}$ 与 $\underset{1}{f}{}^a_{\mu\nu}$，记

$$
\begin{aligned}
&k_{ia}=\frac{1}{2}\varepsilon_{ijk}f^a_{jk},\ K(x)=(k_{ia}(x)),\ L(x)=(f^a_{i0}(x)),\\
&\underset{1}{k}{}_{ia}=\frac{1}{2}\varepsilon_{ijk}\underset{1}{f}{}^a_{jk},\ \underset{1}{K}(x)=(\underset{1}{k}{}_{ia}(x)),\ \underset{1}{L}(x)=(\underset{1}{f}{}^a_{i0}(x)).
\end{aligned}
\tag{3.15}
$$

先讨论 $K(x)$ 与 $L(x)$ 中至少有一个的秩数 $\geqslant 2$ 的情形.不妨取 $K(x)$ 的秩数 $\geqslant 2$. 又由假定，可推知 $\mathscr{F}$ 与 $\mathscr{F}_1$ 的强度和源之间有正交关系，因而必存在 SO_3 中的正交阵 $B(x, t)$ 使

$$
K(x)=\underset{1}{K}(x)B(x, t),\tag{3.16}
$$

由于 $B(x, t)$ 由两对独立的对应向量完全决定，由上式就推出 $B(x, t)$ 与 t 无关，因而可记 $B(x)=(B_{ab}(x))$①，又因 $\mathscr{F}_1$ 和 $\mathscr{F}$ 为同步球对称的，所以若空间经过转动$\tilde{x}=Cx$，

① B_{ab} 即上文中的 B^b_a.

$C \in SO_3$，则

$$K(\tilde{x}) = CK(x)C^{-1},\ \underset{1}{K}(\tilde{x}) = C\underset{1}{K}(x)C^{-1}, \tag{3.17}$$

由(3.16)与(3.17)得

$$K(x)C^{-1} = K(x)B^{-1}(x)C^{-1}B(Cx), \tag{3.17'}$$

因 $K(x)$ 的秩数 $\geqslant 2$，又 C^{-1} 与 $B^{-1}(x)C^{-1}B(Cx)$ 为正交阵，所以对正交阵 B 成立

$$B(Cx) = CB(x)C^{-1}. \tag{3.18}$$

现在要作一个规范场满足三个条件：1. 和 $\mathscr{F}_1$ 等价；2. 本身是静态同步球对称的；3. 和 $\mathscr{F}$ 有相同的强度和源.

记 $\zeta(x)$ 为 SU_2 群中的元素，其伴随变换使 $K \to KB$，那么 $\zeta(x)$ 除一符号外为确定的，又从(3.18) 式可知

$$\zeta(Cx) = \zeta_c \zeta(x) \zeta_c^{-1}, \tag{3.19}$$

以此 ζ_c 的伴随变换使 $K \to KC^{-1}$，记 $\underset{1}{b}{}^i_\lambda(x)X_i = \underset{1}{b}{}_\lambda$，对 $\underset{1}{b}{}_\lambda$ 利用 $\zeta(x)$ 作规范变换，得

$$b'_\lambda(x) = \zeta(x)\underset{1}{b}{}_\lambda \zeta^{-1}(x) - \frac{\partial \zeta(x)}{\partial x^\lambda}\zeta^{-1}(x). \tag{3.20}$$

我们可证明规范势 b'_λ 所对应的规范场 $\mathscr{F}'_1$ 满足所要求的三个条件.现证 $b'_\lambda(x)$ 为同步球对称的，由(3.20)得

$$b'_\lambda(Cx) = \zeta(Cx)\underset{1}{b}{}_\lambda(Cx)\zeta^{-1}(Cx) - \left(\frac{\partial \zeta(x)}{\partial x^\lambda}\right)_{Cx}\zeta^{-1}(Cx),$$

特别

$$\begin{aligned} b'_0(Cx) &= \zeta(Cx)\underset{1}{b}{}_0(Cx)\zeta^{-1}(Cx) = \zeta(Cx)\zeta_c\underset{1}{b}{}_0(x)\zeta_c^{-1}\zeta^{-1}(Cx) \\ &= \zeta_c\zeta(x)\underset{1}{b}{}_0(x)\zeta^{-1}(x)\zeta_c^{-1} = \zeta_c b'_0(x)\zeta_c^{-1}, \end{aligned} \tag{3.21}$$

$$b'_i(Cx) = \zeta(Cx)\underset{1}{b}{}_i(Cx)\zeta^{-1}(Cx) - \left(\frac{\partial \zeta(x)}{\partial x^i}\right)_{Cx}\zeta^{-1}(Cx)$$

$$=C_{ij}\zeta_c\zeta(x)\underset{1}{b}_j(x)\zeta^{-1}(x)\zeta_c^{-1}-C_{ij}\zeta_c\frac{\partial\zeta(x)}{\partial x^j}\zeta^{-1}(x)\zeta_c^{-1}$$

$$=C_{ij}\zeta_c b'_j(x)\zeta_c^{-1},\tag{3.22}$$

这两式表明 $b'_\lambda(x)$ 是同步球对称的，又由 $b'_\lambda(x)$ 的作法可见，它的场强和源

$$\begin{aligned}f'^a_{\lambda\mu}(x)&=\underset{1}{f}{}^b_{\lambda\mu}(x)B_{ba}(x)=f^a_{\lambda\mu}(x),\\ J'^a_\mu(x)&=\underset{1}{J}{}^b_\lambda(x)B_{ba}(x)=J^a_\lambda(x),\end{aligned}\tag{3.23}$$

所以 $\underset{1}{\mathscr{F}}'$ 和 $\mathscr{F}$ 的强度和源相同.

从定理 2 可知 $\mathscr{F}$ 和 $\mathscr{F}'_1$ 等价，因而 $\mathscr{F}$ 和 $\mathscr{F}_1$ 等价.

再讨论 $K(x)$ 和 $L(x)$ 的秩数均 $\leqslant 1$ 的情形.在 $(r,0,0)$ 点作 K 和 L，得

$$K=\begin{pmatrix}Ar^2+C & 0 & 0\\ 0 & C & -Br\\ 0 & Br & C\end{pmatrix},\quad L=\begin{pmatrix}Hr^2+E & 0 & 0\\ 0 & E & -Fr\\ 0 & Fr & E\end{pmatrix},\tag{3.24}$$

由 K 和 L 的秩数不大于 1，所以得

$$B=C=0,\quad E=F=0.\tag{3.25}$$

同样有

$$B_1=C_1=0,\quad E_1=F_1=0.\tag{3.26}$$

这时 A，H 不全为 0，否则 $\mathscr{F}$，$\mathscr{F}_1$ 为显然等价.

如果 $J^a_\mu X_a$ 中包含不和 X_1 平行的同位旋向量，这时 K，L 和 J_μ 中有两个独立的同位旋向量，仍然可以归到上述情况去，如果一切 $J^a_\mu X_a$ 和 X_1 平行，则 $\mathscr{F}$ 和 $\mathscr{F}_1$ 的场强和源的关系仍然是(3.3) 和(3.5)（当然其中有一些方程化为平凡的方程），因而仍然可以用定理 2 的结果而知道 $\mathscr{F}$ 和 $\mathscr{F}_1$ 等价.定理 3 证毕.

附　录

现证明方程组(3.3A)～(3.3F)，(3.5A)～(3.5D)的非平凡解必须是情形(a)或(b).

为此,引入

$$D=Ar^2+C=-r^2V^2-2V-S^2,$$

(3.5A)～(3.5D)成为

$$D(V-V_1)+Br^2(T-T_1)+B(S-S_1)-E(U-U_1)=0, \tag{1}$$

$$Br^2(V-V_1) \qquad\qquad +C(S-S_1) \qquad\qquad =0, \tag{2}$$

$$Cr^2(T-T_1)+(D+C)(S-S_1)+Fr^2(U-U_1)=0, \tag{3}$$

$$-E(V-V_1) \qquad\qquad +F(S-S_1) \qquad\qquad =0. \tag{4}$$

又把(3.3E),(3.3F)改写为

$$r^2U_1(V-V_1)+(1+r^2V)(U-U_1)=0, \tag{5}$$

$$U_1(S-S_1)+S(U-U_1)=0. \tag{6}$$

(1)～(6)连同 $B=B_1$, $C=C_1$, $D=D_1$, $H=H_1$ 便是我们的全系方程,A, B, C, H, E, F 的表达式见(2.11A)～(2.11F).

现分若干情况进行讨论.

(Ⅰ)$U\neq U_1$ 且 $U_1\neq 0$. 由(5)和(6)得

$$S-S_1=-\frac{S}{U_1}(U-U_1),\ V-V_1=-\frac{(1+r^2V)}{r^2U_1}(U-U_1). \tag{7}$$

代入(4)式,又利用 E, F 的表达式得

$$U\{r^2S^2+(1+r^2V)^2\}=0.$$

不妨设 $U\neq 0$(因为 U 和 U_1 是对称的,$U=0$ 的情况也可归到下述的情况 Ⅱ 去)从而得:

$$S=0,\ 1+r^2V=0. \tag{8}$$

从(7)式得 $S_1=0$, $V=V_1$, $1+r^2V_1=0$. 易见此时 $B=B_1=0$, $C=C_1=0$, $E=F=0$, $D=D_1$,余下只要考虑 $H=H_1$,这时,这个方程取形式

$$(U-U_1)+r(U-U_1)'=0.$$

从此得

$$U_1-U=\frac{k}{r}.$$

这样,我们便有解(b)即

$$\begin{cases} V=V_1=-\dfrac{1}{r^2},\ S=S_1=0, \\ T(r),\ T_1(r),\ U(r)\ \text{为}\ r\ \text{的任意函数}, \\ U_1=U+\dfrac{k}{r},\ k\ \text{为任意常数}. \end{cases} \tag{9}$$

但这时有假定 $U_1\neq 0$.

(Ⅱ) $U\neq U_1$,但 $U_1=0$,由(5)(6)仍然得(8)式,代入 $D=D_1$ 就有 $S_1=0$, $1+r^2V_1=0$,和(Ⅰ)一样得到解(b),这时除去了(9)中对 $U_1\neq 0$ 的限制.

(Ⅲ) $U=U_1$, $S=S_1$, $T=T_1$ 的情况:此时必须有 $V-V_1\neq 0$,所以有 $B=B_1=0$.从此得

$$S+r^2T=0,\ \frac{S'}{r}-T=0,$$

因而

$$S_1=S=\frac{l}{r},\ T_1=T=-\frac{l}{r^3}\quad(|\ l\ |<1).$$

又由 $D=D_1=0$ 得

$$V=\pm\frac{1}{r^2}\sqrt{1-l^2}-\frac{1}{r^2},\quad V_1=\mp\frac{1}{r^2}\sqrt{1-l^2}-\frac{1}{r^2}.$$

再由(5)式得 $U_1=0$ 因而 $U=0$,容易见到这时 $f^k_{\lambda\mu}=0$,属于情况(a).

(Ⅳ) $U=U_1$, $S=S_1$, $T\neq T_1$ 的情况.

这时必有 $B=0$,否则由(2)会得到 $V=V_1$,再由(1)会得出 $T=T_1$.

如再设 $V=V_1$,则由 $B=B_1$ 就得出 $V=-\frac{1}{r^2}$,从 $A=A_1$ 得出 $S=0$,容易验证,从此可以得到一组解

$$V=V_1=-\frac{1}{r^2},\ S=S_1=0. \tag{10}$$

而 $U=U_1$, T, T_1 均为任意,这组解属于情况(b).

如 $V\neq V_1$,则由(1)知 $D=0$ 由此得

$$V=-\frac{1}{r^2}\pm\frac{1}{r^2}\sqrt{1-S^2r^2},$$

再利用 $D_1=D=0$ 得

$$V_1=-\frac{1}{r^2}\mp\frac{1}{r^2}\sqrt{1-S^2r^2},$$

从此可见 $0\leqslant S^2r^2<1$,由(3)必有 $C=0$,这和 $A=0$ 等价.设 $S\neq 0$,则

$$T=\frac{1}{S}\left(V^2-\frac{V'}{r}\right),\ T_1=\frac{1}{S}\left(V_1^2-\frac{V_1'}{r}\right).$$

又由(5)得 $U=U_1=0$. 这样,我们就得到一组解,易见它属于情况(a).

当 $S=0$ 时有 $V=0$, $V_1=-\frac{2}{r^2}$,由 $B=0$ 可得 $T=0$,又 $B_1=0$ 得到 $T_1=0$ 因而有 $T_1=T$,这是不合于假设的.

(Ⅴ) $U=U_1$, $S\neq S_1$ 的情形.

由 $F=F_1$ 得知 $U=U_1=0$,这时把(1)(2)(3)看成为 $V-V_1$, $T-T_1$, $S-S_1$ 的线性方程组,计算其系数行列式,为使问题有非平凡解它必须为0,就得

$$D(B^2r^2+C^2)=0. \tag{11}$$

再分别讨论

$(\mathrm{V})_1$ 如 $B=0$,由(2)必有 $C=0$,由(3)必有 $D=0$,这时方程 $D=D_1$, $B=B_1$, $C=C_1$ 是有解的,但这个解所作出的 $f^i_{\lambda\mu}=0$, 因而属于情况(a).

$(\mathrm{V})_2$ $B\neq 0$时,由(11)式必有$D=0$, $T-T_1=-\dfrac{S-S_1}{r^2}$, $V-V_1=-\dfrac{C}{r^2B}(S-S_1)$ 把 $D=D_1$ 写成

$$[2+r^2(2V-(V-V_1))](V-V_1)+(2S-(S-S_1))(S-S_1)=0, \qquad (12)$$

将 $V-V_1$ 用 $-\dfrac{C}{r^2B}(S-S_1)$ 代入,再注意到 $D=0$ 时必有

$$SS'=-V'-rV^2-r^2VV',$$

与

$$SBr^2-(1+r^2V)C=0,$$

就可将(12)化为

$$(S-S_1)^2=0.$$

由此得 $S=S_1$ 和前提矛盾,这时无解.

这样,我们就证明了§3中所需要的结论.

参 考 资 料

[1] T. T. Wu(吴大峻) and C. N. Yang(杨振宁), *Concept of Nonintegrable Phase Factors and Global Formulation of Gauge Fields*, *Phys. Rev.* D12, 3845 (1975).

[2] 谷超豪,电磁场和 U_1 群的整体规范场,Ⅰ、Ⅱ,复旦学报,1975 年,第 4 期.

[3] T. T. Wu(吴大峻) and C. N. Yang(杨振宁), *Unquantized Non-Abelian Gauge Fields*, *Phys. Rev.* D12, 3843 (1975).

[4] G.'t Hooft, *Magnetic Monopoles in Unified Gauge Theories*, TH., 1873, CERN., (1974).

[5] T. T. Wu(吴大峻) and C. N. Yang(杨振宁), *Some Solutions of the Classical Isotopic Gauge*

Field Equations (p. 349 *in Properties of Matter under Unusual Conditions*, Compiled by Fernback and Mark).

[6] 侯伯宇,电与磁对偶的双重协变规范场,科学通报,第20卷,第6期(1975),273-276.

[7] Gu Chao-hao(谷超豪) and Yang Chen-ning(杨振宁), *Some Problems on the Gauge Field Theories*, Scientia Sinica, **18** (1975), 483-501,规范场理论的若干问题,复旦学报,1975年,第2期,27-41.

(本文曾发表于《复旦学报(自然科学版)》1976年第1期,72-81)

关于球对称的 SU_2 规范场

西北大学　侯伯宇　　复旦大学　谷超豪　胡和生

摘　要

本文对一般球对称 SU_2 规范场依其不同特性选取适当规范而得出规范势的简化形式，分析了无源方程的结构，阐明了各种情况下场的某些物理性质.

§1. 引　言

近年来，在不可换群规范场和对偶荷的研究中，人们注意到这项理论和物理学中一些比较基本的问题，如强子结构，真空激发，自旋和同位旋的关系，自旋和统计的关系等，有着一定的联系.因而，有需要对不可换规范场的一种最简单的情况——经典的球对称 SU_2 规范场，作比较详细的讨论.

已经知道，球对称的 SU_2 规范场（也就是对空间任一旋转，均可配上一个适当的规范变换，使规范势保持不变的规范场），只能有三种类型[1]：

Ⅰ. 狭义球对称规范场——其特点是单纯的空间转动就能使其规范势（在适当规范下）保持不变.

Ⅱ. 同步球对称规范场——其特点是空间的转动配上同位旋空间同一旋转使规范势（在适当规范下）保持不变.

Ⅲ′. 化约的球对称规范场.

在类型Ⅲ′中有一部分也属于类型Ⅰ（磁荷为 0 的情形），有一部分也属于类型Ⅱ（具有单位磁荷（正或负）的情形）余下的是磁荷为 $\pm m$ 单位的情形（$m>1$，整数），我们把它划为类型Ⅲ.

Ⅲ. 倍步的球对称规范场——其特点是它内在地联系于一个同位旋向量场(Higgs 场),这个场将单位球面映照到同位旋空间单位球面去,这种映象有$\pm m$ 重复盖的特点,其 Kronecker 指标为$\pm m$[2,3].

在这样的划分下,每一球对称 SU_2 规范场只能属于一种类型.

在本文中,我们经过适当的规范变换,写出各种类型的规范势的简化形式,然后通过强度和源来分析场的性质.如果把球对称 SU_2 规范场解释为由电磁场和带电矢粒子构成,那么这三种情形的某些物理性质可列表如下:

类　别	Higgs 场	磁荷	电　　荷	带电矢粒子	备　　注
狭义球对称	平　　行	0	连续分布加原点点电荷	不化约时出现	无源解只是静电场
同步球对称	径　　向	± 1	同　上	同　上	
倍步球对称	$\pm m$ 重复盖	$\pm m$	同　上	必不出现	无源解只是静电场加静磁场

这里的无源解是指在原点以外满足无源方程的场,原点可以是奇性的点源.还可看出,一切球对称 SU_2 场都不含纵向极化向量的带电矢粒子.对其他有关性质,文中也作了一些讨论.在同步球对称情况下,我们弄清了规范变换对静态无源解的影响,从而使求解的问题得到了简化.

本文所用记号如下:

X 和 $x_i(i=1,\ 2,\ 3)$——空间点及其坐标,

x_0 或 t——时间坐标,

x 和 $x_\mu(\mu=0,\ 1,\ 2,\ 3)$——四维时空中点及其坐标,

$\boldsymbol{e}_a=\dfrac{1}{2i}\sigma_a(a=1,\ 2,\ 3)$—— 同位旋空间标准基,$\sigma_a$ 是 Pauli 阵,

$\boldsymbol{b}_\mu=b_\mu^a\boldsymbol{e}_a$—— 规范势,

$\boldsymbol{f}_{\lambda\mu}=\boldsymbol{b}_{\lambda,\mu}-\boldsymbol{b}_{\mu,\lambda}-\boldsymbol{b}_\lambda\times\boldsymbol{b}_\mu=f_{\lambda\mu}^a\boldsymbol{e}_a$—— 规范场强度,

$\boldsymbol{J}_\mu=\eta^{\sigma\nu}\boldsymbol{f}_{\mu\sigma|\nu}=\eta^{\sigma\nu}(\boldsymbol{f}_{\mu\sigma,\nu}+\boldsymbol{b}_\nu\times\boldsymbol{f}_{\mu\sigma})=J_\mu^a\boldsymbol{e}_a$—— 源(式中“,”表偏导数,“|”表规范

导数，$\times$ 表同位旋空间向量积即换位运算，$\eta^{\sigma\nu}$ 四维明可夫斯基空间度规：$\eta^{00}=-1$，$\eta^{11}=\eta^{22}=\eta^{33}=1$，其余为 0)．

$\boldsymbol{b}_\lambda(x)\to\boldsymbol{b}'_\lambda(x)=(\mathrm{ad}\zeta(x))\boldsymbol{b}_\lambda(x)-\zeta_{,\lambda}\zeta^{-1}$—— 势的规范变换，式中 $\zeta(x)\in SU_2$．

§2. 狭义球对称规范场

(a) 势的简化.已知在适当规范下，势的形式为[1]

$$b_j^a(x)=G^a(r,t)x_j,\ b_0^a(x)=K^a(r,t). \tag{2.1}$$

若记 $\boldsymbol{G}=G^a\boldsymbol{e}_a$，$\boldsymbol{K}=K^a\boldsymbol{e}_a$，则(2.1)可写作

$$\boldsymbol{b}_j=\boldsymbol{G}(r,t)x_j,\ \boldsymbol{b}_0=\boldsymbol{K}(r,t). \tag{2.2}$$

选 $\zeta=\zeta(r,t)\in SU_2$，以之作规范变换，则势仍然具形式(2.1)或(2.2)，但 $\boldsymbol{G}(r,t)$，$\boldsymbol{K}(r,t)$ 受到变换

$$\boldsymbol{G}'=(\mathrm{ad}\zeta)\boldsymbol{G}-r^{-1}\partial_r\zeta\zeta^{-1},\ \boldsymbol{K}'=(\mathrm{ad}\zeta)\boldsymbol{K}-\partial_0\zeta\zeta^{-1}. \tag{2.3}$$

特别，取 $\zeta(r,t)$ 为方程

$$\partial_r\zeta=r\zeta\boldsymbol{G}\ (r>0) \tag{2.4}$$

的属于 SU_2 的一解(利用 $\boldsymbol{G}\in$ 李代数 SU'_2，不难证明这种解必存在).所以由(2.3)得知，势可简化为

$$\boldsymbol{b}_j=0,\ \boldsymbol{b}_0=\boldsymbol{K}(r,t)=K^a(r,t)\boldsymbol{e}_a. \tag{2.5}$$

此外，还可取 $\zeta=\zeta(t)\in SU_2$ 作规范变换，它使(2.5)的形式不改变，但

$$\boldsymbol{K}(r,t)\to(\mathrm{ad}\zeta(t))\dot{\boldsymbol{K}}(r,t)-\partial_0\zeta\zeta^{-1}, \tag{2.6}$$

这里 $\partial_0\zeta\zeta^{-1}$ 可取到任何已给函数 $\boldsymbol{l}(t)\in SU'_2$．

(b) 场的强度.现在可直接算出场的强度，

$$f_{ij}=0,\ f_{0i}=\boldsymbol{K}'(r,\ t)\frac{x_i}{r},\tag{2.7}$$

广义的磁场强度和电场强度为

$$H_l^a=\frac{1}{2}\varepsilon_{ijl}f_{ij}^a=0,\ E_i^a=f_{0i}^a=K^{a'}(r,\ t)\frac{x_i}{r},\tag{2.7$'$}$$

因而,不仅广义的磁场强度为0,而且在每一以原点为中心的球面上,广义电场都具有相互平行的同位旋方向.当 $|K'|=\sqrt{(K^{1'})^2+(K^{2'})^2+(K^{3'})^2}\neq 0$ 时,可以定义出 Higgs 场

$$\boldsymbol{n}=\frac{\boldsymbol{K}'}{|\boldsymbol{K}'|},\tag{2.8}$$

由于 $\boldsymbol{n}$ 在球面上平行,所以无磁荷.

(c) 化约条件.因 $\boldsymbol{f}_{\lambda\mu}$ 取 $\boldsymbol{n}$ 的方向,所以化约条件为 $\nabla_\mu\boldsymbol{n}=0$,[4]写出来就是

$$\boldsymbol{n}'=0,\ \dot{\boldsymbol{n}}+\boldsymbol{K}\times\boldsymbol{n}=0\ \left(\dot{\boldsymbol{n}}=\frac{d\boldsymbol{n}}{dt}\right),\tag{2.9}$$

由此可推出势(2.5)为可化约的充要条件为

$$\boldsymbol{K}(r,\ t)=\boldsymbol{n}(t)\varphi(r,\ t)+\boldsymbol{c}(t)\ (\boldsymbol{n}\text{—— 单位同位旋向量}),\tag{2.10}$$

其中 $\boldsymbol{n}$ 和 $\boldsymbol{c}$ 之间满足

$$\frac{d\boldsymbol{n}}{dt}+\boldsymbol{c}\times\boldsymbol{n}=0.\tag{2.11}$$

由(2.6)可见,可作规范变换使 $\boldsymbol{c}(t)=0$,那么 $\boldsymbol{n}$ 和 t 无关,且 $\boldsymbol{K}(r,\ t)=\boldsymbol{n}\varphi(r,\ t)$.

由此可见,对一般的狭义球对称 SU_2 规范场,(2.10)未必成立,当(2.10)不成立时,除电磁场外,还出现带电矢介子场,但这种场只有时间分量.

(d) 源和无源解.源的表达式为

$$\boldsymbol{J}_0=\boldsymbol{K}''+\frac{2}{r}\boldsymbol{K}',\ \boldsymbol{J}_l=(\dot{\boldsymbol{K}}'+\boldsymbol{K}\times\boldsymbol{K}')\frac{x_l}{r},\tag{2.12}$$

若在 $r>0$ 时满足无源方程，则由 $\boldsymbol{J}_0=0$ 得 $r>0$ 时

$$\boldsymbol{K}=\frac{1}{r}\boldsymbol{g}(t)+\boldsymbol{l}(t),$$

可作规范变换使 $\boldsymbol{l}(t)=0$，再代入 $\boldsymbol{J}_l=0$，就得 $\boldsymbol{g}(t)$ 和 t 无关. 据(c)可见场必为化约，而且只是库仑静电场，从而也见到，非静态的狭义球对称的无源解是不存在的.

§3. 同步球对称的 SU_2 规范场

(a) 规范变换和势的简化. 依定义，在适当规范下，同步球对称 SU_2 规范场的势满足条件：对任一 $A\in SO_3$，必成立

$$\boldsymbol{b}(AX,t)=A(\mathrm{ad}\zeta_A)\boldsymbol{b}(X,t),\ \boldsymbol{b}_0(AX,t)=(\mathrm{ad}\zeta_A)\boldsymbol{b}_0(X,t),\tag{3.1}$$

这里 $\zeta_A\in SU_2$，$\mathrm{ad}\zeta_A$ 在同位旋空间的旋转就是 A，$\boldsymbol{b}$ 表 $\boldsymbol{b}_i(i=1,2,3)$.

现在要引入适当的规范变换使势得到简化，设 $\zeta(X,t)\in SU_2$，它满足如下的条件：对任何 $A\in SO_3$ 均成立

$$\zeta(AX,t)=(\mathrm{ad}\zeta_A)\zeta(X,t).\tag{3.2}$$

那么显然成立

$$\mathrm{ad}\zeta(AX,t)=(\mathrm{ad}\zeta_A)(\mathrm{ad}\zeta(X,t))(\mathrm{ad}\zeta_A^{-1}).\tag{3.3}$$

特别(i)若 $AX=X$，即 A 为使 X 不变的旋转，则由(3.3)可见，$\mathrm{ad}\zeta_A$ 和 $\mathrm{ad}\zeta(X,t)$ 可交换，因而 $\mathrm{ad}\zeta(X,t)$ 是使同位旋向量 $x_a\boldsymbol{e}_a$（和普通空间的向径 OX 具相同的分量）不变的旋转. (ii) 若 $AX=X'$，由(3.3)可见，$\mathrm{ad}\zeta(X't)$ 的转角（绕 $x'_a\boldsymbol{e}_a$）和 $\mathrm{ad}\zeta(X,t)$ 的转角（绕 $x_a\boldsymbol{e}_a$）相同，从而可见：如果用满足(3.3)的 $\zeta(X,t)$ 来定义一个规范变换，那么在每点 X，同位旋空间相应的旋转的转轴和向径同方向[①]，在每一瞬时，在每一以 O 为心的球面上，转角相同.

① 如果把 $x_a\boldsymbol{e}_a$ 和 OX 看成同方向的话.

现证明这种规范变换必定使满足同步球对称条件(3.1)的规范势保持为同步球对称的.事实上,由规范变换的定义和(3.1),(3.2)得知,变后的规范势,$\boldsymbol{b}'_\lambda(X, t)$ 满足

$$\begin{aligned}\boldsymbol{b}'(AX, t) &= (\mathrm{ad}\zeta(AX, t))\boldsymbol{b}(AX, t) - (\nabla\zeta(X, t))_{AX}\zeta^{-1}(AX, t)\\ &= (\mathrm{ad}\zeta_A)(\mathrm{ad}\zeta(X, t))A\boldsymbol{b}(X, t) - A(\mathrm{ad}\zeta_A)(\nabla\zeta(X, t))\zeta^{-1}(X, t)\\ &= (\mathrm{ad}\zeta_A)A\boldsymbol{b}'(X, t),\end{aligned}$$

式中 $(\nabla\zeta(X, t))_{AX}$ 记 $\zeta(X, t)$ 的梯度在 AX 的值,同样也有

$$\boldsymbol{b}'_0(AX, t) = (\mathrm{ad}\zeta_A)\boldsymbol{b}'_0(X, t),$$

这些就是所要证的事项.

再来简化规范势.已知同步球对称规范势的一般形式为[5]

$$\begin{aligned}&b_i^a(X, t) = \varepsilon_{iak}x_k V(r, t) + \delta_{ia}S(r, t) + x_i x_a T(r, t),\\ &b_0^a(X, t) = x_a U(r, t).\end{aligned} \tag{3.4}$$

取 $\zeta(X, t) \in SU_2$,满足(3.2)式,那么 $\mathrm{ad}\zeta(r, 0, 0, t)$ 必使 $\boldsymbol{e}_1$ 不变,所以$\zeta(r, 0, 0, t) = \exp(\varphi(r, t)\boldsymbol{e}_1)$,在$(r, 0, 0, t)$ 处,$(\partial_1\zeta)\zeta^{-1} = \varphi'(r, t)\boldsymbol{e}_1$,因此,

$$b_1^1(r, 0, 0, t) = S(r, t) + T(r, t)r^2 \rightarrow S(r, t) + T(r, t)r^2 - \varphi'(r, t),$$

所以必可通过规范变换使

$$S + Tr^2 = 0. \tag{3.5}$$

规范势 b_i^a 还可以写为下述形式:

$$b_i^a(X, t) = -\varepsilon_{iak}x_k\frac{1}{r^2} + \frac{1}{r^2}\left(\varepsilon_{iak}x_k\phi(r, t) + \left(\delta_{ia} - \frac{x_i x_a}{r^2}\right)\sigma(r, t)\right), \tag{3.6}$$

式中

$$\phi(r, t) = 1 + r^2 V, \ \sigma = Sr. \tag{3.7}$$

还可证明(见附录),能够选取 $\zeta = \zeta(X, t)$,使满足(3.2) 式及 $\partial_r\zeta = 0$,这时(3.6) 经规范

变换后化为

$$b_i^{a'}(X, t) = -\varepsilon_{iak}x_k \frac{1}{r^2} + \frac{1}{r^2}\left(\varepsilon_{iak}x_k \bar{\phi}(r, t) + r\left(\delta_{ia} - \frac{x_i x_a}{r^2}\right)\sigma(r, t)\right), \quad (3.8)$$

这里

$$\begin{aligned} \bar{\phi}(r, t) &= \phi(r, t)\cos\alpha(t) + \sigma(r, t)\sin\alpha(t), \\ \bar{\sigma}(r, t) &= -\phi(r, t)\sin\alpha(t) + \sigma(r, t)\cos\alpha(t). \end{aligned} \quad (3.9)$$

(b) 场的强度.在 $S + r^2 T = 0$ 成立时,[5]中场的强度的表达式化为

$$\begin{aligned} f_{ij}^a &= A\varepsilon_{ijk}x_k x_a + B(\delta_{ai}x_j - \delta_{aj}x_i) + C\varepsilon_{aij}, \\ f_{i0}^a &= Lx_i x_a + M\delta_{ia} + N\varepsilon_{iak}x_k, \end{aligned} \quad (3.10)$$

式中

$$\begin{aligned} &A = \frac{V'}{r} - V^2 - \frac{S^2}{r^2},\ B = \frac{S'}{r} + \frac{S}{r^2},\ C = -(2V + rV'), \\ &L = UV - \frac{U'}{r} - \frac{\dot{S}}{r^2},\ M = -U(1 + r^2 V) + \dot{S},\ N = SU + \dot{V}, \end{aligned} \quad (3.11)$$

f_{i0}^a 即 $-E_i^a$,这里 E_i^a 是广义的电场强度,又广义的磁场强度 H_l^a 为

$$H_l^a = \frac{1}{2}\varepsilon_{ijl}f_{ij}^a = Ax_l x_a + B\varepsilon_{ail}x_i + C\delta_{la}. \quad (3.12)$$

由于同步球对称性,在这种 SU_2 场中可以自然地引入 Higgs 场,它就取 $\pm\frac{x_a}{r}\boldsymbol{e}_a$,相应的磁荷为 ± 1,由 't Hooft 公式所作出的磁场强度就是库仑场的表达式,电场强度 E_i 的表达式[1]化为

$$E_i = \pm(U + rU')\frac{x_i}{r}. \quad (3.13)$$

(c) 强度和同位旋方向平行的场和化约的场.现讨论同步球对称规范场的某些特殊情况,若在每一点强度的同位旋方向都平行于一个同位旋向量(可随点而变化),由球对

称性可见,这方向应和向径平行.在$(r,0,0,t)$点考察,就得到条件

$$M=N=B=C=0, \tag{3.14}$$

从$B=C=0$得

$$\sigma=rS=k_1(t),\ \phi-1=r^2V=k_2(t), \tag{3.15}$$

代入$M=N=0$,得

$$\frac{\dot{k}_1}{r}-U(1+k_2)=0,\ \frac{\dot{k}_2}{r}+Uk_1=0, \tag{3.16}$$

从而得

$$k_1^2+(1+k_2)^2=k=\text{const}. \tag{3.17}$$

当k_1,$1+k_2$不全为0时,k_1和k_2之一可以作为t的任意函数,函数U由(3.16)确定,此时

$$H_l^a=\frac{1-k}{r^4}x_a x_l,\ E_l^a=0, \tag{3.18}$$

k可取任意大于0的常数.这时广义的电磁场只能是静态的,而且广义电场强度为0,广义磁场没有量子化条件.

当k_1和$1+k_2$全取0时,U可为任意函数,$k=0$.

$$H_l^a=\frac{x_a x_l}{r^4},\ E_l^a=-\frac{(rU)'}{r^2}x_a x_l, \tag{3.19}$$

这时广义磁场是静态的,广义电场却有相当大的自由度.

特别,我们考虑到化约条件$\nabla_\mu \boldsymbol{n}=0$,就得出化约的势

$$V=-\frac{1}{r^2},\ S=0,\ U\text{为任意函数}, \tag{3.20}$$

这就是$k_1=(1+k_2)=0$的情形.

(d) 静态无源方程的讨论.设在$r>0$时无源方程满足.我们限于讨论静态的情形.经

计算,源的表达式为

$$J_0^a = \mathscr{A} x_a, \qquad (3.21)$$
$$J_l^a = \mathscr{B} \varepsilon_{lai} x_i + \mathscr{C} \delta_{al} + \mathscr{D} x_a x_l,$$

式中

$$\mathscr{A} = U'' - 2(r^2V^2 + S^2 + 2V)U + \frac{4U'}{r},$$
$$\mathscr{B} = MU + A - \frac{C'}{r} + r^2VA + VC, \qquad (3.22)$$
$$\mathscr{C} = -r^2UN + B'r + 2B + r^2SA + CS,$$
$$\mathscr{D} = UN - \frac{B'}{r} - SA + \frac{SC}{r^2} + 2VB,$$

所以无源方程就是

$$\mathscr{A} = \mathscr{B} = \mathscr{C} = \mathscr{D} = 0. \qquad (3.23)$$

从 $\mathscr{B} = \mathscr{C} = 0$ 得出

$$B(1 + Vr^2) + CS = 0, \qquad (3.24)$$

由于 $\phi = (1 + Vr^2)$, $\sigma = rS$, (3.24)化为

$$\phi'\sigma - \phi\sigma' = 0. \qquad (3.25)$$

若 $\phi = \sigma = 0$, 则得

$$V = -\frac{1}{r^2},\ S = 0,\ U = \frac{c_1}{r} + c_2. \qquad (3.26)$$

为可化约的解,若 ϕ, σ 不全为 0, 就可以令

$$\phi = \phi_1 \cos\alpha,\ \sigma = \phi_1 \sin\alpha,\ G = r^2U. \qquad (3.27)$$

从(3.22)就得到方程

$$r^2\phi''_1=\phi_1(\phi_1^2-1-G^2),\quad r^2G''=2\phi_1^2G. \tag{3.28}$$

应该注意的是，在 $S=0$ 时，(3.28) 是已经导出的无源方程[6, 7].现在我们从一般同步球对称情况(即不假设 $S=0$) 出发，也把求无源解的问题归到方程(3.28) 去.实际上，若置 $\sigma_1=0$，从(3.28) 得出了无源解(ϕ_1, G)，那么，从(3.27) 式得出的无源解(ϕ, σ, G) 并不是实质上新的解，它是可以由解(ϕ_1, G) 经过适当的规范变换而导出的，这一点可以由(3.9)式看出.另一方面，从计算这两个解相应的规范场的强度，可以知道它们的强度之间是能够通过同位旋转动互化的，从而利用[5]中的定理 3，也能推出这个结论.

从而可见，在考虑静态无源解时，即使预先不设定 S, T 为 0，但实际上可考虑 $T=0$, $S=0$ 的情形.

方程(3.28)一般要用数值方法求解，$U=0$ 时已有详细分析[8]，当 U 容许为纯虚时，已有一组解析解[6].

§4. 倍步球对称的 SU_2 规范场

已经知道，这种场一定是可化约的，因而可表为整体的 U_1 规范场.但我们现在要给出它的 SU_2 规范势.为此，先作出一种具 m 重复盖特性的 Higgs 场，设(r, θ, φ)为空间的球面坐标，选

$$\boldsymbol{n}(r, \theta, \varphi, t)=(\sin\theta\cos m\varphi,\ \sin\theta\sin m\varphi,\ \cos\theta), \tag{4.1}$$

式中 $|m|>1$, m 为整数，它确将以 O 为心的球面映照为同位旋空间的单位球面去，并遮盖它 m 次.这种场的规范势由 $\partial_\lambda\boldsymbol{n}+\boldsymbol{b}_\lambda\times\boldsymbol{n}=0$ 确定，特别可取

$$\boldsymbol{b}_i=-\boldsymbol{n}\times\partial_i\boldsymbol{n},\ \boldsymbol{b}_0=U\boldsymbol{n}, \tag{4.2}$$

式中 U 是(t, r)的函数.

通过计算可见普通意义下的磁场强度和电场强度为

$$H_k = m\frac{x_k}{r^3},\ E_k = E(r,\ t)\frac{x_k}{r^3}, \tag{4.3}$$

式中 $\frac{E}{r^2}=U'$. 由于(4.3)已表示了一般的具 m 磁荷的球对称电磁场,所以(4.2)也已是一般(除规范变换外)的规范势了.

如在 $r>0$ 满足无源方程,也就是(4.3)满足无源的 Maxwell 方程,则由 $\operatorname{div}\boldsymbol{E}=0$ 得出 $E(r,\ t)=E(t)$,又由 $\nabla\times\boldsymbol{H}=0$ 得知 $E(t)$ 和 t 无关,所以场化为静电场加静磁场.

附　　录

取 $\zeta(r,\ 0,\ 0,\ t)=\exp(\alpha(t)\boldsymbol{e}_1)$, A 为使 $(r,\ 0,\ 0)$ 变为 X 的旋转, 定义

$$\zeta(X,\ t)=(\mathrm{ad}\zeta_A)\exp(\alpha(t)\boldsymbol{e}_1),$$

这样的 $\zeta(X,\ t)$ 是有确定意义的,因为若 A' 是使 $(r,\ 0,\ 0)$ 变为 X 的另一旋转,则 $A^{-1}A'$ 使 $(r,\ 0,\ 0)$ 不变.

$$(\mathrm{ad}\zeta_{A'})\exp(\alpha(t)\boldsymbol{e}_1)=(\mathrm{ad}\zeta_A)(\mathrm{ad}\zeta_A^{-1}\zeta_{A'})\exp(\alpha(t)\boldsymbol{e}_1)=(\mathrm{ad}\zeta_A)\exp(\alpha(t)\boldsymbol{e}_1).$$

现取 A 为绕 x_3 轴的无限小旋转,使 $(r,\ 0,\ 0)$ 变为 $(r,\ \delta x_2,\ 0)$,则

$$\zeta_A=\exp\left(\frac{\delta x_2}{r}\boldsymbol{e}_3\right),$$

从此可见

$$\zeta(r,\ \delta x_2,\ 0,\ t)=\exp(\alpha\boldsymbol{e}_1)+\frac{\delta x_2}{r}(\boldsymbol{e}_3\exp(\alpha\boldsymbol{e}_1)-\exp(\alpha\boldsymbol{e}_1)\boldsymbol{e}_3).$$

从而得到在 $(r,\ 0,\ 0,\ t)$ 处的

$$(\partial_2\zeta)\zeta^{-1}=\frac{1}{r}\boldsymbol{e}_3-\frac{1}{r}(\mathrm{ad}\exp(\alpha\boldsymbol{e}_1))\boldsymbol{e}_3=\frac{\sin\alpha}{r}\boldsymbol{e}_2+\frac{1-\cos\alpha}{r}\boldsymbol{e}_3,$$

利用这样的 $\zeta(X,\ t)$ 作规范变换将

$$b_2^2(r,0,0,t)=S \longrightarrow \bar{S}=S\cos\alpha - rV\sin\alpha - \frac{\sin\alpha}{r},$$

$$b_2^3(r,0,0,t)=rV \longrightarrow r\bar{V}=S\sin\alpha + rV\cos\alpha - \frac{1-\cos\alpha}{r},$$

这就是

$$\bar{\phi}(r,t)=\phi(r,t)\cos\alpha+\sigma(r,t)\sin\alpha,$$

$$\bar{\sigma}(r,t)=-\phi(r,t)\sin\alpha+\sigma(r,t)\cos\alpha,$$

(3.9)式得证.

参考资料

[1] 谷超豪,胡和生,球对称的 SU_2 规范场和磁单极的规范场描述(未发表).

[2] 侯伯宇,段一士,葛墨林,兰州大学学报自然科学版 1975 年第 2 期,26.

[3] Arafune J., Freund P.G.O., & Goebel C.J., *Jour. Math. Phys.* **16** 433 (1975).

[4] 侯伯宇,SU_2 规范场的结构(未发表).

[5] 胡和生,关于一类 SU_2 群的规范场,复旦学报 1976 年第 1 期,72.

[6] Prased M. K. & Sommerfield C. M., *Phys. Rev. Lett.* **35** 760(1975).

[7] Wu T. T. & Yang C. N. *A static sourceless gauge field* ITP-SB-76-2.

[8] Wu T. T. & Yang C. N., "*Properties of Matter under Unusual Conditions*", 349 (1969).

(本文曾发表于《复旦学报(自然科学版)》1977 年第 1 期,92-99)

局部对偶的黎曼空间和引力瞬子解

复　旦　大　学　谷超豪　李大潜　忻元龙

沈纯理　胡和生

美国纽约州立大学　杨振宁①

摘　　要

四维正定黎曼空间 R_4 能局部地生成两个 SU_2 规范场 $\mathscr{F}^+$ 和 $\mathscr{F}^-$，如果 $\mathscr{F}^+$，$\mathscr{F}^-$ 至少有一个具有自对偶性或反自对偶性，那么空间称为具局部对偶性的.我们证明它们是 Einstein 空间、数量曲率为 0 的共形平坦空间以及 $R^{++}=0$(或 $R^{--}=0$) 的空间.文中得出了 $R^{++}=0(R^{--}\neq 0)$ 的一类黎曼线素.对曲率张量平方可积的情形，作出了规范场作用量，Euler 示性数，Pontrjagin 示性数之间的一个不等式，证明它的等号在而且只在 R_4 具局部对偶性时达到，这结果改进了[7]中关于引力瞬子解的研究，并以 Hitchin 关于四维紧致 Einstein 流形的一个不等式作为特殊情况[8].

§1. 引　　言

近来欧氏空间中具自对偶(反自对偶)性的 SU_2 规范场引起了人们的注意[1~4].对于四维正定的黎曼空间，由于 SO_4 群局部同构于 $SU_2\times SU_2$，所以它的 Christoffel 联络也生成了两个局部的 SU_2 规范场 $\mathscr{F}^+$ 和 $\mathscr{F}^-$.在本文中，我们首先研究这两个规范场在什么时候至少有一个具自对偶(反自对偶) 性的问题.我们证明了：除了熟知的 Einstein 空间

① 杨振宁教授于 1977 年 7 月访问了上海，和复旦大学的教师进行了共同的学术讨论和研究.

之外，这里还有 $R=0$ 的共形平坦空间(包括 't Hooft, Jackiw-Nohl- Rebbi 解[5, 6]) 以及满足 $R^{++}=0$, $R^{--}\neq 0$ 或 $R^{--}=0$, $R^{++}\neq 0$ 的空间，后面这两类空间的 SU_2 规范场 $\mathscr{F}^+$ 与 $\mathscr{F}^-$ 中只有一个有自对偶性(反自对偶性)，我们举出了一类具这种性质的局部黎曼线素.

其次，我们讨论由规范场作用量所定义的引力瞬子解[7]，作出了作用量、Euler 数、Pontrjagin 数应满足的不等式，并且证明，这些不等式的等号在且仅在 $\mathscr{F}^+$ 或 $\mathscr{F}^-$ 有自对偶性(反自对偶性)时成立，此时作用量达到了最小值，从而得出相应各类流形的这两种示性数的关系.这些结果改进了[7]中的研究，并指出了引力瞬子解和 SU_2 瞬子解的关系.又 Hitchin 关于紧致的 Einstein 空间的不等式[8]是这里的特殊情况.

§2. 四维黎曼空间中的局部 SU_2 规范场

设 R_4 是四维正定黎曼空间，线素为 $ds^2=g_{\lambda\mu}dx^\lambda dx^\mu(\lambda, \mu=1, 2, 3, 4)$，黎曼曲率张量为 $R_{\lambda\mu\rho\sigma}$，定义

$$\eta_{\lambda\mu\rho\sigma}=\sqrt{g}\,\varepsilon_{\lambda\mu\rho\sigma}. \tag{1}$$

式中 $g=\det|g_{\lambda\mu}|$，$\varepsilon_{\lambda\mu\rho\sigma}$ 关于下标为反称，$\varepsilon_{1284}=1$.利用 $\eta_{\lambda\mu\rho\sigma}$，可以对二阶反称张量作对偶张量.对于曲率张量 $R_{\lambda\mu\rho\sigma}$，也可以作

$$\begin{aligned}&R^{*\,\cdot}_{\lambda\mu\rho\sigma}=\frac{1}{2}\eta_{\lambda\mu}{}^{\alpha\beta}R_{\alpha\beta\rho\sigma},\ R^{\cdot\,*}_{\lambda\mu\rho\sigma}=\frac{1}{2}\eta_{\rho\sigma}{}^{\alpha\beta}R_{\lambda\mu\alpha\beta},\\&R^{*\,*}_{\lambda\mu\rho\sigma}=\frac{1}{4}\eta_{\lambda\mu}{}^{\alpha\beta}\eta_{\rho\sigma}{}^{\gamma\delta}R_{\alpha\beta\gamma\delta}.\end{aligned} \tag{2}$$

在不致引起误解的情形下，我们把 $R^{*\,\cdot}_{\lambda\mu\rho\sigma}$ 等等的下标略去分别简记为 $R^{*\,\cdot}$ 等等.又 $R_{\lambda\mu\rho\sigma}$ 简记为 $R^{\cdot\,\cdot}$，它可以表示为四个张量 R^{++}, R^{--}, R^{+-}, R^{-+} 之和，这里

$$R^{++}=\frac{1}{4}(R^{\cdot\,\cdot}+R^{*\,\cdot}+R^{\cdot\,*}+R^{*\,*}),$$

$$R^{--}=\frac{1}{4}(R^{\cdot\,\cdot}-R^{*\,\cdot}-R^{\cdot\,*}+R^{*\,*}),$$

$$R^{+-}=\frac{1}{4}(R^{\cdot\cdot}+R^{*\cdot}-R^{\cdot *}-R^{**}),$$
$$R^{-+}=\frac{1}{4}(R^{\cdot\cdot}-R^{*\cdot}+R^{\cdot *}-R^{**}). \tag{3}$$

这四个张量各有其不同的对偶性：＋号表示自对偶性，－号表示反自对偶性，例如 R^{++} 关于前后两对指标均有自对偶性，R^{--} 关于前后两对指标均有反自对偶性，等等.

由于 R_4 的切丛是 SO_4 群的纤维丛，而 SO_4 局部地分解为直积 $SU_2\times SU_2$（或 $SO_3\times SO_3$），所以由 R_4 的克氏记号可以得出两个局部的 SU_2 规范场.我们选一组归一正交标架 e_i，记其变位方程为

$$De_i=\Gamma_{ij\mu}dx^{\mu}e_j. \tag{4}$$

又令

$$\overset{+}{\Gamma}_{ij\mu}=\frac{1}{2}(\Gamma_{ij\mu}+\Gamma^{*}_{ij\mu}),\ \overset{-}{\Gamma}_{ij\mu}=\frac{1}{2}(\Gamma_{ij\mu}-\Gamma^{*}_{ij\mu}), \tag{5}$$

那么 4×4 阵 $\overset{+}{\Gamma}_{\mu}=(\overset{+}{\Gamma}_{ij\mu})$ 和 $\overset{-}{\Gamma}_{\mu}=(\overset{-}{\Gamma}_{ij\mu})$ 就局部地定义了两个 SU_2 规范场 $\mathscr{F}^{+}$ 和 $\mathscr{F}^{-}$ 的规范势. 并成立

引理 1. $\mathscr{F}^{+}$ 和 $\mathscr{F}^{-}$ 的强度分别是

$$R^{+}=R^{++}+R^{+-},\ R^{-}=R^{-+}+R^{--}. \tag{6}$$

【证明】 $\Gamma_{\mu}=(\Gamma_{ij\mu})$ 是 SO_4 规范势，其强度为

$$f_{\lambda\mu}=\Gamma_{\lambda,\mu}-\Gamma_{\mu,\lambda}-[\Gamma_{\lambda},\Gamma_{\mu}]=-(R_{ij\lambda\mu}). \tag{7}$$

右边 $R_{ij\lambda\mu}$ 是曲率张量 $R_{\rho\sigma\lambda\mu}$ 的混合指标表示，也就是把 $R_{\rho\sigma\lambda\mu}$ 的前两个指标转换到正交标架 e_i 后所得到的分量.由于 $\Gamma_{\mu}=\overset{+}{\Gamma}_{\mu}+\overset{-}{\Gamma}_{\mu}$，又 $\overset{+}{\Gamma}_{\mu}$ 和 $\overset{-}{\Gamma}_{\mu}$ 可交换，所以就得到 $\mathscr{F}^{+}$ 和 $\mathscr{F}^{-}$ 的强度分别为

$$f^{+}_{\lambda\mu}=\overset{+}{\Gamma}_{\lambda,\mu}-\overset{+}{\Gamma}_{\mu,\lambda}-[\overset{+}{\Gamma}_{\lambda},\overset{+}{\Gamma}_{\mu}]=-(R^{+}_{ij\lambda\mu}),$$
$$f^{-}_{\lambda\mu}=\overset{-}{\Gamma}_{\lambda,\mu}-\overset{-}{\Gamma}_{\mu,\lambda}-[\overset{-}{\Gamma}_{\lambda},\overset{-}{\Gamma}_{\mu}]=-(R^{-}_{ij\lambda\mu}). \tag{8}$$

式中 $R^{+}_{ij\lambda\mu}=\frac{1}{2}(R_{ij\lambda\mu}+R^{\ \ **}_{ij\lambda\mu})$，$R^{-}_{ij\lambda\mu}=\frac{1}{2}(R_{ij\lambda\mu}-R^{\ \ **}_{ij\lambda\mu})$，由此就得出(6)式,引理证毕.

由此还可见到

引理 2. (i) $\mathscr{F}^{+}$具反自对偶性的充要条件是 $R^{++}=0$，$\mathscr{F}^{-}$具自对偶性的充要条件是 $R^{--}=0$.

(ii) $\mathscr{F}^{+}$具自对偶性的充要条件是 $R^{+-}=0$，$\mathscr{F}^{-}$具反自对偶性的充要条件是 $R^{-+}=0$;又 $R^{+-}=0$ 和 $R^{-+}=0$ 等价.

【证明】 (i)和(ii)的前一结论是由引理 1 及 R^{++}，R^{--}，R^{+-}，R^{-+} 的定义直接给出的.(ii)的后一结论由

$$R^{\ \ *\ \cdot}_{\lambda\mu\rho\sigma}=R^{\ \ \cdot\ *}_{\rho\sigma\lambda\mu} \tag{9}$$

给出.

从此引理可以见到自对偶的 $\mathscr{F}^{+}$和反自对偶的 $\mathscr{F}^{-}$ 必同时出现.

§3. 具对偶性的局部黎曼空间

我们把满足 $R^{++}=0$，$R^{--}=0$ 或 $R^{+-}=0$ 的空间通称为具对偶性的黎曼空间.

已经知道满足 $R^{+-}=0$ 的空间是 Einstein 空间[8].

以下讨论 $R^{++}=0$(或 $R^{--}=0$) 的空间.

引理 3. $R^{++}=0$(或 $R^{--}=0$) 的空间的数量曲率 $R=0$.

【证明】 采用归一化正交标架,由 $R^{++}=0$ 得出

$$R_{1234}+R_{3434}+R_{1212}+R_{3412}=0,$$

轮换指标 2, 3, 4,把所得的式子相加,利用 R_{ijkl} 间熟知的等式,就得到引理的结论.

定理 1. 满足 $R^{++}=0$，$R^{--}=0$ 的空间是数量曲率 $R=0$ 的共形平坦空间.

【证明】 在归一正交标架下,由 $R^{++}=0$，$R^{--}=0$ 能推出

$$R_{1234}=0,\ R_{1212}=-R_{3434},\ R_{2312}=R_{4314} \tag{10}$$

等式子，把它们代入 Weyl 张量

$$C_{hijk}=R_{hijk}+\frac{1}{2}(\delta_{hj}R_{ik}-\delta_{hk}R_{ij}+\delta_{ik}R_{hj}-\delta_{ij}R_{hk})+\frac{R}{6}(\delta_{hk}\delta_{ij}-\delta_{hj}\delta_{ik}).$$

并注意到 $R=0$，就会得出 $C_{hijk}=0$，因此空间为共形平坦.

由此可见，满足 $R^{++}=R^{--}=0$ 的线素必为

$$ds^2=u^2((dx^1)^2+(dx^2)^2+(dx^3)^2+(dx^4)^2). \tag{11}$$

式中 u 是调和函数 $u>0$[4]，因而从局部的观点来看，这种线素是很多的.

现在看满足 $R^{++}=0$，$R^{--}\neq 0$ 的线素是否存在？从局部的观点看，这种线素也是很多的，例如我们考虑

$$ds^2=A^2((dx^1)^2+(dx^2)^2)+B^2((dx^3)^2+(dx^4)^2),$$

令 $A=\ln\alpha$，$B=\ln(\alpha+\rho)$，利用曲率张量的表达式([9]的(37.4))写出 $R^{++}=0$ 的条件，经过一定的演化后，这条件化为

$$\rho_{,24}=\rho_{,13},\ \rho_{,23}=-\rho_{,14}, \tag{12}$$

$$\rho_{,33}+\rho_{,44}-e^{2\rho}(\rho_{,11}+\rho_{,22})=0, \tag{13}$$

$$A_{,33}+A_{,44}+e^{2\rho}(A_{,11}+A_{,22}+2A_{,1}\rho_{,1}+2A_{,2}\rho_{,2}+(\rho_{,1}{}^2+\rho_{,2}{}^2)A)=0. \tag{14}$$

我们并要求 $\rho_{,24}$，$\rho_{,13}$，$\rho_{,23}$，$\rho_{,14}$ 中至少有一个不为 0(否则就会有 $R^{--}=0$).

从(12)易知

$$(\rho_{,11}+\rho_{,22})_{,3}=0,\ (\rho_{,11}+\rho_{,22})_{,4}=0,$$

$$(\rho_{,33}+\rho_{,44})_{,1}=0,\ (\rho_{,33}+\rho_{,44})_{,2}=0.$$

如果 $\rho_{,11}+\rho_{,22}\neq 0$，那么由(13)式可知 ρ 具形状

$$\rho=f(x^1, x^2)-g(x^3, x^4).$$

因此$\rho_{,24}$，$\rho_{,13}$，$\rho_{,23}$，$\rho_{,14}$全部为0，这是不许可的，所以我们必须有

$$\rho_{,11}+\rho_{,22}=0, \rho_{,33}+\rho_{,44}=0. \tag{15}$$

作任意的解析函数$\Phi(x_1-ix_2, x_3+ix_4)$，其实部$\rho(x_1, x_2, x_3, x_4)$能使(12)，(15)成立，从而(13)也成立，方程(14)是关于A的一个线性二阶椭圆型方程，它有很多不等于0的解，因而就得到$R^{++}=0$，$R^{--}\neq 0$的局部黎曼空间.对$R^{--}=0$，$R^{++}\neq 0$的情形也有类似的情况，因而有

定理2. $\mathscr{F}^+$为反自对偶，$\mathscr{F}^-$不为自对偶或反自对偶(或者$\mathscr{F}^-$为自对偶，$\mathscr{F}^+$不为自对偶或反自对偶)的局部黎曼空间是存在的.

现举一特例：令$\Phi=\ln(x_1-ix_2+x_3+ix_4)$，那么

$$\rho=\frac{1}{2}\ln((x_1+x_3)^2+(x_2-x_4)^2).$$

如果设A和x^1，x^2无关，那么(14)就化为

$$A_{,33}+A_{,44}+A=0.$$

取特解

$$A=\sin\frac{x^3}{a}\sin\frac{x^4}{b}\left(\frac{1}{a^2}+\frac{1}{b^2}=1\right).$$

那么就得到$R^{++}=0(R^{--}\neq 0)$的一个线素.

$$ds^2=\sin^2\frac{x^3}{a}\sin^2\frac{x^4}{b}\{(dx^1)^2+(dx^2)^2+[(x_1+x_3)^2+(x_2-x_4)^2]$$
$$((dx^3)^2+(dx^4)^2)\},$$

它在

$$0<x^3<a\pi, 0<x^4<b\pi, (x_1+x_3)^2+(x_2-x_4)^2\neq 0$$

处有定义，是一个不完备的黎曼流形.

§4. 一个积分不等式

现在考虑有关的整体性质.我们考察连通的微分流形 M_4(可以是有边的),研究在 M_4 上定义的黎曼结构(可以是不完备的),我们选取规范场的作用量

$$S_W=\int R_{\alpha\beta\gamma\delta}R^{\alpha\beta\gamma\delta}\sqrt{g}\,d^4x=\int R\cdot R\sqrt{g}\,d^4x, \tag{16}$$

要求它为平方可积.[7]中定义 S_W 取最小值的黎曼结构为引力瞬子解.流形 M_4 的 Euler 数和 Pontrjagin 数 I_E 和 I_P 可定义为①

$$I_E=\int\frac{1}{4}\eta^{\alpha\beta\gamma\delta}\eta^{\lambda\mu\rho\sigma}R_{\alpha\beta\lambda\mu}R_{\gamma\delta\rho\sigma}\sqrt{g}\,d^4x, \tag{17}$$

$$I_P=\int\frac{1}{2}\eta^{\alpha\beta\gamma\delta}R_{\gamma\delta}{}^{\lambda\mu}R_{\alpha\beta\lambda\mu}\sqrt{g}\,d^4x. \tag{18}$$

又记积分 S_W, I_E, I_P 的核分别为 K_W, K_E, K_I.

引理 4. 具有限作用量的四维黎曼流形其 S_W, I_P, I_E 之间必成立等式

$$\begin{aligned}S_W=&\frac{4}{A^2+B^2+C^2+D^2}\int(AR^{++}+BR^{+-}+CR^{-+}+DR^{--})^2\sqrt{g}\,d^4x\\&+\frac{(-2I_P-I_E)A^2+I_EB^2+I_EC^2+(2I_P-I_E)D^2}{A^2+B^2+C^2+D^2}.\end{aligned} \tag{19}$$

这里 A, B, C, D 是任意常数.

【证明】 由定义经直接计算知道

$$\begin{aligned}&K_W=R^{\cdot\cdot}\cdot R^{\cdot\cdot}=R^{*\cdot}\cdot R^{*\cdot}=R^{\cdot *}\cdot R^{\cdot *}=R^{**}\cdot R^{**},\\&K_E=R^{\cdot\cdot}\cdot R^{**}=R^{*\cdot}R^{\cdot *},\\&K_P=R^{\cdot\cdot}\cdot R^{*\cdot}=R^{\cdot\cdot}\cdot R^{\cdot *}=R^{*\cdot}\cdot R^{**}=R^{\cdot *}\cdot R^{**}.\end{aligned} \tag{20}$$

① 我们采用[7]中的表达式,它们和数学中常用的 I_E, I_P 均相差一倍数.

所以从(3)可知

$$\begin{aligned} R^{++}\cdot R^{++} &= \frac{1}{4}(K_W + 2K_P + K_E), \\ R^{+-}\cdot R^{+-} &= \frac{1}{4}(K_W - K_E), \\ R^{-+}\cdot R^{-+} &= \frac{1}{4}(K_W - K_E), \\ R^{--}\cdot R^{--} &= \frac{1}{4}(K_W - 2K_P + K_E). \end{aligned} \tag{21}$$

其他诸如 $R^{++}\cdot R^{+-}$，…，$R^{--}\cdot R^{-+}$ 均为 0.

现在把(19)式右端的积分项乘开来，利用这些关系式就得出了(19).

定理 3. 对具有限作用量的黎曼流形 M_4 必有

(i) 当 $|I_P|\leqslant I_E$ 时，$S_W\geqslant I_E$；

(ii) 当 $I_P\leqslant 0$，$I_P\leqslant -I_E$ 时，$S_W\geqslant -2I_P-I_E$；

(iii) 当 $I_P\geqslant 0$，$I_P\geqslant I_E$ 时，$S_W\geqslant 2I_P-I_E$.

【证明】 据引理 4，成立

$$\begin{aligned} S_W &\geqslant \max_{A^2+B^2+C^2+D^2=1}\left[(-2I_P-I_E)A^2+I_EB^2+I_EC^2+(2I_P-I_E)D^2\right] \\ &= \max\{-2I_P-I_E,\ I_E,\ 2I_P-I_E\}. \end{aligned} \tag{22}$$

所以当 $I_E\geqslant -2I_P-I_E$，$I_E\geqslant 2I_P-I_E$ 时，即当 $|I_P|\leqslant I_E$ 时，$S_W\geqslant I_E$，此即结论(i).

当 $-2I_P-I_E\geqslant I_E$，$-2I_P-I_E\geqslant 2I_P-I_E$ 时，即当 $I_P\leqslant 0$ 和 $-I_P\geqslant I_E$ 时成立 $S_W\geqslant -2I_P-I_E$，此即情形(ii).

当 $2I_P-I_E\geqslant I_E$，$2I_P-I_E\geqslant -2I_P-I_E$ 时，即当 $I_P\geqslant 0$，$I_P\geqslant I_E$ 时成立 $S_W\geqslant 2I_P-I_E$，此即情形(iii).

这定理改进了[7]中所得到的关于 S_W 的不等式.

§5. 引力瞬子解

我们现在看定理 3 中关于 S_W 的不等式中等号成立的情形.这种黎曼线素使作用量

S_W 取到了最小值，因而满足相应的 Euler-Lagrange 方程，这种黎曼线素是[7]中所定义的引力瞬子解.

现设(A, B, C, D)使(22)式右边达到最大值，且使(22)成为等式，那么必须有

$$AR^{++}+BR^{+-}+CR^{-+}+DR^{--}=0.$$

因为 R^{++}, R^{+-}, R^{-+}, R^{--} 具有不同的对偶性，所以成立

$$AR^{++}=0,\ BR^{+-}=0,\ CR^{-+}=0,\ DR^{--}=0. \tag{23}$$

因 A, B, C, D 不全为 0，所以 R^{++}, R^{+-}, R^{--} 至少有一为 0.

反过来，若

(1) $R^{+-}=0$.选 $A=C=D=0$, $B=1$，那么就成立 $S_W=I_E$，再利用不等式(22)就得出

$$I_E\geqslant |I_P|. \tag{24}$$

已经知道 $R^{+-}=0$ 的空间是 Einstein 空间.若成立 $I_E=I_P$，可再选 $A=B=C=0$, $D=1$，而且(22)为等式，从而得 $R^{--}=0$.若成立 $I_E=-I_P$，可再选 $B=C=D=0$, $A=1$，而且(22)为等式，从而得 $R^{++}=0$.特别若 $I_E=I_P=0$，空间为平坦的.当流形为紧致时，(24)式化为[8]中所得到的不等式.

(2) $R^{++}=0$.选 $B=C=D=0$, $A=1$，就成立 $S_W=-2I_P-I_E$. 根据(22)式，就有

$$I_P\leqslant -I_E,\ I_P\leqslant 0. \tag{25}$$

如(25)的第一式等号成立，那么可再选 $A=C=D=0$, $B=1$，由于(22)为等式，从而得 $R^{+-}=0$，这就是(1)中已列出的情形.若(25)的第二式等号成立，可再选 $A=B=C=0$, $D=1$，因(22)为等式，从而得 $R^{--}=0$. 因而空间局部共形平坦，且 $R=0$. 't Hooft, Jackiw-Nohl-Rebbi 解就属于这种情形.

(3) $R^{--}=0$，选 $A=B=C=0$, $D=1$，就成立 $S_W=2I_P-I_E$，根据(22)式就有

$$I_P\geqslant I_E,\ I_P\geqslant 0. \tag{26}$$

若(26)第一式等号成立,就得到(1)中已讨论过的 $R^{+-}=0$. 若(26)式第二式等号成立,就得到(2)中已讨论过的 $R^{++}=R^{--}=0$ 的情形.

总结以上情况,我们得到

定理 4. 使(22)式等号成立的充要条件是黎曼空间 M_4 具局部对偶性,这种黎曼结构必能使 M_4 上的 S_W 达到最小值,其性质列表如下:

R^{++}	R^{+-}	R^{--}	S_W	I_E	I_P	说　明
$\neq 0$	0	$\neq 0$	I_E	>0	$\lvert I_P \rvert < I_E$	Einstein 空间
0	0	$\neq 0$	I_E	>0	$I_P = I_E$	Einstein 空间,成立 $R=0$
$\neq$	0	0	I_E	>0	$I_P = -I_E$	Einstein 空间,成立 $R=0$
0	0	0	0	0	0	平坦空间
0	$\neq 0$	$\neq 0$	$-2I_P - I_E$	$I_E < -I_P$	<0	成立 $R=0$
0	$\neq 0$	0	$-I_E$	<0	0	$R=0$ 的共形平坦空间 't Hooft, Jackiw-Nohl-Rebbi 解属此一类
$\neq 0$	$\neq 0$	0	$2I_P - I_E$	$I_E < I_P$	>0	成立 $R=0$

如何定出这些空间将是值得进一步研究的问题.

参考资料

[1] A. Belavin, A. Polyakov, A. Schwartz, and Y. Tyupkin, *Phys. Lett.*, **59B**, 85 (1975).

[2] C. N. Yang, *Generalization of Dirac's Monopole to SU_2 gauge fields*, ITP-SB-77-20C.

[3] M. F. Atiyah, N. J. Hitchin and I. M. Singer, *Deformations of Instantons*, (preprint) 1977.

[4] 谷超豪、杨振宁等,欧氏空间瞬子解的几何解释,复旦学报,本期.

[5] G. 't Hooft,见[6].

[6] R. Jackiw, C. Nohl, and C. Rebbi, *Phys. Rev.*, **D15**, № 6, 1642-1646 (1977).

[7] A. Belavin and D. E. Burlankov, *Phys. Lett.*, **58A**, № 1, 7-8 (1976).

[8] N. Hitchin, *J. Differential Geometry*, **9** (1974), 435-441.

[9] L. P. Eisenhart, *Riemannian Geometry*, 1949.

(本文曾发表于《复旦学报(自然科学版)》1977 年第 4 期,1-7;英文版曾发表于 *Scientia Sinica*, 1978, 21(4), 475-482)

欧氏空间瞬子解的几何解释

复　旦　大　学　谷超豪　沈纯理　胡和生

美国纽约州立大学　杨振宁①

摘　　要

本文利用黎曼几何学及共形映照下规范场的一些性质，用统一而自然的方式给出 Polyakov，'t Hooft，Jackiw-Nohl-Rebbi 等人所得到的瞬子解，并作出几何解释.

§1. 引　　言

近年来欧氏空间 E_4 中 SU_2 规范场的瞬子解引起了不少研究[1~3]，已经知道 Pontrjagin 示性数为 k 的自对偶(反自对偶)瞬子解依赖于 $8k-3$ 个参数，其中相当一部分是有显式的解析表达式的[1, 2].在本文中我们利用黎曼几何学及共形映照下规范场的一些性质[4]用统一而自然的方式给出这些具有显式表达式的瞬子解并对之作出几何解释.我们证明了，这些瞬子解可以由相应的黎曼空间内在确定，并且它的规范势就是由某些共形欧氏空间的克氏记号分解而形成.

§2. 瞬子解及共形映照

设 E_4 为四维欧氏空间，$b=b_\mu dx^\mu=b_\mu^i(x)dx^\mu X_i(i=1, 2, 3; \mu=1, 2, 3, 4)$ 为

① 杨振宁教授于 1977 年 7 月访问了上海，和复旦大学的教师进行了共同的学术讨论和研究.

SU_2 规范势形式,其强度为

$$f_{\lambda\mu}=b_{\lambda,\mu}-b_{\mu,\lambda}-[b_\lambda, b_\mu], \tag{1}$$

定义对偶强度

$$f^*_{\lambda\mu}=\frac{1}{2}\varepsilon_{\lambda\mu\alpha\beta}f^{\alpha\beta}, \tag{2}$$

若

$$f_{\lambda\mu}=f^*_{\lambda\mu} \text{ 或 } -f^*_{\lambda\mu}, \tag{3}$$

就称 $b=b_\mu dx^\mu$ 为 SU_2 规范场的自对偶解及反自对偶解,它们都是无源解. 又

$$k=\frac{1}{8\pi^2}\{\|f^+\|^2-\|f^-\|^2\}, \tag{4}$$

就是规范场的 Pontrjagin 示性数,这里 $f^+=\frac{1}{2}(f+f^*)$, $f^-=\frac{1}{2}(f-f^*)$, f^+ 具有自对偶性,f^- 具反自对偶性.

如果无源规范势 b_μ 满足 1. 在无穷远处 $f_{\lambda\mu}\to 0$ 相当快;2. 作用量有限;3. 无奇点(或规范变换后无奇点),则称这样的无源解为瞬子解或拟粒子解.自对偶(反自对偶)的瞬子解是其中重要的一类.

同样,我们也可以定义四维黎曼空间无源的 SU_2 规范场和自对偶(或反自对偶)的规范场,并且已经知道,负载规范场的四维黎曼空间当空间受到共形变换而规范势保持不变时,无源规范势保持为无源的[4],自对偶(反自对偶)的规范势保持为自对偶(反自对偶)的,这个事项为求欧氏空间 E_4 的 SU_2 瞬子解提供了一个有效的途径.

§3. 共形欧氏空间的平行移动规范场

每个 n 维黎曼空间内在地决定一个 $SO(n)$的平行移动规范场,其规范势即克氏记号[5],特别,四维共形平坦的黎曼空间 R_4 内在地决定了一个 $SO(4)$平行移动规范场,设

这空间的度规为

$$ds^2=e^{2\sigma}((dx^1)^2+(dx^2)^2+(dx^3)^2+(dx^4)^2). \tag{5}$$

其克氏记号为

$$\begin{Bmatrix}\lambda\\ \mu\nu\end{Bmatrix}=\delta_\mu^\lambda\sigma_{,\nu}+\delta_\nu^\lambda\sigma_{,\mu}-\delta_{\mu\nu}\sigma_{,\lambda}\quad(\lambda,\mu,\nu=1,2,3,4), \tag{6}$$

这就是 R_4 上 SO_4 的规范势在自然标架下的表达式.记 R_4 中和坐标$(x^1,\cdots,x^4)$相应的自然标架的基为 $I_\mu\left(=\dfrac{\partial}{\partial x^\mu}\right)$,那么变位方程为

$$DI_\mu=\begin{Bmatrix}\lambda\\ \mu\nu\end{Bmatrix}dx^\nu I_\lambda. \tag{7}$$

令 $e_\mu=e^{-\sigma}I_\mu$,它们构成 R_4 的归一正交标架,在这正交标架下

$$De_\mu=\Gamma_{\mu\nu}^\lambda dx^\nu e_\lambda,$$

其中

$$\Gamma_{\mu\nu}^\lambda=\delta_\nu^\lambda\sigma_{,\mu}-\delta_{\mu\nu}\sigma_{,\lambda}. \tag{8}$$

它关于 λ, μ 为反称,$b_\nu=(\Gamma_{\mu\nu}^\lambda)(\nu=1,2,3,4)$是 R_4 上的 SO_4 群的平行移动规范场的规范势在正交标架下的表达式.又记 $\overset{*}{b}_\nu=(\overset{*}{\Gamma}{}_{\mu\nu}^\lambda)$,这里 $*$ 是关于指标 λ, μ 的对偶记号,令

$$\overset{+}{b}_\nu=\frac{1}{2}(b_\nu+\overset{*}{b}_\nu), \tag{9}$$

$$\bar{b}_\nu=\frac{1}{2}(b_\nu-\overset{*}{b}_\nu), \tag{10}$$

我们就得到 R_4 上的两个 SU_2 规范势 $\overset{+}{b}_\nu$ 与 $\bar{b}_\nu$.

现在把 R_4 和欧氏空间 E_4 相共形对应,把 $\overset{+}{b}_\nu$, $\bar{b}_\nu$ 看成定义在 E_4 上的两个 SU_2 规范势,记相应的规范场为 $\mathscr{F}^+$ 与 $\mathscr{F}^-$.

利用(8)式,我们可将$\overset{+}{b}_\nu$, $\bar{b}_\nu$写成下述形式

$$\overset{+}{b}_i=-\varepsilon_{ijk}\sigma_{,j}Y_k-\sigma_{,4}Y_i,\ \overset{+}{b}_4=\sigma_{,i}Y_i, \tag{9'}$$

$$\bar{b}_i=-\varepsilon_{ijk}\sigma_{,j}X_k+\sigma_{,4}X_i,\ \bar{b}_4=-\sigma_{,i}X_i, \quad (i,j,k=1,2,3) \tag{10'}$$

这里X_i, Y_j为李代数$SO'(4)$的基

$$X_1=\frac{1}{2}\begin{pmatrix}0&0&0&1\\0&0&-1&0\\0&1&0&0\\-1&0&0&0\end{pmatrix},\ X_2=\frac{1}{2}\begin{pmatrix}0&0&1&0\\0&0&0&1\\-1&0&0&0\\0&-1&0&0\end{pmatrix},\ X_3=\frac{1}{2}\begin{pmatrix}0&-1&0&0\\1&0&0&0\\0&0&0&1\\0&0&-1&0\end{pmatrix},$$

$$Y_1=\frac{1}{2}\begin{pmatrix}0&0&0&-1\\0&0&-1&0\\0&1&0&0\\1&0&0&0\end{pmatrix},\ Y_2=\frac{1}{2}\begin{pmatrix}0&0&1&0\\0&0&0&-1\\-1&0&0&0\\0&1&0&0\end{pmatrix},\ Y_3=\frac{1}{2}\begin{pmatrix}0&-1&0&0\\1&0&0&0\\0&0&0&-1\\0&0&1&0\end{pmatrix}. \tag{11}$$

由于$[X_i, X_j]=\varepsilon_{ijk}X_k$, $[Y_i, Y_j]=\varepsilon_{ijk}Y_k$,因而$X_1$, X_2, X_3与Y_1, Y_2, Y_3分别构成两个SU_2的李代数的基.从(9′)和(10′)就可见到SU_2规范势$\overset{+}{b}_\nu$与$\bar{b}_\nu$实质上分别和't Hooft等人首先引入的Ansatz[2]

$$\overset{+}{b}_\nu=i\tau_{\mu\nu}\sigma_{,\nu}, \tag{12}$$

$$\bar{b}_\nu=i\bar{\tau}_{\mu\nu}\sigma_{,\nu} \tag{13}$$

相同①.所以他们的两个Ansatz可以同时由共形平坦空间的克氏记号自然地导出.

① 在[2]中

$$\bar{\tau}_{ij}=\frac{1}{4i}[\sigma_i\sigma_j]\ \bar{\tau}_{i4}=-\frac{1}{2}\sigma^i,\ \tau_{ij}=\bar{\tau}_{ij},\ \tau_{i4}=-\bar{\tau}_{i4}.$$

式中σ_i为Pauli阵,但我们现在用X_i或Y_i代替$\frac{\sigma_i}{2i}$.

§4. 具自对偶性的解及其几何解释

已知线素为(5)的共形平坦空间的曲率张量 $R_{\lambda\mu\nu\rho}$ 表达式为[6]

$$e^{-2\sigma}R_{\lambda\mu\nu\rho}=\delta_{\lambda\rho}\sigma_{\mu\nu}+\delta_{\mu\nu}\sigma_{\lambda\rho}-\delta_{\lambda\nu}\sigma_{\mu\rho}-\delta_{\mu\rho}\sigma_{\lambda\nu}+(\delta_{\lambda\rho}\delta_{\mu\nu}-\delta_{\lambda\nu}\delta_{\mu\rho})\Delta_1\sigma. \tag{14}$$

式中

$$\sigma_{\lambda\mu}=\sigma_{,\lambda\mu}-\sigma_{,\lambda}\sigma_{,\mu},\ \Delta_1\sigma=\delta^{\alpha\beta}\sigma_{,\alpha}\sigma_{,\beta}. \tag{15}$$

转换到归一正交标架 e_λ，曲率张量为

$$\widetilde{R}_{\lambda\mu\nu\rho}=e^{-4\sigma}R_{\lambda\mu\nu\rho}. \tag{16}$$

将 $\widetilde{R}_{\lambda\mu\nu\rho}$ 关于前两个指标作对偶，得 $\widetilde{R}^*_{\lambda\mu\nu\rho}$，那么

$$\widetilde{R}^+_{\lambda\mu\nu\rho}=\frac{1}{2}(\widetilde{R}_{\lambda\mu\nu\rho}+\widetilde{R}^*_{\lambda\mu\nu\rho}), \tag{17}$$

$$\widetilde{R}^-_{\lambda\mu\nu\rho}=\frac{1}{2}(\widetilde{R}_{\lambda\mu\nu\rho}-\widetilde{R}^*_{\lambda\mu\nu\rho}). \tag{18}$$

所形成的 4×4 阵 $(\widetilde{R}^+_{\lambda\mu\nu\rho})$，$(\widetilde{R}^-_{\lambda\mu\nu\rho})$ 就是 $\mathscr{F}^+$ 和 $\mathscr{F}^-$ 的强度 $f^+_{\nu\rho}$，$f^-_{\nu\rho}$，这里 λ，μ 是矩阵的行和列的指标.从(17)，(18) 式通过直接的计算可知

(1) $\mathscr{F}^-$ 为自对偶或 $\mathscr{F}^+$ 为反自对偶的充要条件是

$$\Delta u=0\ (u>0). \tag{19}$$

这里 $u=e^\sigma$；因此具这样的对偶性的共形平坦空间 R_4 可通过求解 Laplace 方程得出.这样我们就用几何的方法导出了[2]中所得到的解，这种解所相应的 R_4 的线素为

$$ds^2=\left(\sum_{a=0}^{k}\frac{\lambda_a^2}{(x-y_a)^2}\right)^2((dx^1)^2+(dx^2)^2+(dx^3)^2+(dx^4)^2). \tag{20}$$

式中 y_a 是 $k+1$ 个定点，λ_a 是常数.

$$(x-y_a)^2=(x^1-y_a^1)^2+(x^2-y_a^2)^2+(x^3-y_a^3)^2+(x^4-y_a^4)^2. \tag{21}$$

现给出这种 R_4 的一个几何结构的描述.如果将 E_4 添上一个无限远点使它成为 S_4，然后在 S_4 上除去 $k+1$ 个点 $y_a(a=0, \cdots, k)$，那么就得到一个非紧致流形 M_4，R_4 上的线素可以看成是定义在 M_4 上的黎曼结构.利用反演变换，可以见到所添上的无限远点是 R_4 的一个正常点.又若 x 沿测地线 $\rightarrow y_a$ 时，此测地线可无限延伸，也即测地线长度 $\rightarrow \infty$，R_4 在 $x\rightarrow y_a$ 的渐近性质和 E_4 中 $x\rightarrow$ 无限远的渐近性质相同，因此 R_4 是一个非紧致的完备的黎曼空间.特别 $k=0$ 时，如取 y_0 为(0, 0, 0, 0)，作一次坐标反演 $\xi^i=\dfrac{x^i}{x^2}$，我们就得到普通的欧氏线素 $ds^2=(d\xi^1)^2+(d\xi^2)^2+(d\xi^3)^2+(d\xi^4)^2$，从而可见原来的无限远点实际上是正常点，原来的(0, 0, 0, 0)点附近的空间性质与欧氏空间无限远点的性质相同.

(2) $\mathscr{F}^+$为自对偶或$\mathscr{F}^-$为反自对偶的充要条件为

$$\sigma_{\lambda\mu}=0,\ \sigma_{\lambda\lambda}=\sigma_{\mu\mu}(\lambda\neq\mu). \tag{22}$$

令

$$v=\frac{1}{u}=e^{-\sigma}, \tag{23}$$

上式化为

$$V_{,\lambda\mu}=0,\ V_{,\lambda\lambda}=V_{,\mu\mu}(\lambda\neq\mu). \tag{24}$$

从而得到(22)的通介为

$$u=\frac{c}{a+b(x-y)^2}, \tag{25}$$

式中 a, b, c 为常数($c>0$)，y 为一定点.这时线素(5)是常曲率空间的线素.当 $a>0$，$b>0$ 时得出正常曲率空间，这就是 BPST 解[1]的几何解释，当 $\dfrac{a}{b}<0$，我们得到负常曲率空间，这时作用量不是有限的.

因而,我们利用共形欧氏空间导出了那些已具有具体表达式的瞬子解,并给出了它们的几何意义.

§5. 无 源 解

如上所述由无源的共形平坦空间的平行移动规范场①可以导出 E_4 上的两个无源的 SU_2 规范场.由于共形平坦空间成立

$$R_{\lambda\mu;\nu}-R_{\lambda\nu;\mu}+\frac{1}{6}(g_{\lambda\nu}R_{,\mu}-g_{\lambda\mu}R_{,\nu})=0. \tag{26}$$

所以得到共形平坦空间无源的充要条件为 $R=k=\text{const}$,利用 R 的表达式就得到方程

$$\Delta u=ku^3(u>0). \tag{27}$$

当 u 满足(27)时,$\mathscr{F}^+$,$\mathscr{F}^-$ 都是无源的规范场,相反地,若 $\mathscr{F}^+$ 和 $\mathscr{F}^-$ 有一为无源,那么(27)就必须成立,从而另一个也必为无源.事实上,若 $\mathscr{F}^+$ 为无源,则把(26)转换到正交标架后,就应有

$$(\delta_{\lambda\nu}\widetilde{R}_{,\mu}-\delta_{\lambda\mu}\widetilde{R}_{,\nu})+(\delta_{\lambda\nu}\widetilde{R}_{,\mu}-\delta_{\lambda\mu}\widetilde{R}_{,\nu})^*=0.$$

这时 $*$ 是关于指标 μ,ν 而作的,取 $\mu=1$,$\nu=2$,$\lambda=2$ 就有 $\widetilde{R}_{,1}=0$,同样可得 $\widetilde{R}_{,2}=\widetilde{R}_{,3}=\widetilde{R}_{,4}=0$,这就是 $R=\text{const}$.

这样,我们就用几何的方法自然地导出了[7]中关于无源解的方程.

特别当 $k=1$ 时,(27)的解包含了 BPST 解.

如果不计边界条件或容许有奇性,(27)的解是很多的,其中,下述的解

$$u=\frac{1}{\sqrt{(x-y_1)^2(x-y_2)^2}} \tag{28}$$

或许是有意思的,这里 y_1,y_2 是两个定点,$k=-(y_1-y_2)^2$.

---

① 平行移动规范场为无源的定义是 $R_{\lambda\mu;\nu}-R_{\lambda\nu;\mu}=0$.[5]

参 考 资 料

[1] A. Belavin, A. Polyakov, A. Schwartz and Y. Tyupkin, *Phys. Lett.*, **59B**,85 (1975).

[2] R. Jackiw, C. Nohl and C. Rebbi, *Phys. Rev.*, **D15**, 1642 (1977).

[3] M. F. Atiyah, N. J. Hitchin and I. M. Singer, *Deformations of Instantons* (preprint 1977).

[4] C. N. Yang, *Conformal Mapping of gauge field.*, preprint, ITP - SB, 77 - 23.

[5] C. N. Yang, *Phys. Rev. Lett.*, **33**, 445 (1974).

[6] L. P. Eisenhart, *Riemannian Geometry*, 1949.

[7] E. Corrigan, D. B. Fairlie, *Scalar field theory and exact solution to a classical SU(2) gauge theory* (preprint, 1977).

(本文曾发表于《复旦学报(自然科学版)》1977 年第 4 期,8 - 12;英文版曾发表于 *Scientia Sinica*, 1978, 21(6), 767 - 772)

球对称的 SU_2 规范场和磁单极的规范场描述*

谷超豪　胡和生

复旦大学数学系

提　要

本文讨论球对称的 SU_2 规范场，证明了满足最一般的球对称定义的 SU_2 规范场只能有三种基本类型：(1) 同步球对称规范场；(2) 狭义球对称规范场；(3) 化约为 U_1 子群的球对称规范场．文中详细讨论了球对称的带同位旋向量场（Higgs 场）的 SU_2 规范场，完全决定了它们的类型．如果把这种场看成为由电磁场和带电矢介子构成，那么就有如下的结论：如果磁单极所含的磁荷是最小单位的 m 倍，当 $|m|>1$ 时，球对称的带 Higgs 场的 SU_2 规范场只能是纯电磁场，而不能有带电矢介子场出现．但当 $m=0$，± 1 时，球对称的带电矢介子场是可以出现的．从而可见，具有非单位磁荷的磁单极隐含了某种破坏球对称的因素．

一、引　言

在物理学中研究场的理论时，球对称的模型是常见的，球对称的场有着重要的作用．

球对称的标量场、向量场、张量场的一般形式是不难写出的．但对于规范场，因为相互成规范变换的规范场是等价的，所以在定义球对称时就必须考虑到这种规范变换，情

* 1976 年 3 月 24 日收到.

况就较为复杂.近年来,球对称 SU_2 规范场[1]已引起许多注意.吴大峻和杨振宁最先得出两种球对称的无源解[2],1974 年 't Hooft 用某些球对称的 SU_2 规范场来表示磁单极[3],又引起了一系列的研究[4, 5].在这些研究中,球对称的势形式已变得一般一些了,但还没有回答这样的问题:球对称的 SU_2 规范场到底有哪些形式?

本文的目的就在于定出球对称的 SU_2 规范场,我们既讨论 't Hooft 提出的带同位旋向量场 Q① 的 SU_2 规范场,也讨论不带 Q 的场.我们首先要给出球对称的一般定义:一个 SU_2 规范场 $\mathscr{F}$,如其规范势经过空间的任一旋转后,能和原来的规范势规范等价,那么场 $\mathscr{F}$ 就是球对称的.然后得出如下的结果:球对称 SU_2 规范场(取适当规范后)实际上只有三种基本类型:(1) 同步球对称的场,也就是说,其规范势在经受任一转动 A 的同时又经受同位旋空间的同一旋转后是不变的;(2) 狭义球对称的场,即其规范势在任一转动 A 下就是不变的;(3) 化约为 U_1 子群的球对称规范场.每一球对称的 SU_2 规范场总是由这三种类型的场以适当方式组合而成.

如果用 't Hooft 建议的公式表示电磁场强度,s 为偶合常数,则 $\frac{1}{s}$ 为磁荷的最小单位,又把 SU_2 场看成为电磁场和带电矢介子偶合的场,那么类型(1)的场容有磁荷为 $\pm\frac{1}{s}$ 的球对称磁单极,类型(2)的场中无磁荷,在这两种场中还可以有球对称带电矢介子场出现.类型(3)的场和普通的 U_1 群整体规范场无本质区别,可以表示磁荷为 $\frac{m}{s}$ 的磁单极,它只能是纯电磁场,即不能出现矢量介子场.

我们论证的方法是:先设场的势满足一般球对称的定义,然后区分各种情况,找到几个规范变换,在小范围内把规范势变到所需的形式,然后再把这个结果整体化.我们先讨论带 Q 的场,再讨论不带 Q 的场,最后叙述了对于用 SU_2 群规范场描述磁单极的几点看法.

---

① 同位旋向量场 Q 是一种 Higgs 场.

二、定义和主要结果

我们研究四维平坦时空中的球对称 SU_2 规范场.如所知,SU_2 群为 2×2 的单模么正阵

$$\zeta=\begin{pmatrix}\alpha & \beta\\ -\bar{\beta} & \bar{\alpha}\end{pmatrix}\quad(\alpha\bar{\alpha}+\beta\bar{\beta}=1)\tag{1}$$

所成的群,其李代数 SU_2'的基可选为

$$X_1=\frac{1}{2}\begin{pmatrix}0 & -i\\ -i & 0\end{pmatrix},\ X_2=\frac{1}{2}\begin{pmatrix}0 & -1\\ 1 & 0\end{pmatrix},\ X_3=\frac{1}{2}\begin{pmatrix}-i & 0\\ 0 & i\end{pmatrix}.\tag{2}$$

SU_2'中的一般元素可记为 $q=q_1X_i(i=1,2,3)$,称为同位旋向量.(2)式中 X_i 的换位运算为

$$[X_i,X_j]=\varepsilon_{ijk}X_k.\tag{3}$$

SU_2 中的伴随变换 $q\to\zeta q\zeta^{-1}$ 记为 $ad\zeta$.如果行向量(q_1,q_2,q_3)也记为 q,那么 $ad\zeta$ 使q变为

$$q'=qC(\zeta),\tag{4}$$

这里 $C(\zeta)$为 SO_3 中的阵,其元素为

$$\zeta X_i\zeta^{-1}=C_{ij}X_1\tag{5}$$

中的 C_{i1}.同一 SO_3 中的阵可对应于两个 ζ,它们差一正负号.从(4)式还容易得出 $C(\zeta_1\zeta_2)=C(\zeta_2)C(\zeta_1)$.又记由 X_1 生成的子群为 G_1,G_1 的一般元素为

$$\exp(\varphi X_1)=\begin{pmatrix}\cos\frac{\varphi}{2} & -i\sin\frac{\varphi}{2}\\ -i\sin\frac{\varphi}{2} & \cos\frac{\varphi}{2}\end{pmatrix}.\tag{6}$$

SU_2 规范场由一组规范势 $b_\lambda^i(x, t)(\lambda=1, 2, 3, 4)$ 所定义,这里 $x=x_ie_i$ 表空间的点,e_i 为空间直角坐标的基向量,t 或 x_4 表时间,已选取单位使 $c=h=1$.记 $b_\lambda^iX_i=b_\lambda$,变换

$$b_\lambda(x, t)\to\zeta(x, t)b_\lambda(x, t)\zeta^{-1}(x, t)-\partial_\lambda\zeta(x, t)\zeta^{-1}(x, t) \tag{7}$$

称为规范变换,式中 ∂_λ 记 $\dfrac{\partial}{\partial x^\lambda}$. 互为规范变换的规范势称为等价的,它们代表同一个物理的场.我们假设 $b_\lambda(x, t)$, $\zeta(x, t)$ 等均为 x, t (除空间原点外)的有一定光滑性的函数.

't Hooft 在研究磁单极时[3]引进了一个同位旋向量场 $Q(x, t)=q_i(x, t)X_i$,对带 Q 的 SU_2 规范场,在 q_i 不全为 0 之处,他以

$$f_{\lambda\mu}=\frac{1}{|Q|}q_if_{\lambda\mu}^i-\frac{\varepsilon_{abc}}{|Q|^3}q_aD_\lambda q_bC_\mu q_c \text{ ①} \tag{8}$$

作为电磁场张量,式中 $f_{\lambda\mu}^i$ 为场的强度,$D_\lambda Q_b$ 为 Q 的规范导数[6],又

$$|Q|=(q_1^2+q_2^2+q_3^2)^{1/2}.$$

现作出球对称性的定义:

定义 设 $\mathscr{F}$ 为 SU_2 规范场,如果其规范势经过空间的任一旋转后,都能和原来的规范势相等价,则称 $\mathscr{F}$ 为球对称的.

容易见到,这个定义的另一形式是:对空间的任一旋转 A,存在一个适当的 $\zeta(A, x, t)$,使 $\mathscr{F}$ 的势经过 $\zeta(A, x, t)$ 所定义的规范变换和旋转 A 后仍然保持不变,那么场 $\mathscr{F}$ 称为球对称的.现要用式子表示这个定义.以 b 记 b_i, Ab 记 $a_{i1}b_1$,这里 a_{ij} 是 A 的矩阵元,∇ 表示三维空间的梯度,那么这个定义就是说,对任何 $A\in SO_3$,必有 $\zeta(A, x, t)\in SU_2$ 使

---

① 在 't Hooft 的定义中,第二项有因子 $\dfrac{1}{c}$,我们在第八节再引入这个因子.

$$A(\zeta(A, x, t)b(x, t)\zeta^{-1}(A, x, t)-\nabla\zeta(A, x, t)\cdot\zeta^{-1}(A, x, t))=b(Ax, t), \tag{9}$$

$$\zeta(A, x, t)b_4(x, t)\zeta^{-1}(A, x, t)-\partial_4\zeta(A, x, t)\zeta^{-1}(A, x, t)=b_4(Ax, t) \tag{10}$$

成立.事实上,这两个式子左边表示势经过规范变换和旋转后的表达式.因这时 (x, t) 变为 (Ax, t),所以(9),(10) 两式的右边即为原来的势在 (Ax, t) 的表达式,这两边相等,正是球对称的定义.

对带 Q 的场,除(9),(10)式外,显然还要满足

$$\zeta(A, x, t)Q(x, t)\zeta^{-1}(A, x, t)=Q(Ax, t), \tag{11}$$

才称为球对称的,我们假设 $x\neq 0$ 时, $Q\neq 0$.

特别是:

(1) 如果(9),(10)式中的 $\zeta(A, x, t)$ 和 x, t 无关,且 $ad\zeta$ 的阵 $C(\zeta)$ 即为 A^{-1},则场 $\mathscr{F}$ 称为同步球对称,文献[5]中已指出,同步球对称的势的一般形式为

$$b_i^l=\varepsilon_{ilk}x_kV(r, t)+\delta_{1l}S(r, t)+x_ix_lT(r, t),\ b_4^l=x_lU(r, t). \tag{12}$$

(12)式中 $r=(x_1^2+x_2^2+x_3^2)^{\frac{1}{2}}$.当 $V=-\dfrac{1}{r^2}$, $S=0$ 时,场称为特异同步球对称的,否则称为非异的,对带 Q 的场来说,还有

$$q_i=-W(r, t)x_i(W(r, t)\neq 0). \tag{13}$$

在 $W>0$ 时,我们称场为正型的,$W<0$ 时,场为负型的.

(2) 如 $\zeta(A, x, t)$ 是单位阵 I,则场称为狭义球对称的,容易证明

$$b_i^l=R^l(r, t)x_i,\ b_4^l=Y^l(r, t). \tag{14}$$

如带 Q,则 $q_i=q_1(r, t)$.又若经规范变换后,能使 $Q\parallel X_1$ 和 $b_i\parallel X_1$ 同时成立,则称场为特异的,否则就称为非异的.

(3) 在适当规范下能有 $b_\lambda\parallel X_1$,则称场化约为子群 G_1 的规范场,如带 Q,也要求

$Q \parallel X_1$ ①.这种场的球对称性相同于电磁场的球对称性，即电磁场张量应为球对称的，特别是其磁场部分为

$$f_{ij}=n\varepsilon_{ijk}\frac{x_k}{r^3},$$

n 只能取整数(这就是磁量子化条件，来源于场的整体性质)[7]，势函数可取得使 $b_\lambda^2=b_\lambda^3=0$，$b_\lambda^1$ 由电磁场张量依普通的方法定出.后文还将指出，当 $n=\pm 1$ 时，场等价于特异的同步球对称场；$n=0$ 时，场等价于特异的狭义球对称场.

我们的主要结果有

定理 1 带 Q 的球对称规范场必属于下述四种类型之一：

(1) 负型的同步球对称规范场；(2) 正型的同步球对称规范场；(3) 狭义球对称规范场；(4) 化约为子群 G_1 的球对称规范场(满足量子化条件)，且 $|n|>1$.

我们还将注意到，如果只考虑局部的球对称性，即球对称条件只在某角状(立体)区域成立时，量子化条件不出现.

对于不带 Q 的 SU_2 规范场，成立.

定理 2 球对称的 SU_2 规范场有下述三种基本类型：

(1) 非异同步球对称规范场；(2) 非异狭义球对称规范场；(3) 化约为子群 G_1 的球对称规范场(需满足量子化条件).任一球对称的 SU_2 规范场由它们以适当方式组成.

三、带 Q 的场在一点的分类

在讨论整个场之前，我们先要弄清在一点的情况.

设已给了一个带 Q 的球对称场 $\mathscr{F}$，对于任何一点，必可选取一邻域 Ω，使通过适当规范变换后，在此邻域中必有 $Q \parallel X_1$，而且和 X_1 同向.事实上，只要在各点作同位旋空间的适当旋转就可以了，因为 $Q(x, t)$ 是光滑的，在局部范围内，这种旋转也可取为光滑地依

---

① $b_\lambda \parallel X_1$，Q 不平行于 X_1 的情况是可能出现，但它包含于情形(2)之中.

赖于x, t的①.

我们不妨讨论正x_1轴上点(ζ_0, τ)的一个邻域.在Ω中已有

$$q_2(x, t)=q_3(x, t)=0, \ q_1(x, t)>0. \tag{15}$$

当(x, t)和(x_0, τ)相距不远,$A\xi_0$和ξ_0相距不远时(在后面,当我们讨论局部性质时,一般都略去这种限制性的说明),由(11)式得知,$\zeta(A, x, t)\in G_1$,而$C(\zeta(A, x, t))$为

$$C(\zeta(A, x, t))=\begin{pmatrix}1 & 0 & 0\\ 0 & \cos\varphi & \sin\varphi\\ 0 & -\sin\varphi & \cos\varphi\end{pmatrix}, \tag{16}$$

式中$\varphi=\varphi(A, x, t)$. 又记

$$D(\varphi(A, x, t))=\begin{pmatrix}\cos\varphi & \sin\varphi\\ -\sin\varphi & \cos\varphi\end{pmatrix}, \tag{17}$$

$$B(x, t)=\begin{pmatrix}b_1^2(x, t) & b_1^3(x, t)\\ b_2^2(x, t) & b_2^3(x, t)\\ b_3^2(x, t) & b_3^3(x, t)\end{pmatrix}, \ b^1(x, t)=\begin{pmatrix}b_1^1(x, t)\\ b_2^1(x, t)\\ b_3^1(x, t)\end{pmatrix}. \tag{18}$$

(9)式就可写成为

$$B(Ax, t)=AB(x, t)D(\varphi(A, x, t)), \tag{19}$$

$$b^1(Ax, t)=A(b^1(x, t)-\nabla\varphi(A, x, t)). \tag{20}$$

设$R(\theta)$为绕x_1轴的转动,转角为θ,即

$$R(\theta)=\begin{pmatrix}1 & 0 & 0\\ 0 & \cos\theta & -\sin\theta\\ 0 & \sin\theta & \cos\theta\end{pmatrix}. \tag{21}$$

---

① 严格的证明可利用(27)式所定义的正交阵.

取 $x=\xi_0$, $t=\tau$, $A=R(\theta)$,则(19)式化为

$$B(\xi_0,\ \tau)=R(\theta)B(\xi_0,\ \tau)D(\varphi). \tag{22}$$

先证明

引理 1 如果对任何 θ 必有 φ,使(22) 式成立,则 $B(\xi_0,\ \tau)$ 必属于下列四种情形之一:

$$\begin{aligned}
&(1)\ B(\xi_0,\ \tau)=\begin{pmatrix}0 & 0\\ a & b\\ -b & a\end{pmatrix},\ \varphi=\theta;\\
&(2)\ B(\xi_0,\ \tau)=\begin{pmatrix}0 & 0\\ a & b\\ b & -a\end{pmatrix},\ \varphi=-\theta;\\
&(3)\ B(\xi_0,\ \tau)=\begin{pmatrix}a & b\\ 0 & 0\\ 0 & 0\end{pmatrix},\ \varphi=0;
\end{aligned} \tag{23}$$

(4) $B(\xi_0,\ \tau)$ 的一切元素为零.

在情形(1),(2),(3)中 a, b 不全为零.

证明 由(22)式知道,对每一 θ 必有 φ,使

$$\begin{pmatrix}\cos\theta & -\sin\theta\\ \sin\theta & \cos\theta\end{pmatrix}\begin{pmatrix}b_2^2 & b_2^3\\ b_3^2 & b_3^3\end{pmatrix}=\begin{pmatrix}b_2^2 & b_2^3\\ b_3^2 & b_3^3\end{pmatrix}\begin{pmatrix}\cos\varphi & -\sin\varphi\\ \sin\varphi & \cos\varphi\end{pmatrix}, \tag{24}$$

显然有 SO_2 中的阵 S, T 使

$$S\begin{pmatrix}b_2^2 & b_2^3\\ b_3^2 & b_3^3\end{pmatrix}T=\begin{pmatrix}\lambda_1 & 0\\ 0 & \lambda_2\end{pmatrix}. \tag{25}$$

因为 SO_2 中的阵可交换,所以用 S 和 T 左乘和右乘(24)式,就有

$$\begin{pmatrix}\cos\theta & -\sin\theta\\ \sin\theta & \cos\theta\end{pmatrix}\begin{pmatrix}\lambda_1 & 0\\ 0 & \lambda_2\end{pmatrix}=\begin{pmatrix}\lambda_1 & 0\\ 0 & \lambda_2\end{pmatrix}\begin{pmatrix}\cos\varphi & -\sin\varphi\\ \sin\varphi & \cos\varphi\end{pmatrix}. \tag{26}$$

由此易见，当 λ_1，λ_2 不全为零时，必有 $\lambda_1=\pm\lambda_2$，并分别对应 $\theta=\pm\varphi$，从而由(25)式可见

$$b_2^2=\pm b_3^3,\ b_2^3=\mp b_3^2.$$

再由(22)式可见 $b_2^1=b_3^1=0$，因而就有情形(1)和(2).

当 λ_1，λ_2 全为零时，如 b_1^2，b_1^3 不全为零，就有情形(3).

余下来就是情形(4).引理 1 证毕.

这个引理给出了带 Q 的球对称 SU_2 场在一点的分类.我们将分别讨论这四种情形.

四、情形(1)和(2)的讨论

设对 (ξ_0,τ) 出现情形(1)，则对邻近的 (x_0,t) 也如此，这里 x_0 仍记 x_1 轴上的点.

引理 2 阵

$$P(x)=\begin{pmatrix}\dfrac{x_1}{r} & -\dfrac{x_2}{r} & -\dfrac{x_3}{r}\\ \dfrac{x_2}{r} & 1-\dfrac{x_2^2}{r(r+x_1)} & -\dfrac{x_2x_3}{r(r+x_1)}\\ \dfrac{x_3}{r} & -\dfrac{x_2x_3}{r(r+x_1)} & 1-\dfrac{x_3^2}{r(r+x_1)}\end{pmatrix} \tag{27}$$

当 $x\neq -re_1$ 时为解析，属于 SO_3，将 $x_0=re_1$ 变为 $x=x_ie_i$.又 $P(x_0)=E$，这里 E 为单位阵.

证明 这些性质都可以从 $P(x)$ 的表达式直接看出.

现希望由(19)式定出 $\varphi(A,x,t)$.在(19)式中取 A 为 $P(x)$，得

$$B(x,t)=P(x)B(x_0,t)D(\varphi(P(x),x_0,t)). \tag{28}$$

记

$$\varphi(P(x),\ x_0,\ t)=\varphi(x,\ t), \tag{29}$$

则

$$B(Ax,\ t)=P(Ax)B(x_0,\ t)D(\varphi(Ax,\ t)). \tag{30}$$

由(19)式及(28)式可见

$$B(Ax,\ t)=AP(x)B(x_0,\ t)D(\varphi(x,\ t))D(\varphi(A,\ x,\ t)), \tag{31}$$

从而得到

$$B(x_0,\ t)=P^{-1}(Ax)AP(x)B(x_0,\ t)D(\varphi(A,\ x,\ t)+\varphi(x,\ t)-\varphi(Ax,\ t)). \tag{32}$$

$P^{-1}(Ax)AP(x)$ 使 x_0 保持不变,所以等于某一 $R(\theta)$，这里

$$\theta=\theta(A,\ x)=P^{-1}(Ax)AP(x)\text{ 的转角}. \tag{33}$$

由(32)式及引理 1 情况(1)可见

$$\varphi(A,\ x,\ t)=\theta(A,\ x)+\varphi(Ax,\ t)-\varphi(x,\ t). \tag{34}$$

现通过规范变换来简化 $\varphi(A,\ x,\ t)$. 取 $\zeta(x,\ t)\in SU_2$, 使 $ad\zeta$ 的阵 $C(\zeta)$ 即为 $R(-\varphi(x,\ t))$. 利用 $\zeta(x,\ t)$ 作规范变换，规范势 b_λ 变为 $\underset{1}{b}_\lambda$，$\varphi(A,\ x,\ t)$ 变为 $\underset{1}{\varphi}(A,\ x,\ t)$. 这时

$$\underset{1}{B}(x,\ t)=B(x,\ t)D(-\varphi(x,\ t)), \tag{35}$$

从而得

$$\begin{aligned}\underset{1}{B}(Ax,\ t)&=B(Ax,\ t)D(-\varphi(Ax,\ t))\\&=AB(x,\ t)D(\varphi(A,\ x,\ t)-\varphi(Ax,\ t)).\end{aligned} \tag{36}$$

另一方面

$$\begin{aligned}\underset{1}{B}(Ax,\ t)&=A\underset{1}{B}(x,\ t)D(\underset{1}{\varphi}(A,\ x,\ t))\\&=AB(x,\ t)D(\underset{1}{\varphi}(A,\ x,\ t)-\varphi(x,\ t)).\end{aligned} \tag{37}$$

比较(36)和(37)式的右边,又利用(34)式,就得出

$$\underset{1}{\varphi}(A, x, t)=\varphi(A, x, t)-\varphi(Ax, t)+\varphi(x, t)=\theta(A, x). \tag{38}$$

从而可见 $\mathscr{F}$ 等价于一个场 $\mathscr{F}_1$,它的 $\underset{1}{\varphi}(A, x, t)$ 即为(33)式定义的 $\theta(A, x)$. 因而不妨设对 $\mathscr{F}$ 就已经成立

$$\varphi(A, x, t)=\theta(A, x). \tag{39}$$

再证明 $\mathscr{F}$ 等价于同步球对称的规范场. 为此,对任一 x_0 邻近的 x,作 $\zeta(x)$,使 $C(\zeta(x))=P^{-1}(x)$①,用 $\zeta(x)$ 作规范变换,则 Q 化为 $\underset{1}{Q}$,它和向径平行且同向,又 $|\underset{1}{Q}|$ 只是 r 的函数. 记所得的规范势为

$$\underset{1}{b_\lambda}(x, t)=\zeta(x)b_\lambda(x, t)\zeta^{-1}(x)-\partial_\lambda\zeta(x)\zeta^{-1}(x). \tag{40}$$

对任一 $A\in SO_3$,作 $\zeta(A, x)\in SU_2$,使 $C(\zeta(A, x))=R(-\theta(A, x))$. 又作 ζ_A,使 $C(\zeta_A)=A^{-1}$. 因为 $P^{-1}(Ax)AP(x)=R(\theta(A, x))$,所以

$$C(\zeta(A, x))=C(\zeta^{-1}(Ax)\zeta_A\zeta(x)),$$

从而就得

$$\zeta(Ax)\zeta(A, x)\zeta^{-1}(x)=\pm\zeta_A. \tag{41}$$

再从(40)式作

$$\begin{aligned}\underset{1}{b}(Ax, t)&=\zeta(Ax)b(Ax, t)\zeta^{-1}(Ax)-(\nabla\zeta(x))_{Ax}\zeta^{-1}(Ax)\\&=A\zeta(Ax)\zeta(A, x)b(x, t)\zeta^{-1}(A, x)\zeta^{-1}(Ax)\\&\quad-A\zeta(Ax)\,\nabla\zeta(A, x)\zeta^{-1}(A, x)\zeta^{-1}(Ax)-(\nabla\zeta(x))_{Ax}\zeta^{-1}(Ax),\end{aligned}$$

这里 $(\nabla\zeta(x))_{Ax}$ 记 $\nabla\zeta(x)$ 中置 x 为 Ax 的结果,再利用(40)式得

$$b(x, t)=\zeta^{-1}(x)\,\underset{1}{b}(x, t)\zeta(x)-(\nabla\zeta^{-1}(x))\zeta(x).$$

又直接计算可知

① 这种 ζ 除正负号外为唯一确定的,由于我们在局部范围中讨论问题,所以符号取法无关紧要.

$$A\zeta(Ax)\nabla\zeta(A,x)\zeta^{-1}(A,x)\zeta^{-1}(Ax)+(\nabla\zeta(x))_{Ax}\zeta^{-1}(Ax)$$
$$=A\nabla(\zeta(Ax)\zeta(A,x))(\zeta(Ax)\zeta(A,x))^{-1},$$

$$A\zeta(Ax)\zeta(A,x)(\nabla\zeta^{-1}(x))\zeta(x)\zeta^{-1}(A,x)\zeta^{-1}(Ax)$$
$$+A\nabla(\zeta(Ax)\zeta(A,x))(\zeta(Ax)\zeta(A,x))^{-1}=A\nabla\zeta_A\zeta_A^{-}=0,$$

从而可见

$$\underset{1}{b}(Ax,t)=A\underset{1}{b}(x,t)A^{-1}. \tag{42}$$

同样也有

$$\underset{1}{b}_4(Ax,t)=\underset{1}{b}_4(x,t)A^{-1}, \tag{43}$$

$$\underset{1}{Q}(Ax,t)=\underset{1}{Q}(x,t)A^{-1}. \tag{44}$$

这样,我们就证明了:$\mathscr{F}$局部地等价于同步球对称的场.

再证明,这种同步球对称的场必为非异的.先证

引理 3 在进行规范变换(40)式时,

$$\underset{1}{B}(x_0,t)=B(x_0,t)-\begin{pmatrix}0 & 0\\ 0 & \frac{1}{r}\\ -\frac{1}{r} & 0\end{pmatrix}. \tag{45}$$

证明 由 $\zeta(x)$的定义可知

$$\zeta^{-1}(x)X_i\zeta(x)=p_{ij}(x)X_j. \tag{46}$$

这里 $p_{ij}(x)$ 为 $P(x)$ 的矩阵元.将(46)式微分,将所得结果左乘 $\zeta(x)$,右乘 $\zeta^{-1}(x)$,就得到

$$\zeta\partial_\lambda\zeta^{-1}X_i+X_i\partial_\lambda\zeta\zeta^{-1}=\partial_\lambda p_{ij}p_{kj}X_k,$$

或

$$[X_i,\ \partial_\lambda \zeta\zeta^{-1}] = \partial_\lambda p_{ij} p_{kj} X_k. \tag{47}$$

如置

$$\partial_\lambda \zeta\zeta^{-1} = W_\lambda^j X_j, \tag{48}$$

则

$$W_\lambda^l = \frac{1}{2}\varepsilon_{kil} \partial_\lambda p_{i1} p_{kj}. \tag{49}$$

又由(27)式易见

$$(dp_{ij})_{x_0} = \begin{pmatrix} 0 & -\frac{dx_2}{r} & -\frac{dx_3}{r} \\ \frac{dx_2}{r} & 0 & 0 \\ \frac{dx_3}{r} & 0 & 0 \end{pmatrix}, \tag{50}$$

从此就得出

$$(W_i^l) = \begin{pmatrix} 0 & 0 & 0 \\ 0 & 0 & \frac{1}{r} \\ 0 & -\frac{1}{r} & 0 \end{pmatrix},\ W_4^l = 0. \tag{51}$$

从规范变换的定义就得到(45)式,引理 3 证毕.

把(45)式和同步球对称的规范势(12)式相比较,就得

$$S = b_2^2(x_0,\ t),\ V = \frac{1}{r} b_2^3(x_0,\ t) - \frac{1}{r^2},$$

因 $b_2^2(x_0,\ t)$, $b_2^3(x_0,\ t)$ 不全为零,所以

$$S = 0,\ V = -\frac{1}{r^2}$$

不会同时成立，因而场是非异同步球对称的.

从而得到结论：属于情形(1)的场局部等价于负型的、非异的同步球对称场.

情形(2)和情形(1)可通过同位旋空间转动

$$X_1 \to -X_1,\ X_2 \to -X_2,\ X_3 \to X_3 \tag{52}$$

而互化，这也是一种规范变换，但情形(2)化为情形(1)后，同位旋向量场 Q 和 X_1 反向，因而有结论：属于情形(2)的场局部地等价于正型的、非异的同步球对称场.

五、情形(3)和(4)的讨论．局部结构小结

现讨论情形(3)，记四维时空中的半平面 $x_2=x_3=0,\ x_1>0$ 为Π.如果Π上有一点 $(\xi_0,\ \tau)$ 属于情形(3)，那么必有这点的一个Π上的邻域，在其中 $B(x_0,\ t)$ 的秩数 $\geqslant 1$，而且 $b_1^2(x_0,\ t),\ b_1^3(x_0,\ t)$ 不全为零.据引理 1，在这个邻域中每点 $(x_0,\ t)$，$B(x_0,\ t)$ 必须属于情形(3).因而(19)式仍然可以写成为(28)式的形式，照上节相同的方式可以证明，(34)式现在变为

$$\varphi(A,\ x,\ t)=\varphi(Ax,\ t)-\varphi(x,\ t). \tag{53}$$

仍然可以作规范变换，使 $\varphi(A,\ x,\ t)$ 化为零，从而(9)，(10)和(11)式便化为

$$b(Ax,\ t)=Ab(x,\ t),$$

$$b_4(Ax,\ t)=b_4(x,\ t),$$

$$Q(Ax,\ t)=Q(x,\ t), \tag{54}$$

这便是狭义球对称的情形，而且是非异的.从而得知：在情形(3)的点的周围，场 $\mathscr{F}$ 是非异的狭义球对称规范场.

再讨论情形(4).如果在一点 $(\xi_0,\ \tau)$ 出现情形(4).先假设Π上有 $(\xi_0,\ \tau)$ 的一个邻域，其中的点均属于情形(4).(9)，(10)式化为

$$b_\lambda^\alpha(x,\ t)=0 \quad (\alpha=2,\ 3),$$

$$b^1(Ax, t)=A(b^1(x, t)-\nabla\varphi(A, x, t)). \tag{55}$$

如果把 $b_\lambda^1(x, t)$ 看成 G_1 群的规范势，那么场就化为球对称的 G_1 群规范场，即化为电磁场的情况$\left(G_1\right.$ 群和 U_1 群同构，$\exp(\varphi X_1)$ 和 $\exp\left(i\dfrac{\varphi}{2}\right)$ 相对应，关于 φ 的周期是 $\left.4\pi\right)$. 这时场的强度除 $f_{\lambda\mu}^2=f_{\lambda\mu}^3=0$ 外，$f_{\lambda\mu}^1$ 为一球对称的张量，应有

$$H_j=\frac{1}{2}\varepsilon_{ijk}f_{jk}^1=H(r, t)x_i, \tag{56}$$

$$E_i=-f_{i4}^1=E(r, t)x_i. \tag{57}$$

此外，Bianchi 恒等式(即第一套 Maxwell 方程)给出

$$H_i=\frac{hx_i}{r}\ (h=\text{const}). \tag{58}$$

根据(57)和(58)式，在局部范围内可以解出 b_λ^1 作为势函数，并且除了关于 G_1 群的规范变换外，b_λ^1 是唯一确定的，在这种局部的考虑中，对 h 也暂时不出现量子化条件.

如果 (ξ_0, τ) 是Π 上属于情形(1)(或情形(2))的点的极限点，依连续性，在球面 S：$r=r_0$, $t=\tau$ 上，规范场是负型(或正型) 同步特异球对称的，这时应有 $h=-1$(或$+1$)，点(ξ_0, τ) 不可能是属于情形(2)(或情形(1))和(3) 的点的极限点. 又如果(ξ_0, τ) 是Π 上属于情形(3) 的点的极限点时，在这个球面上，场是特异狭义球对称的，因而 $h=0$.

这样，我们在一切可能情况下确定了 (ξ_0, τ) 的一个邻域中球对称场的结构，得到

定理 3 设带 Q 的 SU_2 规范场在(ξ_0, τ) 的一个邻域中为球对称的，那么场在该点附近的结构必为下述情况之一：

(1) 为正型非异同步球对称的；(2) 为负型非异同步球对称的；(3) 为非异狭义球对称的；(4) 为子群 G_1 的球对称规范场；(5) 为正型同步球对称的，在过(ξ_0, τ)的球面 S 上为特异，并为非异情况的极限；(6) 为负型同步球对称的，在 S 上为特异，并为非异情况的极限；(7) 为狭义球对称的，在 S 上为特异，并为特异情况的极限.

六、带 Q 的球对称场的整体结构

现设 $\mathscr{F}$ 是带 Q 的球对称 SU_2 规范场，其势除空间原点外处处为光滑，由于 SU_2 群和 S^3 同胚，其任何连通的二维子流形可以在其中连续收缩为一点，所以在定义 $\mathscr{F}$ 时没有必要用分解域的概念.

记 $S(r, t)$ 为以原点为中心，r 为半径的球面，t 为时间. 先证

引理 4　$S(r, t)$ 上任何两点只能同属于引理 1 所列举的同一种情形.

证明　由定理 3 显然可见，对 $S(r, t)$ 上每点，均可选取一球面上的邻域，使其中各点都属于引理 1 所列举的同一种情形.在 $S(r, t)$ 上任取二点 P_1，P_2，把它们用曲线弧 P_1P_2 连结起来，那么可以有有限个球面区域 Ω_1，…，Ω_s 覆盖弧 P_1P_2，每一 Ω_i 中的点同属于引理 1 的同一种情况，而且 $P_1 \in \Omega_1$，$P_2 \in \Omega_s$，Ω_i 和 Ω_{i+1} 相交，因此 P_1 和 P_2 属于同一种情况.引理证毕.

再证

引理 5　如果 $S(r, t)$ 上有一点 P_0 属于情形(4)，那么在这球面上成立磁量子化条件

$$H_i = \frac{n x_i}{r^3}, \tag{59}$$

这里 n 是整数.

证明　若有点 P_0 的一个四维邻域，其中各点均属于情形(4)，那么在 $S(r, t)$ 的周围，就有一整体的 G_1 群规范场，依照 G_1 群整体规范场的性质[7, 8]，并注意到 $\exp(\varphi X_1)$ 的周期为 4π，就得出(59)式.若 P_0 点为属于情形(1)，(2) 或(3) 的点的极限，那么由 H_i 的连续性也得出(59)式，且 $n=-1$，1 或 0.引理证毕.

四维时空可以看作由 Π 旋转而成.记Π中属于情形(1)，(2)，(3)，(4) 的点的集合为 Σ_1，Σ_2，Σ_3，Σ_4.由第四和第五节的讨论可知 Σ_1，Σ_2，Σ_3 在Π上都是开集，它们的边界无公共点. 我们证明 Σ_1，Σ_2，Σ_3 中至多只能有一个非空. 例如，设 Σ_1 为非空，由函数

$r^2H_1=n$ 的连续性可知,n 在Π上必须为 -1,所以 Σ_2, Σ_3 必为空集,所以当 $n=-1$ 时,$\Pi=\Sigma_1\cup\Sigma_4$,我们就有负型同步球对称场,这就是定理1中的情形(1),当 Σ_4 为空集时,整个场是非异的,当 Σ_1 为空集时,整个场是特异的,同时也等价于 G_1 群的规范场.同理,当 $n=+1$ 或0时,就有定理1中的情形(2)和(3),当 $|n|>1$ 时,必有 $\Pi=\Sigma_4$,有定理1中的情形(4).这样,我们就完全证明了定理1,定出了带 Q 的球对称场的一切情况.

七、不带 Q 的球对称场

引理6 设 $\mathscr{F}$ 为球对称的 SU_2 规范场,$f^a_{\lambda\mu}$ 是它的场强,记

$$h_i^a=\frac{1}{2}\varepsilon_{ijk}f^a_{jk},\tag{60}$$

$$H(x,t)=(h_i^a),\ E(x,t)=(-f^a_{i4}),\tag{61}$$

则对任何 $A\in SO_3$ 成立

$$\begin{aligned}H(Ax,t)&=AH(x,t)C(\zeta(A,x,t)),\\E(Ax,t)&=AE(x,t)C(\zeta(A,x,t)).\end{aligned}\tag{62}$$

证明 为节省篇幅计,这里用一下外微分形式记号.令

$$\omega=b^a_\lambda dx_\lambda X_a\tag{63}$$

为 $\mathscr{F}$ 的势形式[8],记 $Ax=\tilde{x}$,易见(9),(10)式可写为

$$\tilde{\omega}\zeta=\zeta\omega-d\zeta,\tag{64}$$

这里 $\tilde{\omega}=\omega(Ax,dAx)$. 对(64)式进行外微分,得

$$d\tilde{\omega}\zeta-\tilde{\omega}\wedge d\zeta=d\zeta\wedge\omega+\zeta d\omega.$$

利用(64)式消去 $d\zeta$,就得到

$$\zeta(d\omega+\omega\wedge\omega)=(d\tilde{\omega}+\tilde{\omega}\wedge\tilde{\omega})\zeta,\tag{65}$$

或

$$\zeta(d\omega+\omega\wedge\omega)\zeta^{-1}=d\tilde{\omega}+\tilde{\omega}\wedge\tilde{\omega}. \tag{66}$$

写开来,根据场强的定义和(60),(61)式,就推出(62)式.

附注 这个引理适用于任何球对称的规范场,只需把 ζ 的意义改一下就行了.

现在讨论场在一点 (ξ_0, τ) 附近的结构.设在这点 H(或 E)的秩数为 3,由(62)式得

$$C^{-1}(\zeta(A, x, t))=H^{-1}(Ax, t)AH(x, t), \tag{67}$$

由此可见,对 $B\in SO_3$,有

$$C^{-1}(\zeta(AB, x, t))=C^{-1}(\zeta(A, Bx, t)\zeta(B, x, t)), \tag{68}$$

因而在适当地选取符号后,就有

$$\zeta(AB, x, t)=\zeta(A, Bx, t)\zeta(B, x, t). \tag{69}$$

现在证明,可取规范变换,使变换后成立

$$\zeta(P(x), x_0, t)=I. \tag{70}$$

事实上,任取 $\zeta(x, t)$ 作规范变换,$H(x, t)$ 变为 $H(x, t)C(\zeta(x, t))$,从(67)式可见,$\zeta(A, x, t)$ 变为 $\zeta(Ax, t)\zeta(A, x, t)\zeta^{-1}(x, t)$.取 $\zeta(x, t)$ 为 $\zeta^{-1}(P(x), x_0, t)$ 又取 A 为 $P(x)$, $x=x_0$,从而可见 $\zeta(P(x), x_0, t)$ 变为 $\zeta^{-1}(P(x), x_0, t)\zeta(P(x), x_0, t)\zeta(P(x_0), x_0, t)$,但 $\zeta(P(x_0), x_0, t)=I$,所以 $\zeta(P(x), x_0, t)$ 就变为 I.

在(62)式中取 $x=x_0$, $A=R(\theta)$,我们得

$$H(x_0, t)=R(\theta)H(x_0, t)C(\zeta(R(\theta), x_0, t)). \tag{71}$$

考察第一行元素,得

$$(h_1^1, h_1^2, h_1^3)C(\zeta(R(\theta), x_0, t))=(h_1^1, h_1^2, h_1^3). \tag{72}$$

因 $H(x_0, t)$ 为非异,故 (h_1^1, h_1^2, h_1^3) 为非零向量,可作规范变换,使 h_1^2, h_1^3 变为零,可选取这个规范变换的 $\zeta(x, t)$ 使得 $\zeta(Ax_0, t)=\zeta(x_0, t)$,从而不影响(70)式.这时由(72)

式可得 $\zeta(R(\theta), x_0, t) \in G_1$.取 $H(x_0, t)$ 的第二、三列构成矩阵 $B'(x, t)$,那么它也满足(22)形式的关系,从而就可以应用引理 1.由于 H 为非异,故只能出现情形(1) 和(2),而情形(2) 又可通过规范变换变为情形(1),所以我们有

$$C(\zeta(R(\theta), x, t)) = R(-\theta). \tag{73}$$

和第四节相类似,我们有

$$P^{-1}(Ax)AP(x)B'(x_0, t)C(\zeta(A, x_0, t)) = B'(x_0, t). \tag{74}$$

由引理 1 的情形(1),(73)和(33)式关于 $\theta(A, x)$的定义可见

$$C(\zeta(A, x, t)) = R(-\theta(A, x)). \tag{75}$$

从此可见,这里的 $\zeta(A, x)$ 和第四节中(41) 式中所用的 $\zeta(A, x)$ 相同,那时的论述也适用于现在的情况.所以,如在(ξ_0, τ), H(或 E) 的秩数为 3,则有一(ξ_0, τ) 的邻域,在其中场为非异同步球对称的(因为特异同步球对称空间中 H 和E 的秩数为 1),但这时没有正型和负型的区别.

如在一区域中 H(或E) 的秩数为 2, H 中各列向量构成一个二维平面,这平面的法线是转动下的不变同位旋向量场;如 H(或 L) 在一区域中的秩数是 1,则 H 中各列向量共线,也存在转动下的不变向量场,从而带Q的场的讨论在这里都适用.又若 H 和L 在一区域中的秩数均为零,那么场就是平凡的,即可化为 $b^i_\lambda = 0$ 的情形(这时也可以把它看成同步球对称、狭义球对称或化约的 G_1 群的规范场).

分析了这些情况以后,我们就得出了定理 2.

八、关于用带 Q 的 SU_2 场表示球对称磁单极的讨论

取 s 为任一实数,以 sX_i 代替(2)式中的 X_i,那么(6)式就化为

$$\exp(\varphi X_1) = \begin{pmatrix} \cos\frac{s\varphi}{2} & -i\sin\frac{s\varphi}{2} \\ -i\sin\frac{s\varphi}{2} & \cos\frac{s\varphi}{2} \end{pmatrix}, \tag{76}$$

换位运算化为

$$[X_i, X_j]=s\varepsilon_{ijk}X_k, \tag{77}$$

't Hooft 张量化为

$$f_{\lambda\mu}=\frac{1}{|Q|}q_a f^a_{\lambda\mu}-\varepsilon_{abc}\frac{1}{s|Q|^3}q_a D_\lambda q_b D_\mu q_c. \tag{78}$$

$s=e$ 时,就有 Schwinger 量子化条件;$s=2e$ 时,就有 Dirac 量子化条件,$\frac{1}{s}$是磁荷的最小单位.

在同步球对称的情形,可以算出

$$f_{ij}=-\frac{W}{s|W|}\varepsilon_{ijk}\frac{x_k}{r^3}, \tag{79}$$

$$f_{i4}=-\frac{W}{|W|}\left(U+r\frac{\partial U}{\partial r}-\frac{\partial S}{\partial s}-r^2\frac{\partial T}{\partial t}\right)\frac{x_i}{r}. \tag{80}$$

因此,正型同步球对称场的磁荷总是 $\frac{1}{s}$,负型同步球对称场的磁荷总是 $-\frac{1}{s}$①,电磁场强度张量远不能确定 SU_2 规范场.事实上,给了磁荷为 $\pm\frac{1}{s}$ 的电磁场张量之后,$|W|$,U,V,S,T 之中还有三个是任意的,除规范变换外,规范场还包含相当大的自由度.

在狭义球对称的情况,总成立

$$f_{ij}=0.$$

因而磁单极不出现,此时电磁场张量也不能决定规范场.

从上面的论述我们可以见到:

(1) 磁荷为 $\frac{n}{s}$ 的球对称磁单极,其电磁场都可以用带 Q 的 SU_2 规范场来描述,但 n 必须为整数;

① 同步球对称场和狭义球对称场的磁荷,也可以由文献[9,10]的结论中推出.

(2) 磁荷为$\frac{n}{s}$的球对称磁单极，当$|n|>1$时，它相应的带Q的SU_2规范场必化约为G_1群的规范场，这个场由电磁场张量唯一确定；

(3) 同步球对称的带Q的SU_2规范场可用来描述磁荷为$\pm\frac{1}{s}$的球对称磁单极，但这个场不能由电磁场张量唯一确定；

(4) 狭义球对称的带Q的SU_2规范场可用来描述无磁荷的球对称电磁场，但这个场不能由电磁场张量唯一确定；

(5) 用化约的G_1群规范场描述磁单极，实质上就是用U_1群规范场(整体)来描述电磁场，但它可以利用SU_2群的规范变换，化为只需用一个坐标区域来描述[7].

用U_1群整体规范场来描述磁单极，有电磁场和U_1群规范场一一对应的优点.如把带Q的SU_2规范场理解为电磁场和带电矢介子场的偶合，那么由以上论述可以见到，在非单位的磁单极周围出现的球对称的场只能是电磁场，不能再有带电矢介子场.因而具有非单位的磁荷的磁单极隐含了某种破坏球对称性的因素.

究竟是否存在磁单极，磁单极有否球对称性，用什么理论体系来描述它才是正确的，这些问题终将有待于实践去解决，伟大的领袖和导师毛主席指出：**“许多自然科学理论之所以被称为真理，不但在于自然科学家们创立这些学说的时候，而且在于为尔后的科学实践所证实的时候.”** 对于磁单极的学说，也完全是这样.

参考文献

[1] C. N. Yang & R. L. Mills, *Phys. Rev.*, **96** (1954), 191.

[2] T. T. Wu & C. N. Yang, In Properties of Matter under Unusual Conditions (edited by H. Mark and S. Fernbach, Interscience 1969), 349.

[3] G. 't Hooft, *Nucl. Phys.*, **79B** (1974), 276.

[4] 侯伯宇，科学通报，**20** (1975)，273.

[5] 胡和生，复旦学报(自然科学版)，1976，1，72.

[6] 谷超豪、杨振宁，中国科学，1975，471；*Scientia Sinica*，**18** (1975)，483.

[7] T. T. Wu & C. N. Yang，*Phys. Rev.*，**D12** (1975)，3845.

[8] 谷超豪，复旦学报(自然科学版)，1975，4，83.

[9] 侯伯宇、段一士、葛墨林，兰州大学学报(自然科学版)，1975，2，26.

[10] J. Arafune，P. G. O. Freund & C. J. Goebel，*Jour. Math. Phys.*，**16** (1975)，433.

On Spherically Symmetric SU_2 Gauge Fields and Monopoles

Gu Chaohao　Hu Hesheng

(Futan University)

Abstract

In this paper we consider some general properties of spherically symmetric SU_2 gauge fields. We prove that the SU_2 gauge fields which are spherically symmetric in the most general sense may be divided into three basic types only: (1) synchro-spherically symmetric SU_2 gauge fields, (2) strict-spherically symmetric SU_2 gauge fields, (3) spherically symmetric SU_2 gauge fields reducible to U_1 gauge fields.

The spherically symmetric SU_2 gauge fields with a Higgs field are investigated in detail and their types are completely determined. If such a field is regarded as due to the interaction of an electromagnetic field and electrically-charged vector bosons, then the following conclusion is obtained: In the spherically symmetric case there cannot exist an electrically-charged vector boson field around an m-unit monopole except for $m = \pm 1, 0$.

(本文曾发表于《物理学报》1977 年第 26 卷第 2 期，155－168)

关于规范条件的变分问题*

谷超豪　胡和生

复旦大学

一、引　言

规范场可以有各种不同形式的规范条件.近来,对于库伦规范引起了许多讨论[1],我们注意到如对规范场作规范不变的积分,取它的稳定值,就会给出某些种类的规范条件.本文着重讨论其中的一种,其特点是使场的强度形式 $f_{\lambda\mu}dx^{\lambda}\wedge dx^{\mu}$ 成为闭的形式.这种规范条件的可解性归到一个复杂的非线性偏微分方程组,对于解析的 SU_2 规范场,它的局部解是存在的.对于球对称的 SU_2 规范场,我们作了整体的讨论,在这种规范存在的条件下,"磁通量"

$$\frac{1}{2}\int_{s} f_{ij}dx^{i}\wedge dx^{j}$$

具某种"守恒性".此外,我们也讨论了由变分方法导来库伦规范和洛仑兹规范的问题.

二、一种新的规范条件

设 E_3 为欧氏空间,G 为 r 参数紧致李群,$\mathscr{F}$ 为群 G 的规范场,$b_i=b_i^a X_a$ ($i=1, 2, 3$; $a=1, \cdots, r$) 为其规范势.当 $x\to\infty$ 时,$\mathscr{F}$ 为真空,亦即 b 的强度化为 0.我们假设同伦群

* 本文 1978 年 10 月 24 日收到.

$\pi_2(G)=0$,因而可以设 $x\to\infty$ 时 $b_i\to 0$.记 $s(x)$ 为 E_3 到 G 的可微分映照,当 $x\to\infty$ 时 $s(x)\to G$ 的恒等元素 I. $w_i=b_i-s^{-1}\partial_i s$,作如下的积分

$$J=\frac{1}{6}\int(w_i,[w_j,w_k])dx^i\wedge dx^j\wedge dx^k, \tag{1}$$

这里(,)表示群 G 的非退化的伴随不变内积,[,]表示李代数元素的换位运算.我们假定 $x\to\infty$ 时,$b_i\to 0$, $s\to I$ 充分快,使得 J 为有意义的.

作 J 关于 s 的变分,设

$$s\to s'=(I+\varepsilon v)s,$$

这里 v 为取值于李代数的任意函数,ε 为无限小量,在 ε 一次范围内

$$s'^{-1}\partial_i s'=s^{-1}(I-\varepsilon v)\partial_i(I+\varepsilon v)_s=s^{-1}\partial_i s+\varepsilon s^{-1}\partial_i vs, \tag{2}$$

因此

$$\delta w_i=-\varepsilon s^{-1}\partial_i vs. \tag{3}$$

计算 δJ,利用李代数中熟知的公式$(\alpha,[\beta,\gamma])+([\alpha,\gamma],\beta)=0$,易见

$$\delta J=\frac{\varepsilon}{2}\int(s^{-1}\partial_i vs,[w_j,w_k])dx^i\wedge dx^j\wedge dx^k. \tag{4}$$

为使 δJ 对任何 v 均为 0,取 v 使其支集为有限,又注意到

$$b'_j=sw_js^{-1}=sb_js^{-1}-(\partial_j s)s^{-1} \tag{5}$$

为 b_j 经 $s(x)$所定义的规范变换后所得到的规范势,从而得到

$$\delta J=-\frac{\varepsilon}{2}\int(\partial_i v,[b'_j,b'_k])dx^i\wedge dx^j\wedge dx^k$$

$$=\frac{\varepsilon}{2}\int(v,\partial_i[b'_j,b'_k])dx^i\wedge dx^j\wedge dx^k. \tag{6}$$

由于 v 为任意,所以我们得到 Euler-Lagrange 方程

$$\partial_1[b'_2,b'_3]+\partial_2[b'_3,b'_1]+\partial_3[b'_1,b'_2]=0. \tag{7}$$

利用场的强度的定义和 Jacobi 恒等式，容易知道，这就是

$$[b'_1, f'_{23}]+[b'_2, f'_{31}]+[b'_3, f'_{12}]=0, \tag{8}$$

或

$$[b_1-s^{-1}\partial_1 s, f_{23}]+[b_2-s^{-1}\partial_2 s, f_{31}]+[b_3-s^{-1}\partial_3 s, f_{12}]=0. \tag{9}$$

利用 Bianchi 恒等式，由(8)就知道

$$f'_{23,1}+f'_{31,2}+f'_{12,3}=0, \tag{10}$$

这里，表示普通微分.(10)式说明，利用 s 作规范变换后，规范场的强度形式

$$\frac{1}{2}f'_{ij}dx^i \wedge dx^j$$

为 r 个闭形式.从而得

定理 1 设 $s(x)$ 使积分 J 稳定，那么用 $s(x)$ 作规范变换后，规范场的强度形式是闭形式.即

$$f_{ij,k}+f_{jk,l}+f_{ki,j}=0. \tag{11}$$

在条件 $x\to\infty$, $s(x)\to I$ 下，E_3 的无限远可视为一个点，E_3 可视为 S^3, $s(x)$ 为映照 $S^3\to G$. $b_i=0$ 时，J 化为

$$J_1=\frac{1}{6}\int(s^{-1}\partial_i s, [s^{-1}\partial_j s, s^{-1}\partial_k s])dx^i \wedge dx^j \wedge dx^k. \tag{12}$$

它表示 S^3 到 G 这映照的拓扑度(除一常数因子外)，因而不同的 J_1 值标志着各种不相同伦的真空(由 $s^{-1}\partial_i s$ 所确定).

(9)关于 s 是偏微分方程组，有 r 个方程 r 个未知函数，结构相当复杂，对于 SU_2 规范场，在解析情况下，其局部解的存在性虽不能直接用 Cauchy-Kowalewsky 定理来证明，但却可以用 Cartan[2] 的方法予以证明.这就是，我们可以把这方程组化为发甫方程组，考察它的正则积分元素链，可以知道，它的一般解依赖于 3 个两自变数的任意函数(解析).

三、球对称的情形

现考虑 4 维时空中的球对称规范场，这种规范场分为三类[3].

1. 狭义球对称规范场——已知这时 $f_{ij}=0$，所以规范条件(11)恒满足；

2. 化约的球对称规范场——这时 f_{ij} 退化为 U_1 群的规范场的强度，方程(11)也显然成立；

3. 同步球对称规范场——已经知道，这时规范势必能写为[4]

$$b_i^l=\varepsilon_{ilk}x^kV+\delta_{il}S+x^ix^lT, \tag{13}$$

式中 V，S，T 是 r，t 的任意函数，$r=\sqrt{(x^1)^2+(x^2)^2+(x^3)^2}$，又强度的表达式为

$$f_{ij}^l=A\varepsilon_{ijk}x^kx^l+B(\delta_{li}x^j-\delta_{lj}x^i)+C\varepsilon_{lij}, \tag{14}$$

其中

$$A=\frac{1}{r}\frac{\partial V}{\partial r}-\frac{S^2}{r^2}-V^2,\ B=\frac{1}{r}\frac{\partial S}{\partial r}+\frac{S}{r^2},\ C=-\left(2V+\frac{\partial V}{\partial r}r\right). \tag{15}$$

在 A，B，C 的表达式中，我们已利用 $S+r^2T=0$ 这个关系式，因为同步球对称规范场可通过规范变换使这个条件成立.

现在的问题是求 s 使得利用 s 作规范变换后(11)式成立.我们现在限于采用使规范势(13)式保持为同样形式的规范变换[5]，即以

$$s=\frac{1+\mathbf{u}\cdot i\boldsymbol{\sigma}}{\sqrt{1+u^2}},\ u_i=x_if(r,t) \tag{16}$$

来作规范变换，这里 σ_i 为 Pauli 阵，这种规范变换称为同步球对称的规范变换.在以 O 为心，r 为半径的球面上，它在各点所相应的同位旋旋转是以向径为轴的旋转，且点点的转角相同.

由(16)式，我们可以得出

$$s^{-1}\partial_i s=\frac{i\sigma_i}{1+u^2}+\frac{i\varepsilon_{ijk}\sigma_j u_k}{1+u^2}, \tag{17}$$

这里 $u^2=(u_1)^2+(u_2)^2+(u_3)^2$.将(13),(14)及(17)式代入(9)式经过一定的计算发现,所有具 $f'(r, t)$ 的项都自行消去,余下的只是 $f(r, t)$ 的二次方程

$$Pf^2(r, t)+Qf(r, t)+R=0, \tag{18}$$

这里

$$\begin{aligned}&P=2C+(BS-VC)r^2,\\&Q=-2B,\\&R=BS-VC.\end{aligned} \tag{19}$$

又记

$$\Delta=Q^2-4PR, \tag{20}$$

我们得出

定理 2 同步球对称 SU_2 规范场,通过同步球对称的规范变换后,在

(i) $\Delta>0$ 时,有两个规范使 $f_{ij}dx^i\wedge dx^j$ 为闭形式;

(ii) $\Delta<0$ 时,不存在球对称的规范使 $f_{ij}dx^i\wedge dx^j$ 为闭形式;

(iii) $\Delta=0$, $P\neq 0$ 或 $P=0$, $Q\neq 0$ 时有一个规范使 $f_{ij}dx^i\wedge dx^j$ 为闭形式;

(iv) $P=Q=R=0$ 时,任何情况下 $f_{ij}dx^i\wedge dx^j$ 为闭形式.

这四种情形都是可能的.特别当

$$b_i^l=\varepsilon_{ilk}x^k V \tag{21}$$

时,(18)式即为

$$\frac{2C}{1+u^2}f^2(r, t)-VC=0, \tag{22}$$

$C=0$ 时,对任何 $u_i=x_i f(r, t)$, $f_{ij}dx^i\wedge dx^j$ 为闭形式,在 $C\neq 0$ 时

$$f=\pm\sqrt{\frac{V}{2-r^2V}}. \tag{23}$$

因而在 $V>0$ 时,若$\frac{2}{r^2}>V>0$ 成立,从而有二个规范使 $f_{ij}dx^i \wedge dx^j$ 为闭形式,而当 $V<0$ 时,不存在球对称的规范变换使 $f_{ij}dx^i \wedge dx^j$ 为闭形式.

如果我们所讨论的规范条件满足,那么包围原点的任一球面,“磁通量” $\frac{1}{2}\int_s f_{ij}dx^i \wedge dx^j$ 是不因球面的位置而变的,它具拓扑的守恒性.

四、关于库伦规范和洛仑兹规范的变分推导

对已给的规范势 b_λ,作

$$I=\int\Sigma(w_i, w_i)d^3x, \tag{24}$$

对 s 作变分,由 $\delta I=0$ 就得出 E_3 中的库伦规范条件

$$\Sigma b'_{i,i}=0. \tag{25}$$

(24)式是一个正的积分,显然有序列 $s^{(n)}$ 使 I 的极限取得最小值,因而可以预期 $s^{(n)}$ 有收敛的子序列,而且极限函数充分光滑,但严格的数学证明还未能给出.

如所论的空间是 4 维时空 E_4,则对每一时间截面 $t=\text{const.}$ 可以作同样的讨论,只要把 t 看成参数就行了.这时我们得出:如果对每一 t, $s(x, t)$ 使 I 稳定,那么利用 $s(x, t)$ 作规范变换后,在每一时间截面 $t=\text{const.}$ 上,(25)成立.这样,4 维时空中整体库伦规范的存在性问题也化为变分极值问题.

同样,4 维明可夫斯基空间的洛仑兹规范也可由

$$\int\eta^{\lambda\mu}(w_\lambda, w_\mu)d^4x \tag{26}$$

的变分导出,这里 $\eta^{\lambda\mu}$ 是明可夫斯基时空的度规.

如果我们要求 $s^{-1}(x)\partial_i s(x)$ 和 b_i 在 E_3 都是平方可积，那么

$$I=\int \sum_i (s^{-1}\partial_i s-b_i,\ s^{-1}\partial_i s-b_i)d^3x, \tag{27}$$

可理解为规范势 b_i 和强度O的规范势$s^{-1}\partial_i s$（真空）的"距离"平方，从而库伦规范是由和 b_i"距离"最小的真空所定出的. 又积分

$$I_1=\int \sum_i (s^{-1}\partial_i s,\ s^{-1}\partial_i s)d^3x, \tag{28}$$

也可视为 $s=s(x)$ 代表的真空到 $s=\text{const}$ 所代表的真空的一种，但它是取连续数值的，不是一个拓扑不变量.事实上，若

$$I_1(s(x))\neq 0,$$

令 $\bar{s}(x)=s(\tau x)$ $(\tau=\text{const.})$，$\bar{s}(x)$ 和 $s(x)$ 所标志的真空属于同一同伦类，但

$$I_1(\bar{s}')=\frac{1}{\tau}I_1(s),$$

所以 I_1 可取到任何非零数值，且当 $\tau\to\infty$ 时 $I_1(s')\to 0$.

致谢：1978 年暑期在探讨这个问题时，承杨振宁教授提供了宝贵的意见，特此谨表谢意.

参 考 文 献

[1] Gribov, V. N., *SLAC Translation*, **176** (1977).

[2] Cartan, E., *Les systémes différentialo exiérieurs e1 leurs applications geometriques*, Paris, 1945.

[3] 谷超豪、胡和生，物理学报，**26** (1977)，155—168.

[4] 胡和生，复旦学报，1976，1：72.

[5] 侯伯宇、谷超豪、胡和生，复旦大学学报，1977，1：92.

（本文曾发表于《科学通报》1979 年第 24 卷第 11 期，492－495）

关于具有质量的杨-米尔斯方程*

胡和生

复旦大学数学研究所

一、引　　言

关于四维时空中杨-米尔斯方程是否存在具有限能量又无奇点、且场强在无限远处为 0 的静态解，是规范场研究中大家关心的一个问题[1]. 到 1976 年，Deser[2]得出：除 $n=5$ 外，n 维时空（指度规为 $ds^2=-dx_0^2+dx_1^2+dx_2^2+\cdots+dx_{n-1}^2$）中紧致群规范场方程不存在具有限能量又无奇点，且场强在无穷远处适当快地趋于 0 的静态解. 这里，特别令人注意的是四维时空中紧致群规范场不存在满足要求的静态解. 另一方面，由于近年来人们对四维欧氏空间规范场的自对偶解（瞬子解）给予极大的注意，而这些瞬子解都可以看作是五维时空中的静态解，所以 $n=5$ 这个例外情形就显得更为重要[3, 4].

在过去的规范场理论体系中，规范势所表示的矢量粒子无质量，但物理中往往需要有质量的规范粒子（如传播弱相互作用的中间玻色子），为使规范粒子获得质量，物理学家通常是从规范场体系之外加进其他粒子（如 Higgs 粒子）[5]. 最近谷超豪对规范场的概念作了一些分析，把规范场的拉氏密度作了适当的增补，得出规范不变的拉氏密度函数，并且能够从此导出质量项①. 这样，就从理论上证明了在规范场体系之内，规范粒子是可以有质量的.

本文研究这种具有质量的杨-米尔斯方程的静态解问题，我们仍然考虑无外源的情

* 本文 1979 年 1 月 9 日收到.

① 见谷超豪的《关于规范粒子的质量》一文.

形.我们发现,与无质量的规范场时不同,实质量的紧致群的杨-米尔斯方程在 $n \neq 4$ 时都不存在无奇点、有限能量、强度及势在∞处适当快地趋于 0 的静态解,因此,只有 $n=4$ 时有待继续研究.我们也举例说明在 $n>2$ 时虚质量的杨-米尔斯方程存在满足无限远边界条件 $r \to \infty$ 时规范势→0 场强→0 的正则解(即无奇点的解).

二、场方程和能量动量张量

在谷超豪的一文中已经证明,r 参数紧致群 G 的规范场 $\mathscr{F}$ 可以有如下的规范不变的拉氏密度函数,

$$\mathscr{L}=\frac{1}{4}G_{ab}f^{a}_{\lambda\mu}f^{b\lambda\mu}+\frac{m'}{2}G_{ab}(b-\omega^{a}_{\lambda})(b^{b\lambda}-\omega^{b\lambda})$$

$$=\frac{1}{4}(f_{\lambda\mu},\ f^{\lambda\mu})+\frac{m'}{2}(b_{\lambda}-\omega_{\lambda},\ b^{2}-\omega^{\lambda})$$

$$(\lambda,\ \mu=0,\ 1,\ \cdots,\ n-1;\ a,\ b=1,\ 2,\ \cdots,\ r),\tag{1}$$

式中 $b_{\lambda}=b^{a}_{\lambda}X_{a}$ 表示规范势(X_{a} 是 G 的李代数 g 的一组基),$f_{\lambda\mu}=f^{a}_{\lambda\mu}X$ 表示场强,$G_{ab}=(X_{a},\ X_{b})$,这里(,)表示 g 的伴随不变内积,m 为代表质量的偶合常数,ω^{a}_{μ} 由

$$U^{-1}(\partial_{\mu}U)=\omega^{a}_{\mu}X_{a}\tag{2}$$

定义,$U=U(x)$ 为规范势的"参考截面".

$$f^{a\lambda\mu}=\eta^{\lambda\alpha}\eta^{\mu\beta}f^{a}_{\alpha\beta},\tag{3}$$

$$\eta^{\lambda\mu}=\begin{pmatrix}-1 & & & \\ & 1 & & \\ & & \ddots & \\ & & & 1\end{pmatrix}.$$

$\mathscr{L}$ 关于 U 及 b 作变分,可以得出场方程和规范条件,如果就利用这个 U 作规范变换

$$b'_{\lambda}=(a\,d\,U)b_{\lambda}-(\partial_{\lambda}U)U^{-1},\tag{4}$$

又再把 b'_λ 重记为 b_λ，那么 $\mathscr{L}$ 化为

$$\mathscr{L}=\frac{1}{4}(f_{\lambda\mu},\ f^{\lambda\mu})+\frac{m^2}{2}(b_\lambda,\ b^\lambda), \tag{5}$$

场方程可以直接从这个 $\mathscr{L}$ 出发关于 b 变分得出：

$$J_\alpha-m^2b_\alpha=0, \tag{6}$$

这里

$$J_\alpha=\eta^{\lambda\mu}f_{\alpha\lambda|\mu}=\eta^{\lambda\mu}(f_{\alpha\lambda,\mu}+[b_\mu,\ f_{\alpha\lambda}]), \tag{7}$$

这时 b_α 自然地满足 Lorentz 规范条件①

$$\eta^{\lambda\mu}\frac{\partial b_\lambda}{\partial x^\mu}=0 \tag{8}$$

这样做，在经典理论范围之内是完全严格的.

由于能量动量张量与规范无关，因而在这一特殊规范下仍然可采取普通的方法进行推导[6]，我们得出其具体表达式为

$$T_{\alpha\beta}=(f_{\alpha\nu},\ f_\beta{}^\nu)-\frac{1}{4}\eta_{\alpha\beta}(f_{\mu\nu},\ f^{\mu\nu})+m^2(b_\alpha,\ b_\beta)-\frac{m^2}{2}\eta_{\alpha\beta}(b_\lambda,\ b^\lambda). \tag{9}$$

由于 Noether 定理，成立守恒定律

$$\frac{\partial T_\alpha^\beta}{\partial x^\alpha}=0. \tag{10}$$

由 $T_{\alpha\beta}$ 的(9)式即得

$$T_\alpha^\alpha=\eta^{\alpha\beta}T_{\alpha\beta}=\frac{1}{4}(4-n)(f_{\lambda\mu},\ f^{\lambda\mu})-\frac{m^2}{2}(n-2)(b_\lambda,\ b^\lambda), \tag{11}$$

$$T_0^0=\eta^{0\beta}T_{\beta0}=-T_{00}=\frac{1}{2}\left[(f_{0i},\ f^{0i})-\frac{1}{2}(f_{ij},\ f^{ij})\right]$$

---

① 谷超豪一文已证明，这样做法和原来的 $\mathscr{L}$ 出发所得场方程一致.

$$-m^2(b_0, b_0)-\frac{m^2}{2}(b_\lambda, b^\lambda), \tag{12}$$

$$T_i^i=-\frac{1}{2}(n-3)(f_{0i}, f^{0i})+\frac{1}{4}(5-n)(f_{ij}, f^{ij})$$
$$+\frac{m^2}{2}(3-n)(b_\lambda, b^\lambda)+m^2(b_0, b_0). \tag{13}$$

三、$m^2>0$ 时静态解的讨论

这里的静态解是指 b_λ 和 x_0 无关的解.现设 $m^2>0$，我们要求的是

$$r=\sqrt{(x^1)^2+\cdots+(x^{n-1})^2}\to 0$$

时，b_λ^a，$f_{\lambda\mu}^a\to 0$ 适当快的静态解.

先注意到，由于 b_λ^a 和 x^0 无关，就有

$$(b_0, J_0)=(b_0, f_{0i|i})=\partial_i(b_0, f_{0i})-(f_{0i}, f_{0i}). \tag{14}$$

因此

$$0=\int(b_0, J_0-m^2b_0)d^{n-1}x=-\int[(f_{0i}, f_{0i})+m^2(b_0, b_0)]d^{n-1}x. \tag{15}$$

这里积分是沿 $x^0=\text{const.}$所作. 由 $m^2>0$ 得出

$$f_{0i}=0, b_0=0. \tag{16}$$

由此就得出

$$T_{i0}=0, \frac{\partial T_{ij}}{\partial x^j}=0. \tag{17}$$

因而积分

$$\int T_i^i d^{n-1}x=\int\partial_i(x^jT_j^i)d^{n-1}x, \tag{18}$$

利用 $r \to \infty$ 处的条件,可知此积分为 0.利用(13),(16)式得

$$\int\left[\frac{1}{4}(5-n)(f_{ij},\ f^{ij})+\frac{m^2}{2}(3-n)(b_i,\ b^i)\right]d^{n-1}x=0. \tag{19}$$

由此可见

1) $n=3$ 时,$f_{ij}=0$,连同 $f_{0i}=0$ 得出规范场强度 $f_{\lambda\mu}=0$,由场方程就有 $b_\lambda=0$,所以只是平凡解.

2) $n=5$ 时,有 $b_0=b_i=0$, 所以也只是平凡解.

3) $n>5$ 或 $n<3$ 时,仍然只有平凡解.

因此得到

定理 除 $n=4$ 外,n 维时空中具有实质量的紧致群规范场不存在具有限能量、无奇点且场强及势在无限远处适当快地趋近于 0 的静态解.

剩下来是 $n=4$ 的情形,此时是否有满足这些条件的静态解的问题尚未解决,这样的解的存在估计还是有可能的.

与四维欧氏空间的瞬子解相比较,从上述定理我们得到

推论 四维欧氏空间中实质量的杨-米尔斯场没有具有限能量又无奇点且在无限远处强度与势趋于 0 相当快的无源解.

注 1 由定理的证明可见,如果 $r \to \infty$ 时

$$f_{\lambda\mu}=O(r^{-\frac{n+1}{2}-\varepsilon}),\ b_\lambda=O(r^{-\frac{n+1}{2}-\varepsilon}) \quad (\varepsilon>0,\text{任意小}),$$

那么能量有限及 $r \to \infty$ 时强度及势适当快地趋于 0,这两项条件就得到了保证,而

$$f_{\lambda\mu}=O(r^{-\frac{n-1}{2}-\varepsilon}),\ b_\lambda=O(r^{-\frac{n-1}{2}-\varepsilon}),$$

只能保证能量为有限.

注 2 显然,四维欧氏空间的自对偶和反自对偶场都是没有质量的.

四、$m^2<0$ 时静态解的讨论

对于虚质量时与前节的情况不同,我们可以证明在 $n>2$ 时,场方程有满足条件 $r \to$

∞ 时 $b_\lambda \to 0$，$f_{\lambda\mu} \to 0$ 且无奇点的静态解.为此记 $m^2 = -\lambda^2 < 0$，λ 为实，我们先置

$$b_\lambda^1 = 0 \ (a \neq 1),\ b_i^1 = 0, \tag{20}$$

这时

$$f_{0l}^1 = -f_{l0}^1 = \frac{\partial b_0^1}{\partial x^l}, \tag{21}$$

其余 $f'_{\lambda\mu}$ 均为 0，代入无源方程(6)，只留下一个方程

$$\Delta b' + \lambda^2 b_0^1 = 0. \tag{22}$$

设 b_0^1 只依赖于 $r = \sqrt{(x^1)^2 + \cdots + (x^{n-1})^2}$，置 $b_0^1 = u$，$R = \lambda r$，那么 $u(R)$ 满足

$$R\frac{d^2 u}{dR^2} + (n-2)\frac{du}{dR} + Ru = 0. \tag{23}$$

令

$$u = R^{-p}\nu \ \left(p = \frac{n-3}{2} \geqslant 0\right), \tag{24}$$

那么 ν 就满足 p 阶的 Bessel 方程

$$\nu'' + \frac{\nu'}{R} + \left(1 - \frac{p^2}{R^2}\right)\nu = 0. \tag{25}$$

利用此方程的性质可见，在原点和在∞处均正则的解

$$b_0^1(r) = Cr^{-p} J_p(\lambda r). \tag{26}$$

因此，对任何 $n > 2$，具虚质量的杨-米尔斯方程必有 $r \to \infty$ 时 $b_\lambda \to 0$，$f_{\lambda\mu} \to 0$ 的正则解.特别，在 $n = 4$ 时此解即为

$$b_0^1(r) = C\frac{\sin \lambda r}{r}.$$

致谢：作者在此对苏步青教授的鼓励与帮助，表示感谢.

参 考 文 献

[1] Wu, T. T. & Yang, C. N., *in Properties of Matter under Unusual Conditions*, 1968, 349.

[2] Deser, S., *Ref. TH. 2214 CERN*, 1976.

[3] Belavin, A., Polyakov, A., Schwarz. A. & Tyupkin, X., *Phys. Lett.*, **59B** (1975), 85.

[4] Jackiw, R., Nohl, C. & Rebbi, C., *Phys. Rev.*, **D15** (1977), 1642.

[5] 't Hooft, G., *Nucl. Phys.*, **B79** (1974), 276.

[6] Л. Д.朗道,E. M.栗弗席兹,场论(中译本),1959.

(本文曾发表于《科学通报》1980 年第 25 卷第 6 期,241 - 244;英文版曾发表于 *Kexue Tongbao*, 1980, 25(3), 191 - 195)

On the Spherically Symmetric Gauge Fields

Gu Chaohao[1] and Hu Hesheng[2]

1 Institute for Theoretical Physics, State University of New York at Stony Brook, Stony Brook, NY 11794, USA; and Department of Mathematics, Fudan University, Shanghai, China

2 Institute for Mathematics, Fudan University, Shanghai, China

Abstract The spherically symmetric gauge fields with a compact gauge group over 4-dimensional Minkowski space are determined completely. Expressions for the gauge potentials of these fields are obtained.

Part I. Construction of Spherically Symmetric Gauge Fields

1. Introduction

Spherical symmetry is one of the most important symmetry in nature. In a field theory the spherically symmetric fields are very useful for understanding the theory itself. For example, the important role of Schwarzschild solution in Einstein's gravitation theory is well-known. For the gauge theories, the fields with spherical symmetry are also very interesting. It is known that two gauge fields are equivalent if they are related by a gauge transformation. Consequently, in the construction of spherically symmetric gauge fields we have to consider the effect of gauge transformations. Spherically symmetric potentials of SU_2 gauge fields were firstly considered by Wu and Yang [1]. Now the complete classification of spherically symmetric SU_2 gauge fields is known [2]. Some spherically symmetric SU_3 gauge potentials were studied by several authors

[3 - 5]. For the general gauge groups the problem of determining spherically symmetric field may be reduced to solve some system of partial differential equations by using Lie derivatives [6]. But, the results in [6] are local in character and no formulas for the gauge potential were obtained. The same problem was treated in a different way in a previous paper [7], an algebraic method for determining all spherically symmetric gauge fields was proposed. However, in this paper the expressions of the gauge potentials were given only for some special cases and the proof of the general theorem was not complete yet. It should be noted that some special form of spherically symmetric gauge fields were pointed in [8].

In the present paper we develop this method and give a complete determination of general spherically symmetric gauge fields for any compact gauge group. In Part I we state the method of construction and identify a special but important class of such fields. In Part II we prove that all possible spherically symmetric gauge fields can be obtained in this way. The proof is rather long, since we have to prove the completeness.

2. Definitions

Let G be a compact Lie group and $\mathscr{F}$ a gauge field with gauge group G over 4 - dimensional Minkowski space-time R^{3+1}. We assume that the field $\mathscr{F}$ is regular except for $r=0$. Here $r=(x_1^2+x_2^2+x_3^2)^{1/2}$.

If the field $\mathscr{F}$ corresponds to a trivial bundle, then the gauge potential $b(x, dx)=b_\lambda(x)dx_\lambda$ $(\lambda=1, 2, 3, 4)$ is a g - valued 1 - form defined on R^{3+1} except for $r=0$. Here g is the Lie algebra of G. For simplipifying the description we assume that G and g consist of $N\times N$ matrices.

Definition 1. If for each $A\in SO_3$ there exists a G - valued function u_A such that

$$b(Ax, \mathrm{A}dx)=(\mathrm{ad}u_A(x))b(x, dx)-(du_A(x))u_A^{-1}(x), \tag{1}$$

then the field $\mathscr{F}$ is called spherically symmetric and the function u_A is defined as the complementary function associated with A.

If the field $\mathscr{F}$ corresponds a nontrivial bundle we have to cover R^{3+1} by two regions [9] i.e. $R^{3+1}=M^+ \cup M^-$

$$M^+ = R^{3+1} - \{(0, 0, r, t)\},$$
$$M^- = R^{3+1} - \{(0, 0, -r, t)\}, \tag{2}$$
$$(0 \leqq r < \infty, -\infty < t < \infty).$$

Here $(0, 0, r, t)$ and $(0, 0, -r, t)$ are the north pole and south pole of the sphere $x_1^2 + x_2^2 + x_3^2 = r^2$, $x_4 = t$. The field $\mathscr{F}$ should be expressed as the combination of two gauge fields $\mathscr{F}^+$ and $\mathscr{F}^-$ which are defined by two gauge potentials b^+ and b^- over M^+ and M^- respectively. Moreover, b^+ and b^- are related by a gauge transformation

$$b^+(x, dx) = (\mathrm{ad}\zeta(x))b^-(x, dx) - (d\zeta(x))\zeta^{-1}(x) \tag{3}$$

in $M^+ \cap M^-$, where $\zeta(x)$ is a G-valued function.

Definition 2. $\mathscr{F}^+$ (or $\mathscr{F}^-$) is called almost spherically symmetric if the definition 1 holds true for all x and A, satisfying $x \in M^+$ (or M^-) and $Ax \in M^+$ (or M^-).

Definition 3. If $\mathscr{F}$ consists of $\mathscr{F}^+$ and $\mathscr{F}^-$, and $\mathscr{F}^+$, $\mathscr{F}^-$ are almost spherically symmetric, then $\mathscr{F}$ is called a spherically symmetric field.

Remark. If $\mathscr{F}$ corresponds to a trivial bundle it can also be represented as a combination of $\mathscr{F}^+$ and $\mathscr{F}^-$. In this case it is ready seen that definition 3 agrees with Definition 1.

3. Construction of Spherically Symmetric Gauge Fields

We give here a description of the method of constructing spherically symmetric gauge fields. The main procedure is to construct $\mathscr{F}^+$ and $\mathscr{F}^-$, then connect them by a gauge transformation. In order to obtain $\mathscr{F}^+$ (and $\mathscr{F}^-$), we construct the complementary function u_A corresponding to each rotation A satisfying $Ax_0 \neq x_0'$ where $x_0 = (0, 0, r,$

t), $x_0'=(0, 0, -r, t)$. Then, we determine the possible gauge potential of x_0, and further we obtain the gauge potential b^+ at any arbitrary points of $\mathscr{F}^+$, the gauge potential b^- is constructed similarly. Moreover, we need a quantization condition to combine the b^+ and b^-.

Let (ψ, θ, ϕ) be the Eulerian angles for the rotation A, i. e. $A = C_3(\psi)C_1(\theta)C_3(\phi)$, where $C_a(\alpha)$ is the rotation around x_a - axis through an angle α.

(a) The first step is to take an element Y of g such that the quantization condition

$$\exp(4\pi Y)=e \tag{4}$$

be satisfied, where e is the unit element of G. Let the value of a complementary function $u_{C_3(\phi)}$ at x_0 be

$$u_{C_3(\phi)}(x_0)=\exp(\phi Y). \tag{5}$$

The condition (4) is a crucial one for combining $\mathscr{F}^+$ and $\mathscr{F}^-$ in a global gauge field $\mathscr{F}$.

(b) In order to obtain $u_A(x_0)$ for general A we set

$$u_{C_3(\psi)C_1(\theta)}(x_0)=u_{C_3(\psi)}(x_0), \quad 0\leqq\theta<\pi. \tag{6}$$

In Part II we shall prove that any other possible choices of $u_{C_3(\psi)C_1(\theta)}(x_0)$ will be equivalent to (6) via a gauge transformation. Then, the value of the complementary function $u_A(x)$ at $x=x_0$ is

$$u_A(x_0)=u_{C_3(\psi)C_1(\theta)C_3(\phi)}(x_0)=u_{C_3(\psi)}(x_0)u_{C_3(\phi)}(x_0)=u_{C_3(\psi+\phi)}(x_0) \quad (\theta<\pi). \tag{7}$$

(c) Let $x=Bx_0$, $B\in SO_3$, $x\neq(0, 0, -r, t)$. It is seen in Part II that the formula

$$\begin{aligned} u_A(x) &= u_{C_3(\psi_1)C_1(\theta_1)C_3(\phi_1)}(Bx_0) \\ &= u_{C_3(\psi_1)C_1(\theta_1)C_3(\phi_1)B}(x_0)u_B^{-1}(x_0)=u_{AB}(x_0)u_B^{-1}(x_0) \end{aligned}$$

defines the value of a complementary function $u_{C_3(\psi_1)C_1(\theta_1)C_3(\phi_1)}$ at the point x. It is easy to verify that this value is independent upon the choice of B. Thus, we obtain the complementary function $u_A(x)$.

(d) To obtain the possible gauge potential at the point x_0, we put $x=x_0$ and $A=C_3(\phi)$ in Eq. (1) and obtain

$$b(x_0, C_3(\phi)dx)=(\text{ad}\exp(\phi Y))b(x_0, dx). \tag{8}$$

For simplicity one can choose a gauge such that $x_i b_i(x)=0$, then

$$b_3(x_0)=0. \tag{9}$$

By differentiating Eq. (8) with respect to ϕ, we obtain a system of linear equations

$$-b_2(x_0)+[Y, b_1(x_0)]=0, \quad b_1(x_0)+[Y, b_2(x_0)]=0, \tag{10}$$

$$[Y, b_4(x_0)]=0. \tag{11}$$

For given G and Y, it is quite easy to solve these equations. Thus, (9)-(11) give $b_\lambda(x_0)$ or

$$b(x_0, dx)=b_\lambda(x_0)dx_\lambda=b_\lambda(0, 0, r, t)dx_\lambda \quad (r>0). \tag{12}$$

For abbreviation, we write

$$b_\lambda(r, t)=b_\lambda(0, 0, r, t)=b_\lambda(x_0), \quad b(x_0, dx)=b(r, t, dx). \tag{12'}$$

(e) Using (1) and (6), after some calculation we obtain the expression of b^+ at any point of M^+:

$$b^+(C_3(\psi)C_1(\theta)x_0, dx)=(\text{ad}\exp(\psi Y))b(r, t, C_1(-\theta)C_3(-\psi)dx) - \frac{r-x_3}{r}Yd\psi, \tag{13}$$

where $C_3(\psi)C_1(\theta)x_0(\pi>\theta\geqq 0)$ is a general point of M^+.

This is the formula for gauge potentials of the field $\mathscr{F}^+$. In Part II, we shall verify that it is almost spherically symmetric.

(f) Similarly, the field $\mathscr{F}^-$ is defined by the potential

$$b^-(C_3(\psi)C_1(\theta-\pi)x_0',\ dx)=(\text{ad}\ \exp(-\psi Y))b(r,\ t,\ C_1(\pi-\theta)C_3(-\psi)dx)$$
$$+\frac{r+x_3}{r}Yd\psi, \tag{14}$$

where $x_0'=(0,\ 0,\ -r,\ t)$ and $b(x_0',\ dx)=b(x_0,\ dx)=b(r,\ t,\ dx)$.

(g) It is readily verified that

$$b^+(x,\ dx)=\text{ad}\ (\exp(2\psi Y))b^-(x,\ dx)-(d\exp(2\psi Y))\exp(-2\psi Y). \tag{15}$$

Hence, $\mathscr{F}^+$ and $\mathscr{F}^-$ can be combined together as a spherically symmetric gauge field $\mathscr{F}$ if (4) holds.

Thus the construction of spherically symmetric gauge fields is accomplished. In short, we have the theorem

Main Theorem. *All the spherically symmetric gauge fields of a compact gauge group are the combination of two fields* $\mathscr{F}^+$ *and* $\mathscr{F}^-$ *which are defined over* M^+ *and* M^- *with gauge potentials* (13) *and* (14) *respectively. In* (13) *and* (14) Y *is an arbitrary element of* g *satisfying* $\exp(4\pi Y)=e$ *and* $b(r,\ t,\ dx)$ *is determined by* (9)-(11).

4. Simple Examples

We shall give some examples to illustrate the general construction in Sect. 3.

(a) U_1 gauge fields. When G is U_1, the Lie algebra g is generated by i, thus we have to take $Y=\frac{m}{2}i$, where m is an integer. From Eq. (9)-(11), the general solution for $b_\lambda(x_0)$ is

$$b_1(x_0)=b_2(x_0)=b_3(x_0)=0,\quad b_4(x_0)=\sigma(r,\ t)i.$$

Consequently, from (13), (14) we have

$$b^{+}(x, dx)=-\frac{r-x_3}{r}Yd\psi+i\sigma(r, t)dt=-\frac{1-\cos\theta}{2}mid\psi+\sigma(r, t)idt,$$
$$b^{-}(x, dx)=\frac{r+x_3}{r}Yd\psi+i\sigma(r, t)dt=\frac{1+\cos\theta}{2}mid\psi+\sigma(r, t)idt. \tag{16}$$

So the field consists of a standard m - monopole [9] and a spherically symmetric electromagnetic field with scalar potential $\sigma(r, t)$.

(b) SU_2 gauge fields. Without loss of generality we may take

$$Y=i\begin{bmatrix}\frac{m}{2} & 0\\ 0 & -\frac{m}{2}\end{bmatrix} \quad (m \text{ is an integer}). \tag{17}$$

From (10) we see that

$$b_1(x_0)+[Y, [Y, b_1(x_0)]]=0. \tag{18}$$

If $m\neq 0, 1$, we have

$$b_1(x_0)=b_2(x_0)=b_3(x_0)=0,$$
$$b_4(x_0)=\sigma\begin{bmatrix}i & 0\\ 0 & -i\end{bmatrix}.$$

The field is reduced to a U_1 field and the potential has the form (15), the only change is that i should be replaced by $\begin{pmatrix}i & 0\\ 0 & -i\end{pmatrix}$.

If $m=0$ or 1, we obtain the strictly spherically symmetric SU_2 gauge field and the synchro-spherically symmetric SU_2 gauge fields [2]. We shall discuss them in the following section.

5. Spherically Symmetric Gauge Fields of Proper Type

Definition 4. Let $\mathscr{F}$ be a spherically symmetric gauge field corresponding to a trivial bundle. If there exists a gauge such that the complementary function u_A does not depend on x, then it is called a spherically symmetric gauge field of proper type [7]. It has been shown that if a spherically symmetric field $\mathscr{F}$ is regular everywhere, then it is of proper type. In fact, in this case the bundle is trivial. Let 0, $0'$, x be the points $(0, 0, 0, 0)$, $(0, 0, 0, t)$, (x_1, x_2, x_3, t) respectively. Choose a gauge such that the phase factors for each segment $00'$ and $0'x$ are all equal to e. The phase factor for the arc $xx+dx$ is $e-b(x, dx)$ where $b(x, dx)$ is the gauge potential [10]. This quantity is the phase factor for the infinitessimal loop $00'xx'\,0''0$, where $x'=x+dx$, $0''=(0, 0, 0, t+dt)$ [11]. The spherical symmetry implies the existence of u_A such that

$$b(Ax, dAx) = (\mathrm{ad}\, u_A) b(x, dx). \tag{19}$$

Here u_A is independent of x.

Evidently, a spherically symmetric gauge field $\mathscr{F}$ of proper type can be singular at $r=0$. The construction of spherically symmetric gauge fields of proper type can be accomplished in the following way. Let τ: $A \to u_A$ be an homomorphism of SO_3 to G and τ_1 the corresponding homomorphism of the Lie algebra so_3 to g, X_1, X_2, X_3 be a set of standard base of so_3 and $Y_i = \tau_1 X_i (i=1, 2, 3)$. It is easily seen that

$$u_{C_3(\phi)} = \exp(\phi Y_3). \tag{20}$$

Moreover, $u_{C_3(\psi)C_1(\theta)}$ is given by the homorphism τ, i.e. $u_{C_3(\psi)C_1(\theta)} = \tau C_3(\psi) C_1(\theta)$, $b_\lambda(x_0)$ are still determined by (9)-(11) with $Y=Y_3$ and the gauge potential is given by

$$b(Ax_0, \mathrm{Ad}x) = (\mathrm{ad}\, u_A) b(x_0, dx). \tag{21}$$

The gauge potential obtained this way is equivalent to the gauge potential obtained through the method in Sect. 3 with the same $u_{C_3(\phi)(x_0)} = \exp(\phi Y_3)$ and $b(x_0, dx)$. This is a consequence of Lemma 7. However, in the present case the potential corresponds to a trivial bundle and admits a simpler explicit expression. Besides, we can conclude that a spherically symmetric gauge field $\mathscr{F}$ is of proper type if and only if the element Y of the Lie algebra, described in the main theorem, be $\tau_1 X_3$.

For SU_2 case, we know that the representations of SO_3 in a two dimensional space are either a trivial representation $A \to e$ or the spin representation $A \to u_A$ with $(\mathrm{ad}\, u_A) = A$, if a suitable base for SU_2 is chosen. Then Y_3 is 0 or $\frac{i}{2}\begin{pmatrix}1 & 0\\ 0 & -1\end{pmatrix}$. The SU_2 gauge field obtained are strictly spherically symmetric and synchro-spherically symmetric (see the end of Appendix II) respectively.

In the appendix we shall list all possible gauge potential for SU_3 gauge field. A class of SU_N gauge potentials is also presented.

Part II. Proof of the Main Theorem

6. Sets of Complementary Functions

At first we shall analyze in detail the properties of complementary functions defined in Definition 1. The analysis is made for the case that $\mathscr{F}$ corresponds to a trivial bundle, but all results are valid for the fields $\mathscr{F}^+$ and $\mathscr{F}^-$ with some obvious modifications.

Because there may be a set of complementary functions $\{u_A\}$ associated with each $A \in SO_3$, we use the following notations:

$U_A = \{u_A\}$— the set of complementary functions associated with A.

$u_A(x)$— the value of u_A at x.

$U_A(x) = \{u_A(x)\}$— the set of values at x of complementary functions associated with A.

Lemma 1. *Let* $A, B \in SO_3$. *Then*

$$\{u \mid u = u_A(Bx) u_B(x)\} = U_{AB}(x). \tag{22}$$

Proof. Using (1), we see that

$$\begin{aligned} b(ABx, ABdx) &= \text{ad}\, u_A(Bx) b(Bx, Bdx) - du_A(Bx) u_A^{-1}(Bx) \\ &= \text{ad}\, u_A(Bx) \text{ad}\, u_B(x) b(x, dx) - \text{ad}\, u_A(Bx) du_B(x) u_B^{-1}(x) \\ &\quad - du_A(Bx) u_A^{-1}(Bx) \\ &= \text{ad}(u_A(Bx) u_B(x)) b(x, dx) - d(u_A(Bx) u_B(x)) \\ &\quad (u_A(Bx) u_B(x))^{-1}. \end{aligned}$$

Hence $u_A(Bx)u_B(x) \in U_{AB}(x)$. Similarly we have

$$u_B^{-1}(x) \in U_{B^{-1}}(Bx). \tag{23}$$

Consequently, for each $u_{AB}(x) \in U_{AB}(x)$ and $u_B(x) \in U_B(x)$ we have

$$u_{AB}(x) u_B^{-1}(x) = u_{AB}(x) u_{B^{-1}}(Bx) \in U_A(Bx). \tag{24}$$

Thus (22) is proved.

For simplicity we write (22) as

$$U_A(Bx) U_B(x) = U_{AB}(x). \tag{25}$$

Lemma 2. *Under the gauge transformation*

$$b'(x, dx) = (\text{ad}\, \zeta(x)) b(x, dx) - (d\zeta(x)) \zeta^{-1}(x), \tag{26}$$

the complementary function u_A *becomes* u_A' *defined by*

$$u'_A(x) = \zeta(Ax) u_A(x) \zeta^{-1}(x). \tag{27}$$

Proof. Using (1), (26), and (27), it is easily seen that

$$b'(Ax, \text{Ad}\, x) = (\text{ad}\, u'_A(x)) b'(x, dx) - (du'_A(x)) u'^{-1}_A(x).$$

This is the conclusion of Lemma 2.

Let x_0 be a fixed point, say $(0, 0, r, t)$, and E the unit element of SO_3.

Lemma 3. $U_E(x_0)$ *is a closed subgroup of* G.

Proof. From Lemma 1 we see that $U_E(x_0)$ is a subgroup. If $\alpha \in U_E(x_0)$, then there is a G-valued function $u \in U_E$ such that $u(x_0)=\alpha$. The function u is the solution of the differential equation

$$du = ub - bu \tag{28}$$

with the initial condition $u(x_0)=\alpha$. Suppose that $\{\alpha_n\}$ is a sequence of elements in $U_E(x_0)$ and $\alpha_n \to \alpha_0$ as $n \to \infty$. For each α_n the differential equation (28) has a solution $u_n(x)$ and $u_n(x_0)=\alpha_n$. The value $u_n(x)$ can be obtained by solving a system of ordinary equation

$$\frac{du}{d\sigma} = u(\sigma) b\left(x(\sigma), \frac{dx(\sigma)}{d\tau}\right) - b\left(x(\sigma), \frac{dx(\sigma)}{d\sigma}\right) u(\sigma) \tag{29}$$

with initial condition $u(0)=\alpha_n$, where $x(\sigma)$ $0 \leqq \sigma \leqq 1$ is a smooth arc connecting x_0 and x. Moreover, the value of $u_n(x)$ is independent of the choice of the arc $x(\sigma)$. Consequently, $u_n(x)$ converges to a solution $u(x)$ of (29) with initial condition $u(x_0)=\alpha_0$ and $u(x)$ is independent of the choice of $x(\sigma)$, so it is a solution of (28). Hence $\alpha_0 \in U_E(x_0)$. This proves that the group $U_E(x_0)$ is closed.

Lemma 4. *Suppose that the Lie algebra g does not contain a nontrivial center. If $U_E(x_0)$ is a nondiscrete proper subgroup of G, then the field is reducible to a gauge field whose gauge group has Lie algebra $g_1 \neq g$.*

Proof. Suppose that $u \in U_E$. Then u satisfies (28). The integrability condition of (28) is

$$(\mathrm{ad}\, u(x)) f_{\lambda\mu}(x) = f_{\lambda\mu}(x). \tag{30.0}$$

Here $f_{\lambda\mu}(x)$ is the field strength. The successive gauge derivative $f_{\lambda\mu/\nu}(x)$, ... satisfy

$$(\mathrm{ad}\, u(x)) f_{\lambda\mu/\nu}(x) = f_{\lambda\mu/\nu}(x), \tag{30.1}$$

$$(\mathrm{ad}\, u(x)) f_{\lambda\mu/\nu\sigma}(x) = f_{\lambda\mu/\nu\sigma}(x). \tag{30.2}$$

Let the subalgebra generated by $f_{\lambda\mu}(x)$, $f_{\lambda\mu/\nu}(x)$, ... be $\Sigma(x)$. From (30.0), (30.1), ..., it is seen that each element of $\Sigma(x)$ remains unchanged under ad $u(x)$ with $u(x) \in U_E(x)$. Moreover, by their construction $\Sigma(x)$ are parallel along any path with respect to the gauge potential b. Consequently, the loop phase factors at point x_0 (or holonomy group at that point) keep $\Sigma(x_0)$ unchanged. From the hypothesis on G and $U_E(x)$ we see that $\Sigma(x_0) \neq g$ and the loop phase factors at the point x_0 belong to a subgroup G_1 of lower dimension. Consequently, the field is reducible to a G_1 gauge field. Lemma 4 is proved.

If g contains a nontrivial center, then G is decomposed to the direct product $U_1 \times U_1 \times \cdots \times U_1 \times G'$, where the Lie algebra of G'' does not contain nontrivial center. The gauge potential is also decomposed. Further, spherically symmetric U_1 gauge fields are readily constructed (see Sect. 4). Hence it remains to consider the fields with gauge group G'.

Consequently, without loss of generality we can assume that $U_E(x_0)$ is discrete and g does not contain nontrivial center.

Lemma 5. *The mapping* $A \to U_A(x_0)$ *is a smooth mapping from* SO_3 *into the coset space* $G/U_E(x_0)$ *where* $x_0=(0, 0, r, t)$.

Proof. The integrability condition of (1) is

$$f_{\lambda_1\lambda_2}(Ax)a^{\lambda_1}_{\mu_1}\, a^{\lambda_2}_{\mu_2}\, u_A(x) - u_A(x) f_{\mu_1\mu_2}(x) = 0,$$

$$A = (a^{\lambda}_{\mu}) \in SO_3. \tag{31.0}$$

Differentiation gives

$$f_{\lambda_1\lambda_2/\lambda_3}(Ax)a^{\lambda_1}_{\mu_1}\, a^{\lambda_2}_{\mu_2}\, a^{\lambda_3}_{\mu_3}\, u_A(x) - u_A(x) f_{\mu_1\mu_2/\mu_3}(x) = 0. \tag{31.1}$$

Let $x = x_0$ and consider $u(x_0)$ as unknowns. We solve these equations firstly. The independent equations in (31.0), (31.1), ... must be finite in number, for otherwise, the solution $u_A(x_0)$ does not exist. Evidently, when $A = E$, $u_A = e$ satisfies these equations.

We use the implicit function theorem to prove the smoothness of $u_A(x_0)$ near E. Let $u = u(\alpha_1, \cdots, \alpha_r)$ be a parametric representation of the group G near e and $u(0, \cdots, 0) = e$. The system (31.0), (31.1), ... with $x = x_0$ can be considered as equations of $\alpha_1, \cdots, \alpha_r$. Differentiating these equations with respect to α_a $(a = 1, \cdots, r)$ and letting $A = E$, $\alpha_a = 0$, we get

$$f_{\lambda_1\lambda_2}(x_0)u_a - u_a f_{\lambda_1\lambda_2}(x_0) = 0,$$

$$f_{\lambda_1\lambda_2/\lambda_3}(x_0)u_a - u_a f_{\lambda_1\lambda_2/\lambda_3}(x_0) = 0,$$

where $u_a = \left.\dfrac{\partial u}{\partial \alpha_a}\right|_{\alpha_a = 0} \in g$. From these equations we should have $u_a = 0$. Otherwise, $\Sigma(x_0)$ would not be equal to g. The implicit function theorem implies that $u_A(x_0)$ with $u_E(x_0) = e$ is smooth with respect to A near E.

Consider the solution of (31.0), (31.1), ... near C. Let $A = CB$, $w = u_C^{-1}(Bx_0)u_A(x_0)$. Noting

$$f_{\lambda_1\lambda_2/\lambda_3\cdots\lambda_s}(Ax_0)a^{\lambda_1}_{\mu_1}\cdots a^{\lambda_s}_{\mu_s} = u_C(Bx_0)f_{\nu_1\nu_2/\nu_3\cdots\nu_s}(Bx_0)b^{\nu_1}_{\mu_1}\cdots b^{\nu_s}_{\mu_s}u_C^{-1}(Bx_0),$$

$$B = (b^{\lambda}_{\mu}),$$

we obtain

$$f_{\lambda_1\lambda_2}(Bx_0)b^{\lambda_1}_{\mu_1}b^{\lambda_2}_{\mu_2}w(x_0) - w(x_0)f_{\mu_1\mu_2}(x_0) = 0,$$

$$f_{\lambda_1\lambda_2/\lambda_3}(Bx_0)b^{\lambda_1}_{\mu_1}b^{\lambda_2}_{\mu_2}b^{\lambda_3}_{\mu_3}w(x_0) - w(x_0)f_{\mu_1\mu_2/\mu_3}(x_0) = 0,$$

and hence w is smooth with respect to B near E. Consequently $u_A(x_0)$ is smooth with respect to A near arbitrary C. The Eq. (1) is integrable for $u_A(x)$ if we take the obtained $u_A(x_0)$ as initial condition.

7. Determination of the Potentials

Let (ψ, θ, ϕ) be the Eulerian angles for rotation A, i.e. $A = C_3(\psi)C_1(\theta)C_3(\phi)$. From

Lemmas 1 and 5 we may write

$$u_{C_3(\psi)C_1(\theta)C_3(\phi)}(x_0)=u_{C_3(\psi)C_1(\theta)}(x_0)u_{C_3(\phi)}(x_0) \tag{32}$$

with

$$u_{C_3(\phi)}(x_0)=\exp(\phi Y), \quad u_{C_3'(2\pi)}(x_0)\in U_E(x_0), \tag{33}$$

where $Y=Y(r, t)$ is a g - valued smooth function. Moreover, $u_{C_3(\psi)C_1(\theta)}(x_0)$ is smooth with respect to (ψ, θ) and

$$u_{C_3(\psi)C_1(0)}(x_0)=u_{C_3(\psi)}(x_0), \tag{34}$$

$$u_{C_3(2\pi)C_1(\theta)}(x_0)=u_{C_3(0)C_1(\theta)}(x_0)u_{C_3(2\pi)}(x_0). \tag{35}$$

When $\mathscr{F}$ corresponds to a trivial bundle we have also

$$u_{C_3(\psi)C_1(\pi)}(x_0)(x_0)=u_{C_1(\pi)}(x_0)u_{C_3(-\psi)}(x_0). \tag{36}$$

From Lemma 1 it is seen that

$$u_A(Bx_0)=u_{AB}(x_0)u_{B^{-1}}(Bx_0), \tag{37}$$

so $u_A(x)$ can be determined by the whole set of $u_A(x_0)$.

Lemma 6.

$$\left.\frac{\partial u_{C_3(\psi)C_1(\theta)}(x_0)}{\partial\theta}\right|_{\theta=0} u^{-1}_{C_3(\psi)}(x_0)=\alpha\cos\psi+\beta\sin\psi, \tag{38}$$

where α, β *take values in* g *and are independent of* ψ.

Proof. From Lemma 5 and (37) we have

$$u_{C_3(\psi_1)C_1(\theta_1)}(x)=u_{C_3(\psi_1)C_1(\theta_1)C_3(\psi)C_1(\theta)}(x_0)u^{-1}_{C_3(\psi)C_1(\theta)}(x_0), \tag{39}$$

where x denotes the point $C_3(\psi)C_1(\theta)x_0$. For simplicity we neglect the symbols r and t. Let $C_3(\psi_1)C_1(\theta_1)C_3(\psi)C_1(\theta)=C_3(\psi')C_1(\theta')C_3(\phi')$. From the matrices in both sides it is easily seen that

$$\cos\theta' = -\sin\theta_1 \cos\psi \sin\theta + \cos\theta_1 \cos\theta,$$

$$\operatorname{ctg}\phi' = \operatorname{ctg}\psi \cos\theta + \operatorname{ctg}\theta_1 \frac{\sin\theta}{\sin\psi},$$

$$\operatorname{ctg}\psi' = -\frac{\sin\psi_1 \sin\psi \sin\theta - \cos\psi_1 \cos\theta_1 \cos\psi \sin\theta - \cos\psi_1 \sin\theta_1 \cos\theta}{\cos\psi_1 \sin\psi \sin\theta + \sin\psi_1 \cos\theta_1 \cos\psi \sin\theta + \sin\psi_1 \sin\theta_1 \cos\theta}.$$

From these equations and

$$\theta' \mid_{\theta=0} = \theta_1, \quad \phi' \mid_{\theta=0} = \psi, \quad \psi' \mid_{\theta=0} = \psi_1, \tag{40}$$

we have

$$\left.\frac{\partial\theta'}{\partial\theta}\right|_{\theta=0} = \cos\psi, \quad \left.\frac{\partial\phi'}{\partial\theta}\right|_{\theta=0} = -\sin\psi \operatorname{ctg}\theta_1, \quad \left.\frac{\partial\psi'}{\partial\theta}\right|_{\theta=0} = \frac{\sin\psi}{\sin\theta_1}. \tag{41}$$

Consequently,

$$\begin{aligned}\left.\frac{\partial u_{C_3(\psi_1)C_1(\theta_1)}(x)}{\partial\theta}\right|_{\theta=0} = \sin\psi\Bigg(&\frac{1}{\sin\theta_1}\frac{\partial u_{C_3(\psi_1)C_1(\theta_1)}(x_0)}{\partial\psi_1} - \\ &\frac{\cos\theta_1}{\sin\theta_1} u_{C_3(\psi_1)C_1(\theta_1)}(x_0)\frac{\partial u_{C_3(\psi)}(x_0)}{\partial\psi} u^{-1}_{C_3(\psi)}(x_0) + \\ &\cos\psi \frac{\partial u_{C_3(\psi_1)C_1(\theta_1)}(x_0)}{\partial\theta_1} - \\ &u_{C_3(\psi_1)C_1(\theta_1)}(x_0)\left.\frac{\partial u_{C_3(\psi)C_1(\theta)}(x_0)}{\partial\theta}\right|_{\theta=0} u^{-1}_{C_3(\psi)}(x_0)\Bigg).\end{aligned} \tag{42}$$

On the other hand, if $f(x)$ is any differentiable function of x at $x = x_0$, we have $dx_1 = \sin\psi d\theta$, $dx_2 = -\cos\psi d\theta$ at x_0. Thus

$$\begin{aligned}\left.\frac{\partial f}{\partial\theta}\right|_{\substack{\psi=\text{const}\\ \theta=0}} d\theta &= \left.\frac{\partial f}{\partial x_1}\right|_{x=x_0} dx_1 + \left.\frac{\partial f}{\partial x_2}\right|_{x=x_0} dx_2 \\ &= \left(\left.\frac{\partial f}{\partial x_1}\right|_{x=x_0} \sin\psi - \left.\frac{\partial f}{\partial x_2}\right|_{x=x_0} \cos\psi\right) d\theta.\end{aligned} \tag{43}$$

Comparing to (42), we obtain the conclusion of Lemma 6.

If $u_{C_2(\phi)}(x_0)$ is given, we may use a special $u_{C_3(\psi)C_1(\theta)}(x_0)$ satisfying (34), (35), and (39) to construct a set of complementary functions $u_A(x)$. Of course we may use another $u'_{C_3(\psi)C_1(\theta)}(x_0)$ satisfying the same condition to construct another set of complementary functions $u'_A(x)$. However, we have

Lemma 7. *$u'_A(x)$ and $u_A(x)$ are related by (26) with a suitable $\zeta(x)$.*

Proof. Let

$$\zeta(x) = u'_{C_3(\psi)C_1(\theta)}(x_0)u^{-1}_{C_3(\psi)C_1(\theta)}(x_0). \tag{44}$$

It is easily verified that $\zeta(x)$ is a G - valued function on M^+ and is regular at x_0.

Let $A = C_3(\psi_1)C_1(\theta_1)C_3(\phi_1)$, $A\,C_3(\psi)C_1(\theta) = C_3(\psi')C_1(\theta')C_3(\phi')$, then $C_3(\psi')C_1(\theta')x_0 = AC_3(\psi)C_1(\theta)x$, and

$$\begin{aligned} u'_A(x) &= u'_{C_3(\psi')C_1(\theta')}(x_0)u_{C_3(\phi')}(x_0)u'^{-1}_{C_3(\psi)C_1(\theta)}(x_0) \\ &= \zeta(Ax)u_{C_3(\psi')C_1(\theta')}(x_0)u_{C_3(\phi')}(x_0)u^{-1}_{C_3(\psi)C_1(\theta)}(x_0)\zeta^{-1}(x) \\ &= \zeta(Ax)u_A(x)\zeta^{-1}(x). \end{aligned}$$

Lemma 7 follows from Lemma 2.

From Lemma 7 it is seen that without loss of generality we may take

$$u_{C_3(\psi)C_1(\theta)}(x_0) = u_{C_3(\psi)}(x_0) \quad (\theta \neq \pi). \tag{45}$$

Lemma 8. *If $b(x_0, dx)$ satisfies*

$$b(x_0, u_{C_3(\phi)}(x_0)dx) = (\mathrm{ad}\,u_{C_3(\phi)}(x_0))b(x_0, dx) - du_{C_3(\phi)}(x)\,|_{x=x_0}\, u^{-1}_{C_3(\phi)}(x_0), \tag{46}$$

then

$$b(x, dx) = (\mathrm{ad}\,u_A(x_0))b(x_0, A^{-1}dx) - du_A(x)\Big|_{\substack{x=x_0 \\ dx=A^{-1}dx}} u_A^{-1}(x_0)$$

is a spherically symmetric gauge potential, where $x = Ax_0$.

Proof. We first prove that $b(x, dx)$ does not depend on the choice of A. Let $Ax_0 = Bx_0 = x$,

$$\underset{1}{b}(Ax_0, dAx) = (\mathrm{ad}u_A(x_0))(x_0, dx) - du_A(x)\mid_{x=x_0} u_A^{-1}(x),$$

$$\underset{2}{b}(Bx_0, dBx) = (\mathrm{ad}u_B(x_0))b(x_0, dx) - du_B(x)\mid_{x=x_0} u_B^{-1}(x).$$

We want to prove that $\underset{1}{b}(Ax_0, dx) = \underset{2}{b}(Bx_0, dx)$.

By a direct calculation this is equivalent to

$$b(x_0, dB^{-1}Ax) = (\mathrm{ad}\, u_B^{-1}(x_0)u_A(x_0))b(x_0, dx) + u_B^{-1}(x_0)du_B(x)\Big|_{\substack{x=x_0 \\ dx=B^{-1}Adx}}$$
$$- u_B^{-1}(x_0)du_A(x)\mid_{x=x_0} u_A^{-1}(x_0)u_B(x_0). \tag{47}$$

It is easily seen that $B^{-1}A = C_3(\phi)$ for some ϕ and

$$u_B^{-1}(x_0)u_A(x_0) = u_{B^{-1}}(Bx_0)u_A(x_0) = u_{B^{-1}}(Ax_0)u_A(x_0) = u_{B^{-1}A}(x_0).$$

Moreover, differentiating $u_{B^{-1}A}(x) = u_{B^{-1}}(Ax)u_A(x)$ and setting $x = x_0$, we have

$$du_{B^{-1}A}(x)\mid_{x=x_0} = du_{B^{-1}}(Ax)\mid_{x=x_0} u_B(x_0) + u_{B^{-1}}(Ax_0)du_A(x)\mid_{x=x_0}$$
$$= du_B^{-1}(B^{-1}Ax)\mid_{x=x_0} u_B(x_0) + u_B^{-1}(x_0)du_A(x)\mid_{x=x_0}.$$

Thus we see that (47) is equivalent to (46). So we proved that $b(x, dx)$ is independent of the choice of A.

Similarly, a direct calculation gives that

$$b(ABx_0, dABx_0) = (\mathrm{ad}u_A(Bx_0))b(Bx_0, dBx) - du_A(Bx)\mid_{x=x_0} u_A(Bx_0).$$

Lemma 8 is proved.

Thus the problem is reduced to construct $b(x_0, dx)$ such that (46) is satisfied.

It is easily seen that (46) is equivalent to (10), (11) and

$$[Y, b_3(x_0)] = 0, \tag{48}$$

if we take $u_{C_3(\phi)}(x_0) = \exp(\phi Y)$, since (10), (11), and (48) are the equivalent formulas of (46) in the Lie algebra. For the time being Y can depend on x_0.

Now we are going to construct the potential. Let $b(x_0, dx)$ be a solution of (10),

(11), and (48). From (42) and (45) it follows that

$$\left.\frac{\partial u_{C_3(\psi_1)C_1(\theta_1)}(\psi, \theta)}{\partial \theta}\right|_{\theta=0} = \sin\psi \frac{1-\cos\theta_1}{\sin\theta_1} Y. \tag{49}$$

Noting that

$$r\sin\psi\, d\theta = dx_1,$$

we have

$$b^+(Ax_0, dAx) = (\mathrm{ad}\, u_3(\psi_1))b(x_0, dx) - \frac{1}{r}\left(\frac{1-\cos\theta_1}{\sin\theta_1}\right)Ydx_1, \tag{50}$$

where $A = C_3(\psi_1)C_1(\theta_1)$. Replacing (ψ_1, θ_1) by (ψ, θ) and noticing that

$$\begin{bmatrix} x_1 \\ x_2 \\ x_3 \end{bmatrix} = A \begin{bmatrix} 0 \\ 0 \\ r \end{bmatrix} = \begin{bmatrix} r\sin\psi\sin\theta \\ -r\cos\psi\sin\theta \\ \cos\theta \end{bmatrix}.$$

$$A^{-1} = \begin{bmatrix} \cos\psi & \sin\psi & 0 \\ -\sin\psi\cos\theta & \sin\psi\cos\theta & \sin\theta \\ \cos\psi\sin\theta & -\cos\psi\sin\theta & \cos\theta \end{bmatrix},$$

we obtain (13). Here we use (48) instead of $b_3(x_0) = 0$. Thus all possible almost spherically symmetric fields $\mathscr{F}^+$ are constructed.

By the symmetry we can choose a suitable gauge such that the potential of $\mathscr{F}^-$ is (14). Since x_0 and x_0' lie in opposite direction, we have to replace ψ by $-\psi$ in the construction of $u_{C_3(x)(x_0')}$. Moreover, it is noted that

$$C_3(\psi)C_1(\theta)x_0 = C_3(\psi)C_1(\theta - \pi)x_0', \tag{51}$$

We prove formula (4). Let $\zeta(x) = \exp(2\psi Y)$. It is easily seen that

$$b^+(x, dx) = (\mathrm{ad}\, \zeta(x))b^-(x, dx) - (d\zeta(x))\zeta^{-1}(x), \tag{52}$$

so $\mathscr{F}^+$ and $\mathscr{F}^-$ are equivalent at each point of $M_+ \cap M_-$. It is required that $\zeta(x)$ be a single valued function of $M_+ \cap M_-$. Hence we must have (4).

In order to simplify $Y(r, t)$ we shall use the following lemma.

Lemma 9. *If* $Y(r, t)$ *is a* g *-valued smooth function, satisfying* $\exp(4\pi Y(r, t)) = e$, *then there exists a* G *-valued smooth function* $\kappa(r, t)$ *and an element* $Y_0 \in g$ *such that*

$$Y(r, t) = (\mathrm{ad}\,(\kappa(r, t)))Y_0. \tag{53}$$

Proof. The local existence of a continuous $\kappa(r, t)$ is known [12]. In addition we shall prove that $\kappa(r, t)$ can be a smooth function of r and t. Let H be the subgroup of G

$$H = \{\alpha \in G \mid (\mathrm{ad}\,\alpha)Y_0 = Y_0\},$$

h the Lie algebra of H and $h^\perp$ the orthogonal complementary of h. For any given r_1, t_1, we have

$$Y(r_1, t_1) = (\mathrm{ad}\,\lambda)Y_0.$$

Consider the equation

$$Y(r, t) = \mathrm{ad}(\lambda \exp(\kappa(r, t)))Y_0$$

with $k(r, t)$ *as unknown function taking values in* $h^\perp$. The equation is satisfied by $r = r_1$, $t = t_1$, and $k = 0$. Differentiating the equation and set $r = r_1$, $t = t_1$, and $k = 0$ we obtain

$$[dk, Y_0] = 0.$$

Since $dk \in h^\perp$ we have $dk = 0$. From the implicit function theory we obtain the existence of a smooth solution $\kappa(r, t)$ near an arbitrary point (r_1, t_1). So the set K: $0 < r < \infty$, $-\infty < t < \infty$ may be covered by a system of neighborhoods $\{U_\alpha\}$ such that

$$Y(r, t) = (\mathrm{ad}\,\kappa_\alpha(r, t))Y_0$$

in U_α. In $U_\alpha \cap U_\beta$ define $g_{\beta\alpha} = \kappa_\beta \kappa_\alpha^{-1} \in H$. We have H bundle over K. Since K is homeomorphic to R^2 the H bundle is a trivial bundle. Hence for each U_α there exists $\psi_\alpha \in H$ such that $g_{\beta\alpha} = \psi_\beta \psi_\alpha^{-1}$. Define

$$\kappa = \kappa_\alpha \psi_\alpha^{-1}$$

which is equal to $\kappa_\beta \psi_\beta^{-1}$ in $U_\alpha \cap U_\beta$. We obtain the conclusion of Lemma 9.

Consequently, in the construction of gauge potentials we may assume $Y(r, t)$ be independent of (r, t).

Further, we define a G-valued function $\zeta(r, t)$ by

$$\frac{d\zeta}{dr} = \zeta(r, t) b_3(r, t), \quad r(r_0, t) = e, \quad (r_0 \neq 0).$$

By the gauge transformation via $\zeta(r, t)$ we obtain $b_3(r, t) = 0$ instead of (48).

Thus the proof of the main theorem is accomplished.

Appendix I

Spherically Symmetric SU$_3$ Gauge Fields

For the SU$_3$ case we may assume

$$Y = \frac{i}{2} \begin{bmatrix} m_1 & 0 & 0 \\ 0 & m_2 & 0 \\ 0 & 0 & m_3 \end{bmatrix},$$

where m_1, m_2, m_3 are integers with $m_1 + m_2 + m_3 = 0$. The spherically symmetric gauge potentials are classified as follows.

I. Proper Type

(a) $m_1 = 2$, $m_2 = 0$, $m_3 = -2$. Y is the generator Y_3 of the 3-dimensional irreducible representation of SO$_3$. This is the case considered in [5] (see also Appendix II).

(b) $m_1=1$, $m_2=0$, $m_3=-1$. Y is the generator Y_3 of the reducible representation which is the 2 - dimensional irreducible representation of SO_3 acturely. We have

$$b(x, dx)=\begin{bmatrix} b'(x, dx) & & 0 \\ & & 0 \\ 0 & 0 & 0 \end{bmatrix}+i\begin{bmatrix} \lambda(r, t) & 0 & 0 \\ 0 & \lambda(r, t) & 0 \\ 0 & 0 & -2\lambda(r, t) \end{bmatrix} dx_{\psi},$$

where $b'(x, dx)$ is a syncro-spherically symmetric potential and $\lambda(r, t)$ is an arbitrary function.

(c) $m_1=m_2=m_3=0$. We have

$$b_i=0,$$

if we take the gauge $x^i b_i=0$ and

$$b_4=(a_{ij}(r, t)),$$

where $(a_{ij}(r, t))$ is an arbitrary function valued in su_3. In this case the "magnetic part" of the potential vanishes

II. Improper Type

(a) $m_i \neq m_j$, $m_i \neq m_j \pm 2$ $(i, j=1, 2, 3, i \neq j)$. We have

$$b_i^+ dx^i=-\frac{r-x_3}{r} Y d\psi, \quad b_i^- dx^i=\frac{r+x_3}{r} Y d\psi$$

$$b_4^-=b_4^+=i\begin{bmatrix} \lambda(r, t) & 0 & 0 \\ 0 & \mu(r, t) & 0 \\ 0 & 0 & -\lambda(r, t)-\mu(r, t) \end{bmatrix}.$$

(b) $m_1=m_2 \neq m_3$, $m_3 \neq m_1 \pm 2$.

b_i^+ and b_i^- are same as II(a).

$$b_4^+ = b_4^- = i\begin{bmatrix} \lambda(r, t) & z(r, t) & 0 \\ \bar{z}(r, t) & \mu(r, t) & 0 \\ 0 & 0 & -\lambda(r, t) - \mu(r, t) \end{bmatrix}.$$

(c) $m_2 = m_1 - 2$, $m_3 \neq m_1, m_2, m_1 + 2, m_2 - 2$.

$b_4^{\pm}$ are same as II(a). $b_i^{\pm}$ are equal to those of II(a) with the additional terms which are gauge potentials of a syncro-spherically symmetric field in the gauge considered in Sect. 3.

Remark. Except the cases I(a) and I(c) the fields are reducible.

Appendix II

A Special Class of Spherically Symmetric SU_N Gauge Fields

Let

$$Y = i\begin{bmatrix} l & & & & \\ & l-1 & & & \\ & & \ddots & & \\ & & & -l+1 & \\ & & & & -l \end{bmatrix} \quad (N = 2l + 1)$$

be a diagonal matrix. Y is a generator Y_3 of the irreducible representation of SO_3.

Let

$$D_{mn}^l(\psi, \theta, 0) = T_{mn}^l(\phi, \theta),$$

where D_{mn}^l are the generalized spherical functions and $\phi = \psi^{-\frac{\pi}{2}}$. So (θ, ϕ) are the spherical coordinates of a unit sphere. Using the method of Sect. 3 we obtain the matrix expressions of the gauge potentials

$$b_1(x) = \left[\begin{array}{l} \sum_{s=-l+1}^{l} \{(T^l_{ms}\overline{T}^l_{ns-1} - T^l_{ms-1}\overline{T}^l_{ns})(-d_s \sin\phi + e_s \cos\theta\cos\phi) \\ \quad -i(T^l_{ms}\overline{T}^l_{ns-1} + T^l_{ms-1}\overline{T}^l_{ns})(d_s\cos\theta\cos\phi + e_s\sin\phi)\} \end{array}\right],$$

$$b_2(x) = \left[\begin{array}{l} \sum_{s=-l+1}^{l} \{(T^l_{ms}\overline{T}^l_{ns-1} - T^l_{ms-1}\overline{T}^l_{ns})(d_s \cos\phi + e_s \cos\theta\sin\phi) \\ \quad +i(T^l_{ms}\overline{T}^l_{ns-1} + T^l_{ms-1}\overline{T}^l_{ns})(-d_s\cos\theta\sin\phi + e_s\cos\phi)\} \end{array}\right], \tag{A}$$

$$b_3(x) = \left[\begin{array}{l} \sum_{s=-l+1}^{l} \{(T^l_{ms}\overline{T}^l_{ns-1} - T^l_{ms-1}\overline{T}^l_{ns})(-e_s \sin\theta) + i(T^l_{ms}\overline{T}^l_{ns-1} \\ \quad + T^l_{ms-1}\overline{T}^l_{ns}d_s\sin\theta)\} \end{array}\right],$$

$$b_4(x) = \left[i \sum_{s=-l+1}^{l} c_s T_{ms}\overline{T}_{ns}\right],$$

where m, n are the indices for the elements of $N\times N$ matrices, c_s, d_s, e_s are arbitrary functions of (r, t).

In particular, for $N=2$ (A) are equavalent to the general form of the synchro-spherically symmetric potentials

$$b_i^a = \varepsilon_{iak}x_kV(r,t) + \delta_{ia}S(r,t) + x_ix_aT(r,t),$$

$$b_4^a = x_aU(r,t)$$

with the gauge condition

$$S + r^2T = 0.$$

For $N=3$ the expression (A) are consistent with the expression obtained in [15].

Note. From a letter from Prof. R. Jackiw dated March 31, 1980, we became aware that he solved the same problem, using the method in [6]. He stated the results and the method in his lectures in February 1980 at Schladming, Austria.

The results are consistent with ours. Our proof is more complicated, but we do not assume a priori that the complementary function $U_A(x)$ is single-valued in the local sense and differentiable with respect to A. We are grateful to Prof. Jackiw for his letter and for telling us some negligence in the Appendix I of our preprint ITP - SB - 100 - 79.

Acknowledgements. We are very grateful to Prof. C.N. Yang for valuable discussions. We are also benifited from the discussions with Prof. C. T. Yang and Prof. A. S. Goldhaber. We express our thanks to the State University of New York at Stony Brook and the University of California at Berkeley for their hospitality.

References

[1] Wu, T. T., Yang, C. N.: In properties of matter under unusual conditions (ed. Mark H, Fernbach, S.), p.349. New York, London: Interscience 1969.

[2] Gu, C.H., Hu, H.S.: Acta Phys. Sin. **26**, 155 - 168 (1977).

[3] Wu, A.C.T., Wu, T.T.: J. Math. Phys. **15**, 53 - 54 (1974).

[4] Marciano, W.J., Pagels, S.: Phys. Rev. D **12**, 1093 - 1095 (1975).

[5] Chakrabarti, A.: Ann. Inst. Henri Poincaré A **23**, 235 - 249 (1975).

[6] Forgacs, P., Manton, M.S.: Commun. Math. Phys. **72**, 15 - 35 (1980).

[7] Gu, C.H.: Fudan J. **2**, 30 - 36 (1977).

[8] Wilkinson, D., Goldhaber, A.S.: Phys. Rev. D **16**, 1221 - 1231 (1977).

[9] Wu, T.T., Yang, C.N.: Phys. Rev. D **12**, 3845 - 3857 (1975).

[10] Yang, C.N.: Phys. Rev. Lett. **33**, 445 - 447 (1974).

[11] Gu, C.H.: Phys. Energ. Fort. Phys. Nucl. **2**, 97 - 108 (1978).

[12] Montgomery, D., Zippin, L.: Topological transformation groups. New York, London: Interscience 1955.

Communicated by E. Brezin

Received December 26, 1979

（本文曾发表于 *Communications in Mathematical Physics*, 1981, **79**, 75－90）

On the Static Solutions of Massive Yang-Mills Equations*

Hu Hesheng (H.S. Hu)

Institute of Mathematics, Fudan University

Dedicated to Professor Su Bu-chin on the Occasion of his 80th Birthday and his 50th Year of Educational Work

§ 1. Introduction

Usually, a pure Yang-Mills field over Minkowski spacetime $R^{1,n-1}$ is considered as a field of massless particles. Its action integral is[1]

$$L=\int -\frac{1}{4}(f_{\lambda\mu},\ f^{\lambda\mu})d^n x. \tag{1}$$

Here $f_{\lambda\mu}$ is the strength of gauge field with a compact gauge group G and (,) denotes the cartan's inner product of the Lie algebra g of G. However, many particles in nature are not massless and hence it is a problem of general interest to consider the massive Yang-Mills fields. It has been proved in [2] and [3], from different points of view, that the following gauge invariant functional

$$L_m=\int\left[-\frac{1}{4}(f_{\lambda\mu},\ f^{\lambda\mu})-\frac{m^2}{2}(b_\lambda-\omega_\lambda,\ b^\lambda-\omega^\lambda)\right]d^n x \tag{2}$$

* Manuscript received May 12, 1981.

may be considered as the action integral of the massive Yang-Mills field. Here b_λ is the gauge potential, ω_λ is defined by

$$\omega_\lambda = U^{-1}\partial_\lambda U \tag{3}$$

and U is a G - valued function which is a section of the product bundle $R^{n-1} \wedge G$.

One may think that the choice of U is a choice of gauge and that the gauge is a reference system of measuring the generalized phase of a gauge field. Let U be the variational variables as well as b_λ, the Euler equations of the action integral (1) and (2) are the massless Yang-Mills equations and massive Yang-Mills equations respectively.

We mentioned that the massive Yang-Mills field is also attractive for its relationship with harmonic mapping. The functional (2) is nothing else than the coupling of the pure Yang-Mills functional and the following action integral of the harmonic maps from $R^{1,n-1}$ to the gauge group G

$$S(U) = \int (\omega_\lambda, \omega^\lambda) d^n x. \tag{4}$$

There are guite a lot of papers devoted to the solutions to Yang-Mills equations. One problem of considerable interest is whether there exists any static solution to the Yang-Mills equations such that it has finite energy and no singularities. Recently the following facts concerning the nonexistence of the global solutions are discovered.

(a) if $n \neq 5$, the pure Yang-Mills equations on an n - dimensional spacetime $R^{1,n-1}$ do not admit any static solution which has (i) finite energy (ii) no singularities and (iii) the field strength approaching to zero sufficiently fast at infinity. (Deser, S.[5])

Thus, for $n=4$, i.e. on the real spacetime, there does not exist such solution. For $n=5$, solutions do exist[6, 7], since the instantons in 4 - dimensional Euclidean space may be regarded as static solutions in 5 - dimensional spacetime.

(b) In an n - dimensional spacetime with $n \neq 4$, the massive Yang-Mills equations with real mass do not admit any static solution which has (i) finite energy (ii) no singularities and (iii) the field strength and potential approaching to zero sufficiently fast at infinity (Hu, H.S.[8]).

Comparing these two results, we discovered that there is a "discontinuity" as $m \to 0$ in 5 - dimensional spacetime, i.e. for $n=5$ and $m \neq 0$, no such solution, but when $m = 0$ such solutions do exist. Deser, S. and Isham, C.J. in a recent paper[9] wrote that this is the first explicit example which make us recognize that there exists a classical "discontinuity". In their paper, the results are extended to the gauge field with "soft" mass, i. e. Yang-Mills-Higgs-Kibble field. For $n = 5$, the "discontinuity" holds in general.

In the present paper, we will show that in the results (a) and (b) not only condition (iii) can be removed, but also the finite energy condition (ii) can be weakened. In other words, when the total energy within the sphere of radius r approaches to infinity quite slowly as $r \to \infty$, the above nonexistence theorem holds true also. In the proof we use a certain technique used in[10] with some improvement. Since finite energy and infinite energy is essentially different in physics, this new discovery may be of interest in physics.

The method of proving the main theorem of the present paper is utilizable for more general case. For example, in the case of the Yang-Mills-Higgs-Kibble field, the results for "soft" mass is improved similarly.

§ 2. Massive Yang-Mills fields

By choosing the Lorentz gauge, the gauge in variant functional becomes

$$L=-\int\left\{\frac{1}{4}(f_{\lambda\mu},\ f^{\lambda\mu})+\frac{m^2}{2}(b_\lambda,\ b^\lambda)\right\}d^{n-1}x\quad(\lambda,\ \mu=0,\ 1,\ \cdots,\ n-1).\tag{5}$$

Here the field strength is

$$f_{\lambda\mu}=b_{\lambda,\ \mu}-b_{\mu,\ \lambda}-[b_\lambda,\ b_\mu]\quad\left(b_{\lambda,\ \mu}=\frac{\partial b_\lambda}{\partial x^\mu}\right)\tag{6}$$

and the metric of spacetime is

$$ds^3=\eta_{\lambda\mu}dx^\lambda dx^\mu=-dx^{0^2}+dx^{1^2}+\cdots+dx^{n-1^2}.\tag{7}$$

The massive Yang-Mills equation become

$$J_\alpha-m^2b_\alpha=0,\tag{8}$$

$$\eta^{\lambda\mu}b_{\lambda,\ \mu}=0.\tag{9}$$

Here

$$J_\alpha=\eta^{\lambda\mu}f_{\alpha\lambda|\mu}=\eta^{\lambda\mu}(f_{\alpha\lambda,\ \mu}+[b_\mu,\ f_{\alpha\lambda}]),\tag{10}$$

moreover, it is interesting to note that (9) is a consequence of (8), so we only need to consider equation (8).

We always assume the gauge group G is a compact group. Under the Lorentz gauge the energy momentum tensor

$$T_{\alpha\beta}=(f_{\alpha\nu},\ f_\beta^\nu)-\frac{1}{4}\eta_{\alpha\beta}(f_{\mu\nu},\ f^{\mu\nu})+m^2(b_\alpha,\ b^\alpha)-\frac{m^2}{2}\eta_{\alpha\beta}(b_\lambda,\ b^\lambda)\tag{11}$$

and we have the conservation law

$$\frac{\partial T_\alpha^\beta}{\partial x^\alpha}=0.\tag{12}$$

Especially, the energy density is

$$T_{00}=\frac{1}{2}\left[(f_{0i},\ f_{0i})+\frac{1}{2}(f_{ij},\ f_{ij})\right]+\frac{m^2}{2}(b_0,\ b_0)+\frac{m^2}{2}(b_i,\ b_i)$$

$$(i, j = 1, 2, \cdots, n-1) \tag{13}$$

and

$$T_{ii} = T^{\alpha}_{\alpha} - T^{0}_{0} = -\frac{1}{2}(n-3)(f_{0i}, f^{0i}) + \frac{1}{4}(5-n)(f_{ij}, f^{ij})$$
$$+\frac{m^2}{2}(3-n)(b_\lambda, b^\lambda) + m^2(b_0, b_0). \tag{14}$$

For a static gauge field, b_λ is independent of x^0.

The total energy of the field is

$$\int_{R^{n-1}} T_{00} d^{n-1} x, \tag{15}$$

where R^{n-1} is $x^0 = \text{const}$. In the previous works, one often assume that the total energy is finite, now the condition is weakened to

$$\int \frac{T_{00}}{\psi(r)} d^{n-1} x < \infty, \tag{16}$$

where $\psi(r)$ is a positive, unbounded, continuous function of r satisfying

$$\int_R^\infty \frac{dr}{r\psi(r)} = \infty \quad (R > 0). \tag{17}$$

If $\psi(r) = 1$, the energy is finite. But the energy may be infinite, for example, in the case $\psi(r) = O(\log r)$ (as $r \to \infty$). Hence when (16) holds true, the total enegy may be either finite or infinite. In this paper the energy is called "slowly divergent energy" if $\int T_{00} d^{n-1} x = \infty$ and (16) holds.

§ 3. Nonexistence of the static solution

In the following we give the precise statement and the proof of the main theorem concerning the massive Yang-Mills equations.

Theorem. *In an n-dimensional spacetime* $R^{1,n-1}$ *with* $n\neq 4$, *the compact group Yang-Mills field with real mass does not possess any non-trivial static solution which is free of singularities and has finite or "slowly divergent" energy.*

Proof From the expression (10) for J_0 and the static condition $b_{\alpha,0}=0$, we have

$$\begin{aligned}(b_0, J_0) &= (b_0, f_{0i|i}) = (b_0, f_{0i})_{,i} - (b_{0,i}, f_{0i}) + (b_0, [b_i, f_{0i}]) \\ &= (b_0, f_{0i})_{,i} - (b_{0,i}, f_{0i}) - ([b_i, b_0], f_{0i}) \\ &= (b_0, f_{0i})_{,i} + (f_{0i}, f_{0i}).\end{aligned} \tag{18}$$

Consider the integral

$$0=\int_0^\infty \omega(r)dr\int_{|x|\leqslant r}(J_0-m^2b_0, b_0)d^{n-1}x, \tag{19}$$

where $|x|=\{(x^1)^2+\cdots+(x^{n-1})^2\}^{\frac{1}{2}}$ and $\omega(r)$ will be defined later. Using (18), we have

$$0=\int_0^\infty \omega(r)dr\int_{|x|\leqslant r}Kd^{n-1}x+\int_0^\infty \omega(r)dr\int_{|x|=r}(b_0, f_{0i})\frac{x^2}{r}dS, \tag{20}$$

where

$$K=-(f_{0i}, f_{0i})-m^2(b_0, b_0)\leqslant 0, \tag{21}$$

The equality in (21) holds if and only if $f_{0i}=b_0=0$. If K does not equal zero identically, then there exists a constant $R>0$ and a positive constant ε such that

$$\int_{|x|\leqslant R_1}Kd^{n-1}x<-\varepsilon \quad (R_1\leqslant R). \tag{22}$$

Choose

$$\omega(r)=\begin{cases}0, & r<R,\\ \dfrac{1}{r\psi(r)}, & R\leqslant r\leqslant R_1,\\ 0, & r>R_1,\end{cases} \tag{23}$$

where $\psi(r)$ is positive unbounded, continuous functon of r satisfying

$$\int_R^\infty \frac{dr}{r\psi(r)}=\infty. \tag{24}$$

Then, from (21), we have

$$0<-\varepsilon\int_R^{R_1}\frac{dr}{r\psi(r)}+\int_R^{R_1}\frac{dr}{r\psi(r)}\int_{|x|=r}\{(b_0, b_0)+(f_{0i}, f_{0i})\}dS$$

$$<-\varepsilon\int_R^{R_1}\frac{dr}{r\psi(r)}+\frac{1}{R}\int_{|x|<R_1}\frac{(b_0, b_0)+(f_{0i}, f_{0i})}{\psi(r)}d^{n-1}x. \tag{25}$$

Choose R_1 sufficiently large, it is easily seen that the right side should be negative. This is a contradiction. Consequently, we should have $K=0$ identically, i.e.

$$b_0=0, \ f_{0i}=0. \tag{26}$$

Thus, we have

$$T_{ii}=\frac{1}{4}(5-n)(f_{ij}, f_{ij})+\frac{m^2}{2}(3-n)(b_i, b_i) \tag{27}$$

and (12) is reduced to

$$T_{ij,i}=0. \tag{28}$$

Consider the integral

$$0=\int_0^\infty \omega(r)dr\int x^j T_{ij,i}d^{n-1}x=\int_0^\infty \omega(r)dr\int_{|x|\leqslant r}\{(x^jT_{ij})_{,i}-T_{ii}\}d^{n-1}x$$

$$=\int_0^\infty \omega(r)dr\int_{|x|=r}(x^jT_{ij})\frac{x^i}{r}-\int_0^\infty \omega(r)dr\int_{|x|\leqslant r}T_{ii}d^{n-1}x. \tag{29}$$

It is easily seen that there exists a constant A such that

$$|T_{ij}|\leqslant AT_{00}. \tag{30}$$

Moreover, from (27), we have

(a) If $n\geqslant 5$, then $T_{ii}\leqslant 0$

and the equality holds only when $b_i=0$.

(b) If $n\leqslant 3$, then $T_{ii}\geqslant 0$

and the equality holds only when $b_i=0$.

In either case, if $T_{ii}\not\equiv 0$, we have $T_{ii}<0$ (or $T_{ii}>0$) in some region. Hence there exist two constant $R>0$ and $\varepsilon>0$ such that

$$\int_{|x|\leqslant R_1} T_{ii} d^{n-1}x < -\varepsilon \quad (\text{or} > \varepsilon) \quad (R_1 \geqslant R). \tag{31}$$

Choosing the same $\omega(r)$ as in (23), for the case (a), we have

$$0 < -\varepsilon \int_R^{R_1} \frac{dr}{r\psi(r)} + A\int_0^\infty \frac{T_{00}}{\psi(r)} dS dr. \tag{32}$$

By the assumption that the energy is finite or "slowly divergent", we can choose R_1 sufficiently large, and it is easily seen that the right side of equation (32) should be negative. This gives a contradiction again. Consequently, we should have

$$T_{ii}=0.$$

For the case (b), the situation is quite similar. Consequently

(i) when $n\neq 3, 4, 5$, we have $f_{ij}=0$, $b_i=0$;

(ii) when $n=5$, we have $b_i=0$, hence $f_{ij}=0$;

(iii) when $n=3$, from the field equation (8) and $f_{ij}=0$ we have $b_i=0$.

In other words, when $n\neq 4$, the solution should be a trivial one. Thus the Theorem is proved completely.

Remarks.

1. For the massless case $m=0$. Deser's Theorem is also improved similarly.

2. Consider the Yang-Mills-Higgs-Kibble field (the gauge field with "soft" mass)

$$I=\int\left(-\frac{1}{4}F_{\mu\nu}F^{\mu\nu}-\frac{m^2}{2}b_\mu b^\mu-\frac{1}{4}\nabla_\mu\phi\nabla^\mu\phi-V(\phi)\right), \tag{33}$$

where ϕ is a scalar invariant and $V(\phi)$ is the potential. By using the same method, the result of [9] can be improved and extended to the case of "slowly divergent" energy and the classical "discontinuity" holds also for $n=5$.

In the following section, we shall specialize that the condition for the energy in our theorem cannot be omitted, because for any dimensional spacetime in massive and massless Yang-Mills field we can find static regular solutions with energy diverges sufficiently fast. In the meanwhile we obtain all the static solutions of strictly spherically symmetric gauge field.

§ 4. Static solutions of strictly spherically symmetric gauge field

Suppose the gauge potential of stricty spherically symmetric static gauge fields are in the canonical form[11]

$$b_i(x)=\phi(r)x_i, \quad b_0=\sigma(r), \tag{34}$$

where $\phi(r)$, $\sigma(r)$ are g - valued functions, depending only on r. In order to solve the massive Yang-Mills equations, we substitute (34) in (9) and we have

$$r\phi'(r)+(n-1)\phi(r)=0. \tag{35}$$

Hence we obtain

$$\phi(r)=\frac{\text{const}}{r^{n-1}}. \tag{36}$$

The requirement of regularity at the origin implies that $\phi(r)=0$. From (8) we obtain

$$\Delta\sigma(r)-m^2\sigma(r)=0 \tag{37}$$

or

$$\frac{d^2\sigma}{dr^2}+\frac{(n-2)}{r}\frac{d\sigma}{dr}-m^2\sigma=0. \tag{38}$$

Let

$$mr=R, \quad \sigma=R^{-p}q(R) \quad \left(p=\frac{n-3}{2}\right). \tag{39}$$

We obtain

$$q''+\frac{q'}{R}-\left(1+\frac{p^2}{R^2}\right)q=0, \tag{40}$$

this is the modified. Bessel equation[12]. It is known that this equation admits the following solutions which are everywhere regular.

$$q=q_0 I_p(R), \tag{41}$$

where q_0 is an element of g, $I_p(R)$ is the Bessel function with purely imaginary argument. Hence the equation (8) posses the following everywhere regular solutions

$$b_i=0, \quad b_0=q_0(mr)^{-\frac{n-3}{2}}I_p(mr). \tag{42}$$

Since when $r \to \infty$

$$I_p(mr) \sim \frac{e^{n+r}}{(2\pi mr)^{\frac{1}{2}}}, \tag{43}$$

the energy of such solutions is infinite and is not "slowly divergent".

From the above discussion we conclude that for any n, the condition of energy in the theorem cannot be omitted. That is to say, the massive Yang-Mills equation admits infinite many static regular solution whose energy is infinite and divergent sufficienty fast.

At the conclusion of the present paper we give two open problems.

1. In the case $n=4$, does there exist a static regular solution of massive (or massless i. e. pure) Yang-Mills equation with finite energy or "slowly divergent" energy?

2. Are there static solutions of the massless Yang-Mills equations in R^{4+1} with $b_0 \neq 0$? This problem arises due to the fact that the instantons of R^4 are static solutions in R^{4+1} with $b_0 = 0$ and the solutions in R^{4+1} with $b_0 \neq 0$ may be consider as some solutions of a certain coupled Yang-Mills equations in 4 - dimensional Euclidean space.

References

[1] Yang, C.N., Integral formulism of Gauge fields, *Phys. Rev. Letters*, **33** (1974), 445 - 447.

[2] Shizuya, K., *Nucl. Phys.* **B94** (1975), 260.

[3] Gu Chaohao, On the mass of gauge particles, a report at the seminar of modern physics of Fudan University, 1978.

[4] Gu Chaohao, The loop phase-factor approach to gauge field, *Physica Energiae Fortis et Physica Nuclearis*, **2** (1978), 97 - 108.

[5] Deser, S., Absence of static solutions in source-free Yang-Mills theory, Ref. TH. 2214 CERN, 1976.

[6] Jackiw, R., Nohl, C., Rebbi, C., *Phys. Rev.*, **D15** (1977), 1642 - 1645.

[7] Gu Chaohao, Hu Hesheng & Shen Chunli, A geometrical interpretation of instanton solutions in Euclidean space, *Scientia Sinica*, **21** (1978), 767 - 772.

[8] Hu Hesheng, On equations of Yang-Mills gauge fields with mass, *Research Reports of Mathematics of Fudan Univ.* (1979, Spring), *Kexue Tongbao* **25** (1980), 191 - 195.

[9] Deser, S., Isham, C.J., Static solution of Yang-Mills-Higgs-Kibble system, *Kexue Tongbao*, **25** (1980), 773 - 776.

[10] Weder, R., *Phys. Lett.*, **85B** (1979), 249.

[11] Gu Chaohao & Hu Hesheng, On spherically symmetric SU_2 gauge fields and monopoles, *Acta Physica Sinica*, **26** (1977), 155 - 168.

[12] Whittaker, E.T. & Watson, G.N., *A course of modern analysis*, (1955), 360 - 361.

关于具有质量的 Yang-Mills 方程的静态解

胡和生
(复旦大学数学研究所)

摘　要

关于 Yang-Mills 方程的静态解,Deser, S[5].证明了:当 $n \neq 5$ 时,无质量的紧致群 Yang-Mills 方程不存在满足条件(i)无奇性(ii)能量有限(iii)当 $r \to \infty$ 时,场强 $f_{\lambda\mu} \to 0$ 足够快的静态解.又已知当 $n = 5$ 时,正则静态解确实存在[6].

对于具实质量的紧致群 Yang-Mills 方程,作者[8]证明了:当 $n \neq 4$ 时,不存在满足条件(i)无奇性(ii)能量有限(iii)当 $r \to \infty$ 时规范势 b_λ 与场强 $f_{\lambda\mu} \to 0$ 足够快的静态解.从而发现在 $n = 5$, 当质量 $m \to 0$ 时,对 Yang-Mills 方程的可解性问题而言,在性质上有一种"不连续性".物理学家认为这是存在着经典的不连续性的第一个明确的例子,并对包括 Higgs 场的情况作了推广的研究[9].

本文进一步证明了上述两个结果中不仅条件(iii)可以取消,而且条件(ii)也可减弱.即能量为无限,但当以 r 为半径的球体的总能量趋于无限相当慢时定理仍成立.这时经典的"不连续性"也仍成立.由于能量有限与能量无限在物理上有根本的不同,所揭示的现象是值得注意的.

文中又证明,如果取消总能量趋于无限相当慢这个条件,定理的结论就不成立.

这里的证明方法,可用于更一般的情况.例如包括 Higgs 场的情况,从而[9]中的结果也得到改进.

(本文曾发表于 *Chinese Annals of Mathematics*, 1982, 3(4), 519-526)

Sine-Laplace Equation, Sinh-Laplace Equation and Harmonic Maps

Hu Hesheng*

Institute of Mathematics, Fudan University, Shanghai, China

The relationship between harmonic maps from R^2 to S^2, H^2, $S^{1,1}(+1)$, $S^{1,1}(-1)$ and the $\mp$ sinh-Laplace, $\mp$ sine-Laplace equation is found respectively. Existence theorems of some boundary value problems for the above harmonic maps are obtained. In the cases of H^2, $S^{1,1}(+1)$, $S^{1,1}(-1)$ the results are global.

1. Introduction

The relationship between the harmonic map and the famous sine-Gordon equation was pointed out by K. Pohlmeyer in [1]. In fact, from the harmonic map of the Minkowski plane $R^{1,1}$ to a sphere $S^2 \subset R^3$ one can construct a Lax pair and its integrability condition is the sine-Gordon equation. Conversely, the solution of the sine-Gordon equation can be used to construct a harmonic map from $R^{1,1}$ to S^2 in general by solving a system of completely integrable linear partial differential equations. Moreover, Gu developed this idea, established the global existence theorem for the Cauchy problems of the harmonic map from $R^{1,1}$ to any n - dimensional complete Riemannian manifold

* Research supported partially by the Institute for Applied Mathematics, Sonderforschungsbereich 72 of the University of Bonn.

M_n and gave a geometrical interpretation of the single soliton solutions of the sine-Gordon equation in [2].

Recently [3] Gu found the relationship between the sinh-Gordon equation and the harmonic map of $R^{1,1}$ to the sphere $S^{1,1}$ with indefinite metric in a 3 dimensional Minkowski space $R^{2,1}$, furthermore he proved an existence theorem of global solutions of the harmonic map with suitable initial conditions. It should be mentioned that the sinh-Gordon equation has been geometrically interpretated through timelike surfaces of positive constant curvature in $R^{2,1}$ by S.S. Chern in [4].

Thus, the following problem arises naturally: Whether there exists similar relationship between the nonlinear elliptic equation

$$\frac{\partial^2 u}{\partial t^2}+\frac{\partial^2 u}{\partial x^2}=\mp \sinh u, \quad (1),(2)$$

$$\frac{\partial^2 u}{\partial t^2}+\frac{\partial^2 u}{\partial x^2}=\mp \sin u, \quad (3),(4)$$

and the harmonic maps. We call equations (1)~(4) ∓ sinh-Laplace and ∓ sine-Laplace equations respectively.

Let S^2, H^2, $S^{1,1}(+1)$ and $S^{1,1}(-1)$ be the sphere, hyperbolic plane, the "spheres" with indefinite metric of constant curvature $+1$ and -1 respectively. In this paper we shall show the intimate relationship between equation (1)~(4) and the harmonic maps from R^2 to S^2, H^2 and the indefinite $S^{1,1}(1)$, $S^{1,1}(-1)$ respectively. From the harmonic maps, we can obtain the Lax pairs with one of the equations (1)~(4) as their integrability conditions respectively, and from each solution of (1)~(4) by using the solution of the Lax pairs, we can construct the corresponding harmonic maps.

Further, we consider certain boundary value problems for harmonic maps from R^2

to the above mentioned 2 dimensional manifolds of constant curvature. Owing to the known properties of non-linear elliptic equations, we obtained existence theorems of the solutions. In the cases of H^2, $S^{1,1}(+1)$ and $S^{1,1}(-1)$, the results are global.

2. The case of S^2

Consider a unit sphere S^2 in 3 dimensional Euclidean space R^3 and choose the parameter (t, x) such that the metric of S^2 can be written in the form

$$ds^2 = \cosh^2 \frac{\alpha}{2} dt^2 + \sinh^2 \frac{\alpha}{2} dx^2 \quad (\alpha \neq 0) \tag{5}$$

locally. Considering the formula of curvature

$$K = \frac{R_{1212}}{g_{11} g_{22} - g_{12}^2}, \tag{6}$$

we obtain

$$1 = \frac{\dfrac{\partial^2 \alpha}{\partial t^2} + \dfrac{\partial^2 \alpha}{\partial x^2}}{-\sinh \alpha}, \tag{7}$$

this is the negative sinh-Laplace equation (1). Consequently, if the metric of $S^2 \subset R^3$ is locally written in the form (5), then α must be the solution of the negative sinh-Laplace equation (1). Evidently, the converse is also true, if $\alpha \neq 0$.

Let l be the radius vectors of the points on S^2, equation (5) can be written as

$$\begin{aligned} l_t &= \cosh \frac{\alpha}{2} m, \\ l_x &= \sinh \frac{\alpha}{2} n, \end{aligned} \tag{8}$$

where m, n are unit vectors and l, m, n are mutually orthogonal. For definiteness, we assume that det $(l, m, n)=1$.

Hence we can write

$$\begin{aligned} & l_t=\cosh \frac{\alpha}{2} m, && l_x=\sinh \frac{\alpha}{2} n, \\ & m_t=-\cosh \frac{\alpha}{2} 1-\sigma n, && m_x=-\lambda n, \\ & n_t=\sigma m, && n_x=-\sinh \frac{\alpha}{2} l+\lambda m. \end{aligned} \tag{9}$$

From the integrability condition $l_{xt}=l_{tx}$, we have

$$\lambda=-\frac{1}{2} \alpha_t, \quad \sigma=\frac{1}{2} \alpha_x. \tag{10}$$

Substituting (10) into (9) and calculating $m_{tx}=m_{xt}$, $n_{tx}=n_{xt}$, we can verify easily that these integrability conditions of (9), (10) are the negative sinh-Laplace equation

$$\alpha_{xx}+\alpha_{tt}=-\sinh \alpha. \tag{1}'$$

Consequently, equation (9) and (10) constitute a Lax pair of the group SO(3) with negative sinh-Laplace equation as its integrability condition. On the other hand, from (8) we have

$$l_t^2-l_x^2=1, \quad l_t \cdot l_x=0. \tag{11}$$

By differentiation, it is easily seen that $l(t, x)$ satisfies

$$l_{tt}+l_{xx}+(l_t^2+l_x^2) l=0. \tag{12}$$

This is exactly the equation of harmonic maps from R^2 to S^2. Consequently, $l(t, x)$ is a harmonic map from R^2 to S^2. In this paper, a normalized harmonic map is defined as a harmonic map $l(t, x)$ which satisfies the condition (11).

Moreover, let $\alpha(t, x)$ be a given solution to the negative sinh-Laplace equation on a simply connected region $\Omega \subset R^2$. By solving the Lax pair with the initial condition (t,

$x) = (t_0, x_0)$, $(l, m, n) = (l_0, m_0, n_0)$, where $(t_0, x_0) \in \Omega$, $(l_0, m_0, n_0) \in$ SO(3), and extracting $l(t, x)$ from the solution, we obtain a normalized harmonic map $l(t, x)$ from Ω to the sphere. Thus, we can construct a normalized harmonic map from a solution of the negative sinh-Laplace equation.

Now there arises another question. Whether all of the harmonic maps $l(t, x)$ from R^2 to S^2 can be obtained through solving $\alpha(t, x)$ from the negative sinh-Laplace equation. In the following we prove that in general a harmonic map from a region $\Omega \subset R^2$ to S^2 can always be constructed through a normalized harmonic map together with a conformal map.

Let $l(t, x)$ be a harmonic map from $\Omega \subset R^2$ to S^2. Then $l(t, x)$ satisfies equation (12) and the equation $l^2 = 1$. Since (12) is a system of elliptic equations with analytic coefficients, the regular solution $l(t, x)$ must be analytic. Hence the vector-valued function $l(t, x)$ can be complexified to $l(\tau, \zeta)$ where τ, ζ are complex variables and $l(\tau, \zeta)$ is a complex analytic function defined in a region $\Omega_C (\supset \Omega)$ in $\mathbb{C}^2$, $l(\tau, \zeta)$ still satisfies the equation

$$l_{\tau\tau} + l_{\zeta\zeta} + (l_\tau^2 + l_\zeta^2) l = 0. \tag{13}$$

Introduce

$$\xi = \frac{\tau + i\zeta}{2}, \quad \eta = \frac{\tau - i\zeta}{2}, \tag{14}$$

as two new independent variables, then τ, ζ take real values if and only if ξ and η are conjugate. Using (14), we can rewrite (13) into

$$l_{\xi\eta} + (l_\xi \cdot l_\eta) l = 0. \tag{15}$$

Since $l^2 = 1$, we obtain

$$(l_\xi^2)_\eta = 0, \quad (l_\eta^2)_\xi = 0, \tag{16}$$

and hence

$$l_\xi^2=k(\xi),\quad l_\eta^2=h(\eta), \tag{17}$$

where $k(\xi)$ and $h(\eta)$ are all analytic functions. We assume

$$\begin{aligned} l_\xi^2&=(l_t-il_x)^2=l_t^2-l_x^2-2il_t\cdot l_x\neq 0,\\ l_\eta^2&=(l_t+il_x)^2=l_t^2-l_x^2+2il_t\cdot l_x\neq 0, \end{aligned} \tag{18}$$

in the region Ω. Then we can solve the differential equations

$$\frac{\partial\xi'}{\partial\xi}=2k^{\frac{1}{2}}(\xi),\quad \frac{\partial\eta'}{\partial\eta}=2h^{\frac{1}{2}}(\eta), \tag{19}$$

and obtain the solutions

$$\xi'=\beta(\xi),\quad \eta'=\bar{\beta}(\eta), \tag{20}$$

in an open set $\Omega_1(\Omega\subset\Omega_1\subset\Omega_C)$ such that ξ' and η' are conjugate when ξ and η are conjugate. Denote

$$\xi'=t'+ix',\quad \eta'=t'-ix',$$

then

$$t'+ix'=\beta(t+ix) \tag{21}$$

is a conformal mapping T from the region Ω to some region Ω'. Let $l(t',x')$ be the composition of T^{-1} and $l(t,x)$ (i.e. $1\circ T^{-1}$), then we have $l_{\xi'}^2=1/4$ or

$$l_{t'}^2-l_{x'}^2=1,\quad l_{t'}\cdot l_{x'}=0, \tag{22}$$

on Ω'. Hence $l(t',x')$ is a normalized harmonic map and can be determined from a solution of the negative sinh-Laplace equation. Consequently, when the condition (18) is satisfied, a harmonic map from Ω to S^2 can be constructed by solving the negative sinh-Laplace equation, and solving the Lax pair (9), (10), and then combining the solution with a conformal map.

Theorem 1. Let $\alpha(t, x)$ be a given solution to the negative sinh-Laplace equation (1) on a simply connected region $\Omega \subset R^2$, we can determine a normalized harmonic map $l(t, x)$ from Ω to S^2 through solving the Lax pair (9), (10). Conversely, any harmonic map from $\Omega \subset R^2$ to S^2 satisfying (18) equals to the product of a conformal map of region Ω and a normalized harmonic map constructed by solving the negative sinh-Laplace equation and Lax pair (9), (10).

Remark. As is known, in 2 dimensional case, the product of a conformal map and a harmonic map is also a harmonic map.

Remark. If $l_\xi^2 = l_\eta^2 = 0$ identically, l is a conformal map.

3. Other cases

Let l, m be vectors in space $R^{2,1}$, their scalar product is defined by

$$l \cdot m = l_1 m_1 + l_2 m_2 - l_3 m_3. \tag{23}$$

$S^{1,1}(+1)$ is the locus of points in $R^{2,1}$ satisfying the equation

$$l^2 = 1. \tag{24}$$

If the metric of $S^{1,1}(+1)$ is written in the following form locally

$$ds^2 = \cos^2 \frac{\alpha}{2} dt^2 - \sin^2 \frac{\alpha}{2} dx^2 \quad (\alpha \neq 0), \tag{25}$$

calculating the condition $K = 1$, we have

$$\frac{\partial^2 \alpha}{\partial t^2} + \frac{\partial^2 \alpha}{\partial x^2} = -\sin \alpha, \tag{3$'$}$$

this is just the negative sine-Laplace equation (3). Consequently, in order that the metric of $S^{1,1}(+1) \subset R^{2,1}$ can be written in the form (25), the necessary and sufficient condition is that α must be the solution of the negative sine-Laplace equation (3) and

$\alpha \neq 0$.

Hence $l = l(t, x)$ satisfies

$$l_t = \cos\frac{\alpha}{2} m, \quad l_x = \sin\frac{\alpha}{2} n, \tag{26}$$

Here $m^2 = 1$, $n^2 = -1$, $m \cdot n = 0$, and $\det(l, m, n) > 0$. The Lax pair of the group SO(2, 1) takes the following form

$$\begin{aligned} l_t &= \cos\frac{\alpha}{2} m, & l_x &= \sin\frac{\alpha}{2} n, \\ m_t &= -\cos\frac{\alpha}{2} l - \frac{\alpha_x}{2} n, & m_x &= \frac{\alpha_t}{2} n, \\ n_t &= -\frac{\alpha_x}{2} m, & n_x &= \sin\frac{\alpha}{2} l + \frac{\alpha_t}{2} m. \end{aligned} \tag{27}$$

Calculating $m_{xt} = m_{tx}$, $n_{xt} = n_{tx}$ we obtain the negative sine-Laplace equation again. Similar to the section 2 we have

THEOREM 2. For any solution of the negative sine-Laplace equation (3′) in a simply connected region Ω, we can always construct a normalized harmonic map $l(t, x)$ from $\Omega \subset R^2$ to $S^{1,1}(+1)$ through solving the Lax pair (26), (27). Conversely, any harmonic map $l(t, x)$ from $\Omega \subset R^2$ to $S^{1,1}(+1)$ satisfying (18) equals to the product of a conformal map of region Ω and a normalized harmonic map constructed by solving the negative sine-Laplace equation and Lax pair (27).

We have the similar results on the sinh-Laplace equation and sine-Laplace equation, corresponding to the harmonic map from $\Omega \subset R^2$ to $H^2 \subset R^{2,1}$ and to $S^{1,1}(-1) \subset R^{1,2}$ respectively. For simplicity, we write down here the metrics and the Lax pairs only.

If we write the metric of H^2 in the form

$$ds^2 = \cosh^2\frac{\alpha}{2} dt^2 + \sinh^2\frac{\alpha}{2} dx^2, \tag{28}$$

we will obtain the sinh-Laplace equation

$$\alpha_{tt}+a_{xx}=\sinh\alpha.$$

Here H^2 is considered as the upper component of

$$l^2=-1 \tag{29}$$

in $R^{2,1}$. The corresponding Lax pair is

$$\begin{aligned} &l_t=\cosh\frac{\alpha}{2}m, && l_x=\sinh\frac{\alpha}{2}n,\\ &m_t=\cosh\frac{\alpha}{2}l-\frac{1}{2}\alpha_x n, && m_x=\frac{1}{2}\alpha_t n,\\ &n_t=\frac{1}{2}\alpha_x m, && n_x=\sinh\frac{\alpha}{2}l-\frac{1}{2}\alpha_t m, \end{aligned} \tag{30}$$

with $l^2=-1$, $m^2=n^2=1$, $l\cdot m=1\cdot n=m\cdot n=0$ and $\det(l, m, n)=1$.

If we write the metric of $S^{1,1}(-1)$ in the form

$$ds^2=\cos^2\frac{\alpha}{2}dt^2-\sin^2\frac{\alpha}{2}dx^2. \tag{31}$$

Here $S^{1,1}(-1)$ is the sphere

$$l^2=-1, \tag{32}$$

in the space $R^{1,2}$ with $l^2=-l_1^2-l_2^2+l_3^2$, we will obtain the sine-Laplace equation

$$\alpha_{tt}+\alpha_{xx}=\sin\alpha.$$

The related Lax pair is

$$\begin{aligned} &l_t=\cos\frac{\alpha}{2}m, && l_x=\sin\frac{\alpha}{2}n,\\ &m_t=\cos\frac{\alpha}{2}l-\frac{1}{2}\alpha_x n, && m_x=\frac{1}{2}\alpha_t n,\\ &n_t=-\frac{1}{2}\alpha_x m, && n_x=-\sin\frac{\alpha}{2}l+\frac{1}{2}\alpha_t m, \end{aligned} \tag{33}$$

with $l^2=-1$, $m^2=1$, $n^2=-1$, $l\cdot m=l\cdot n=m\cdot n=0$, $\det(l, m, n)=1$.

4. A boundary value problem for normalized harmonic maps

From now on we will consider a boundary value problem for normalized harmonic maps. By using the properties of the elliptic equations (1)～(4), some existence theorems can be obtained.

We can pose the following boundary value problem: Let Ω be a simply connected region with C^2 boundary and $v=(v_1, v_2)$ be a field of unit vectors on the boundary $\partial\Omega$. We wish to find the normalized harmonic map from Ω to S^2 (resp. $S^{1,1}(+1)$, H^2, $S^{1,1}(-1)$) such that l_v^2 (the square length of the v-direction derivative of l) along $\partial\Omega$ equals to a given function σ.

For the S^2 case, from equation (8) we have

$$l_v=l_t v_1+l_x v_2=v_1\cosh\frac{\alpha}{2}m+v_2\sinh\frac{\alpha}{2}n. \tag{34}$$

The boundary condition $l_v^2=\sigma$ means that

$$v_1^2\cosh^2\frac{\alpha}{2}+v_2^2\sinh^2\frac{\alpha}{2}=\sigma. \tag{35}$$

It follows that

$$\cosh^2\frac{\alpha}{2}=\sigma+v_2^2. \tag{36}$$

Suppose that

$$\sigma+v_2^2>1, \tag{37}$$

we can get two boundary values α defined by

$$\alpha=\pm 2\cosh^{-1}\sqrt{\sigma+v_2^2}, \tag{38}$$

for the negative sinh-Laplace equation. Using the known result of the nonlinear elliptic equations [5], it follows that the Dirichlet problem for the negative sinh-Laplace equation admits solutions if Ω is small enough (or the variation of α is small enough). With no loss of generality, we can take + sign in (38), since the negative sign can be compensated by a symmetry. Hence the harmonic map is uniquely determined except an orthogonal transformation in R^3.

For the case $S^{1,1}(+1)$ and $S^{1,1}(-1)$, the condition (37) should be replaced by

$$0<\sigma+v_2^2<1, \tag{40}$$

and for the case of H^2, we still have the condition (37).

It is interesting to point out that the Dirichlet problem for positive and negative sine-Laplace equation and positive sinh-Laplace equation are solvable globally [5] since the right hand sides satisfy

$$|\sin\alpha|\leqslant 1 \quad \text{or} \quad \frac{d}{d\alpha}\sinh\alpha=\cosh\alpha>0.$$

Hence for the cases H^2, $S^{1,1}(+1)$, $S^{1,1}(-1)$ the region Ω can be arbitrary large.

Thus we have

Theorem 3. If (37) is satisfied for the cases of S^2 and H^2, or (40) is satisfied for the cases of $S^{1,1}(+1)$ and $S^{1,1}(-1)$. There exists a normalized harmonic map from a simply-connected region Ω to S^2 (or H^2, $S^{1,1}(+1)$, $S^{1,1}(-1)$) such that at the boundary $l_v^2=\sigma$ hold. In the S^2 case, Ω should be small enough. In the cases of H^2, $S^{1,1}(+1)$ and $S^{1,1}(-1)$, the results are global. The maps are determined except for a rotation in R^3, $R^{2,1}$ or $R^{1,2}$ respectively.

The results of this section can be listed as follows:

Target manifold	boundary condition	equation for α	normalized harmonic map
$S^2 \subset R^3$	$l_v^2 = \sigma$ with$\sigma + v_2^2 > 1$	$\Delta\alpha = -\text{sh}\,\alpha$	exist for small Ω
$H^2 \subset R^{2,1}$	$l_v^2 = \sigma$ with $\sigma + v_2^2 > 1$	$\Delta\alpha = \text{sh}\,\alpha$	exist globally
$S^{1,1}(+1) \subset R^{2,1}$	$l_v^2 = \sigma$ with $0 < \sigma + v_2^2 < 1$	$\Delta\alpha = -\sin\alpha$	exist globally
$S^{1,1}(-1) \subset R^{1,2}$	$l_v^2 = \sigma$ with $0 < \sigma + v_2^2 < 1$	$\Delta\alpha = \sin\alpha$	exist globally

Acknowledgements. The author is grateful to Prof. S. Hildebrandt for his support and concerns.

Bibliography

[1] Pohlmeyer, K.: Integrable Hamiltonian System and Interaction Through Quadratic Constraints. Comun.Math. Phys. **46**, 207 - 221 (1976).

[2] Gu, C.H.: On the Cauchy Problem for Harmonic Map Defined on Two-Dimensional Minkowski Space. Comun. on Pure and Appl.Math. **33**, 727 - 737 (1980).

[3] Gu, C.H.: On the Harmonic Maps from $R^{1,1}$ to $S^{1,1}$. to appear.

[4] Chern, S.S.: Geometrical Interpretation of Sinh-Gordon Equations. Annales Polinici Math. **39**, 63 - 69 (1981).

[5] Courant, R. and Hilbert, D.: Methods of Mathematical Physics. vol. Ⅱ, 369 - 374 Interscience Publishers, (1966).

(Received July 20, 1982)

(本文曾发表于 *Manuscripta Mathematica*, 1982, **40**, 205 - 216)

On the Massive Yang-Mills Equations

Hu Hesheng (胡和生)
Institute of Mathematics, Fudan University, Shanghai, China

The massive Yang-Mills theory may be derived from the action integral

$$L_m = \int_R 1, \ n-1 \left[-\frac{1}{4}(f_{\lambda\mu}, f^{\lambda\mu}) - \frac{m^2}{2}(b_\lambda - w_\lambda, b^\lambda - w^\lambda) \right] d^n x \tag{1}$$

$$(\lambda, \mu = 0, 1, \cdots, n-1).$$

Here $R^{1,n-1}$ is the Minkowski space-time of demension n, b_λ and $f_{\lambda\mu}$ are the gauge potential and the field strength of the gauge field respectively. Moreover, $w_\lambda = U^{-1}\partial_\lambda U$ is defined via a G - valued function U, where G, being a compact Lie group, is the gauge group of the field. As usual, (,) denotes the Cartan's inner product of the Lie algebra of G. The first variation of L_m with respect to b_λ and U gives the massive Yang-Mills equations

$$J_\lambda + m^2(b_\lambda - w_\lambda) = 0 \tag{2}$$

$$\eta^{\mu\nu}(b_{\mu,\nu} - w_{\mu,\nu}) = 0 \tag{3}$$

with $\eta_{00} = -1$, $\eta_{11} = \cdots = \eta_{n-1\,n-1} = 1$, $\eta_{\lambda\mu} = 0 (\lambda \neq \mu)$. Here the comma stands for the partial derivatives and

$$J_\lambda = \eta^{\mu\nu} f_{\lambda\mu|\nu} = \eta^{\mu\nu}(f_{\lambda\mu,\nu} + [b_\nu, f_{\lambda\mu}]) \tag{4}$$

perform the gauge transformation via the G - valued function U, the massive Yang-

Mills equations are reduced to

$$J_\lambda + m^2 b_\lambda = 0 \tag{5}$$

$$\eta^{\mu\nu} b_{\mu,\nu} = 0. \tag{6}$$

It is noticed that (6) is a consequence of (5).

The energy momentum tensor of the field has the expression

$$T_{\alpha\beta} = (f_{\alpha\nu}, f_\beta^\nu) - \frac{1}{4}\eta_{\alpha\beta}(f_{\mu\nu}, f^{\mu\nu}) + m^2(b_\alpha, b^\alpha) - \frac{m^2}{2}\eta_{\alpha\beta}(b_\lambda, b^\lambda), \tag{7}$$

which obeys the conservation law

$$T^\beta_{\alpha,\beta} = 0. \tag{8}$$

Especially, the energy density is

$$T_{00} = \frac{1}{2}[(f_{0i}, f_{0i}) + \frac{1}{2}(f_{ij}, f_{ij})] + \frac{m^2}{2}(b_0, b_0) + \frac{m^2}{2}(b_i, b_i) \tag{9}$$

$$(i, j = 1, 2, \cdots, n-1)$$

and

$$\begin{aligned} T_{ii} &= T^\alpha_\alpha - T^0_0 \\ &= -\frac{1}{2}(n-3)(f_{0i}, f^{0i}) + \frac{1}{4}(5-n)(f_{ij}, f^{ij}) + \\ &\quad \frac{m^2}{2}(3-n)(b_\lambda, b^\lambda) + m^2(b_0, b_0). \end{aligned} \tag{10}$$

For a static gauge field, b_λ is independent of x^0.

The total energy of the field is

$$\int_{R^{n-1}} T_{00} d^{n-1} x, \tag{11}$$

where R^{n-1} is $x^0 = \text{const}$.

For the massless case, i.e. $m = 0$, it is found that if $n \neq 5$, the pure Yang-Mills eguations on an n - dimensional spacetime $R^{1,n-1}$ do not admit any static solution which has (i) finite energy (ii) no singularities and (iii) the field strngth approaching to zero sufficiently fast at infinity. (Deser, S. [1]) Thus, for $N=4$, i.e. on the real spacetime, there does not exist such solution. For $n=5$, solutions do exist[2, 3], since the instantons in 4 - dimensional Euclidean space may be regarded as static solutions in 5 - dimensional spacetime. For the massive case i. e. $m \neq 0$, we have proved that in an n - dimensional spacetime with $n \neq 4$, the massive Yang-Mills equations with real mass do not admit any static solution which has (i) finite energy (ii) no singularities and (iii) the field strength and potential approaching to zero sufficiently fast at infinity (Hu, H.S.[4]).

Comparing these two results, we discovered that there is a "discontinuity" as $m \rightarrow 0$ in 5 - dimensional spacetime, i.e. for $n=5$ and $m \neq 0$, no such solution, but when $m=0$ such solutions do exist. Deser, S. and Isham, C.J. in a recent paper[5] wrote that this is the first explicit example which make us recognize that there exists a classical "discontinuity". In their paper, the results are extended to the gauge field with "soft" mass, i.e. Yang-Mills-Higgs-Kibble field. For $n=5$, the "discontinuity" holds in general.

Now we will show that in the results (a) and (b) not only condition (iii) can be removed, but also the finite energy condition (ii) can be weakened.

Let $\psi(r)$ be a positive, unbounded and continuous function of r on $[0, \infty)$ satisfying

$$\int_R^{\infty} \frac{dr}{r\psi(r)} = \infty \quad (R>0). \tag{12}$$

We introduce the following condition

$$\int \frac{T_{00}}{\psi(r)} d^{n-1}x < \infty. \tag{13}$$

If $\psi(r)=1$, the energy is finite. But the energy may be infinite, for example, in the case $\psi(r)=0(\log r)$ (as $r\to\infty$). Hence when (13) holds true, the total energy may be either finite or infinite. In this paper the energy is called "slowly divergent energy" if $\int T_{00}d^{n-1}x=\infty$ and (13) holds.

We have the following

Theorem. In an n - dimensional spacetime $R^{1,n-1}$ with $n\neq 4$, the compact group Yang-Mills field with real mass does not possess any non-trivial static solution which is free of singularities and has finite or "slowly divergent" energy.

Proof From the expression (4) for J_0 and the static condition $b_{\alpha,0}=0$, we have

$$(b_0, J_0)=(b_0, f_{0i})_{,i}+(f_{0i}, f_{0i}). \tag{14}$$

Consider the integral

$$0=\int_0^\infty \omega(r)dr\int_{|x|\leqslant r}(J_0-m^2b_0, b_0)d^{n-1}x, \tag{15}$$

where $|x|=\{(x^1)^2+\cdots+(x^{n-1})^2\}^{1/2}$ and $\omega(r)$ will be defined later. Using (14), we have

$$0=\int_0^\infty \omega(r)dr\int_{|x|\leqslant r}Kd^{n-1}x+\int_0^\infty \omega(r)dr\int_{|x|=r}(b_0, f_{0i})\frac{x^i}{r}dS, \tag{16}$$

where

$$K=-(f_{0i}, f_{0i})-m^2(b_0, b_0)\leqslant 0, \tag{17}$$

The equality in (17) holds if and only if $f_{0i}=b_0=0$. If K does not equal zero identically, then there exists a constant $R>0$ and a positive constant ε such that

$$\int_{|x|\leqslant R_1}Kd^{n-1}x<-\varepsilon \quad (R_1\leqslant R). \tag{18}$$

Choose

$$\omega(r)=\begin{cases}0, & r<R, \\ \dfrac{1}{r\psi(r)}, & R\leqslant r\leqslant R_1, \\ 0, & r>R_1,\end{cases} \tag{19}$$

where $\psi(r)$ is positive unbounded, continuous functon of r satisfying

$$\int_R^{\infty}\frac{dr}{r\psi(r)}=\infty.$$

Then, from (17), we have

$$\begin{aligned}0&<-\varepsilon\int_R^{R_1}\frac{dr}{r\psi(r)}+\int_R^{R_1}\frac{dr}{r\psi(r)}\int_{|x|=r}\{(b_0, b_0)+(f_{0i}, f_{0i})\}dS \\ &<-\varepsilon\int_R^{R_1}\frac{dr}{r\psi(r)}+\frac{1}{R}\int_{|x|<R_1}\frac{(b_0, b_0)+(f_{0i}, f_{0i})}{\psi(r)}d^{n-1}x.\end{aligned} \tag{20}$$

Choose R_1 sufficiently large, it is easily seen that the right side should be negative. This is a contradiction. Consequenty, we should have $K=0$ identically, i.e.

$$b_0=0,\ f_{0i}=0. \tag{21}$$

Thus, we have

$$T_{ii}=\frac{1}{4}(5-n)(f_{ij}, f_{ij})+\frac{m^2}{2}(3-n)(b_i, b_i) \tag{22}$$

and (8) is reduced to

$$T_{ij,i}=0. \tag{23}$$

Consider the integal

$$\begin{aligned}0&=\int_0^{\infty}\omega(r)dr\int x^jT_{ij,i}d^{n-1}x \\ &=\int_0^{\infty}\omega(r)dr\int_{|x|\leqslant r}\{(x^jT_{ij})_{,i}-T_{ii}\}d^{n-1}x\end{aligned}$$

$$=\int_0^\infty \omega(r)dr\int_{|x|\leqslant r}(x^jT_{ij})\frac{x^i}{r}-\int_0^\infty \omega(r)dr\int_{|x|\leqslant r}T_{ii}d^{n-1}x. \tag{24}$$

It is easily seen that there exists a constant A such that

$$|T_{ij}|\leqslant AT_{00}. \tag{25}$$

Moreover, from (22), we have

(a) if $n\geqslant 5$, then $T_{ii}\leqslant 0$

and the equality holds only when $b_i=0$.

(b) If $n\leqslant 3$, then $T_{ii}\geqslant 0$

and the equality holds only when $b_i=0$.

In either case, if $T_{ii}\not\equiv 0$, we have $T_{ii}<0$ or $T_{ii}>0$ in some region. Hence there exist two constant $R>0$ and $\varepsilon>0$ such that

$$\int_{|x|<R_1}T_{ii}d^{n-1}x<-\varepsilon \quad (\text{or}>\varepsilon) \quad (R_1\geqslant R). \tag{26}$$

Choosing the same $\omega(r)$ as in (19), for the case (a), we have

$$0<-\varepsilon\int_R^{R_1}\frac{dr}{r\psi(r)}+A\int_0^\infty\int\frac{T_{00}}{\psi(r)}dSdr. \tag{27}$$

By the assumption that the energy is finite or "slowly divergent", we can choose R_1 sufficiently large, and it is easily seen that the right side of equation (27) should be negative. This gives a contradiction again. Consequently, we should have

$$T_{ii}=0.$$

For the case (b), the situation is quite similar. Consequently

(i) when $n\neq 3, 4, 5$, we have $f_{ij}=0$, $b_i=0$,

(ii) when $n=5$, we have $b_i=0$, hence $f_{ij}=0$,

(iii) when $n=3$, from the field equation (8) and $f_{ij}=0$ we have $b_i=0$.

In other words, when $n\neq 4$, the solution should be a trivial one. Thus the theorem is proved completely.

Remark 1. For the massless case $m=0$. Deser's theorem is also improved similarly.

Remark 2. Consider the Yang-Mills-Higgs-Kibble field (the gauge field with "soft" mass)

$$I=\int\left(-\frac{1}{4}F_{\mu\nu}F^{\mu\nu}-\frac{m^2}{2}b_\mu b^\mu-\frac{1}{4}\nabla_\mu\phi\nabla^\mu\phi-v(\phi)\right), \tag{28}$$

where ϕ is a scalar invariant and $v(\phi)$ is the potential. By using the same method, the result of [5] can be improved and extended to the case of "slowly divergent" energy.

Remark 3. The condition for the energy in our theorem cannot be omitted, because for any dimensional spacetime in massive and massless Yang-Mills field we can find static regular solutions with energy diverges sufficiently fast.

We would like to point out the following open problems.

1. Are there any static regular solution to the massive Yang-Mills equations with finite energy or "slowly divergent" energy in the case $n=4$?

2. Are there static solutions of the massless Yang-Mills equations in R^{4+1} with $b_0\neq 0$? This problem arises due to the fact that the instantons of R^4 are static solutions in R^{4+1} with $b_0=0$ and the solutions in R^{4+1} with $b_0\neq 0$ may be consider as some solutions of a certain coupled Yang-Mills equations in 4-dimensional Euclidean space.

References

[1] Deser, S., Absence of static solutions in source-free Yang-Mills theory, Ref. TH. 2214 CERN, 1976.

[2] Jackiw, R., Nohl, C., Rebbi, C., Phys. Rev., D15 (1977), 1642-1645.

[3] Gu Chaohao, Hu Hesheng & Shen Chunli, A geometrical interpretation of instanton solutions in Euclidean space, Scientic Sinica, 21 (1978), 767-772.

[4] Hu Hesheng, On equations of Yang-Mills gauge field with mass, Research Reports of Mathematics of Fudan Univ. (1979, Spring), Kexue Tongbao, 25 (1980), 191 - 195.

[5] Deser, S., Isham, C.J., Static solution of Yang-Mills-Higgs-Kibble system, Kexue Tongbao, 25 (1980), 773 - 776.

(本文曾发表于《1981 年“双微”上海讨论会文集》,科学出版社出版,1984 年,115 - 122)

Some Nonexistence Theorems for Massive Yang-Mills Fields and Harmonic Maps

Hu Hesheng (H.S. Hu)

Institute of Mathematics, Fudan University, Shanghai, China

Introduction

The Yang-Mills theory and the theory of harmonic maps between Riemannian manifolds are two important subjects of differential geometry. They have some common features, such as they are both variational theories, being very important in theoretical physics and they both rely on the theory of non-linear partial differential equations, having almost same principal parts. Moreover, the harmonic maps from 2-dimensional space are quite similar to the Yang-Mills (Y - M) fields over 4-dimensional space, since in such cases they are both conformally invariant theories.

There are quite a lot of papers considering the solution to the Y - M equations and equations of harmonic maps. An important problem is to investigate the global existence or nonexistence of certain gauge fields or certain harmonic maps.

In the present paper we will give some nonexistence theorems. We give the concepts of the Y - M fields and harmonic maps in § 1 briefly. Besides, it is emphasized that a massive Y - M field is the coupling of a pure Y - M fields and a harmonic map from the space-time to the gauge group. In § 2 we condider a nonexistence theorem for massive Y - M fields on the Minkowski space-time $R^{n-1,1}$ and its generalization to some curved manifolds. In §3 we consider a nonexistence theorem for harmonic maps from Euclidean space to any Riemannian manifold and its generalizations.

§1. Pure Yang-Mills Fields and Harmonic Maps

We begin with a brief sketch of the two concepts.

(a) Y-M fields (gauge field)

Let G be a Lie group, usually being compact and linear, and g its Lie algebra. A gauge field over a Riemannian (or Lorentzian) manifold M, mathematically is defined by a connection on the principal bundle $P(M, G)$.

We shall consider gauge fields on the Minkowski space-time $R^{1,n-1}$ mainly. The metric of $R^{1,n-1}$ is

$$ds^2 = y_{\lambda\mu} dx^\lambda dx^\mu = -(dx^0)^2 + (dx^1)^2 + \cdots + (dx^n)^2$$
$$(\lambda, \mu = 0, 1, \cdots, n-1). \tag{1}$$

A gauge field is defined by its gauge potential (connection)

$$b = b_\lambda(x) dx^\lambda, \tag{2}$$

which is a l-form valued in g. The field strength (curvature) is

$$F = \frac{1}{2} f_{\lambda\mu} dx^\lambda \wedge dx^\mu, \tag{3}$$

with

$$f_{\lambda\mu} = \frac{\partial b_\lambda}{\partial x^\mu} - \frac{\partial b_\mu}{\partial x^\lambda} - [b_\lambda, b_\mu]. \tag{4}$$

The Y-M functional L or the action integral of the pure Y-M theory is defined as[1]

$$L(b) = -\frac{1}{4}\int (f_{\lambda\mu}, f^{\lambda\mu}) d^n x, \tag{5}$$

Here (,) stands for the Cartan inner product.

A pure Y - M field is a critical point of the Y - M functional, i. e. its gauge potential satisfies the Euler equations of the Y - M functional (5)

$$J_\alpha = y^{\lambda\mu} f_{\alpha\lambda|\mu} = y^{\lambda\mu}(f_{\alpha\lambda,\mu} + [b_\mu, f_{\alpha\lambda}]) = 0. \tag{6}$$

These equations are called pure Y - M equations.

Let S be a G - valued function. The gauge potential

$$b' = (ads)b - (ds)s^{-1} \tag{7}$$

is called the gauge transformation of b. Two gauge potentials related by a gauge transformation are considered as equivalent mathematically and physically.

(b) Harmonic maps

Let M, N be Riemannian manifolds or Lorentzian manifolds and $\phi: M \to N$ a C^2 - map.

The energy integral of ϕ is

$$E(\phi) = \int_M e(\phi) dv_M. \tag{8}$$

Here

$$e(\phi) = \underset{N}{g}_{\alpha\beta}(\phi)\frac{\partial\phi^\alpha}{\partial x^i}\frac{\partial\phi^\beta}{\partial x^j}\underset{M}{g}^{ij}(x) \qquad (i, j = 1, \cdots, m; \alpha, \beta = 1, \cdots, n) \tag{9}$$

is the expression of the energy density in local coordinates.

A map ϕ is called harmonic if it is a critical point of $E(\phi)$, i.e. a solution to the Euler equations of $E(\phi)$

$$\underset{M}{g}^{ij}\left(\frac{\partial^2\phi^\alpha}{\partial x^i x^j} - \underset{M}{\Gamma}^k_{ij}\frac{\partial\phi^\alpha}{\partial x^k} + \underset{N}{\Gamma}^\alpha_{\beta r}\frac{\partial\phi^\beta}{\partial x^i}\frac{\partial\phi^\alpha}{\partial x^j}\right) = 0. \tag{10}$$

Systems of PDEs (10) is elliptic or hyperbolic, respectively, if M is Riemannian or Lorentzian [2].

(c) Massive Y - M fields

The action integral of massive Y - M fields on $R^{1,n-1}$ is a coupling of the pure Y - M functional (5) and the energy integral for harmonic maps from $R^{1,n-1}$ to the gauge group G,

$$L_m(b, U)=\int\left[-\frac{1}{4}(f_{\lambda\mu}, f^{\lambda\mu})-\frac{m^2}{2}(b_\lambda-\omega_\lambda, b^\lambda-\omega^\lambda)\right]d^n x$$

$$(\lambda=0, 1, \cdots, n-1). \tag{11}$$

Here

$$\omega_\lambda=U^{-1}\frac{\partial U}{\partial x^\lambda},$$

U is a G - valued function which is a display of "gauge" and the coupling constant m is the mass of gauge particles[3].

The Euler equations of the action integral of $L_m(b, U)$ are

$$\begin{aligned} &J_\alpha-m^2(b_\alpha-\omega_\alpha)=0,\\ &y^{\alpha\beta}\left(\frac{\partial b_\alpha}{\partial x^\beta}-\frac{\partial \omega_\alpha}{\partial x^\beta}-[b_\beta, \omega_\alpha]\right)=0 \end{aligned} \tag{12}$$

and the gauge transformation is

$$\begin{aligned} &b'=\omega b\omega^{-1}-(d\omega)\omega^{-1},\\ &U'=U\omega^{-1}, \end{aligned} \tag{13}$$

where $W\in G$. The action integral (11) is a gauge invariant.

§ 2. Nonexistence Theorems for the Static Massive Yang-Mills Fields

In the theory of Y - M fields, one problem of considerable interest is whether there exist any nontrivial static solutions to the Y - M equations with finite energy and free of singularities over the whole $R^{1,n-1}$.

The 1st nonexistence theorem[4] was discovered by S. Deser in 1976 as follows: If $n \neq 5$, the pure Y - M equations of a compact group in n - dimensional spacetime $R^{1,n-1}$ does not admit any nontrivial static solution which has (i) no singularities (ii) finite energy (iii) the field strength approaching to zero sufficiently fast at infinity.

It is noticed that

(a) For $n=4$, there does not exist such static solution on the real space-time $R^{1,3}$.

(b) If $n = 5$, such static solution do exist, since the instantons on the 4 - dimensional Euclidean space R^4 may be considered as regular static solutions in $R^{1,4}$.

(c) We will show later (Remark of Theorem 2), condition (iii) can be removed and condition (ii) can be weakened.

Now we turn to the same problem for massive Y - M fields.

A massive Y - M field is called static, if through a gauge transformation, (b, U) is equivalent to (b', U') which is independent of the time variable x^0.

No loss of generalities, we assume that (b, U) is independent of x^0. Taking $W=U$ in (13), U is reduced to the unit element of G and then $L_m(b, U)$ becomes

$$L_m(b) = -\int \left[\frac{1}{4}(f_{\lambda\mu}, f^{\lambda\mu}) + \frac{m^2}{2}(b_\lambda, b^\lambda)\right] d^n x, \tag{14}$$

and the equations (12) are reduced to

$$\begin{gathered} J_\alpha - m^2 b_\alpha = 0, \\ y^{\lambda\mu} \frac{\partial b_\lambda}{\partial x^\mu} = 0. \end{gathered} \tag{15}$$

The second set of equations (15) mean that the potential b satisfies the Lorentz gauge condition.

The energy momentum tensor is

$$T_{\alpha\beta}=(f_{\alpha\nu},f_{\beta}^{\nu})-\frac{1}{4}y_{\alpha\beta}(f_{\mu\nu},f^{\mu\nu})+m^2(b_\alpha,b^\alpha)-\frac{m^2}{2}y_{\alpha\beta}(b_\lambda,b^\lambda). \tag{16}$$

In particular,

$$T_{00}=\frac{1}{2}\left[(f_{0i},f_{0i})+\frac{1}{2}(f_{ij},f_{ij})\right]+\frac{m^2}{2}(b_0,b_0)+\frac{m^2}{2}(b_i,b_i) \tag{17}$$

$$(i,j=1,2,\cdots,n-1)$$

and

$$T_{ii}=T_\alpha^\alpha-T_0^0=-\frac{1}{2}(n-3)(f_{0i},f^{0i})+\frac{1}{4}(5-n)(f_{ij},f^{ij})+$$

$$\frac{m^2}{2}(3-n)(b_\lambda,b^\lambda)+m^2(b_0,b_0). \tag{18}$$

Moreover, the conservative law

$$\frac{\partial T_\alpha^\beta}{\partial x^\beta}=0 \tag{19}$$

holds.

The total energy of the field is

$$E=\int_{R^{n-1}}T_{00}d^{n-1}x. \tag{20}$$

In 1979 the author obtained the following result[5].

Theorem 1. In an n - dimensional spacetime with $n\neq 4$, the massive Y - M equations with real mass do not admit any static solution which has (i) finite energy (ii) no singularity (iii) the field strength and potential approaching to zero sufficiently fast at infinity.

Comparing Theorem 1 with Deser's result we discovered that there is a "discontinuity" as $m\to 0$ in 5 - dimensional spacetime, i.e. for $n=5$ and $m\neq 0$, no such solution, but when $m=0$ such solution do exist.

Afterwards, S. Deser & C. J. Isham in their paper [6] wrote that this is the first explicit example which make us recognize that there exists a classical "discontinuity". Besides, in their paper the result is extended to the gauge field with "soft" mass.

In 1982 the author found that the finite energy condition can be weakened.

Let $\psi(r)$ be a positive, continuous and unbounded function, defined on $(0, \infty)$ and satisfying

$$\int_a^\infty \frac{dr}{r\psi(r)} = \infty \qquad (a>0). \tag{21}$$

If the energy density T_{00} satisfies

$$\int_{R^{n-1}} T_{00} d_x^{n-1} = \infty,$$

and

$$\int_{R^{n-1}} \frac{T_{00}}{\psi(r)} d^{n-1}x < \infty, \tag{22}$$

we say that the energy is slowly divergent. It means that the total energy within the sphere of radius r approaches to infinity quite slowly as $r \to \infty$.

We have[7]

Theorem 2. Let G be a compact Lie group, m real number $(m \neq 0)$, $n \neq 4$. The massive Y-M equations on $R^{1,n-1}$ do not admit any non-trivial static solution which is free of singularities and has finite or "slowly divergent" energy.

Proof. From the expression for J_0 and static condition $b_{\alpha,0}=0$, we have

$$(b_0, J_0) = (b_0, f_{0i})_{,i} - (f_{0i}, f_{0i}). \tag{23}$$

Consider the integral

$$0=\int_0^\infty \omega(r)dr\int_{|x|\leqslant r}(J_0-m^2b_0,\ b_0)d^{n-1}x,$$

where $|x|=((x^1)^2+\cdots+(x^{n-1})^2)^{1/2}$ and $w(r)$ will be defined later. Using (23), we have

$$0=\int_0^\infty \omega(r)dr\int_{|x|\leqslant r}Kd^{n-1}x+\int_0^\infty \omega(r)dr\int_{|x|=r}(b_0,\ f_{0i})\frac{x^i}{r}dS,$$

where

$$K=-(f_{0i},\ f_{0i})-m^2(b_0,\ b_0)\leqslant 0. \tag{24}$$

The equality in (24) holds if and only if $f_{0i}=b_0=0$. If K does not equal zero identically, then there exists a constant $R>0$ and a positive constant ε such that

$$\int_{|x|\leqslant R_1}Kd^{n-1}x<-\varepsilon \qquad (R_1\geqslant R). \tag{25}$$

Choose

$$\omega(r)=\begin{cases}0, & r<R,\\ \dfrac{1}{r\psi(r)}, & R\leqslant r\leqslant R_1,\\ 0, & r>R_1,\end{cases} \tag{26}$$

where $\psi(r)$ is a positive, unbounded, continuous function of r satisfying

$$\int_R^\infty \frac{dr}{r\psi(r)}=\infty. \tag{27}$$

Then, we have

$$\begin{aligned}0&<-\varepsilon\int_R^{R_1}\frac{dr}{r\psi(r)}+\int_R^{R_1}\frac{dr}{r\psi(r)}\int_{|x|=r}\{(b_0,\ b_0)+(f_{0i},\ f_{0i})\}dS\\&<-\varepsilon\int_R^{R_1}\frac{dr}{r\psi(r)}+\frac{1}{R}\int_{|x|\leqslant R_1}\frac{(b_0,\ b_0)+(f_{0i},\ f_{0i})}{\psi(r)}d^{n-1}x.\end{aligned}$$

Choose R_1 sufficiently large, it is easily seen that the right side should be negative. This is a contradiction. Consequently, we should have $K=0$ identically, i.e.

$$b_0=0, \quad f_{0i}=0. \tag{28}$$

Thus we have

$$T_{ii}=\frac{1}{4}(5-n)(f_{ij}, f_{ij})+\frac{m^2}{2}(3-n)(b_i, b_i) \tag{29}$$

and the conservation law becomes

$$T_{ij,i}=0. \tag{30}$$

Consider the integral

$$0=\int_0^\infty \omega(r)dr \int_{|x|\leqslant r} x^j T_{ij,i} d^{n-1}x=\int_0^\infty \omega(r)dr \int_{|x|\leqslant r} \{(x^j T_{ij})_{,i}-T_{ii}\} d^{n-1}x$$
$$=\int_0^\infty \omega(r)dr \int_{|x|=r} (x^j T_{ij})\frac{x^i}{r} dS-\int_0^\infty \omega(r)dr \int_{|x|\leqslant r} T_{ii} d^{n-1}x. \tag{31}$$

It is easily seen that there exists a constant A such that

$$|T_{ij}|\leqslant A T_{00}. \tag{32}$$

Moreover from (29) we have

(a) If $n\leqslant 3$, then $T_{ii}\geqslant 0$ and the equality holds only when $b_i=0$;

(b) If $n\geqslant 5$, then $T_{ii}\leqslant 0$ and the equality holds only when $b_i=0$.

In either case, if $T_{ii}\not\equiv 0$ we have $T_{ii}>0$ (or $T_{ii}<0$) in some region. Hence there exist two constant $R>0$ and $\varepsilon>0$ such that

$$\int_{|x|\leqslant R_1} T_{ii} d^{n-1}x>\varepsilon \quad (\text{or} <-\varepsilon) \qquad (R_1\geqslant R). \tag{33}$$

Choosing the same $\omega(r)$ as that in (26), for the case (a) we have

$$0<-\varepsilon\int_R^{R_1}\frac{dr}{r\psi(r)}+A\int_0^\infty \frac{T_{00}}{\psi(r)} dS dr. \tag{34}$$

By the assumption that the energy is finite or "slowly divergent", we can choose R_1 sufficiently large, and it is easily seen that the right side of equation (34) should be negative. This gives a contradiction again. Consequently, we should have

$$T_{ii}=0.$$

For the case (b) the situation is quite similar. Consequently we have $f_{ij}=0$ and $b_i=0$. In other words, when $n\neq 4$, the solution should be a trivial one. Thus Theorem 2 is proved completely.

Remark 1. The condition for the energy in Theorem 2 cannot be omitted, because for any dimensional space-time in massive and massless Y-M field we can find static regular solutions with energy diverges sufficiently fast.

Remark 2. Consider the Yang-Mills-Higgs- Kibble field (the gauge field with "soft" mass)

$$I=\int\left(-\frac{1}{4}f_{\mu\nu}f^{\mu\nu}-\frac{m^2}{2}b_\mu b^\mu-\frac{1}{4}\nabla_\mu\phi\nabla^\mu\phi-V(\phi)\right),$$

where ϕ is a scalar invariant and $V(\phi)$ is the potential. By using the same method the result of [6] can be improved and extended to the case of "slowly divergent" energy and the classical "discontinuity" hold also for $n=5$.

Remark 3. Open problem. In the case $n=4$, does there exist a static regular solution of massive Y-M equation with finite energy or "slowly divergent" energy?

Remark 4. For the massless case $m=0$, Using (16) and starting from (31), $f_{\mu\nu}=0$ can be obtained. Thus Deser's Theorem is also easily improved.

Recently we consider the massive Y-M field over curved space-time.

Let $R\times C^{n-1}$ be a curved space-time with metric

$$ds^2=-(dx^0)^2+e^{2\rho}((dx^1)^2+\cdots+(dx^{n-1})^2).$$

Here ρ is a function of $x^1, x^2, \cdots, x^n$ satisfying the conditions

$$\text{(i)} \quad 1+L(\rho)=1+x^i\rho_i \geqslant 0, \quad \rho_i=\partial\rho/\partial x^i \qquad (i=1, \cdots, n-1), \tag{35}$$

$$\text{(ii)} \quad 0<c_1<\rho<c_2.$$

The following theorem is obtained by the author and Y.L. Pan[8].

Theorem 3. In an n - dimensional curved space-time $R \times C^{n-1}$ with $n \neq 4$, the compact group Y - M field with real mass does not possess any nontrivial static solution which is free of singularities and has finite and slowly divergent energy.

As is pointed out by Sealey[9], the condition (i) has a geometric significance, i.e. the mean curvature normal of S_r is never pointing away from zero, where S_r is the level surface $(x^1)^2+\cdots+(x^{n-1})^2=r^2$.

For the massless case, a similar result also holds. Same as the flat space-time case, the exceptional dimension is $n=5$ also.

As a consequence of Theorem 3, we have[8]

Theorem 4. If M^n is a Riemannian manifold with metric $ds^2=e^{2\rho}((dx^1)^2+\cdots+(dx^n)^2)$, where ρ satisfies $1+L(\rho)\geqslant 0$, then M^n does not possess any nontrivial massive Y - M gauge field which has finite action or slowly divergent action.

The corresponding theorem for massless case[8] improves a result obtained by Sealey[9].

§ 3. Nonexistence Theorems for Harmonic Map with Finite or Slowly Divergent Energy

It is known that the harmonic maps from Euclidean space R^n $(n>2)$ to any m - dimensional Riemannian space M_m with finite energy must be a constant map, i.e. the image of R^n is a fixed point[2][10].

For each harmonic map ϕ, we have a stress-energy tensor which satisfy a conservative law. Using the conservative law together with the technique in § 2, the author proved that[1]

Theorem 5. Let ϕ: $R^n \to M_m$ be a harmonic map of n - dimensional ($n \neq 2$) Euclidean space R^n into an m - dimensional Riemannian manifold M_m. Suppose that the energy $E(\phi)$ of ϕ is finite or slowly divergent, then ϕ is a constant map.

In theoretical physics, the Chiral field or the nonlinear σ - model on n - dimensional Minkowski space-time $R^{1,n-1}$ is just a harmonic map ϕ from $R^{1,n-1}$ to a homogeneous Riemannian manifold M_m. If the field is static, then ϕ is a harmonic map from R^{n-1} to M_m. Hence the physical significance of Theorem 5 can be expressed as[11]

Theorem 5'. In $n+1(n>2)$ dimensional Minkowski space-time $R^{n,1}$, there does not exist any static nontrivial Chiral field with finite energy or slowly divergent energy.

Remark 1. By using sterographic projection from $S^2 \to R^2$, we will obtain nontrivial harmonic maps from $R^2 \to M$ with finite energy. So $n=2$ is actually an exceptional case.

Remark 2. The energy condition in our theorem cannot be omitted, because we can find many regular harmonic maps from R^n whose energy does not diverge so slowly.

Remark 3. Since a solution to the Ernst equations is a harmonic map from R^n to the hyperbolic plane with metric

$$ds^2 = \frac{1}{\Phi^2}(d\Phi^2 + d\Psi^2),\ \Phi > 0$$

in Poincare representation, we obtain another physical meaning of Theorem 5[11], i.e. a nontrivial solution to the Ernst equation with $n > 2$ must have infinite energy. Furthermore, the energy cannot be slowly divergent. Here the energy is in the sense of harmonic maps.

On the other hand, H.C.J. Sealey in [9] proved the theorem: Let $M^n(n>2)$ be a conformal flat space with metric form $ds^2 = f^2(x)((dx^1)^2 + (dx^2)^2 + \cdots + (dx^n)^2)$. If

$L(f)=\sum_i x^i \partial \log f/\partial x^i \geqslant -1$, then any harmonic map with finite energy from M^n to any Riemannian manifold must be a constant map.

Recently, Pan and the author[12] obtain the following more general theorem.

Theorem 6. Let M^n $(n>2)$ be a Riemannian manifold with metric form $ds^2 = f_1^2(x)(dx^1)^2+f_2^2(x)(dx^2)^2+\cdots+f_n^2(x)(dx^n)^2$ satisgying the following conditions

(i) $L(f_i)=\sum_j x^j \partial \log f_i/\partial x^j \geqslant -1$,

(ii) There exists a positive constant K such that $\max\limits_{1\leqslant i,\, j\leqslant n} \dfrac{f_i}{f_j} \leqslant K$,

(iii) For any index i, and any set of $(n-2)$ indices $j_1, \cdots, j_{n-2} (\neq i)$,

$$\sum_{k=1}^{n-2}(1+L(f_{jk}))\geqslant 1+L(f_i).$$

Then, any harmonic map ϕ with finite or slowly divergent energy from M^n to any Riemannian manifold must be a constant map.

Remark 1. In the case $f_1=\cdots=f_n=f$ the conditions (ii) and (iii) are trivial. Hence Theorem 5 and the above mentioned result of Sealey are special cases of Theorem 6.

Remark 2. We point out that the condition (i) also has the geometric meaning as that in Sealey's case.

Remark 3. Theorem 6 includes essentially the case where M^n is a direct product manifold of p conformal flat manifolds $M_1 x \cdots x M_p$.

Because of the limitation of space, we will not give the proof of Theorem 6 here. Instead, we give the proof only for the special case. We assume M^n be a conformally flat space C^n with metric

For each harmonic map ϕ, we have a stress-energy tensor

$$S_{ij}=\frac{1}{2}e(\phi)\underset{M}{g_{ij}}-\underset{N}{g_{\alpha\beta}}\phi_i^\alpha\phi_j^\beta, \tag{37}$$

which satisfies the conservative law

$$S^{i}_{j,i}=0, \tag{38}$$

here; denotes the covariant derivative with respect to the metric of C^n. Since C^n is conformally flat,

$$S^{j}_{i}=e^{-2\rho}S_{ij}, \tag{39}$$

$$\underset{M}{\Gamma}{}^{i}_{jk}=\delta^{i}_{j}\rho_k+\delta^{i}_{k}\rho_j-\rho_i\delta_{jk} \tag{40}$$

hold true. Consider the integral

$$\begin{aligned} O&=\int_{|x|\leqslant r} x^j S^{j}_{i;j}dV=\int\{(x^iS^{j}_{i})_{;j}-x^{i}_{;j}S^{j}_{i}\}dV \\ &=\int_{|x|\leqslant r}(x^iS^{j}_{i})_{;j}dV-\int_{|x|\leqslant r}S^{i}_{i}dV-\int \underset{M}{\Gamma}{}^{i}_{jk}x^kS^{j}_{i}dV. \end{aligned} \tag{41}$$

Using (40) and Stoke's theorem, we have

$$O=\int_{|x|=r}\frac{e^{-\rho}S_{ij}x^ix^j}{r}dh-\int_{|x|\leqslant r}S^{i}_{i}(1+L\rho)dV,$$

where dh is the volume element of $|x|=r$ and $dV=e^{\rho}dr\ dh$. From $S_{ii}=e^{2\rho}\frac{n-2}{2}e(f)\geqslant 0$ and $1+L\rho\geqslant 0$, we see that if ϕ is not a constant map, there exist positive constants R_1 and ε such that

$$\int_{|x|\leqslant r}S_{ii}(1+L\rho)dV>\varepsilon>0 \qquad (r\geqslant R_1). \tag{42}$$

Hence for $r\geqslant R_1$

$$0<\int_{|x|=r}\frac{e^{-\rho}S_{ij}x^ix^j}{r}dh-\varepsilon<\frac{1}{2}\int_{|x|=r}re^{\rho}e(\phi)dh-\varepsilon \tag{43}$$

multiplying the above inequality by $w(r)$ defined in (26), and integrate, we have

$$0 < \frac{1}{2}\int_0^R \omega(r)dr \int_{|x|=r} re^{\rho}e(\phi)dh - \varepsilon\int_0^R \omega(r)dr$$

$$= \frac{1}{2}\int_{R_1 \leqslant |x| \leqslant R} \frac{e(\phi)}{\psi(r)}dV - \varepsilon\int_{R_1}^R \frac{dr}{r\psi(r)}. \tag{44}$$

It is easily seen that if R is sufficiently large, the right side of (44) should be negative. This is a contradiction. Hence ϕ must be a constant map.

This work is partially supported by the Chinese National Foundation of Natural Sciences.

References

[1] C.N.Yang & R.Mills, Isotopic spin conservation and a generalized gauge invariance, Phys. Rev. 96 (1954). 191 - 195.

[2] J.Eells & L.Lemaire, A report on harmonic maps, Bull. London Math. Soc. 10 (1978). 1 - 68.

[3] C.H.Gu (Gu Chao-hao), On classical Yang-Mills Fields, Physics Reports 80 (1981). 251 - 337.

[4] S.Deser, Absence of static solutions in source-free Yang-Mills theory, Phys. Lett. 64B (1976). 463 - 465.

[5] H.S.Hu (Hu He-sheng), On equations of Yang-Mills gauge fields with mass, Kexue Tongbao 25 (1980). 191 - 195.

[6] S.Deser & C.J.Isham, Static solution of Yang-Mills-Higgs-Kibble system, Kexue Tongbao 25 (1980). 773 - 776.

[7] H.S.Hu, On the static solutions of massive Yang-Mills equations, Chinese Annals of Math. 3 (1982). 519 - 526.

[8] H.S.Hu & Y.L. Pan (Pan Yang-lian), Vanishing theorems on the static solutions of massive Yang-Mills field, Preprint of Fudan Univ. (1984).

[9] H.C.J.Sealey, Some conditions ensuring the vanishing of harmonic differential forms with applications to harmonic maps and Yang-Mills theory, Math. Proc. Camb. Phil. Soc. 91(1982).

441－452.

[10] S.Hildebrandt, Nonlinear elliptic systems and harmonic mappings, Preceedings of 1980 Beijing Symposium on Differential Geometry and Differential Equations, Vol.1, 481－615.

[11] H.S.Hu, A nonexistence theorem for harmonic maps with slowly divergent energy. Chinese Annals of Math. 5 (1984).

[12] H.S.Hu & Y.L.Pan, A theorem on Liouville's type on harmonic maps with finite or slowly divergent energy, Preprint of Fudan Univ. (1984).

（本文曾发表于 *Springer Lecture Notes in Physics*, 1984, **212**, 107－116）

The Construction of Hyperbolic Surfaces in 3 Dimensional Minkowski Space and Sinh-Laplace Equation

Hu Hesheng (胡和生)
Institute of Mathematics, Fudan University
Received Mar.28, 1984

§ 1. Introduction

The relationship between surfaces of negative constant curvature in the Euclidean space R^3 and the sine-Gordon equation was known long time ago. Owing to the significance of the sine-Gordon equation in physical sciences, the geometrical properties of the surfaces of negative constant curvature become more attractive[1].

In the study of harmonic maps from the Minkowski plane $R^{1,1}$ or from the Euclidean plane R^2 to various kinds of "spheres" in R^3 or $R^{3,1}$, the following equations

$$\alpha_{tt} - \alpha_{xx} = \pm \sin\alpha \quad \text{(sine-Gordon)}, \tag{1)(2}$$

$$\alpha_{tt} - \alpha_{xx} = \pm \sinh\alpha \quad \text{(sinh-Gordon)}, \tag{3)(4}$$

$$\alpha_{tt} + \alpha_{xx} = \pm \sin\alpha \quad \text{(sine-Laplace)}, \tag{5)(6}$$

$$\alpha_{tt} + \alpha_{xx} = \pm \sinh\alpha \quad \text{(sinh-Laplace)} \tag{7)(8}$$

play important roles[2, 3, 4].

In the present paper we shall elucidate the relationship between the above mentioned PDEs and the construction of various kinds of surfaces of constant curvature in R^3 or $R^{2,1}$. The metrics of these surfaces may be indefinite if the ambient space is $R^{2,1}$. As is known, if the asymptotic lines are real, there exists a special parametrization of the surfaces of constant curvature, i.e. the Tchebysheff coordinates of the surfaces[5]. In the present paper, we extend the Tchebysheff coordinates to the case that the asymptotic lines are imaginary. Using the Tchebysheff coordinates for each case, we find the relation between the above PDEs and various kinds of surfaces of constant curvature. For example, to each solution of the sinh-Laplace equation there corresponds a space-like surface of curvature -1 in $R^{2,1}$.

Using these relations and the results obtained in the previous paper [4], we gave a general method for the construction of surfaces of constant curvature in the present work. The procedure is to solve the corresponding PDEs together with the integration of a completely integrable system of linear PDEs. For example, we use a class of rather simple solutions of the sinh-Laplace equation to get a class of surfaces of curvature -1 in $R^{2,1}$. Among these surfaces there are surfaces found by Hano and Nomizu in proving the nonrigidity of the simply connected complete hyperbolic plane in $R^{2,1}$; besides, there are a family of surfaces of curvature -1 which are complete graphs, but are not complete with respect to their Riemannian metrics. This shows a distinction between the surfaces in $R^{2,1}$ and R^3.

Thus, we obtain complete graph solutions defined on the whole plane to the Monge-Ampère equations

$$rt-s^2=(1-p^2-q^2)^2$$

or

$$F_{x_1x_1}F_{x_1x_2}-F_{x_1x_2}^2=(1-F_{x_1}^2-F_{x_2}^2)^2$$

through the sinh-Laplace equation.

§ 2. Tchebysheff Coordinates for Surfaces of Negative Constant Curvature in $R^{2,1}$

For the surfaces of constant curvature with real asymptotic directions, the existence of Tchebysheff coordinates can be found in [5]. Using these coordinates, the relations between the sine-Gordon equation and surfaces of negative constant curvature in R^3 and those between the sinh-Gordon equation and time-like surfaces of positive constant curvature in $R^{2,1}$ were found respectively[5].

It is also easy to see that the asymptotic directions are imaginary on the space-like and time-like surfaces of negative curvature in $R^{2,1}$.

Let S be a space-like surface of curvature -1 in $R^{2,1}$. As is well-known, a surface $S \subset R^{2,1}$ is called space-like if the induced metric on S is positively definite. We define Tchebysheff coordinates on S as follows.

Definition 1. *A coordinate system* (t, x) *of the space-like surface* S *is called Tchebysheff coordinates if the metric of* S *is in the form*

$$ds^2 = \cosh^2 \frac{\alpha}{2} dt^2 + \sinh^2 \frac{\alpha}{2} dx^2, \tag{9}$$

and the second fundamental form is

$$\mathrm{II} = \cosh \frac{\alpha}{2} \sinh \frac{\alpha}{2} (dt^2 + dx^2). \tag{10}$$

Theorem 1. *If a space-like surface* $S \subset R^{2,1}$ *is of constant curvature* -1 *and free of umbilici, then* (i) *S can be covered by charts with Tchebysheff coordinates,* (ii) *the function* $\alpha(t, x)$ *in Tchebysheff coordinates satisfies sinh-Laplace equation* (7).

Proof. Let $\{P, e_i\}$ $(i = 1, 2, 3)$ be a field of orthonormal frames of S such that

e_1, e_2 are unit tangent vectors to the lines of curvature and e_3 are the normal vectors to the surface. Hence $e_1^2 = e_2^2 = -e_3^2 = 1$, the fundamental equations of the surface S is

$$dP = \omega^a e_a \quad (a, b = 1, 2), \tag{11}$$

$$de_a = \omega_a^b e_b + \omega_a^3 e_3, \tag{12}$$

$$de_3 = \omega_3^a e_a \tag{13}$$

with

$$\omega_b^a = -\omega_a^b, \quad \omega_3^a = \omega_a^3. \tag{14}$$

Since e_1, e_2 are principal tangent vectors, we have

$$\omega_1^3 = A\omega^1, \ \omega_2^3 = C\omega^1. \tag{15}$$

Using the property that the curvature of S is -1, we have

$$AC = 1. \tag{16}$$

Choose coordinates such that

$$\omega^1 = \lambda dt, \ \omega^2 = \mu dx, \tag{17}$$

and let

$$\omega_2^1 = -\omega_1^2 = \gamma dt + \delta dx.$$

Then, from the integrability condition of (11) we obtain

$$\mu\gamma = \lambda_x, \ \lambda\delta = -\mu_t.$$

From the integrability condition of (13) we obtain

$$A_x\lambda + (A - C)\lambda_x = 0, \ C_t\mu + (C - A)\mu_t = 0. \tag{18}$$

Since the surface is free of umbilici, $A \neq \pm 1$. Set $A = \tanh\frac{\alpha}{2}$, $C = \mathrm{cth}\frac{\alpha}{2}$. From (18) we have

$$(\ln\lambda)_x = \left(\ln\cosh\frac{\alpha}{2}\right)_x,$$

$$(\ln\mu)_t = \left(\ln\sinh\frac{\alpha}{2}\right)_t.$$

Hence

$$\lambda = h(t)\cosh\frac{\alpha}{2},$$

$$\mu = g(t)\sinh\frac{\alpha}{2}.$$

By changing t and x we have

$$\lambda = \cosh\frac{\alpha}{2},\ \mu = \sinh\frac{\alpha}{2}. \tag{19}$$

Hence

$$ds^2 = \cosh^2\frac{\alpha}{2}dt^2 + \sinh^2\frac{\alpha}{2}dx^2,$$

$$\mathrm{II} = \cosh\frac{\alpha}{2}\sinh\frac{\alpha}{2}(dt^2 + dx^2). \tag{20}$$

Thus, the existence of Tchebysheff coordinates is proved. The second part of the theorem follows immediately from Gauss equation

$$d\omega_2^1 = k\omega^1 \wedge \omega^2 \quad (k = -1).$$

The similar argument is valid for time-like surfaces of curvature -1 in $R^{2,1}$. As we have mentioned, the Tchebysheff coordinates for surfaces with real asymptotic lines were known. Now we have

Theorem 2. *The surfaces of constant curvature with imaginary asymptotic direction and free of umbilici in 3 dimensional space R^3 or $R^{2,1}$ can be covered by charts with Tchebysheff coordinates in which the fundamental forms are as follows.*

Surfaces	Signature	Curvature	ds^2 in Tch.coord.	II in Tch.coord.	Related equation
$S\subset R^{2,1}$	++	-1	$\mathrm{ch}^2\frac{\alpha}{2}dt^2+\mathrm{sh}^2\frac{\alpha}{2}dx^2$	$\mathrm{ch}\frac{\alpha}{2}\mathrm{sh}\frac{\alpha}{2}(dt^2+dx^2)$	$\Delta\alpha=\mathrm{sh}\,\alpha$
$S\subset R^3$	++	$+1$	$\mathrm{ch}^2\frac{\alpha}{2}dt^2+\mathrm{sh}^2\frac{\alpha}{2}dx^2$	$\mathrm{ch}\frac{\alpha}{2}\mathrm{sh}\frac{\alpha}{2}(dt^2+dx^2)$	$\Delta\alpha=-\mathrm{sh}\,\alpha$
$S\subset R^{2,1}$	+−	-1	$\cos^2\frac{\alpha}{2}dt^2-\sin^2\frac{\alpha}{2}dx^2$	$\cos\alpha\sin\alpha(dt^2+dx^2)$	$\Delta\alpha=\sin\alpha$

As a consequence of the above theorem, we have

Corollary. *The only minimal surfaces* ($H=0$) *of constant curvature in* 3 *dimensional Minkowski space* $R^{2,1}$ *are planes.*

Proof. At first we notice that every point on the minimal surface ($H=0$) of constant curvature must be an umbilicus, otherwise, in a neighborhood of nonumbilic point we can use Tchebysheff coordinates and hence $H\neq 0$. Furthermore, it is easily seen that a totally umbilic surface is a minimal surface if it is a plane.

§ 3. Construction for Surfaces of Constant Curvature −1 in $R^{2,1}$

The following theorem provides a method for constructing a space-like surface of curvature -1 in $R^{2,1}$ from a solution to the sinh-Laplace equation. We would like to emphasize that the procedure of the construction is linear provided that the solution of sinh-Laplace equation is given.

Theorem 3. *Let* $a\ (t,\ x)$ *be a solution of sinh-Laplace equation on a simply connected region* $\Omega\subset R^2$ *and* $\alpha\neq 0$. *Then there exists a space-like surface* $S\subset R^{2,1}$ *of constant curvature* -1 *with* $(t,\ x)$ *as its Tchebysheff coordinates, i.e.* (9) *and* (10) *hold true.*

Proof. Firstly, we solve e_1, e_2, e_3 from (12), (13). These equations are nothing else than the following Lax pair of the sinh-Laplace equation

$$\begin{aligned}
&l_t = \sinh\frac{\alpha}{2}m, && l_x = \cosh\frac{\alpha}{2}n, \\
&m_t = \sinh\frac{\alpha}{2}l - \frac{1}{2}\alpha_x n, && m_x = \frac{1}{2}\alpha_t n, \\
&n_t = \frac{1}{2}\alpha_x m, && n_x = \cosh\frac{\alpha}{2}l - \frac{1}{2}\alpha_t m,
\end{aligned} \tag{21}$$

which appeared in my previous paper [4]. But in the present paper the notations t, x are interchanged. Here l, m, n are e_3, e_1, e_2 respectively.

The integrability condition of the Lax pair (21) is the sinh-Laplace equation. The orthonormal solution (e_1, e_2, e_3) of (12), (13) is defined on the whole Ω since the system (12), (13) is linear in e_1, e_2, e_3. After solving e_1, e_2, e_3, we solve the equation

$$dP = \omega^1 e_1 + \omega^2 e_2 = \cosh\frac{\alpha}{2}dtm + \sinh\frac{\alpha}{2}dxn, \tag{22}$$

which is also completely integrable. In short, for a solution of sinh-Laplace equation, the system (12), (13) and (21) is complete integrable and the solution is defined on the whole Ω. Choosing any initial condition $(t, x) = (t_0, x_0) \in \Omega$, and $(P, e_1, e_2, e_3) = (P_0, e_1^0, e_2^0, e_3^0)$ which is an orthonormal frame of $R^{2,1}$, we obtain a surface $P = P(t, x)$ expressed in Tchebysheff coordinates.

Remark 1. In the previous paper we already proved that $e_3 = e_3(t, x)$ is a harmonic map from the Minkowski plane (t, x) to the "sphere" H^2. Moreover, $e_3 = e_3(t, x)$ is now the Gauss map $S \to H^2$. Hence we have the following diagram.

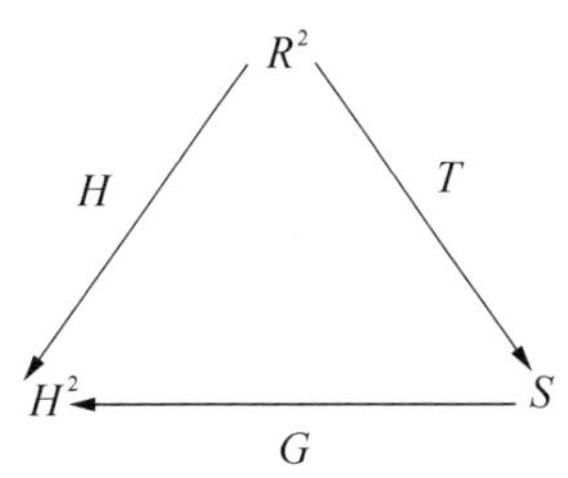

Here T, G, H represent the Tchebysheff coordinate, the Gauss map and the harmonic map respectively.

Remark 2. A similar fact holds for time-like surfaces of negative constant curvature in $R^{2,1}$. Actually, the method in this section is valid for constructing all surfaces of constant curvature which is free of umbilici in R^3 and $R^{2,1}$, and the construction is linear provided that a solution to any one of the equations (1)~(8) is given.

§ 4. Some Hyperbolic Surfaces in $R^{2,1}$

As an application of Theorem 3, we choose the following set of solutions

$$\alpha = 2\sinh^{-1}\left(-\frac{1}{\sinh(\lambda t+\mu x)}\right)(\lambda^2+\mu^2=1,\ \lambda>0,\ \lambda t+\mu x<0) \tag{23}$$

of the sinh-Laplace equation (7) for constructing the hyperbolic surfaces in $R^{2,1}$. Here λ and μ are constants.

It is easily seen that

$$\operatorname{sh}\frac{\alpha}{2} = -\frac{1}{\operatorname{sh}(\lambda t+\mu x)},\quad \operatorname{ch}\frac{\alpha}{2} = -\frac{\operatorname{ch}(\lambda t+\mu x)}{\operatorname{sh}(\lambda t+\mu x)}. \tag{24}$$

We shall solve the 1-st part of the Lax pair (21). Now it can be written as

$$l_t = \operatorname{sh}\frac{\alpha}{2}m,\quad m_t = \operatorname{sh}\frac{\alpha}{2}(l-\mu n),\quad n_t = \mu\operatorname{sh}\frac{\alpha}{2}m. \tag{25}$$

Let

$$a = \mu l - n,\quad b = l + \lambda m - \mu n,\quad c = -l + \lambda m + \mu n. \tag{26}$$

(25) is equivalent to

$$a_t = 0,\quad b_t = \lambda\operatorname{sh}\frac{\alpha}{2}b,\quad c_t = -\lambda\operatorname{sh}\frac{\alpha}{2}c. \tag{27}$$

Integrating (27) we obtain

$$a = a_0(x),\quad b = -b_0(x)\operatorname{cth}\frac{\lambda t+\mu x}{2},\quad c = -c_0(x)\operatorname{th}\frac{\lambda t+\mu x}{2}, \tag{28}$$

where $a_0(x)$, $b_0(x)$, $c_0(x)$ are vector functions independent of t.

Substituting (28) in the 2 - nd part of the Lax pair (21), through complicated calculations we obtain

$$a_0' = -b_0(x) + \frac{1}{2}c_0(x), \quad b_0' = -a_0(x), \quad c_0' = a_0(x), \tag{29}$$

where "1" denotes the differentiation with respect to x.

The general solutions of (29) are

$$\begin{aligned} a_0(x) &= \sigma \operatorname{ch} x + \nu \operatorname{sh} x, \\ b_0(x) &= \tau - \sigma \operatorname{sh} x - \nu \operatorname{ch} x, \\ c_0(x) &= \tau + \sigma \operatorname{sh} x + \nu \operatorname{ch} x, \end{aligned} \tag{30}$$

where ν, σ, τ are constant vectors.

Let $x = 0$ and $t \to -\infty$. From (28), (30) we obtain

$$a_0(0) = \sigma, \; b_0(0) = \tau - \nu, \; c_0(0) = \tau + \nu. \tag{31}$$

Considering (26) we obtain

$$\begin{aligned} a_0(0) &= \mu l_0 - n, \\ b_0(0) &= l_0 + \lambda m_0 - \mu n_0, \\ c_0(0) &= -l_0 + \lambda m_0 + \mu n_0, \end{aligned} \tag{32}$$

where l_0, m_0, n_0 form an orthonormal set of vectors in $R^{2,1}$. Knowing the expression for a, b, c we obtain the general solution of the Lax pair

$$\begin{aligned} l = {} & l_0\left(\frac{\mu}{\lambda^2}(\operatorname{sh} x - \mu \operatorname{ch} x) + \frac{1}{\lambda^2}\operatorname{cth}\xi(\mu \operatorname{sh} x - \operatorname{ch} x)\right) + \\ & m_0\left(-\frac{1}{\lambda}\frac{1}{\operatorname{sh}\xi}\right) + n_0\left(\frac{\mu}{\lambda^2}(\operatorname{ch} x - \mu \operatorname{sh} x) + \frac{1}{\lambda^2}\operatorname{cth}\xi(-\operatorname{sh} x + \mu \operatorname{ch} x)\right), \end{aligned} \tag{33}$$

$$m = l_0\left(\frac{\mu}{2\lambda}\operatorname{sh} x\,\frac{2}{\operatorname{sh}\xi} + \frac{1}{2\lambda}\operatorname{ch} x\left(-\frac{2}{\operatorname{sh}\xi}\right)\right) - m_0 \operatorname{cth}\xi +$$

$$n_0\left(-\frac{1}{2\lambda}\operatorname{sh}x\,\frac{1}{\operatorname{sh}\xi}+\frac{\mu}{\lambda}\operatorname{ch}x\,\frac{1}{\operatorname{sh}\xi}\right), \tag{34}$$

$$n=l_0\left(\frac{1}{\lambda^2}(\operatorname{sh}x-\mu\operatorname{ch}x)+\operatorname{cth}\xi(\mu\operatorname{sh}x-\operatorname{ch}x)\frac{\mu}{\lambda^2}\right)+$$
$$m_0\left(-\frac{\mu}{\lambda}\,\frac{1}{\operatorname{sh}\xi}\right)+n_0\left(\frac{1}{\lambda^2}(\operatorname{ch}x-\mu\operatorname{sh}x)+\operatorname{cth}\xi\,\frac{\mu}{\lambda^2}(\mu\operatorname{ch}x-\operatorname{sh}x)\right), \tag{35}$$

where $\xi=\lambda t+\mu x$.

We turn to the integration of the equation

$$dP=\operatorname{ch}\frac{\alpha}{2}dtm+\operatorname{sh}\frac{\alpha}{2}dxn.$$

Integrating

$$P_t=\operatorname{ch}\frac{\alpha}{2}m,$$

we have

$$P=P_0(x)+l_0\left(-\frac{\mu}{\lambda^2}\operatorname{sh}x+\frac{1}{\lambda^2}\operatorname{ch}x\right)\left(-\frac{1}{\operatorname{sh}\xi}\right)+m_0\,\frac{1}{\lambda}(\xi-\operatorname{cth}\xi)+$$
$$n_0\left(\frac{1}{\lambda^2}\operatorname{sh}x-\frac{\mu}{\lambda^2}\operatorname{ch}x\right)\left(-\frac{1}{\operatorname{sh}\xi}\right). \tag{36}$$

From

$$P_x=\operatorname{sh}\frac{\alpha}{2}n$$

we obtain

$$P_0'+m_0\,\frac{\mu}{\lambda}=0,$$

and then

$$P_0=-m_0\,\frac{\mu}{\lambda}x. \tag{37}$$

Thus the general expression for P is

$$P = -\frac{1}{\operatorname{sh}\xi}\operatorname{ch}x\left(\frac{l_0}{\lambda^2} - \frac{\mu n_0}{\lambda^2}\right) + m_0\frac{1}{\lambda}(\xi - \operatorname{cth}\xi - \mu x) +$$

$$\frac{1}{\operatorname{sh}\xi}\operatorname{sh}x\left(\frac{\mu l_0}{\lambda^2} - \frac{n_0}{\lambda^2}\right). \tag{38}$$

Without loss of generality we may choose

$$(l_0, m_0, n_0) = \left(\frac{1}{\lambda}(l_0 - \mu n_0), m_0, -\frac{1}{\lambda}(\mu l_0 - n_0)\right), \tag{39}$$

and (32) takes the following form

$$S_\lambda:\quad \begin{aligned} x_1 &= t - \frac{1}{\lambda}\operatorname{cth}(\lambda t + \mu x), \\ x_2 &= -\frac{1}{\lambda\operatorname{sh}(\lambda t + \mu x)}\operatorname{sh}x, \\ x_3 &= -\frac{1}{\lambda\operatorname{sh}(\lambda t + \mu x)}\operatorname{ch}x. \end{aligned} \tag{40}$$

In particular, setting $\lambda = 1$, $\mu = 0$ in (40) we obtain the following surface

$$\begin{aligned} x_1 &= t - \operatorname{cth}t, \\ x_2 &= -\frac{1}{\operatorname{sh}t}\operatorname{sh}x, \\ x_3 &= -\frac{1}{\operatorname{sh}t}\operatorname{ch}x. \end{aligned} \tag{41}$$

Actually, it is the surface of revolution obtained in [6] by a different approach. Besides, all the surfaces found in [6] can be obtained by our procedure.

We can show that the surfaces S_λ are complete graphs. From the equation $x_2 = -\frac{1}{\lambda\operatorname{sh}\xi}\operatorname{sh}x$ we see that for fixed $\xi < 0$, the range of x_2 is $(-\infty, \infty)$ if x varies from $-\infty$ to $+\infty$. Since x_1 can be expressed as

$$x_1 = \frac{\xi - \mu \operatorname{sh}^{-1}(-x_2 \lambda \operatorname{sh}\xi) - \operatorname{cth}\xi}{\lambda},$$

it is easy to see that

$$\frac{dx_1}{d\xi} > 0 \quad (x_2 \in (-\infty, +\infty), \ \xi \in (-\infty, 0)).$$

Moreover, for fixed x_2, the range of x_1 is $(-\infty, \infty)$ if ξ varies from $-\infty$ to 0. Hence $(t, x) \to (x_1, x_2)$ is a diffeomorphism from the half-plane $\xi < 0$ in (t, x) plane to the whole plane (x_1, x_2). Consequently, the surfaces S_λ are complete graphs.

It is interesting to see that these surfaces are not complete Riemannian manifolds if $\lambda \neq 1$ (i.e. $\mu \neq 0$). Choose $t = t_0$, $x = x_0$ such that $\xi < 0$. In case $\mu < 0 (\mu > 0)$, consider the curve on the surface $S_\lambda (\lambda \neq 0)$ with parameters $t = t_0$, $x_0 < x < \infty (-\infty < x < x_0)$. The point (x_1, x_2, x_3) on the curve approaches ∞ as $x \to \infty$. However, the length of the curve

$$S = -\int_{x_0}^{\infty} \frac{dx}{\operatorname{sh}(\lambda t_0 + \mu x)} = \frac{1}{\mu} \ln\left(-\operatorname{th}\frac{\lambda t_0 + \mu x_0}{2}\right)$$

is finite. So we have

Theorem 4. *In $R^{2,1}$ there are complete graphs (e.g. $S_\lambda (\lambda \neq 1)$ defined by (40)) which are space-like surfaces of curvature -1 and noncomplete with respect to their metrics.*

Remark. The Monge-Ampère equation

$$F_{x_1 x_1} F_{x_2 x_2} - F_{x_1 x_2}^2 = (1 - F_{x_1}^2 - F_{x_2}^2)^2$$

admits a family of solutions (40) defined over the whole (x_1, x_2) plane as well as the obvious solution $x_3 = \sqrt{1 + x_1^2 + x_2^2}$. Besides we found more solutions to the Monge-Ampère equation. In the present method the nonlinearity of the Monge-Ampère equation is strongly reduced, since the nonlinearity of sinh-Laplace equation is much lower than the nonlinearity of the Monge-Ampère equation, and the system of equations (21) and (22) is linear.

Similarly, from the solution of the sinh-Gordon equation we can construct many surfaces in $R^{2,1}$ with indefinite metric and of curvature -1.

References

[1] Scott, A.C., Chu F.Y. and Mclaugulin, D.W., The Soliton, A new concept in applied science, *Proc. IEEE*, 61 (1973), 1443 - 1483.

[2] Gu Chao-hao (Gu C.H.), On the Cauchy problem for harmonic map defined on two dimonsional Minkowski space, *Comun. on Pure and Appl. Math.*, 33 (1980), 727 - 737.

[3] Gu Chao-hao, On the harmonic maps from $R^{1,1}$ to $S^{1,1}$, *Crelles Jour.*, 346 (1984), 101 - 109.

[4] Hu He-sheng (Hu H.S.), Sine-Laplace equation, sinh-Laplace equation and harmonic maps, *Manuscripta Math.*, 40 (1982), 205 - 216.

[5] Chern, S.S., Geomotrical interpretation of sinh-Gordon equation, *Annales Polinici Math.*, 39 (1981), 63 - 69.

[6] Hano, J. and Nomizu, K., On the isometric immersion of hyperbolic plane into the Lorentz-Minkowski space and the Monge-Ampère equation of a certain type. Preprint (1982), Bonn.

(本文曾发表于 *Acta Mathematica Sinica*, *New Series*, 1985, 1(1), 79 - 86)

A Theorem of Liouville's Type on Harmonic Maps with Finite or Slowly Divergent Energy①

Hu Hesheng(胡和生) Pan Yanglian(潘养廉)
Institute of Mathematics, Fudan University, Shanghai, China

Abstract Some theorems of Liouville's type on harmonic maps from Euclidean space of conformal flat space with finite or slowly divergent energy have been obtained by the first-named author and H.C.J. Sealey, respectively. In this paper, a more general theorem is proved, which includes their results as special cases. The technique is to use a conservation law for harmonic maps.

§ 1. Introduction

In [1], the first-named author proved the theorem: Let φ: $R^n \to M^m$ be a harmonic map of $n(n \neq 2)$ - dimensional Euclidean space R^n into an m - dimensional Riemannian manifold M^m. Suppose that the energy $e(\varphi)$ of φ is finite or slowly divergent. Then φ is a constant map. Here "slowly divergent energy" means that $\int_{R^n} e(\varphi) d^n(x) = \infty$ and $\int_{R^n} \frac{e(\varphi)}{\psi(r)} d^n x < \infty$, where $\psi(r)$ is a positive, continuous function of r satisfying

① Manuscript received April 12, 1984.

$$\int_a^\infty \frac{dr}{r\psi(r)} = \infty \quad (\text{for } a \text{ certain constant } a > 0). \tag{1}$$

On the other hand, H.C.J. Sealey in [2] proved the theorem: Let M^n $(n \geqslant 3)$ be a conformal flat space with metric form $ds^2 = f^2(x)\ (dx'^2 + \cdots + dx^{n^2})$. If $L(f) \equiv \sum_i x^i \frac{\partial \log f}{\partial x^i} \geqslant -1$, then any harmonic map with finite energy from M^n into any Riemannian manifold must be a constant map.

Sealey has pointed out that the condition $L(f) \geqslant -1$ has a geometric significance. In fact, if S_r denotes the level surface $\{x \in R^n \mid \sum (x^i)^2 = r^2\}$, then $L(f) \geqslant -1$ holds if and only if the mean curvature normal of S_r with respect to $ds = f^2(x) \sum_i (dx^i)^2$ is never pointing away from zero.

In this paper, using a similar technique as in [1], we will prove the following more general theorem which includes both the above theorems as special cases.

Main theorem. *Let M^n $(n \geqslant 3)$ be a Riemannian manifold with metric form $ds^2 = f_1^2(x)(dx^1)^2 + f_2^2(x)(dx^2)^2 + \cdots + f_n^2(x)(dx^n)^2$ satisfying the following conditions:*

(A) $L(f_i) = \sum_j x^j \frac{\partial \log f_i}{\partial x^j} \geqslant -1, \quad i = 1, \cdots, n.$

(B) *There exists a positive constant K such that*

$$\max_{1 \leqslant i,\, j \leqslant n} \frac{f_i}{f_j} \leqslant K.$$

(C) *For any index $1 \leqslant i \leqslant n$, and index $j_1 \neq j_2 \neq \cdots \neq j_{n-2}$,*

$$\sum_{k=1}^{n-2} (1 + L(f_{j_k})) \geqslant 1 + L(f_i).$$

Then, any harmonic map φ with finite or slowly divergent energy from M^n into any Riemannian manifold must be a constant map, where "slowly divergent energy" *means that* $\int_{M^n} e(\varphi)dV=\infty$ *and* $\int_{M^n}\frac{e(\varphi)}{\psi(r)}dV<\infty$, *where* $\psi(r)$ *is a positive, continuous function of* r *satisfying*

$$\int_a^{\infty}\frac{dr}{r\psi(r)}=\infty \quad (\text{for a certain constant } a>0).$$

Remark 1. In the case of $f_1=\cdots=f_n=f$, the conditions (B) and (C) are trivial and this theorem is a generalization of Sealey's result as well as Hu's previous result.

Remark 2. We point out that the condition (A) also has the geometric significance as that in Sealey's case.

Remark 3. Theorem 1 includes essentially the case where M^n is a direct product manifold of p conformal flat manifolds $M_1\times\cdots\times M_p$.

§ 2. Preliminary

Let M^n be as above and S_r the level surface $\{x\in M^n \mid \sum_i (x^i)^2=r^2\}$.

Since there exists at least an $x^i\neq 0$ on S_r, say $x^n\neq 0$, we denote the induced metric of S_r from M^n by $g'_{ab}dx^a dx^b$, where $a, b, c, \cdots=1, \cdots, n-1$, and the volume element of S_r by dh. A straightforward computation shows

$$g'_{ab}=f_a^2\delta_{ab}+\left(\frac{f_n}{x^n}\right)^2 x^a x^b. \tag{2}$$

Thus, it is easy to show that

$$\det(g'_{ab})=\left(\prod_{i=1}^n f_i\right)\cdot\frac{\Phi}{(x^n)^2}=\det(g_{ij})\frac{\Phi}{(x^n)^2}, \tag{3}$$

where $\Phi=\sum_{i=1}^n\left(\frac{x^i}{f_i}\right)^2$.

Since $dx^n = d\sqrt{r^2 - \sum_a (x^a)^2} = \frac{rdr - \sum_a x^a dx^a}{x^n}$, and the volume element dV of M^n is $\sqrt{\det(g_{ij})}\, dx^1 \wedge \cdots \wedge dx^{n-1} \wedge dx^n$, we have from (3) the following Lemma 1.

Lemma 1. *On* S_r, *it holds that* $dV = \frac{r}{\sqrt{\Phi}} dh \wedge dr$.

Now suppose that φ is a harmonic map from M^n into any Riemannian manifold $(N^m, \tilde{g})$. The stress-energy tensor S of φ is a (1, 1)-type tensor with components $S_i^j = e(\varphi)\delta_i^j - \sum_{l,\alpha,\beta} g^{jl} \tilde{g}_{\alpha\beta}\varphi^\alpha_{,i}\varphi^\beta_{,l}$, where $e(\varphi)$ is the energy density of φ and $\alpha, \beta, \gamma = 1, \cdots, m$ (cf. [3]).

It is well known that the divergence of S vanishes, i.e.,

$$\sum_j S^j_{i,j} = 0. \tag{4}$$

Here the comma stands for the covariant derivative.

Lemma 2. *It holds that* $\sum_{i,j,k} \left\{ \begin{matrix} i \\ jk \end{matrix} \right\} x^k S_i^j = \sum_{i,k} x^k \frac{\partial \log f_i}{\partial x^k} S_i^i$, *where* $\left\{ \begin{matrix} i \\ jk \end{matrix} \right\}$ *is the second Christoffel symbol of* M^n.

Proof Since $g_{ij} = f_i^2 \delta_{ij}$, from computation we have

$$\left\{ \begin{matrix} i \\ jk \end{matrix} \right\} = \frac{1}{2}\left(\delta_j^i \frac{\partial \log f_i f_j}{\partial x^k} + \delta_k^i \frac{\partial \log f_i f_k}{\partial x^j} - \delta_{jk} \frac{1}{f_i^2} \frac{\partial f_j f_k}{\partial x^i}. \right) \tag{5}$$

Thus

$$\sum_{i,j,k} \left\{ \begin{matrix} i \\ jk \end{matrix} \right\} x^k S_i^j = \sum_{i,k} x^k \frac{\partial \log f_i}{\partial x^k} S_i^i + \sum_{i,j} \frac{x^i}{f_i} \frac{\partial f_i}{\partial x^j} S_i^j - \sum_{i,j} \frac{f_j}{f_i^2} \frac{\partial f_j}{\partial x^i} x^j S_i^j. \tag{6}$$

But, on the other hand, we have

$$\sum_{i,j} \frac{x^i}{f_i} \frac{\partial f_i}{\partial x^j} S_i^j = \frac{1}{2} e(\varphi) \sum_k \frac{x^k}{f_k} \frac{\partial f_k}{\partial x^k} - \sum_{i,j} \frac{x^i}{f_i f_j^2} \frac{\partial f_i}{\partial x^j} \tilde{g}_{\alpha\beta}\varphi^\alpha_{,i}\varphi^\beta_{,j}, \tag{7}$$

$$-\sum_{i,j}\frac{f_j}{f_i^2}\frac{\partial f_j}{\partial x^i}x^j S_i^j=-\frac{1}{2}e(\varphi)\sum_k\frac{x^k}{f_k}\frac{\partial f_k}{\partial x^k}+\sum_{i,j}\frac{x^j}{f_j f_i^2}\frac{\partial f_j}{\partial x^i}\tilde{g}_{\alpha\beta}\varphi^\alpha_{,j}\varphi^\beta_{,i}. \tag{8}$$

From (6), (7) and (8), the lemma is obvious.

Lemma 3. *If condition* (A) *and* (C) *are satisfied, then*

$$\sum_{i=1}^n(1+L(f_i))S_i^i\geqslant 0. \tag{9}$$

Proof For simplicity, we denote $a_i=(1+L(f_i))$. Thus, $a_i\geqslant 0$. Let p be a point in M^n. If, at the point p, $S_i^i\geqslant 0$ for any index i, then (9) holds obviously. Otherwise, since $S_i^i=\frac{1}{2}\Big(\sum_{j\neq i}\frac{1}{f_j^2}\tilde{g}_{\alpha\beta}\varphi^\alpha_{,j}\varphi^\beta_{,j}-\frac{1}{f_i^2}\tilde{g}_{\alpha\beta}\varphi^\alpha_{,i}\varphi^\beta_{,i}\Big)$, it is clear that $S_i^i<0$ holds only for one index i, say, $S_n^n<0$, and in this case we have

$$\frac{1}{f_n^2}\tilde{g}_{\alpha\beta}\varphi^\alpha_{,n}\varphi^\beta_{,n}>\sum_{i=1}^{n-1}\frac{1}{f_i^2}\tilde{g}_{\alpha\beta}\varphi^\alpha_{,i}\varphi^\beta_{,i}. \tag{10}$$

Without loss of generality, we can assume $0\leqslant a_1\leqslant\cdots\leqslant a_{n-1}$ at point p. Now, we have

$$\sum_{i=1}^n(1+L(f_i))S_i^i=\frac{1}{2}\Big\{\sum_{i=1}^{n-1}\Big(\sum_{j\neq i}\frac{a_i}{f_j^2}\tilde{g}_{\alpha\beta}\varphi^\alpha_{,j}\varphi^\beta_{,j}-\frac{a_i}{f_i^2}\tilde{g}_{\alpha\beta}\varphi^\alpha_{,i}\varphi^\beta_{,i}\Big)$$
$$+\sum_{j=1}^{n-1}\frac{a_n}{f_j^2}\tilde{g}_{\alpha\beta}\varphi^\alpha_{,j}\varphi^\beta_{,j}-\frac{a_n}{f_n^2}\tilde{g}_{\alpha\beta}\varphi^\alpha_{,n}\varphi^\beta_{,n}\Big\}. \tag{11}$$

In the case $a_n<a_{n-1}$, we have

$$\text{RHS } of \text{ (11)}=\frac{1}{2}\Big\{\sum_{i=1}^{n-2}\sum_{j\neq i}\frac{a_i}{f_j^2}\tilde{g}_{\alpha\beta}\varphi^\alpha_{,j}\varphi^\beta_{,j}+\underbrace{\sum_{j=1}^{n-2}\frac{a_{n-1}}{f_j^2}\tilde{g}_{\alpha\beta}\varphi^\alpha_{,j}\varphi^\beta_{,j}-\sum_{i=1}^{n-2}\frac{a_i}{f_i^2}\tilde{g}_{\alpha\beta}\varphi^\alpha_{,i}\varphi^\beta_{,i}}_{\geqslant 0}$$
$$\underbrace{-\frac{a_{n-1}}{f_{n-1}^2}\tilde{g}_{\alpha\beta}\varphi^\alpha_{,n-1}\varphi^\beta_{,n-1}+\sum_{j=1}^{n-1}\frac{a_n}{f_j^2}\tilde{g}_{\alpha\beta}\varphi^\alpha_{,j}\varphi^\beta_{,j}}_{\geqslant 0}$$

$$+\frac{a_{n-1}}{f_n^2}\tilde{g}_{\alpha\beta}\varphi^\alpha_{,n}\varphi^\beta_{,n}-\underset{\substack{\vee\\0}}{\frac{a_n}{f_n^2}}\tilde{g}_{\alpha\beta}\varphi^\alpha_{,n}\varphi^\beta_{,n}\Bigg\}$$

$$\geqslant\frac{1}{2}\left(\sum_{i=1}^{n-2}a_i-a_{n-1}\right)\frac{1}{f_{n-1}^2}\tilde{g}_{\alpha\beta}\varphi^\alpha_{,n-1}\varphi^\beta_{,n-1}\geqslant 0. \tag{12}$$

When $a_n \geqslant a_{n-1}$, we have

$$\text{RHS } of\ (11)=\frac{1}{2}\Bigg\{\sum_{i=1}^{n-1}\sum_{j\neq i,n}\frac{a_i}{f_j^2}\tilde{g}_{\alpha\beta}\varphi^\alpha_{,j}\varphi^\beta_{,j}-\underset{\substack{\vee\\0}}{\sum_{i=1}^{n-1}}\frac{a_i}{f_i^2}\tilde{g}_{\alpha\beta}\varphi^\alpha_{,i}\varphi^\beta_{,i}+\underset{\substack{\vee\\0}}{\sum_{j=1}^{n-1}}\frac{a_n}{f_j^2}\tilde{g}_{\alpha\beta}\varphi^\alpha_{,j}\varphi^\beta_{,j}$$

$$+\sum_{i=1}^{n-1}\frac{a_i}{f_n^2}\tilde{g}_{\alpha\beta}\varphi^\alpha_{,n}\varphi^\beta_{,n}-\frac{a_n}{f_n^2}\tilde{g}_{\alpha\beta}\varphi^\alpha_{,n}\varphi^\beta_{,n}\Bigg\}$$

$$\geqslant\frac{1}{2}\left(\sum_{i=1}^{n-1}a_i-a_n\right)\frac{1}{f_n^2}\tilde{g}_{\alpha\beta}\varphi^\alpha_{,n}\varphi^\beta_{,n}\geqslant 0. \tag{13}$$

Thus the lemma is proved.

From the above proof, it is not difficult to prove the following lemma.

Lemma 4. *If conditions* (A) *and* (C) *are satisfied, then* $\sum_i(1+L(f_i))S^i_i\equiv 0$ *holds if and only if* φ *is a constant map.*

§ 3. Proof of Main Theorem

In this section, we use Einstein summation convention. Let $B_r=\{x\in M^n \mid \sum_i(x^i)^2\leqslant r^2\}$. From (4), we have

$$0=\int_{B_r}x^iS^j_{i,j}dV=\int_{B_r}(x^iS^j_i)_{,j}dV-\int_{B_r}\left[\frac{\partial x^i}{\partial x^j}+\begin{Bmatrix} j\\ jk\end{Bmatrix}x^k\right]S^j_idV. \tag{14}$$

Noting that the unit outward normal vector of S_r is

$$W=\frac{1}{r}\sum_{i=1}^{n}\frac{x^i}{f_i}\frac{\partial}{\partial x^i},$$

and using the integral formula

$$\int_{B_r}\operatorname{div}X\,dV=\int_{S_r}\langle X,W\rangle dh$$

and Lemma 2, (14) is reduced to

$$0=\frac{1}{r}\int_{S_r}\sum_{i,j}f_jS_i^jx^ix^jdh-\int_{B_r}\sum_i(1+L(f_i))S_i^idV. \tag{15}$$

By using Schwartz inequality, we have

$$\begin{aligned}\sum_{i,j}f_jS_i^jx^ix^j&=e(\varphi)\sum_j f_j(x^j)^2-\sum_{j,k}\frac{1}{f_j}\tilde{g}_{\alpha\beta}(\varphi^\alpha_{,j}x^j)(\varphi^\beta_{,k}x^k)\\&\leqslant e(\varphi)\sum_j f_j(x^j)^2\leqslant e(\varphi)r^2\sqrt{\sum_i f_i^2}.\end{aligned} \tag{16}$$

From (15) and (16), we obtain

$$r\int_{Sr}e(\varphi)\sqrt{\sum_i f_i^2}dh\geqslant\int_{B_r}\sum_i(1+L(f_i))S_i^idV. \tag{17}$$

If φ is not a constant, from Lemma 3 and Lemma 4, we claim that there exist two positive numbers R_0 and ε such that, for $r\geqslant R_0$,

$$\int_{B_r}\sum_i(1+L(f_i))S_i^idV>\varepsilon. \tag{18}$$

Let $\psi(r)$ be a positive continuous function of r satisfying

$$\int_a^\infty\frac{dr}{r\psi(r)}=\infty\quad(\text{for a certain constant } a>0). \tag{19}$$

In consequence of Lemma 1 and (18), integrating (17), we have

$$\int_{R_0}^R\frac{\varepsilon}{r\psi(r)}dr\leqslant\int_{R_0}^R\int_{S_r}\frac{e(\varphi)}{\psi(r)}\sqrt{\sum_i f_i^2}dhdr\leqslant\int_0^R\int_{S_r}\frac{e(\varphi)}{\psi(r)}\sqrt{\sum_i f_i^2}dhdr$$

$$=\int_0^R\int_{S_r}\frac{e(\varphi)}{\psi(r)}\frac{\sqrt{\sum_i f_i^2}\sqrt{\Phi}}{r}\cdot\frac{r}{\sqrt{\Phi}}dh\,dr$$

$$=\int_{B_R}\frac{e(\varphi)}{\psi(r)}\frac{\sqrt{\sum_i f_i^2}\sqrt{\Phi}}{r}dV,\ \text{for } R>R_0. \tag{20}$$

Furthermore, from condition (B), we have

$$\frac{\sqrt{\sum_i f_i^2}\sqrt{\Phi}}{r}=\sqrt{\left(\sum_i\left(\frac{f_i}{r}\right)^2\left(\sum_j\left(\frac{x^j}{f_j}\right)^2\right)\right)}\leqslant\sqrt{\left(\sum_i f_i^2\right)\left(\sum_j\frac{1}{f_j^2}\right)}\leqslant\sqrt{n^2K^2}\leqslant nK.$$

Thus, (20) reduces to

$$\int_{R_0}^R\frac{\varepsilon}{r\psi(r)}dr\leqslant nK\int_{B_R}\frac{e(\varphi)}{\psi(r)}dV,\ \text{for } R>R_0. \tag{21}$$

Letting $R\to\infty$ in (21), the left hand side of (21) approaches infinite, but the right hand side of (21) is finite, since φ is of finite energy or with slowly divergent energy. This contradiction proves our theorem.

References

[1] Hu Hesheng (H.S. Hu), An Nonexistence Theorem for Harmonic Maps with Slowly Divergent Energy, *Chin. Ann. of Math.*, **5B**: 4 (1984), 737-740.

[2] Sealey, H.C.J., Some Conditions Ensuring the Vanishing of Harmonic Differential Forms with Applications to Harmonic Maps and Yang-Mills Theory, *Math. Proc. Camb. Phil. Soc.*, **91** (1982), 441-452.

[3] Baird P., and Eells, J., A Conservation Law for Harmonic Maps, Lecture Notes in Mathematics 838, Springer Verlag.

(本文曾发表于 *Chinese Annals of Mathematics*, 1986, 7B(3), 345-349)

Harmonic Maps and a Pinching Theorem for Positively Curved Hypersurfaces

H. S. Hu, Y. L. Pan and Y. B. Shen*

Abstract In this paper, we establish a theorem of Liouville type for stable harmonic maps in sufficiently pinched, positively curved hypersurfaces of a space form with nonnegative constant curvature. Similar results for the Euclidean sphere S^n have been proved by Y. L. Xin and P. F. Leung, respectively.

1. Statement of result. As is well known, for the Euclidean $n(\geqslant 3)$-sphere S^n, there are no nonconstant stable harmonic maps either from S^n to any Riemannian manifold [**6**] or from any compact manifold to S^n [**3**]. Its generalizations to submanifolds have been attempted by many authors [**3**, **4**, **5**], where the additional conditions imposed on submanifolds are extrinsic. It is natural to find some condition intrinsic for submanifolds. In this paper, we prove the following theorem.

THEOREM. *Let $\widetilde{M}^{n+1}(c)$ be an $(n+1)$-dimensional simply connected space form with constant sectional curvature $c \geqslant 0$. Suppose that M^n $(n \geqslant 3)$ is a compact hypersurface in $\widetilde{M}^{n+1}(c)$ of which the sectional curvature* Riem^M *satisfies the following pinching condition:*

* Received by the editors October 1, 1985.
1980 *Mathematics Subject Classification* (1985 *Revision*). Primary 58E20; Secondary 53C40.
Key words and phrases. Harmonic map, hypersurface, stable, pinching theorem.

(1) $$c+a^2/[(n-2)c+(n-1)a]\leqslant \mathrm{Riem}^M < c+a$$

for some constant $a>0$. *Then there is no nonconstant stable harmonic map between* M^n *and any compact Riemannian manifold* N, *i.e., either from* M^n *to* N *or from* N *to* M^n.

COROLLARY 1. *If* M^n $(n\geqslant 3)$ *is a compact hypersurface in the Euclidean space* E^{n+1} *with pinching condition:* $a/(n-1)\leqslant \mathrm{Riem}^M < a$ *for some constant* $a>0$, *then there is no nonconstant stable harmonic map between* M^n *and any compact Riemannian manifold.*

COROLLARY 2. *If* M^n $(n>4)$ *is a compact hypersurface in* E^{n+1} *with* $\frac{1}{4}\leqslant \mathrm{Riem}^M<1$, *then the same conclusion as in Corollary 1 holds.*

REMARK. The condition (1) is sharp in some sense; see an example below.

2. Preliminaries. Let M and N be Riemannian manifolds of dimensions n and m, respectively, and ϕ: $M\to N$ a smooth map. Throughout this paper, we agree on the following ranges of indices unless otherwise stated:

$$1\leqslant i, j, k, \dots\leqslant n; \qquad 1\leqslant \alpha, \beta, \gamma, \dots \leqslant m.$$

We choose local fields of orthonormal frames $\{e_i\}$ and $\{e'_\alpha\}$ in M and N, respectively, and let $\{\omega_i\}$ and $\{\omega'_\alpha\}$ be the fields of dual frames. Under the map ϕ, we have $\phi^*\omega'_\alpha=\sum_i a_{\alpha i}\omega_i$. The energy of ϕ is defined by

$$E(\phi)=\frac{1}{2}\int_M \sum_{\alpha, i}(a_{\alpha i})^2 * 1,$$

where $*1=\omega_1\wedge\cdots\wedge\omega_n$. The map ϕ is *harmonic* if its tension field $\tau=\sum_{\alpha, i} a_{\alpha i i}e'_\alpha$ vanishes identically, where $a_{\alpha ij}$ is the covariant derivative of $a_{\alpha i}$ [**1**].

For any deformation vector field along ϕ, $V=\sum_\alpha v_\alpha e'_\alpha$, the second variation of $E(\phi)$ is [**2**, **4**]

(2) $$E''(\phi)=-\int_M\Big[\sum_\alpha v_\alpha\big(\Delta_M v_\alpha+\sum_{i,\beta,\gamma,\delta}v_\delta a_{\beta i}a_{\gamma i}R'_{\beta\alpha\gamma\delta}\big)\Big]*1,$$

where $R'_{\alpha\beta\gamma\delta}$ is the curvature tensor of N and Δ_M is the Laplacian on M. If $E''(\phi)\geqslant 0$ for any V along ϕ, then the harmonic map ϕ is said to be *stable*.

Now let $\widetilde{M}^{n+1}(c)$ be an $(n+1)$-dimensional simply connected space form with constant sectional curvature $c\geqslant 0$, and M^n a hypersurface in $\widetilde{M}^{n+1}(c)$. We choose a local field of orthonormal frames $e_1,\ldots,e_{n+1}$ in $\widetilde{M}^{n+1}(c)$ such that, restricted to M^n, the vector e_{n+1} is normal to M^n. Let $\omega_1,\ldots,\omega_{n+1}$ be the field of dual frames relative to the frame field chosen above. Then, restricted to M, we have [**1**]

$$\omega_{n+1}=0,\ \omega_{n+1,i}=\sum_j h_{ij}\omega_j,\quad h_{ij}=h_{ji}.$$

The Gauss equation of M^n in $\widetilde{M}^{n+1}(c)$ is

(3) $$R_{ijkl}=c(\delta_{ik}\delta_{jl}-\delta_{il}\delta_{jk})+h_{ik}h_{jl}-h_{il}h_{jk},$$

where R_{ijkl} is the curvature tensor of M^n.

By [**4** and **5**], we have the following lemma, whose proof is omitted here.

LEMMA 1. *Let M^n $(n\geqslant 3)$ be a compact hypersurface in $\widetilde{M}^{n+1}(c)$ with $c\geqslant 0$. If the second fundamental tensor h_{ij} of M^n satisfies pointwisely*

(4) $$\sum_k(2h_{ik}h_{jk}-h_{kk}h_{ij})<(n-2)c\delta_{ij}$$

for an orthonormal basis $\{e_i\}$ on M^n, then there is no nonconstant stable harmonic map between M^n and any compact Riemannian manifold.

3. The proof of the Theorem. Let P be an arbitrary point in M^n and $\{\lambda_i\}$ eigenvalues of the matrix (h_{ij}). It follows from (3) that

(5) $$R_{ijij}=c+\lambda_i\lambda_j\qquad(i\neq j)$$

at $P\in M$. The pinching condition (1) together with (5) implies that all of $\{\lambda_i\}$ are nonzero and have the same sign. So, without loss of generality, we may assume that at P

(6) $$0 < \lambda_i \leqslant \lambda_2 \leqslant \cdots \leqslant \lambda_n$$

and set

(7) $$H = \sum_i \lambda_i.$$

Thus, (4) becomes

$$2\lambda_i^2 - H\lambda_i - (n-2)c < 0$$

for all i, which together with (6) yields that

$$\lambda_i \leqslant \lambda_n < \frac{1}{4}[H + \sqrt{H^2 + 8(n-2)c}\,].$$

Hence, we have the following lemma.

LEMMA 2. *Under the assumption* (6), *the condition* (4) *is equivalent to*

(8) $$\lambda_n < \frac{1}{4}[H + \sqrt{H^2 + 8(n-2)c}\,]$$

at $P \in M$.

We now suppose that

(9) $$b^2 \leqslant \lambda_i \lambda_j < B^2 \qquad (i \neq j)$$

for some $0 < b < B$ at $P \in M$.

LEMMA 3. *If* $\lambda_1 \geqslant b$ *and* B, b *in* (9) *satisfy*

(10) $$B^2 - Lb + (n-1)b^2 = 0,$$

where

(11) $$L = \frac{1}{2}[3(n-1)b + \sqrt{(n-1)^2 b^2 + 4(n-2)c}\,],$$

then the inequality (8) *holds.*

Proof. Since $\lambda_1 \geqslant b$, it follows from (9) and (10) that

$$\lambda_n < B^2/\lambda_1 \leqslant B^2/b = L - (n-1)b. \tag{12}$$

On the other hand, we have from (11)

$$[3L - 4(n-1)b]^2 = L^2 + 8(n-2)c,$$

i.e.,

$$L - (n-1)b = \frac{1}{4}[L + \sqrt{L^2 + 8(n-2)c}]. \tag{13}$$

Substituting (13) into (12), we get

$$\lambda_n < \frac{1}{4}[L + \sqrt{L^2 + 8(n-2)c}]. \tag{14}$$

If $H \geqslant L$, (14) implies (8). So we need only consider the case that $H < L$. Putting

$$L - H = K > 0, \tag{15}$$

we see from (6), (7) and (15)

$$\lambda_n = H - \sum_{j \neq n} \lambda_j \leqslant H - (n-1)b = L - (n-1)b - K,$$

which together with (13) yields

$$\lambda_n \leqslant \frac{1}{4}[H + \sqrt{L^2 + 8(n-2)c} - 3K]. \tag{16}$$

On the other hand, from the fact that $K > 0$ and $c \geqslant 0$ it is easy to see that

$$[\sqrt{(L-K)^2 + 8(n-2)c} + 3K]^2 > L^2 + 8(n-2)c,$$

i.e.,

$$\sqrt{H^2 - 8(n-2)c} > \sqrt{L^2 + 8(n-2)c} - 3K. \tag{17}$$

Substituting (17) into (16), we complete the proof of Lemma 3.

LEMMA 4. *If* $n \geqslant 3$, *and* (10) *holds, then* (8) *is true.*

PROOF. By Lemma 3, we need only consider the case that $\lambda_1 < b$. Since $n \geqslant 3$, there exist λ_1 and $\lambda_2 (\leqslant \lambda_3)$ such that, by (9), $\lambda_1 \lambda_2 \geqslant b^2$.

Construct

$$\lambda'_1 = \lambda'_2 = \frac{1}{2}(\lambda_1 + \lambda_2). \tag{18}$$

Then,

$$(\lambda'_1)^2 = \frac{1}{4}(\lambda_1 + \lambda_2)^2 \geqslant \lambda_1 \lambda_2 \geqslant b^2,$$

so that $\lambda'_1 \geqslant b$. Since (18) implies that $\lambda'_1 = \lambda'_2 \leqslant \lambda_3$ and $\lambda'_1 + \lambda'_2 + \sum_{k=3}^{n} \lambda_k = H$, then we can apply Lemma 3 to the case where $0 < \lambda'_1 = \lambda'_2 \leqslant \lambda_3 \leqslant \cdots \leqslant \lambda_n$ and conclude that (8) is true. Thus, Lemma 4 is proved completely.

Now, we can easily give the proof of the Theorem.

It is clear, by (5), that (1) becomes

$$a^2/[(n-2)c + (n-1)a] \leqslant \lambda_i \lambda_j < a \qquad (i \neq j) \tag{19}$$

at $P \in M$.

Taking $b = a/\sqrt{(n-2)c + (n-1)a}$, we have

$$a = \frac{1}{2} b[(n-1)b + \sqrt{(n-1)^2 b^2 + 4(n-2)c}\,]. \tag{20}$$

Combining (19) with (20), we obtain (9) and (10) at P. By Lemmas 2 and 3, we conclude that (4) holds at P. Since $P \in M$ is arbitrary, the Theorem follows immediately from Lemma 1.

EXAMPLE. In order to employ Lemma 1, the inequality $\mathrm{Riem}^M < c + a$ in (1) cannot be replaced by $\mathrm{Riem}^M \leqslant c + a$. In such a sense, our pinching condition (4) is sharp. In fact, we may set

$$(21)\quad \lambda_1=\cdots=\lambda_{n-1}=b,\qquad \lambda_n=\frac{1}{2}[(n-1)b+\sqrt{(n-1)^2b^2+4(n-2)c}\,],$$

where

$$b=a/\sqrt{(n-2)c+(n-1)a}$$

for some constant $a>0$. In this case, we have (6) and

$$(22)\quad b^2\leqslant\lambda_i\lambda_j\leqslant\frac{1}{2}b[(n-1)b+\sqrt{(n-1)^2b^2+4(n-2)c}\,]\qquad(i\neq j)$$

which together with (5) implies that

$$c+a^2/[(n-2)c+(n-1)a]\leqslant\mathrm{Riem}^M\leqslant c+a.$$

On the other hand, from (21) and (11) we can see that

$$H=(n-1)b+\lambda_n=\frac{1}{2}[3(n-1)b+\sqrt{(n-1)^2b^2+4(n-2)c}\,]=L,$$

which together with (13) yields that

$$\lambda_n=\frac{1}{4}[H+\sqrt{H^2+8(n-2)c}\,],$$

i.e., (8) is not true. Thus, Lemma 1 cannot be applied to this case.

Acknowledgement

The third-named author would like to thank Professor S. S. Chern and the Mathematical Sciences Research Institute for their hospitality.

References

[1] S. S. Chern and S. I. Goldberg, *On the volume decreasing property of a class of real harmonic mappings*, Amer. J. Math. **97** (1975), 133-147.

[2] J. Eells and L. Lemaire, *A report on harmonic maps*, Bull. London Math. Soc. **10** (1978), 1-68.

[3] P. F. Leung, *On the stability of harmonic maps*, Lecture Notes in Math., vol.949, Springer-Verlag, pp. 122-129.

[4] Y. L. Pan and Y. B. Shen, *Stability of harmonic maps and minimal immersions*, Proc. Amer. Math. Soc. **93** (1985), 111-117.

[5] Y. L. Xin, *Topology of certain submanifolds in the Euclidean sphere*, Proc. Amer. Math. Soc. **82** (1981), 643-648.

[6] ——, *Some results on stable harmonic maps*, Duke Math. J. **47** (1980), 609-613.

Department of Mathematics, Fudan University, Shanghai,
People's Republic of China (Current address of H. S. Hu and Y. L. Pan)
Department of Mathematics, Hangzhou University, Hangzhou,
People's Republic of China (Current address of Y. B. Shen)

(本文曾发表于 *Proceedings of the American Mathematical Society*, 1987, 99 (1), 182-186)

Nonexistence Theorems for Yang-Mills Fields and Harmonic Maps in the Schwarzschild Spacetime

Hu Hesheng*

Forschungsschwerpunkl Geometrie, Universität Heidelberg, Mathematisches Institut, SFB 123, IM Neuenheimer Feld 254, D-6900 Heidelberg, West Germany

(Received: 10 July 1987)

Abstract The nonexistence of static solutions to pure Yang-Mills equations and nonconstant harmonic maps defined on the Schwarzschild spacetime outside the black hole ($r > 2M$) is considered. Nonexistence theorems for pure Yang-Mills equations and harmonic maps in the region $r \geqslant 5M$ and $r \geqslant 3M$ are obtained, respectively.

1. Introduction

There are several nonexistence theorems in physical theories. As is well-known, there are no well-behaved source-free static solutions to the Maxwell or Einstein equations in four dimensions. Furthermore, for the Yang-Mills (YM) theory [1], it has been proved by S. Deser that there are no regular static finite energy solutions going to zero fast enough at spatial infinity in the $n(n\neq 5)$-dimensional Minkowski space $R^{n-1,1}$[2]. The author found that the energy condition can be weakened to a 'slowly divergent'

---

* Permanent address: Institute of Mathematics, Fudan University, Shanghai, People's Republic of China.
Research supported by the Chinese Fund, the Institute of Physical Mathematics of Dijon University, France, and the Institute of Mathematics, Heidelberg University, West Germany.

energy case and the condition at spatial infinity can be removed [3,4]. For the Yang-Mills fields with mass, we have a similar result when $n \neq 4$ [4 - 7]. It is natural to consider such kind of problems in a curved spacetime [4,10].

In this Letter, we consider the nonexistence of the static solutions to pure Yang-Mills equations and nonconstant harmonic maps defined on the Schwarzschild spacetime outside the black hole ($r > 2M$). We obtain nonexistence theorems for pure Yang-Mills fields and harmonic maps in the region $r \geqslant 5M$ and $r \geqslant 3M$, respectively. It is reasonable to have the following conjecture: the above nonexistence theorems in the Schwarzschild spacetime will hold true also when $r \geqslant 2M$. It should be noted that a static YM field or static harmonic map is meaningful only when $r \geqslant 2M$ for Schwarzschild spacetime.

In Section 2 we list our notations and state the main results. Sections 3 and 4 are devoted to the proof of the case of Yang-Mills fields and the case of harmonic maps, respectively.

2. Notations and Results

In this Letter we consider the static solutions of Yang-Mills fields and nonlinear σ-models in the Schwarzschild spacetime.

2.1 Schwarzschild Spacetime

The metric of the Schwarzschild spacetime V_4 is written as

$$ds^2 = g_{\alpha\beta} dx^\alpha dx^\beta = -\left(1 - \frac{2M}{r}\right) dt^2 + \frac{1}{1-(2M/r)} dr^2 + r^2 (d\theta^2 + \sin^2\theta d\varphi^2)$$

$$(\alpha, \beta = 0, 1, 2, 3), \tag{1}$$

where $r = 2M$ is the Schwarzschild radius, the region inside the sphere with the

Schwarzschild radius is the black hole.

The volume element of the sphere in (1) is

$$\mathrm{d}V = \frac{r^2 \sin\theta}{\sqrt{1-(2M/r)}} \mathrm{d}r\,\mathrm{d}\theta\,\mathrm{d}\varphi, \tag{2}$$

whereas the area element of the spherical surface is

$$\mathrm{d}S = r^2 \sin\theta\,\mathrm{d}\theta\,\mathrm{d}\varphi. \tag{3}$$

2.2 Yang-Mills Fields

Let G be a Lie group usually being compact and linear, and g its Lie algebra. A gauge field over a Riemannian (or Lorentzian) manifold M, mathematically is defined by a connection on the principal bundle $P(M, G)$. Here, M is V_A with metric (1).

The gauge potential b (connection) and field strength F (curvature) of YM fields are denoted by

$$b = b_\lambda(x)\mathrm{d}x^\lambda \tag{4}$$

and

$$F = \frac{1}{2} f_{\lambda\mu} \mathrm{d}x^\lambda \wedge \mathrm{d}x^\mu \quad (\lambda, \mu = 0, 1, 2, 3), \tag{5}$$

respectively. Here

$$f_{\lambda\mu} = \frac{\partial b_\lambda}{\partial x^\mu} - \frac{\partial b_\mu}{\partial x^\lambda} - [b_\lambda, b_\mu]. \tag{6}$$

b and F are valued in g. For convenience, we use the notation x^0, x^1, x^2, x^3 instead of t, r, θ, φ.

The action integral of YM theory is defined as

$$L = -\frac{1}{4}\int (f_{\lambda\mu}, f^{\lambda\mu})\mathrm{d}V. \tag{7}$$

The Euler equations of the YM functional (7) are the YM equations

$$J_\alpha = g^{\lambda\mu}(f_{\alpha\lambda;\mu} + [b_\mu, f_{\alpha\lambda}]) = 0, \tag{8}$$

where ; denotes the covariant derivative w.r.t. the metric (1). If $F=0$, the field is called trivial.

The energy momentum tensor of YM fields is

$$T_{\alpha\beta} = (f_{\alpha\nu}, f_{\beta\varepsilon})g^{\nu\varepsilon} - \frac{1}{4}g_{\alpha\beta}(f_{\mu\nu}, f_{\gamma\varepsilon})g^{\mu\gamma}g^{\nu\varepsilon}. \tag{9}$$

If

$$\int_{r>R_0} T_{00}\,\mathrm{d}V < \infty, \tag{10}$$

we say the energy of the field on the region $r > R_0$ is finite. If

$$\int_{r>R_0} T_{00}\,\mathrm{d}V = \infty \tag{11}$$

and

$$\int_{r>R_0} \frac{T_{00}}{\psi(r)}\,\mathrm{d}V < \infty, \tag{12}$$

for a certain positive, continuous and unbounded function $\psi(r)$, satisfying

$$\int_{R_0}^{\infty} \frac{\mathrm{d}r}{r\psi(r)} = \infty. \tag{13}$$

We say the energy is slowly divergent.

2.3 Harmonic Maps

Let M, N be Riemannian manifolds or Lorentz manifolds and $\phi: M \rightarrow N$ a C^2-map. In this Letter, M is the Schwarzschild spacetime outside the black hole.

The energy integral of the map ϕ is

$$E(\phi)=\int_M e(\phi)\mathrm{d}V_M, \tag{14}$$

where

$$e(\phi)=\underset{N}{g}_{st}(\phi)\frac{\partial\phi^s}{\partial x^\alpha}\frac{\partial\phi^t}{\partial x^\beta}\underset{M}{g}^{\alpha\beta}(x),\quad \alpha,\ \beta=0,\ 1,\ 2,\ 3;\ s,\ t=1,\ \ldots,\ n, \tag{15}$$

is the expression of the energy density in local coordinates.

A harmonic map ϕ is a solution to the Euler equation of $E(\phi)$

$$\underset{M}{g}^{\alpha\beta}\left(\frac{\partial^2\phi^s}{\partial x^\alpha\partial x^\beta}-\underset{M}{\Gamma}{}^{\gamma}_{\alpha\beta}\frac{\partial\phi^s}{\partial x^\gamma}+\underset{N}{\Gamma}{}^{s}_{tp}\frac{\partial\phi^t}{\partial x^\alpha}\frac{\partial\phi^p}{\partial x^\beta}\right)=0. \tag{16}$$

The stress-energy tensor of the harmonic map is

$$T_{\alpha\beta}=\frac{1}{2}\underset{N}{g}_{st}\phi^s_\gamma\phi^t_\delta\underset{M}{g}^{\gamma\delta}\underset{M}{g}_{\alpha\beta}-\underset{N}{g}_{st}\phi^s_\alpha\phi^t_\beta=\frac{1}{2}e(\phi)\underset{M}{g}_{\alpha\beta}-\underset{N}{g}_{st}\phi^s_\alpha\phi^t_\beta. \tag{17}$$

Here ϕ^s_α denotes $\partial\phi^s/\partial x^\alpha$.

If N is a Lie group, or more generally, a homogeneous manifold, then physically, ϕ is called a nonlinear σ-model or Chiral field. We will consider the static harmonic map from Schwarzschild spacetime outside the black hole to any Riemannian manifold N.

Some nonexistence theorems have been obtained for a finite energy and slowly divergent energy case [8 - 10].

2.4 Results

Theorem 1. *Let G be a compact Lie group. The pure YM equations on the Schwarzschild spacetime (with $r\geqslant 5M$) do not admit any nontrivial regular static solution with finite or 'slowly divergent' energy, and satisfying the boundary condition*

$$\int_{r=5M}T^1_1\mathrm{dS}\geqslant 0\quad \textit{when } r=5M. \tag{18}$$

For the harmonic maps from Schwarzschild spacetime, we have

Theorem 2. *Let ϕ be a static harmonic map from the Schwarzschild spacetime (with $r\geqslant 3M$) into any n-dimensional Riemannian manifold N. If the energy of ϕ is finite or slowly divergent and the boundary value on $r=3M$ satisfies*

$$\int_{r=3M} T_1^1 \mathrm{d}S \geqslant 0, \tag{19}$$

then ϕ must be a constant map. Here the term 'static' means the map ϕ does not depend on 'time' t.

In proving the above theorems, we use the following two Lemmas.

LEMMA 1. *If $T_{\alpha\beta}$ is any symmetric tensor on a m-dimensional Riemannian or Lorentzian manifold M and satisfies the conservation law*

$$T^{\alpha}_{\beta;\,\alpha}=0, \tag{20}$$

then for any fixed j

$$\begin{gathered}(T^i_j x^j)_{,\,i}=-T^0_{j;\,0}x^j+(-T^{\lambda}_j\Gamma^i_{\lambda i}+T^i_{\lambda}\Gamma^{\lambda}_{ji})x^j+T^j_j \\ (\lambda=0,\,1,\,\ldots,\,m-1;\,i,\,j=1,\,\ldots,\,m-1)\end{gathered} \tag{21}$$

holds true in any coordinate system. Here j is not a summation index.

Proof. By using the definition of covariant derivatives and the conservation law of T^{α}_{β}, it is easy to get (21).

LEMMA 2. *If $T_{\alpha\beta}$ is a static symmetric tensor defined on a region $t=const$, $R_0\leqslant r\leqslant R$ of the Schwarzschild spacetime and satisfies the conservation law (20), then we have*

$$\begin{aligned}&\int_{r=R} T^1_1 \frac{r}{\sqrt{1-(2M/r)}}\mathrm{d}S-\int_{r=R_0} T^1_1 \frac{r}{\sqrt{1-(2M/r)}}\mathrm{d}S \\ &=\int_{R_0\leqslant r\leqslant R}(T^0_0-2T^1_1)\frac{M}{r-2M}\mathrm{d}V+\int_{R_0\leqslant r\leqslant R}T^i_i\,\mathrm{d}V.\end{aligned} \tag{22}$$

Here i is a summation index and

$$T_i^j = T_{il} g^{lj}. \tag{23}$$

Proof. We put $j=1$ in Equation (21) of Lemma 1. By using the assumption that $T_{\alpha\beta}$ is static, we have

$$(T_1^i r)_{,i} = (-T_1^\lambda \Gamma_{\lambda 0}^0 + T_\lambda^0 \Gamma_{10}^\lambda - T_1^\lambda \Gamma_{\lambda i}^i + T_\lambda^i \Gamma_{1i}^\lambda) r + T_1^1, \tag{24}$$

where, i denotes the derivative with respect to x^i.

Integrate the left-hand side of (24) on $x^0=$const., in which θ integrates from 0 to π, φ integrates from 0 to 2π. Considering the periodic property of the integrands, we have

$$\iint \left(T_1^1 \frac{r^3 \sin\theta}{\sqrt{1-(2M/r)}} \right) \Bigg|_{r=R} \mathrm{d}\theta \mathrm{d}\varphi - \iint \left(T_1^1 \frac{r^3 \sin\theta}{\sqrt{1-(2M/r)}} \right) \Bigg|_{r=R_0} \mathrm{d}\theta \mathrm{d}\varphi -$$
$$\iiint T_1^1 \left(\frac{r^2}{\sqrt{1-(2M/r)}} \right)_{,r} r \sin\theta \mathrm{d}r \mathrm{d}\theta \mathrm{d}\varphi - \iiint T_1^2 r \cot\theta \mathrm{d}V. \tag{25}$$

For integration of the right-hand side of (24), by the property of the Christoffel symbol of Schwarzschild spacetime, we obtain

$$\iiint (-T_1^1 + T_0^0) \frac{M}{r-2M} \mathrm{d}V + \iiint (-T_1^2 r \cot\theta) \mathrm{d}V +$$
$$\iiint [(T_1^1 + T_2^2 + T_3^3) - 2T_1^1] \mathrm{d}V. \tag{26}$$

Since expression (25) is equal to expression (26), we obtain the identity (22). Lemma 2 is proved.

3. Nonexistence Theorem for Static Yang-Mills Fields in Schwarzschild Spacetime outside the Black Hole

For the pure YM fields, the energy momentum tensor $T_{\alpha\beta}$ is expressed in the form (9).

Hence, we have

$$T_0^0 = g^{00} T_{00} = (f_{0\nu}, f^{0\nu}) - \frac{1}{4}(f_{\mu\nu}, f^{\mu\nu}), \tag{27}$$

$$T_\alpha^\alpha = 0, \tag{28}$$

$$T_i^i = -T_0^0 = -(f_{0\nu}, f^{0\nu}) + \frac{1}{4}(f_{\mu\nu}, f^{\mu\nu}), \tag{29}$$

$$T_1^1 = g^{11}(f_{1\nu}, f_{1\varepsilon}) g^{\nu\varepsilon} - \frac{1}{4}(f_{\mu\nu}, f^{\mu\nu}). \tag{30}$$

Using (27)-(30) the right side of (22)

$$\begin{aligned}
&\int_{R_0 \leqslant r \leqslant R} (T_0^0 - 2T_1^1) \frac{M}{r-2M} \mathrm{d}V + \int_{R_0 \leqslant r \leqslant R} T_i^i \mathrm{d}V \\
&= \int \left[(f_{0\nu}, f^{0\nu}) - \frac{1}{4}(f_{\mu\nu}, f^{\mu\nu}) - 2g^{11}(f_{1\nu}, f_{1\varepsilon}) g^{\nu\varepsilon} + \frac{1}{2}(f_{\mu\nu}, f^{\mu\nu}) \right] \times \\
&\quad \frac{M}{r-2M} \mathrm{d}V + \int \left[(-f_{0\nu}, f^{0\nu}) + \frac{1}{4}(f_{\mu\nu}, f^{\mu\nu}) \right] \mathrm{d}V \\
&= \int \left[\frac{-r+3M}{r-2M}(f_{0\nu}, f^{0\nu}) + \frac{r-M}{4(r-2M)}(f_{\mu\nu}, f^{\mu\nu}) - \right. \\
&\quad \left. 2g^{11}(f_{1\nu}, f_{1\varepsilon}) g^{\nu\varepsilon} \frac{M}{r-2M} \right] \mathrm{d}V \\
&= \int_{R_0 \leqslant r \leqslant R} \left[\frac{r-M}{4(r-2M)}(f_{ab}, f^{ab}) + \frac{r-5M}{2(r-2M)}(f_{1a}, f_{1a}) g^{11} g^{aa} + \right. \\
&\quad \left. \frac{r(r-5M)}{2(r-2M)^2}(f_{0a}, f_{0a}) g^{aa} + \frac{r-M}{2(r-2M)}(f_{10}, f_{10}) \right] \mathrm{d}V,
\end{aligned} \tag{31}$$

where a, $b = 2, 3$ are summation indices. Hence, when $r \geqslant 5M$ the right-hand side of (31) is $\geqslant 0$, and the equality holds only if $f_{\mu\nu} = 0$.

Writing the integrand of the right-hand side of (31) as $\widetilde{T}(r, f]$, then (22) becomes

$$0=\int_{r=R} T_1^1 \frac{r}{\sqrt{1-(2M/r)}}\mathrm{d}S-\int_{r=R_0} T_1^1 \frac{r}{\sqrt{1-(2M/r)}}\mathrm{d}S-$$

$$\int_{R_0\leqslant r\leqslant R} \widetilde{T}[r,f]\mathrm{d}V. \tag{32}$$

If $f_{\mu\nu}\neq 0$, there exist constants $R_1>0$ and $\varepsilon>0$ such that

$$\int_{R_0\leqslant r\leqslant R_2} \widetilde{T}[r,f]\mathrm{d}V>\varepsilon \quad (\text{for all } R_2\geqslant R_1). \tag{33}$$

Choosing

$$\omega(R)=\begin{cases} 0, & R_0\leqslant R<R_1\ (R_0\geqslant 5M),\\ \dfrac{1}{R\psi(R)}, & R_1\leqslant R\leqslant R_2,\\ 0, & R>R_2,\end{cases} \tag{34}$$

where $\psi(R)$ has been defined above (see (12), (13)).

Multiplying (32) by $\omega(R)$ and integrating it, we obtain

$$0=\int_{R_1}^{R_2}\omega(R)\mathrm{d}R\int_{r=R} T_1^1 \frac{r}{\sqrt{1-(2M/r)}}\mathrm{d}S-$$

$$\int_{R_1}^{R_2}\omega(R)\mathrm{d}R\int_{r=R_0} T_1^1 \frac{r}{\sqrt{1-(2M/r)}}\mathrm{d}S-$$

$$\int_{R_1}^{R_2}\omega(R)\mathrm{d}R\int_{R_0\leqslant r\leqslant R} \widetilde{T}[r,f]\mathrm{d}V. \tag{35}$$

Then we have

$$0=\int_{R_1}^{R_2}\frac{\mathrm{d}R}{\psi(R)}\int_{r=R}\frac{T_1^1}{\sqrt{1-(2M/r)}}\mathrm{d}S-\int_{R_1}^{R_2}\omega(R)\mathrm{d}R\int_{r=R_0} T_1^1\times$$

$$\frac{r}{\sqrt{1-(2M/r)}}\mathrm{d}S-\int_{R_1}^{R_2}\frac{\mathrm{d}R}{R\psi(R)}\int_{R_0\leqslant r\leqslant R}\widetilde{T}[r,f]\mathrm{d}V. \tag{36}$$

Owing to the boundary condition, the second term is nonpositive. Now we consider the first term. It is easily seen that

$$T_1^1 < A T_{00}, \tag{37}$$

where A is a suitable positive constant. Hence the first term is less than

$$A\int_{R>R_2} \frac{T_{00}}{\psi(R)} \mathrm{d}V. \tag{38}$$

From (33), the third term is less than

$$-\varepsilon\int_{R_1}^{R_2} \frac{\mathrm{d}R}{R\psi(R)}. \tag{39}$$

Hence, we have

$$0 < -\varepsilon\int_{R_1}^{R_2} \frac{\mathrm{d}R}{R\psi(R)} + A\int_{R_2>R>R_1} \frac{T_{00}}{\psi(R)} \mathrm{d}V, \tag{40}$$

for every $R_2 > R_1$.

Let $R_2 \to \infty$, we get a contradiction, since

$$\int_{R>R_0} \frac{T_{00}}{\psi(R)} \mathrm{d}V \text{ is finite and } -\varepsilon\int_{R_1}^{R_2} \frac{\mathrm{d}R}{R\psi(R)} \text{ tends to } -\infty.$$

Hence, we have $f_{\mu\nu} = 0$.

Theorem 1. is proved.

Evidently, if $T_1^1 \geqslant 0$ at $r = 5M$, the boundary condition is satisfied. Now we give the physical meaning of $T_1^1 \geqslant 0$.

For gauge theories in Minkowski space, in a Lorentz frame, the energy density of the field is

$$\frac{1}{2}\left(\frac{1}{2}\sum_{i,j=1}^{3}(f_{ij}, f_{ij}) + \sum_{i=1}^{3}(f_{i0}, f_{i0})\right), \tag{41}$$

the first term belongs to the 'magnetic part' and the second term belongs to the 'electric part'.

In the present case: referring to the orthonormal frame $\{e_1, e_2, e_3\}$ where e_1, e_2,

e_3 are parallel to r, θ, φ direction, respectively, $T_1^1 \geqslant 0$ can be written in

$$(f_{12}, f_{12})+(f_{13}, f_{13})+(f_{02}, f_{02})+(f_{03}, f_{03}) \geqslant (f_{23}, f_{23})+(f_{10}, f_{10}), \tag{42}$$

which means that the energy in the radius direction is not larger than that in the transversal direction.

4. Nonexistence Theorems for Harmonic Maps from Schwarzschild Spacetime to any Riemannian Manifold

Now we turn to prove Theorem 2. In this case, the stress-energy tensor $T_{\alpha\beta}$ takes the form (17), we have

$$T_\alpha^\alpha = g^{\alpha\beta} T_{\alpha\beta} = e(\phi), \tag{43}$$

$$T_0^0 = g^{00} T_{00} = \frac{1}{2} e(\phi), \tag{44}$$

$$T_1^1 = g^{11} T_{11} = \frac{1}{2} e(\phi) - \underset{N}{g}_{st} \phi_1^s \phi_1^t g^{11}. \tag{45}$$

Since $T_{\alpha\beta}$ satisfies the conservation law and we consider static harmonic maps, therefore, from Lemma 2, Equation (22) holds true. By using (43)-(45), we obtain

$$(T_0^0 - 2T_1^1)\frac{M}{r-2M} + T_i^i = \frac{r-3M}{2(r-2M)} e(\phi) + \frac{2M}{r-2M} \underset{N}{g}_{st} \phi_1^s \phi_1^t g^{11}, \tag{46}$$

and (22) becomes

$$\int_{r=R} T_1^1 \frac{r}{\sqrt{1-(2M/r)}} \mathrm{d}S - \int_{r=R_0} T_1^1 \frac{r}{\sqrt{1-(2M/r)}} \mathrm{d}S$$
$$= \int_{R_0 \leqslant r \leqslant R} \left(\frac{r-3M}{2(r-2M)} e(\phi) + \frac{2M}{r-2M} \underset{N}{g}_{st} \phi_1^s \phi_1^t g^{11} \right) \mathrm{d}V, \tag{47}$$

when $R_0 \geqslant 3M$, the right-hand side of (47) is $\geqslant 0$ and it can be written in the following form:

$$=\int_{R_0\leqslant r\leqslant R}\left[\frac{r-3M}{2(r-2M)}\left(\underset{N}{g}_{st}\phi_2^s\phi_2^t g^{22}+\underset{N}{g}_{st}\phi_3^s\phi_3^t g^{33}\right)+\frac{r+M}{2(r-2M)}\underset{N}{g}_{st}\phi_1^s\phi_1^t g^{11}\right]\mathrm{d}V.$$

Hence we have the right-hand side of (47) equals zero when and only when ϕ is a constant map.

Using a similar procedure to that in Section 3, we obtain the proof of Theorem 2.

Acknowledgement

The main results for the Yang-Mills case were essentially achieved at Dijon University and Paris Ⅵ University. The present version of this paper was completed at Heidelberg University. The author is grateful to Profs. Moshé Flato and Choquet Bruhat for their beneficial discussions. The author is indebted to these universities for their hospitality and support.

References

[1] Yang, C. N. and Mills, R., Isotopic spm conservation and a generalized gauge invariance, *Phys. Rev.* **96**, 191-195 (1954).

[2] Deser, S., Absence of static solution in source-free Yang-Mills theory, *Phys. Lett.* **64B**, 463-465 (1976).

[3] Hu, H. S. (Hu Hesheng), On the static solutions of massive Yang-Mills equations, *Chinese Ann. Math.* **3**, 519-526 (1982).

[4] Hu, H. S., Some nonexistence theorems for massive Yang-Mills fields and harmonic maps, *Springer Lecture Notes in Physics* **212**, 107-116 (1984).

[5] Gu, C. H. (Gu Chaohao), On classical Yang-Mills fields, *Phys. Rep.* **80**, 251 - 337 (1981).

[6] Hu, H. S., On equations of Yang-Mills gauge fields with mass, *Kexue Tongbao* **25**, 191 - 195 (1980).

[7] Deser, S. and Isham, C. J., Static solution of Yang-Mills-Higgs-Kibble system, *Kexue Tongbao* **25**, 773 - 776 (1980).

[8] Hildebrandt, S., Nonlinear elliptic systems and harmonic mappings, *Proceedings of 1980 Beijing Symposium on Differential Geometry and Differential Equations*, Vol.**1**, pp.481 - 615.

[9] Hu, H. S., A nonexistence theorem for harmonic maps with slowly divergent energy, *Chinese Ann. Math.* **5B**, 737 - 740 (1984).

[10] Hu, H. S. and Pan, Y. L. (Pan Yanglian), A theorem on Liouville's type on harmonic maps with finite or slowly divergent energy, *Chinese Ann. Math.* **7B**, 345 - 349 (1986).

(本文曾发表于 *Letters in Mathematical Physics*, 1987, **14**, 253 - 262)

Nonexistence Theorems for Yang-Mills Fields and Harmonic Maps in the Schwarzschild Spacetime (Ⅱ)*

Hu Hesheng (H. S. HU) and Wu Siye (S. Y. WU)
Institute of Mathematics, Fudan University, Shanghai, China
(Received: 20 August 1987)

Abstract We continue the study of the nonexistence of static pure Yang-Mills fields and harmonic maps defined on the Schwarzschild spacetime outside the black hole. Both the conditions on the regions and on the energy density are improved. For the case of harmonic maps, the region is precisely the best one, i.e., $r \geqslant 2M$, as was conjectured in *Lett. Math. Phys.* **14**, 253 (1987).

1. Introduction

As is well-known, there are no well-behaved source-free static solutions to the Maxwell and Einstein equations. Similar results have been obtained for Yang-Mills fields and harmonic maps [2 - 9].

In a previous Letter [1], the first author obtained some nonexistence theorems for pure Yang-Mills fields and harmonic maps defined on Schwarzschild spacetime. The regions considered are $r \geqslant 5M$ (for Yang-Mills fields) and $r \geqslant 3M$ (for harmonic

* Research supported by the Science Fund of the Chinese Academy of Sciences, Forschungsschwerpunkt Geometrie and SFB 123 of Heidelberg University.

maps), respectively. It is conjectured that the above nonexistence theorems in the Schwarzschild spacetime will hold true for $r \geqslant 2M$, i.e., outside the black hole.

In this Letter, for the harmonic map case the conjecture is proved, and for the Yang-Mills case, the limitation of regions is improved from $r \geqslant 5M$ to $r \geqslant 3M$. Moreover, the condition on the speed of divergence of energy is also improved. Besides, we give an example showing that the condition on the boundary values cannot be removed in the case of Yang-Mills fields.

The main results are stated in Section 2, the proofs of the theorem of Yang-Mills fields and harmonic maps are given in Sections 3 and 4, respectively.

2. Main Results

We consider the static solution of YM fields and harmonic maps defined on the Schwarzschild spacetime whose metric is written as

$$\begin{aligned} ds^2 &= g_{\alpha\beta} dx^\alpha dx^\beta \\ &= -\left(1-\frac{2M}{r}\right) dt^2 + \frac{1}{1-(2M/r)} dr^2 + \\ &\quad r^2 (d\theta^2 + \sin^2\theta d\varphi^2) \\ &\quad (\alpha, \beta = 0, 1, 2, 3). \end{aligned} \tag{1}$$

It should be noticed that a static YM field or static harmonic map is meaningful only when $r \geqslant 2M$, since in the black hole $r < 2m$ the metric itself is nonstatic.

The equations of pure YM fields are

$$J_\alpha = g^{\lambda\mu} (f_{\alpha\lambda;\mu} + [b_\mu, f_{\alpha\lambda}]) = 0, \tag{2}$$

where; denotes the covariant derivative.

The energy momentum tensor of YM fields is

$$T_{\alpha\beta}=(f_{\alpha\nu},f_{\beta\varepsilon})g^{\nu\varepsilon}-\frac{1}{4}g_{\alpha\beta}(f_{\mu\nu},f_{\gamma\varepsilon})g^{\mu\gamma}g^{\nu\varepsilon}. \tag{3}$$

The equations of harmonic map ϕ from $M\to N$ are

$$\underset{M}{g}^{\alpha\beta}\left(\frac{\partial^2\phi^s}{\partial x^\alpha\partial x^\beta}-\underset{M}{\Gamma}{}^{\gamma}_{\alpha\beta}\frac{\partial\phi^s}{\partial x^\gamma}+\underset{N}{\Gamma}{}^{s}_{tp}\frac{\partial\phi^t}{\partial x^\alpha}\frac{\partial\phi^p}{\partial x^\beta}\right)=0. \tag{4}$$

The stress-energy tensor of the harmonic map is

$$T_{\alpha\beta}=\frac{1}{2}\underset{N}{g}_{st}\phi^s_\gamma\phi^t_\delta\underset{M}{g}^{\gamma\delta}\underset{M}{g}_{\alpha\beta}-\underset{N}{g}_{st}\phi^s_\alpha\phi^t_\beta. \tag{5}$$

If

$$\int_{r>R_0}T_{00}\,\mathrm{d}V<\infty, \tag{6}$$

we say the energy of the field on the region $r>R_0$ is finite. If

$$\int_{r>R_0}T_{00}\,\mathrm{d}V=\infty \tag{7}$$

and

$$\int_{r>R_0}\frac{T_{00}}{\psi(r)r^{1-\delta}}\,\mathrm{d}V<\infty \tag{8}$$

for a certain $\delta\leqslant 1$, where $\psi(r)$ is a positive, continuous, and unbounded function satisfying

$$\int_{R_0}^{\infty}\frac{\mathrm{d}r}{r\psi(r)}=\infty, \tag{9}$$

we say the energy is 'mildly divergent' of index δ.

It is noted that the 'slowly divergent' energy case in [1] corresponds to the case $\delta=1$ in Equation (8). Evidently, the smaller the δ is in (8), the quicker the divergent of the energy will be. Hence, mildly divergent energy of index $\delta<1$ may be not slowly divergent.

For pure YM fields we obtain the following theorem.

THEOREM 1. *Let G be a compact Lie group. The pure YM equation on the Schwarzschild spacetime with* $r \geqslant 3M$ *do not admit any nontrivial regular static solution whose energy is finite or 'mildly divergent' of index* $\delta > 0$ *however small and satisfying the following boundary condition*

$$\int_{r=3M} T_1^1 \mathrm{d}S \geqslant 0. \tag{10}$$

For the harmonic maps from Schwarzschild spacetime to any Riemannian manifold, we have

THEOREM 2. *Let* ϕ *be a static harmonic map from the Schwarzschild spacetime outside the black hole, i.e.,* $r \geqslant 2M$ *into any n-dimensional Riemannian manifold N. If the energy of* ϕ *is finite or mildly divergent of index* $\delta > 0$ *however small and the boundary value on* $r = 2M$ *satisfies*

$$\lim_{R_0 \to 2M}\left[\left(1 - \frac{2M}{R_0}\right)^{1/2} \int_{r=R_0>2M} T_1^1 \mathrm{d}S\right] \geqslant 0. \tag{11}$$

then ϕ *must be a constant map.*

Thus, the conjecture of [1] for the nonexistence of the harmonic maps from Schwarzschild spacetime is proved.

Besides, the energy condition for both cases is improved. In the next section we will show that the boundary condition (10) of Theorem 1 cannot be removed.

3. The Case of Yang-Mills Fields

In order to prove Theorems 1 and 2, firstly we point out the following formula which holds true both for the energy momentum tensor (3) and stress-energy tensor (5).

$$(T_1^i f(r)\sqrt{G})_{,i} = (T_\sigma^\mu \Gamma_{1\mu}^\sigma)\sqrt{G} f(r) + T_1^1 \sqrt{G} f'(r), \tag{12}$$

where

$$T_\sigma^\mu = T^{\mu\nu} T_{\sigma\nu}, \qquad \sqrt{G} = \frac{r^2 \sin\theta}{\sqrt{1-(2M/r)}}. \tag{13}$$

Formula (12) is proved by using the conservation law and the static property of $T_{\alpha\beta}$.

Now we begin to prove Theorem 1. Substituting the Christoffel symbols of the metric and expression (3) of the energy momentum tensor $T_{\alpha\beta}$ to Equation (12) and calculating, we obtain

$$(T_1^i f(r)\sqrt{G})_{,i} = \left[\left(2 - \frac{rf'(r)}{f(r)}\right)P + \left(\frac{-2M}{r-2M} + \frac{rf'(r)}{f(r)}\right)Q\right]\frac{f(r)}{r}\sqrt{G}, \tag{14}$$

where

$$P = \frac{1}{2}\left[\frac{1}{2}(f_{ab}, f^{ab}) - (f_{01}, f^{01})\right] \geqslant 0,$$
$$Q = \frac{1}{2}[(f_{1a}, f^{1a}) - (f_{0a}, f^{0a})] \geqslant 0 \tag{15}$$

$(a, b = 2, 3)$ and the following relations

$$T_0^0 = -(P+Q),$$
$$T_i^i = P + Q \quad (i \text{ summation index}), \tag{16}$$
$$T_1^1 = -P + Q$$

hold.

Integrating (14) on $x^0 = \text{const.}$ we have

$$\int_0^{2\pi}\int_0^\pi\int_{R_0}^R (T_1^i f(r)\sqrt{G})_{,i}\, dr d\theta d\varphi$$
$$= \int_0^{2\pi}\int_0^\pi\int_{R_0}^R \left[\left(2 - \frac{rf'(r)}{f(r)}\right)P + \left(-\frac{2M}{r-2M} + \frac{rf'(r)}{f(r)}\right)Q\right] \times$$

$$\frac{f(r)}{r}\sqrt{G}\,\mathrm{d}r\mathrm{d}\theta\mathrm{d}\varphi. \tag{17}$$

Considering the property of the integrands, we have

$$\int_0^{2\pi}\int_0^{\pi}\left(T_1^1 f(r)\frac{r^2\sin\theta}{\sqrt{1-(2M/r)}}\right)\Bigg|_{r=R}\mathrm{d}\theta\mathrm{d}\varphi -$$

$$\int_0^{2\pi}\int_0^{\pi}\left(T_1^1 f(r)\frac{r^2\sin\theta}{\sqrt{1-(2M/r)}}\right)\Bigg|_{r=R_0}\mathrm{d}\theta\mathrm{d}\varphi$$

$$=\int_0^{2\pi}\int_0^{\pi}\int_{R_0}^{R}\left[\left(2-\frac{rf'(r)}{f(r)}\right)P+\left(-\frac{2M}{r-2M}+\frac{rf'(r)}{f(r)}\right)Q\right]\times$$

$$\frac{f(r)}{r}\sqrt{G}\,\mathrm{d}r\mathrm{d}\theta\mathrm{d}\varphi. \tag{18}$$

Let

$$f(r)=r^{\delta}\left(1-\frac{2M}{r}\right)^{1-(\delta/2)},$$

the inequalities

$$2-\frac{rf'(r)}{f(r)}=\frac{r-3M}{r-2M}(2-\delta)>0, \tag{19}$$

$$\frac{-2M}{r-2M}+\frac{rf'(r)}{f(r)}=\delta\frac{r-3M}{r-2M}>0 \tag{20}$$

hold for $r>3M$. Hence, the integrand of the right-hand side of (18) is nonnegative for $r>3M$ and it is zero if and only if $P=Q=0$, i.e., $f_{\lambda\mu}=0$.

Denoting

$$W=\left(\frac{r-3M}{r-2M}(2-\delta)P+\delta\frac{r-3M}{r-2M}Q\right)r^{\delta-1}\left(1-\frac{2M}{r}\right)^{1-(\delta/2)} \tag{21}$$

then (18) becomes

$$\int_{r=R}T_1^1 r^{\delta}\left(1-\frac{2M}{r}\right)^{\frac{1}{2}(1-\delta)}\mathrm{d}S-\int_{r=R_0}T_1^1 r^{\delta}\left(1-\frac{2M}{r}\right)^{\frac{1}{2}(1-\delta)}\mathrm{d}S$$

$$=\int_{R_0\leqslant r\leqslant R} W\mathrm{d}V. \tag{22}$$

If $f_{\mu\nu}\neq 0$, there exist two constants $R_1>0$ and $\varepsilon>0$ such that

$$\int_{R_0\leqslant r\leqslant R_2} W\mathrm{d}V>\varepsilon \tag{23}$$

holds for all $R_2\geqslant R_1$.

Choosing

$$\omega(R)=\begin{cases} 0, & 3M\leqslant R_0\leqslant R<R_1, \\ \dfrac{1}{R\psi(R)}, & R_1\leqslant R\leqslant R_2, \\ 0, & R>R_2 \end{cases} \tag{24}$$

where $\psi(r)$ has been defined in Section 2. Multiplying (22) by $\omega(R)$ and substituting (24) in it and then integrating, we have

$$\int_{R_1}^{R_2}\frac{\mathrm{d}R}{R\psi(R)}\int_{r=R} T_1^1 r^{\delta}\left(1-\frac{2M}{r}\right)^{\frac{1}{2}(1-\delta)}\mathrm{d}S-$$

$$\int_{R_1}^{R_2}\frac{\mathrm{d}R}{R\psi(R)}\int_{r=R_0} T_1^1 r^{\delta}\left(1-\frac{2M}{r}\right)^{\frac{1}{2}(1-\delta)}\mathrm{d}S$$

$$=\int_{R_1}^{R_2}\frac{\mathrm{d}R}{R\psi(R)}\int_{R_0\leqslant r\leqslant R} W\mathrm{d}V. \tag{25}$$

It is easily seen that for a suitable constant A

$$|T_1^1|<AT_{00}. \tag{26}$$

Hence, the first term of the left-hand side of Equation (25) is less than

$$A\int_{R_2>R>R_1}\frac{T_{00}}{\psi(R)R^{1-\delta}(1-(2M/R))^{(\delta/2)-1}}\mathrm{d}V \tag{27}$$

and the 2nd term of the left-hand side is nonpositive due to the boundary condition (10). The term on the right-hand side of (25) is greater than

$$\varepsilon\int_{R_1}^{R_2}\frac{\mathrm{d}R}{R\psi(R)}. \tag{28}$$

Then, from (25) we have

$$A\int_{R_2>R>R_1}\frac{T_{00}}{\psi(R)R^{1-\delta}(1-(2M/R))^{(\delta/2)-1}}\mathrm{d}V>\varepsilon\int_{R_1}^{R_2}\frac{\mathrm{d}R}{R\psi(R)}. \tag{29}$$

Since the energy is mildly divergent, the left-hand side of the above inequality remains finite as $R_2\to\infty$, but according to Equation (9), the right-hand side tends to ∞ as $R_2\to\infty$. It is a contradiction. Hence, we have

$$f_{\mu\nu}=0.$$

Theorem 1 is proved.

Remark 1. Condition (10) holds true if we have

$$T_1^1\geqslant 0 \tag{30}$$

at $r=3M$. The meaning of (30) is explained in [1].

Remark 2. In the following we give a contrary example to show that the boundary condition in Theorem 1 cannot be removed.

Let G be an Abelian group, the YM equations (2) become

$$g^{\lambda\mu}f_{\alpha\lambda;\mu}=0\quad(\alpha,\lambda,\mu=0,1,2,3). \tag{31}$$

It is easy to prove that (31) admits the following solution.

$$b_0=\frac{1}{r},\qquad b_1=b_2=b_3=0. \tag{32}$$

From this solution we get

$$T_{00}=\frac{1}{2r^4}\left(1-\frac{2M}{r}\right)$$

and the total energy is finite since

$$\int_{R_0\leqslant r\leqslant R} T_{00}\,\mathrm{d}V=\frac{2\pi}{3M}\left\{\left(1-\frac{2M}{R}\right)^{3/2}-\left(1-\frac{2M}{R_0}\right)^{3/2}\right\}.$$

For this solution, we have

$$T_1^1=-\frac{1}{2}(f_{01}, f_{01})=\frac{1}{2r^4}<0, \tag{33}$$

hence, condition (10) is violated.

Thus, the pure YM field does admit a static solution which is regular and with finite energy outside the black hole, but does not satisfy the boundary condition. This shows that boundary condition (10) cannot be removed.

Remark 3. By using the same technique it is easily seen that the energy condition of the nonexistence theorems of the pure YM field in [2-4] can be improved too.

4. The Case of Harmonic Maps

Now we consider the harmonic maps from Schwarzschild spacetime to any Riemannian manifold N and prove Theorem 2. The stress-energy tensor $T_{\alpha\beta}$ of the harmonic maps take the form (5). Substituting it in (12), we have

$$(T_1^i f(r)\sqrt{G})_{,i}=\Big[\frac{M}{r(r-2M)}T_0^0+\Big(-\frac{M}{r(r-2M)}+\frac{f'(r)}{f(r)}-\frac{1}{r}\Big)T_1^1+\frac{1}{r}(T_1^1+T_2^2+T_3^3)\Big]f(r)\sqrt{G}. \tag{34}$$

Since we consider static harmonic maps, then

$$\begin{aligned}&T_0^0=P+Q,\\&T_1^1=-P+Q,\\&T_i^i=T_\alpha^\alpha-T_0^0=P+Q\quad(i=1, 2, 3; \alpha=0, 1, 2, 3),\end{aligned} \tag{35}$$

where i, α are summation indexes, and

$$\begin{aligned} P &= \frac{1}{2} \underset{N}{g}_{st} \phi_1^s \phi_1^t g^{11}, \\ Q &= \frac{1}{2} (\underset{N}{g}_{st} \phi_2^s \phi_2^t g^{22} + \underset{N}{g}_{st} \phi_3^s \phi_3^t g^{33}). \end{aligned} \tag{36}$$

Substituting (35), (36) in (34), we get

$$(T_1^i f(r) \sqrt{G})_{,i} = \left[\left(\frac{2(r-M)}{r-2M} - \frac{rf'(r)}{f(r)} \right) P + \frac{rf'(r)}{f(r)} Q \right] \frac{f(r)}{r} \sqrt{G}. \tag{37}$$

Integrating (37) on $x^0 = \text{const.}$, we have

$$\begin{aligned} &\int_0^{2\pi} \int_0^{\pi} \left(T_1^1 f(r) \frac{r^2 \sin\theta}{\sqrt{1-(2M/r)}} \right) \Bigg|_{r=R} \mathrm{d}\theta \mathrm{d}\varphi - \\ &\quad \int_0^{2\pi} \int_0^{\pi} \left(T_1^1 f(r) \frac{r^2 \sin\theta}{\sqrt{1-(2M/r)}} \right) \Bigg|_{r=R_0} \mathrm{d}\theta \mathrm{d}\varphi \\ &= \int_{R_0 \leqslant r \leqslant R} \left[\left(\frac{2(r-M)}{r-2M} - \frac{rf'(r)}{f(r)} \right) P + \frac{rf'(r)}{f(r)} Q \right] \frac{f(r)}{r} \mathrm{d}V. \end{aligned} \tag{38}$$

Let $f(r) = r^\delta (1 - 2M/r)$, where δ is an arbitrary small positive number. The right-hand side of (38) is equal to

$$\int_{R_0 \leqslant r \leqslant R} \left[(2-\delta) P + \left(\delta + \frac{2M}{r-2M} \right) Q \right] r^{\delta-1} \left(1 - \frac{2M}{r} \right) \mathrm{d}V$$

when $r > 2M$, the integrand of the above integral is nonnegative and vanishes if and only if $P = Q = 0$, i.e., ϕ is a constant.

Using a similar procedure to that in Section 3 and considering the limiting procedure of $R_0 \to 2M$, we obtain Theorem 2.

In theoretical physics, the chiral field (or nonlinear σ-model) in 4 - dimensional spacetime is just a harmonic map ϕ from the spacetime to a homogeneous Riemannian manifold. Hence, the physical significance of Theorem 2 can be expressed as follows.

THEOREM 2′. *In Schwarzschild spacetime, there does not exist any static nontrivial chiral field with finite energy, slowly divergent energy, or mildly divergent energy.*

Remark. We should emphasize that in the proof of Theorems 1 and 2 we only use the conservation law

$$T^{\alpha}_{\beta;\alpha}=0$$

of the energy momentum tensor (3) of YM fields and the stress-energy tensor (5) of harmonic maps respectively, instead of the pure Yang-Mills equations and equations of harmonic maps themselves. So the results are valid for more general cases. For example, a relative harmonic map ϕ is a generalization of the harmonic map. It is defined in [10], and obeys the conservation law of the stress-energy tensor. Hence, Theorem 2 also holds true for relative harmonic maps.

Acknowledgements

This paper is completed at Heidelberg University. The first author is grateful to Profs W. Jäger and F. Tomi for their hospitality and beneficial discussions. The second author is grateful to Prof. Chaohao Gu for his encouragement and beneficial discussions. The authors also express their thanks to Prof. M. Flato for beneficial discussions.

References

[1] Hu, H. S. (Hu Hesheng), Nonexistence theorems for Yang-Mills fields and harmonic maps in the Schwarzschild spacetime (I). *Lett. Math. Phys.* **14**, 253 - 262 (1987).

[2] Deser, S., Absence of static solution in source-free Yang-Mills theory, *Phys. Lett.* **64B**, 463 - 465 (1976).

[3] Hu, H. S., Some nonexistence theorems for massive Yang-Mills fields and harmonic maps, *Springer Lecture Notes of Physics* **212**, 107-116 (1984).

[4] Hu, H. S., On the static solutions of massive Yang-Mills equations, *Chinese Ann. Math.* **3**, 519-526 (1982).

[5] Hu, H. S., A nonexistence theorem for harmonic maps with slowly divergent energy, *Chinese Ann. Math.* **5B**, 737-740 (1984).

[6] Hu, H. S. and Pan, Y. L. (Pan Yang-lian), A theorem on Liouville's type on harmonic maps with finite or slowly divergent energy, *Chinese Ann. Math.* **7B**, 345-349 (1986).

[7] Eell, J. and Lemaire, L., A report on harmonic maps, *Bull. London Math. Soc.* **10**, 1-68 (1978).

[8] Sealey, H. C. J., Thesis, Warwick University (1980).

[9] Hildebrandt, S., Nonlinear elliptic systems and harmonic mappings, *Proc. 1980 Beijing Symposium on Differential Geometry and Differential Equations*, Vol.**1**, pp.481-615.

[10] Ishihara, T. and Ishihara, S., Notes on relatively harmonic immersion, *Hokkaido Math. J.* **4**, 234-246 (1975).

(本文曾发表于 *Letters in Mathematical Physics*, 1987, **14**, 343-351)

On the Classical Lump of Yang-Mills Fields

Hu Hesheng*
Institute of Mathematics, Fudan University, Shanghai 200433, People's Republic of China
(Received: 18 January 1991; revised version: 16 May 1991)

Abstract It is shown that for pure Yang-Mills fields there is no lump phenomena if the total energy is infinite and diverges mildly in a certain sense. This improves the well-known classical result on the absence of a lump for the finite total energy case. Some exact lump solutions are obtained showing that the energy condition cannot be removed. The results are valid for more general fields.

AMS subject classification (1991). 81T13.

1. Introduction

It is well known that a finite energy solution to the pure Yang-Mills equation in 3+1-dimensional Minkowski spacetime does not admit lump phenomena [1-3]. In particular, there is no static solution with finite energy [4, 5]. It has been found that the assumption on the finiteness of the energy is not necessary for the nonexistence of a static solution. In fact, there are a series of results showing that the nonexistence theorem holds if the finite energy condition is weakened to a 'slowly or mildly

* This work was supported by the Chinese Natural Science Foundation, the Chinese Fund of Doctor Programs, and the University of Paris VI, and University of Bourgogne.

divergent energy condition' and the spacetime can be generalized to some curved spacetime, say, the outside of the black hole of the Schwarzschild spacetime [6, 7]. Some related results on harmonic maps have also been obtained [8]. Moreover, we have no reason to assume, *à priori*, that the total energy of a physical system should be finite. Hence, it is worthwhile considering the nonexistence or existence of lump phenomena in the infinite energy case.

In this Letter, the following results are proved.

(1) For pure Yang-Mills fields, there is no lump phenomenon if the energy diverges slowly or mildly in certain senses (Section 3). In other words, if interesting lump phenomena do appear, then the total energy in the ball $|x|<R$ should increase to ∞ sufficiently fast as $R\to\infty$.

(2) The energy condition in the above statement cannot be removed. We will show this in Section 4 by giving an exact solution of the pure Yang-Mills fields admitting lump phenomena.

(3) The above result holds true not only for the Yang-Mills fields but also for the gauge fields whose energy-momentum tensor obeys the conservation laws. Some lump solutions which satisfy the conservation laws, but do not satisfy the Yang-Mills equation, will be given in Section 5.

2. Notations and Definitions

Let

$$L=\frac{1}{4}\int(f_{\mu\nu},\ f^{\mu\nu})\mathrm{d}^4x\quad(\mu,\ \nu=0,\ 1,\ 2,\ 3)\tag{2.1}$$

be the Yang-Mills functional of a compact Lie group G on $3+1$ - dimension Minkowski spacetime. Here,

$$f_{\mu\nu}=\partial_{\mu}b_{\nu}-\partial_{\nu}b_{\mu}+[b_{\mu}, b_{\nu}] \tag{2.2}$$

is the field strength and b_μ the potential. $f_{\mu\nu}$ and b_μ are g-valued where g is the Lie algebra of the group G, and the Yang-Mills equations

$$J_{\nu}=\eta^{\lambda\mu}(\partial_{\lambda}f_{\mu\nu}+[b_{\lambda}, f_{\mu\nu}])=0 \tag{2.3}$$

are satisfied.

The energy momentum tensor is

$$T_{\alpha\beta}=(f_{\alpha\nu}, f_{\beta}^{\nu})-\frac{1}{4}\mu_{\alpha\beta}(f_{\mu\nu}, f^{\mu\nu}). \tag{2.4}$$

Here $\eta_{\alpha\beta}$ is the metric tensor and

$$[\eta_{\alpha\beta}]=\begin{bmatrix}-1 & & & \\ & +1 & & \\ & & +1 & \\ & & & +1\end{bmatrix}. \tag{2.5}$$

The integral

$$\int_{|x|<R} T_{00}(t, x)\mathrm{d}^3x \tag{2.6}$$

is the energy in the ball

$$|x|<R\ (r=|x|=\sqrt{(x^1)^2+(x^2)^2+(x^3)^2})\quad \text{at } x_0=t.$$

DEFINITION. If there exist positive constants R, ε, t_0 such that

$$\int_{|x|<R} T_{00}\mathrm{d}^3x>\varepsilon \tag{2.7}$$

holds for every $t>t_0$, we say that the fields admits lump phenomena.

From the expression of $T_{\alpha\beta}$, we have

$$T_{\alpha}^{\alpha}=0 \tag{2.8}$$

and

$$\frac{\partial}{\partial x^{\alpha}} T^{\alpha}_{\beta} = 0 \quad \text{or} \quad T^{\alpha}_{\beta,\alpha} = 0. \tag{2.9}$$

Equation (2.9) is the conservation law of the energy momentum tensor.

3. A Theorem of the Absence of Classical Lumps

In this section, we shall prove the following theorem which improves the results of Coleman [1] and Weder [2].

THEOREM. *There are no lumps of a pure Yang-Mills fields in* R^{3+1} *whose energy satisfies*

$$\int_{R^3} \frac{T_{00}(x, t)}{r^{\lambda}} \mathrm{d}^3 x < B, \tag{3.1}$$

where $0 \leqslant \lambda < 1$, *and* B *is a constant.*

This theorem weakens the finite energy condition in [1, 2].

First, we give the following lemma.

LEMMA. *The energy-momentum tensor* $T_{\alpha\beta}$ *of the Yang-Mills fields in* R^{3+1} *satisfies*

$$| T_{\alpha\beta} | \leqslant T_{00} \quad (\alpha, \beta = 0, 1, 2, 3). \tag{3.2}$$

Using (2.4) and (2.8), by direct calculation, we obtain (3.2).

Now, we start to prove the above theorem.

Proof. Let $f(r)$ be a C^1 function of r. We have

$$\int_{|x| \leqslant \rho} \sum_{i,j} \partial_i (f(r) x^j T_{ij}) \mathrm{d}^3 x = \int_{|x|=\rho} f(r) T_{ij} \frac{x^i x^j}{r} \mathrm{dS} \quad (i, j = 1, 2, 3). \tag{3.3}$$

On the other hand, the left-hand side of the above equation equals

$$\int_{|x|\leqslant\rho} f(r)\sum_{i} T_{ij}\,\mathrm{d}^3 x + \int_{|x|\leqslant\rho} f'(r)\sum_{i,j}\frac{x^i x^j}{r}T_{ij}\,\mathrm{d}^3 x +$$

$$\int_{|x|\leqslant\rho} f(r)\sum_{i,j} x^j \partial_i T_{ij}\,\mathrm{d}^3 x$$

$$=\int_{|x|\leqslant\rho} f(r) T_{00}\,\mathrm{d}^3 x + \int_{|x|\leqslant\rho} f'(r)\sum_{i,j}\frac{x^i x^j}{r}T_{ij}\,\mathrm{d}^3 x +$$

$$\int_{|x|\leqslant\rho} f(r)\sum x^j \frac{\partial T_{0j}}{\partial x^0}\,\mathrm{d}^3 x. \tag{3.4}$$

From (3.3) and (3.4), we have

$$\frac{\partial}{\partial x^0}\int_{|x|\leqslant\rho} f(r)\sum x^j T_{0j}\,\mathrm{d}^3 x$$

$$= -\int_{|x|\leqslant\rho} f(r) T_{00}\,\mathrm{d}^3 x - \int_{|x|\leqslant\rho} f'(r)\sum_{i,j}\frac{x^i x^j}{r}T_{ij}\,\mathrm{d}^3 x +$$

$$\int_{|x|\leqslant\rho} f(r)\sum_{i,j}\frac{x^i x^j}{r}T_{ij}\,\mathrm{d}S. \tag{3.5}$$

Suppose that the lump phenomena exist, i.e. there are positive constants R, ε, t_0 such that, for $t > t_0$

$$\int_{|x|\leqslant R_1} T_{00}(x, t)\,\mathrm{d}^3 x \geqslant \varepsilon \quad (t > t_0,\ R_1 > R). \tag{3.6}$$

Let

$$f(r) = r^{-\lambda}, \quad \lambda > 0.$$

From (3.6), we have

$$\int_{|x|\leqslant R_1} \frac{T_{00}(x, t)}{r^{\lambda}}\,\mathrm{d}^3 x > \varepsilon_1, \tag{3.7}$$

where $\varepsilon_1 = R_1^{-\lambda}\varepsilon$.

Take

$$\mu(r)=\begin{cases}\dfrac{1}{r}, & R\leqslant r\leqslant R_1,\\ 0, & \text{elsewhere.}\end{cases} \tag{3.8}$$

R_1 will be chosen later.

Let

$$w(t)=-\int_0^\infty \mu(\rho)\,\mathrm{d}\rho\int_{|x|\leqslant\rho} r^{-\lambda}\sum_j x^j T_{0j}\,\mathrm{d}^3x. \tag{3.9}$$

Using the lemma and the assumption (3.1), we have

$$|w(t)|\leqslant 3\int_0^\infty \rho\mu(\rho)\,\mathrm{d}\rho\int_{|x|\leqslant\rho} r^{-\lambda}T_{00}\,\mathrm{d}^3x < 3B(R_1-R). \tag{3.10}$$

On the other side,

$$\frac{\mathrm{d}w(t)}{\mathrm{d}t}=-\int_0^\infty \mu(\rho)\,\mathrm{d}\rho\int_{|x|\leqslant\rho} r^{-\lambda}\sum_j x^j T_{0j,0}\,\mathrm{d}^3x. \tag{3.11}$$

From (3.5), we have

$$\begin{aligned}\frac{\mathrm{d}w(t)}{\mathrm{d}t}=&\int_0^\infty \mu(\rho)\,\mathrm{d}\rho\int_{|x|\leqslant\rho} r^{-\lambda}T_{00}\,\mathrm{d}^3x-\\ &\lambda\int_0^\infty \mu(\rho)\,\mathrm{d}\rho\int_{|x|\leqslant\rho} r^{-\lambda-1}\sum_{i,j}\frac{x^ix^j}{r}T_{ij}\,\mathrm{d}^3x-\\ &\int_0^\infty \mu(\rho)\,\mathrm{d}\rho\int_{|x|=\rho} r^{-\lambda}\sum_{i,j}\frac{x^ix^j}{r}T_{ij}\,\mathrm{d}S.\end{aligned} \tag{3.12}$$

From the fact

$$\begin{aligned}&\left|\lambda\int_0^\infty \mu(\rho)\,\mathrm{d}\rho\int_{|x|\leqslant\rho} r^{-\lambda-1}\sum\frac{x^ix^j}{r}T_{ij}\,\mathrm{d}^3x\right|\\ &\leqslant\lambda\int_0^\infty \mu(\rho)\,\mathrm{d}\rho\int_{|x|\leqslant\rho} r^{-\lambda-1}\frac{\left|\sum x^ix^jT_{ij}\right|}{r}\mathrm{d}^3x\\ &<\lambda\int_0^\infty \mu(\rho)\,\mathrm{d}\rho\int_{|x|\leqslant\rho} r^{-\lambda}T_{00}\,\mathrm{d}^3x.\end{aligned} \tag{3.13}$$

Hence, the sum of the first two terms in the right-hand side of (3.12) is larger than

$$(1-\lambda)\int_0^\infty \mu(\rho)\mathrm{d}\rho\int_{|x|\leqslant\rho} r^{-\lambda}T_{00}\mathrm{d}^3x$$
$$> (1-\lambda)(\ln R_1 - \ln R)\varepsilon_1 \quad (t > t_0). \tag{3.14}$$

Moreover, for the last term of (3.12)

$$\left|\int_0^\infty \mu(\rho)\mathrm{d}\rho\int_{|x|=\rho} r^{-\lambda}\sum \frac{x^i x^j}{r}T_{ij}\mathrm{d}S\right| < 3B. \tag{3.15}$$

Hence,

$$\frac{\mathrm{d}w}{\mathrm{d}t} > (1-\lambda)(\ln R_1 - \ln R)\varepsilon_1 - 3B \quad (t > t_0). \tag{3.16}$$

Take R_1 such that

$$(1-\lambda)(\ln R_1 - \ln R)\varepsilon_1 - 3B = K > 0. \tag{3.17}$$

Then we have $w \to \infty$ as $t \to \infty$. This contradicts (3.10), thus, the theorem is proved.

Remark. If we take $f(r) = 1$, this is the result of [1] which corresponds to the finite energy case.

The energy condition (3.1) of Theorem 1 contains the slowly divergent energy case also as a special case. The condition of slowly divergent energy was defined in [4, 5].

4. Lump Solutions to the Pure Yang-Mills Equations

From Section 3, we see that lump phenomena cannot occur if the energy condition (3.1) is satisfied. In this section, we will give some exact solutions of the pure Yang-Mills equations in which lumps occur. The results also show that for the absence of lumps, the energy condition in our Theorem cannot be removed.

Let G be SU (2) and X_1, X_2, X_3 the standard base of its Lie algebra with

$$[X_1, X_2]=X_3, \qquad [X_2, X_3]=X_1, \qquad [X_3, X_1]=X_2. \tag{4.1}$$

Take

$$b_1=b_1(t)X_1, \qquad b_2=b_2(t)X_2, \qquad b_3=b_3(t)X_3, \qquad b_0=0. \tag{4.2}$$

After calculations, we have

$$\begin{aligned} &f_{0t}=-\frac{\mathrm{d}b_t}{\mathrm{d}t}X_t, \\ &f_{12}=-b_1b_2X_3, \qquad f_{23}=-b_2b_3X_1, \qquad f_{31}=-b_3b_1X_2 \end{aligned} \tag{4.3}$$

(where $i=1, 2, 3$ and i is not a summation index) and the Yang-Mills equations are reduced to

$$\begin{aligned} &J_0=0, \\ &J_1=\left(\frac{\mathrm{d}^2b_1}{\mathrm{d}t^2}+b_1(b_2^2+b_3^2)\right)X_1=0, \\ &J_2=\left(\frac{\mathrm{d}^2b_2}{\mathrm{d}t^2}+b_2(b_1^2+b_3^2)\right)X_2=0, \\ &J_3=\left(\frac{\mathrm{d}^2b_3}{\mathrm{d}t^2}+b_3(b_1^2+b_2^2)\right)X_3=0. \end{aligned} \tag{4.4}$$

In the case of

$$b_0=0, \qquad b_1=b_2=b_3=b(t). \tag{4.5}$$

Equations (4.4) reduce to a single equation of b

$$\frac{\mathrm{d}^2b}{\mathrm{d}t^2}+2b^3=0. \tag{4.6}$$

Hence,

$$\left(\frac{\mathrm{d}b}{\mathrm{d}t}\right)^2+b^4=\mathrm{const}=k^4$$

and

$$\frac{\mathrm{d}b}{\mathrm{d}t}=\pm(k^4-b^4)^{1/2}$$

or

$$\int_0^b \frac{\mathrm{d}b}{(k^4-b^4)^{1/2}}=\pm t.$$

Consequently, b is an elliptic function of t [7]

$$b(t)=k\ \operatorname{sinlemn}(kt). \tag{4.7}$$

In this case,

$$T_{00}=\frac{3}{2}\left(\left(\frac{\mathrm{d}b}{\mathrm{d}t}\right)^2+b^4\right)=\frac{3}{2}k^4>0.$$

Hence, (4.5), (4.7) are lump solutions which are nonstatic and periodic with respect to t.

5. Gauge Field Satisfying Conservation Laws

From the proof of the Theorem in Section 3, it is seen that instead of the full Yang-Mills equations, we only use the conservation laws

$$\frac{\partial T_\beta^\alpha}{\partial x_\alpha}=0. \tag{5.1}$$

This was pointed out in [2] for the finite energy case.

It is interesting to see that gauge fields do exist which satisfy the conservation laws (5.1) and the Yang-Mills equations are not satisfied.

First, we start to prove the following

Lemma. *For a gauge field* F, *the conservation laws* (5.1) *are equivalent to the equations*

$$(J_\nu, f_{\beta\varepsilon})\eta^{\nu\varepsilon}=0. \tag{5.2}$$

Proof. From the conservation laws (5.1), by using Bianchi identity, and expressions (2.3) and (2.4) of J_v and $T_{\alpha\beta}$, respectively, we have

$$\begin{aligned}
T_{\alpha\beta|\alpha} &= (f_{\alpha\nu|\alpha}, f^\nu_\beta)+(f_{\alpha\nu}, f^\nu_{\beta|\alpha})-\frac{1}{4}(f_{\mu\nu|\beta}, f^{\mu\nu})-\frac{1}{4}(f_{\mu\nu}, f^{\mu\nu}_{|\beta}) \\
&= (-J_\nu, f^\nu_\beta)+(f^{\alpha\nu}, f_{\beta\nu|\alpha})-\frac{1}{2}(f^{\mu\nu}, f_{\mu\nu|\beta}) \\
&= (-J_\nu, f^\nu_\beta)+\left(f^{\mu\nu}, f_{\beta\nu|\mu}-\frac{1}{2}f_{\mu\nu|\beta}\right) \\
&= (-J_\nu, f^\nu_\beta)+\frac{1}{2}(f^{\mu\nu}, -f_{\nu\beta|\mu}-f_{\beta\mu|\nu}-f_{\mu\nu|\beta}) \\
&= -(J_\nu, f_{\beta\varepsilon})\eta^{\nu\varepsilon}.
\end{aligned}$$

The lemma is proved. Here ',' and '|' stand for derivatives and gauge derivatives, respectively.

Now let us construct the following examples.

We still use the Ansatz (4.2). From the expressions of J_0, J_1, J_2, J_3 in (4.4), Equations (5.2) are reduced to

$$\eta^{\lambda\mu}(J_\lambda, f_{0\mu})=0, \tag{5.3}$$

since the equations

$$\eta^{\lambda\mu}(J_\lambda, f_{i\mu})=0 \quad (i=1, 2, 3) \tag{5.4}$$

automatically hold.

Obviously, the class of solutions to (5.3) is much larger than to (4.4). For example, let b_2, b_3 be constants k_2, $k_3(\neq 0)$ respectively, Equation (5.3) becomes

$$\frac{\mathrm{d}b_1}{\mathrm{d}t}\frac{\mathrm{d}^2 b_1}{\mathrm{d}t^2}+k^2 b_1\frac{\mathrm{d}b_1}{\mathrm{d}t}=0.$$

Here $k^2 = b_2^2 + b_3^2 \neq 0$. Hence,

$$\left(\frac{\mathrm{d}b_1}{\mathrm{d}t}\right)^2 + k^2 b_1^2 = k_1^2. \tag{5.5}$$

Thus, we get the following solution of the conservation laws (5.1)

$$b_1 = \frac{k_1}{k}\sin(kt+\alpha), \qquad b_2 = k_2, \qquad b_3 = k_3. \tag{5.6}$$

Besides, it is interesting to see that $J_2 \neq 0$, $J_3 \neq 0$. Hence, the Yang-Mills equations are not satisfied while as the conservation law holds.

It is also noted for the constructed fields (5.6) that we have

$$T_{00} = \text{const} \neq 0. \tag{5.7}$$

Hence, it is a lump solution. The solution is nonstatic and periodic with respect to t.

We have a similar example in electromagnetic dynamics.

Let the potential b_α be

$$b_0 = \cos t \sum_{i=1}^{3} a_i x_i + \sin t \sum_{i=1}^{3} b_i x_i, \quad b_i = 0 \quad (i=1, 2, 3).$$

Here

$$a^2 = b^2 \neq 0, \qquad a \cdot b = 0 \quad (a = (a_1, a_2, a_3), \, b = (b_1, b_2, b_3)). \tag{5.8}$$

We see that

$$\begin{aligned} f_{0i} &= -(a_i \cos t + b_i \sin t), \quad f_{ij} = 0, \\ J_i &= b_i \cos t - a_i \sin t \neq 0, \quad J_0 = 0, \end{aligned} \tag{5.9}$$

and

$$T_{00} = \frac{1}{2}\sum f_{0i}^2 = a^2 = \text{const}.$$

The conservation laws holds, since

$$f_{\beta\gamma}J^{\nu}=0. \tag{5.10}$$

So (5.8) are nonstatic lump solutions which satisfy the conservation laws of electromagnetic fields but do not satisfy the Maxwell equations.

Remark. It is interesting to find lump solution for which $T_{00}\to 0$ as $r\to\infty$.

Acknowledgement

This work was done mainly in France when the author was visiting the University Paris VI and the University of Bourgogne. The author is grateful to Profs Y. Choquet-Bruhat and M. Flato for their hospitality and valuable discussions.

References

[1] Coleman, S, There are no classical glueballs, *Comm. Math. Phys.* **55**, 113 - 116 (1977).

[2] Weder, R., Absence of classical lumps, *Comm. Math. Phys.* **57**, 161 - 164 (1977).

[3] Glasscy, R. T. and Strauss, W. A., Decay of classical Yang-Mills fields, *Comm. Math. Phys.* **65**, 1 - 13 (1979).

[4] Deser, S., Absence of static solutions in sourcefree Yang-Mills theory, *Phys. Lett.* **64B**, 463 - 466 (1976).

[5] Hu Hesheng (H. S. Hu), On the static solution of massive Yang-Mills equations, *Chinese Ann. Math.* **3**, 519 - 525 (1982).

[6] Hu Hesheng (H. S. Hu), Nonexistence theorems for Yang-Mills fields and harmonic maps in the Schwarzschild spacetime (I), *Lett. Math. Phys.* **14**, 253 - 262 (1987).

[7] Hu Hesheng and Wu Siye (S. Wu), Nonexistence theorems for Yang-Mills fields and harmonic maps in the Schwarzschild spacetime (II), *Lett. Math. Phys.* **14**, 343 - 351 (1987).

[8] Hu Hesheng, Some nonexistence theorems for massive Yang-Mills fields and harmonic maps,

Springer Lecture Notes in Physics No.212, pp.107 – 116 (1984).

[9] Whittaker, E. T and Watson, G. N., A Course of Modern Analysis, Cambridge University Press, Fourth Edition 1958.

(本文曾发表于 *Letters in Mathematical Physics*, 1991, **22**, 267 – 275)

Darboux Transformation of Su-chain

Hu Hesheng*

Institute of Mathematics, Fudan University, Shanghai, China

Dedicated to Prof. Su Buchin on his 90th birthday

§ 1. Introduction

In projective differential geometry, the Laplace sequences of period 4, which consists of 4 congruences and 4 focal surfaces, is an important subject. P. S. Finikoff and Su Buchin investigated the Laplace sequence of period 4 with a stratifiable couple of diagonal congruences and obtained a series of interesting results ([1], [2]). This kind of Laplace sequence is called Finikoff configuration by Su etc. and is called Su chain later by G. Bol etc.. Su proved that the determination of Su chain is reduced to solve sinh-Gordon equation together with a system of completely integrable linear PDSs [2]. In recent terminology, the later is the Lax pair of sinh-Gordan equation. However, there is no explicit examples appeared in the literature. Recently, the Darboux transformation, which is a new form of Bäcklund transformation, appeared brilliantly in the soliton theory. Owing to the Darboux transformation, one can use the fundamental solution of the Lax pair to construct the new solution [3], [4]. The algorithm is purely algebraic and can be continued successively for infinite times.

---

* Partially supported by NNSFC and SFECC.

In the present paper, by using the method of the Darboux transformation to investigate Su chain, we obtain the following main results:

(1) We give a general method to obtain a new Su chain from a known one.

(2) By using Darboux transformations successively, we can obtain infinite sequence of Su chains.

(3) There exist sequences of Su chains with period $2n$ ($n\geqslant 2$).

(4) Starting from the trivial solution of sinh-Gordan equation, we construct a Su chain explicitly which contains 2 quadratic surfaces and the next Su chain is constructed explicitly as well.

We give a sketch of proofs of these facts, and their details will be published elsewhere.

§ 2. Darboux Transformation

Let

$$N_i = N_i(u, v) \qquad (i=1, 2, 3, 4) \tag{1}$$

be 4 surfaces which form a Su chain in the projective space P_3, where u, v are the parameters of the conjugate nets on each surface.

In order to use the Darboux method for constructing new Su chain from a known one, we first rewrite the fundamental equations of Su chain, then we simplify the fundamental equations by rescaling and introduces a spectral parameter λ. We have:

$$\frac{\partial}{\partial u}\begin{pmatrix} N_1 \\ N_2 \\ N_3 \\ N_4 \end{pmatrix} = \lambda \begin{pmatrix} 0 & 0 & e^{-4\phi} & 0 \\ 0 & 0 & 0 & e^{-4\phi} \\ 0 & e^{4\phi} & 0 & 0 \\ e^{4\phi} & 0 & 0 & 0 \end{pmatrix} \begin{pmatrix} N_1 \\ N_2 \\ N_3 \\ N_4 \end{pmatrix},$$

$$\frac{\partial}{\partial v}\begin{pmatrix} N_1 \\ N_2 \\ N_3 \\ N_4 \end{pmatrix} = \begin{pmatrix} -2\phi_v & 0 & 0 & \frac{1}{\lambda} \\ 0 & -2\phi_v & \frac{1}{\lambda} & 0 \\ -\frac{1}{\lambda} & 0 & 2\phi_v & 0 \\ 0 & \frac{1}{\lambda} & 0 & 2\phi_v \end{pmatrix} \begin{pmatrix} N_1 \\ N_2 \\ N_3 \\ N_4 \end{pmatrix}, \tag{2}$$

where ϕ is the solution of sinh-Gordan equation

$$\phi_{uv} = \sinh 4\phi. \tag{3}$$

(3) is just the integrability condition of (2). i.e. (2) is the Lax pair of (3).

Denote

$$\Phi = \begin{pmatrix} N_1 \\ N_2 \\ N_3 \\ N_4 \end{pmatrix}, \quad U = \begin{pmatrix} 0 & e^{-4\phi}a \\ e^{4\phi}b & 0 \end{pmatrix}, \tag{4}$$

$$C = \begin{pmatrix} -2\phi_v I & 0 \\ 0 & 2\phi_v I \end{pmatrix}, \quad D = \begin{pmatrix} 0 & b \\ a & 0 \end{pmatrix}$$

where C, D are 4×4 matrices, and

$$a = \begin{pmatrix} -1 & 0 \\ 0 & 1 \end{pmatrix}, \quad b = \begin{pmatrix} 0 & 1 \\ 1 & 0 \end{pmatrix}, \quad 0 = \begin{pmatrix} 0 & 0 \\ 0 & 0 \end{pmatrix}, \quad I = \begin{pmatrix} 1 & 0 \\ 0 & 1 \end{pmatrix}. \tag{5}$$

Then, (2) can be written in the form

$$\Phi_u = \lambda U \Phi, \quad \Phi_v = \left(C + \frac{1}{\lambda} D\right) \Phi \tag{6}$$

here Φ depends upon u, v, λ, i.e. $\Phi = \Phi(u, v, \lambda)$.

Let ϕ and $\Phi(u, v, \lambda)$ be a known solution of (3) and the Lax pair (6) respectively. The main idea of the Darboux transformation is to construct a 4×4 matrix (called Darboux matrix)

$$S=I+\lambda\alpha \tag{7}$$

such that

$$\Phi_1=S\Phi=(I+\lambda\alpha)\Phi \tag{8}$$

satisfies (6) i.e.

$$\Phi_{1u}=\lambda U_1\Phi_1, \quad \Phi_{1v}=\left(C_1+\frac{D}{\lambda}\right)\Phi_1 \tag{6'}$$

where

$$U_1=\begin{pmatrix}0 & e^{-4\phi_1}a\\ e^{4\phi_1}b & 0\end{pmatrix}, \quad C_1=\begin{pmatrix}-2\phi_{1v}I & 0\\ 0 & 2\phi_{1v}I\end{pmatrix} \tag{4'}$$

and ϕ_1 is a new solution of (3).

From (6′) and (4′), we see that α should take the form

$$\alpha=\begin{pmatrix}0 & \mu a\\ \sigma b & 0\end{pmatrix} \tag{9}$$

where μ, σ satisfy

$$\mu_u=\mu\sigma^{-1}e^{4\phi}-e^{-4\phi}, \quad \sigma_u=\mu^{-1}\sigma e^{-4\phi}-e^{4\phi} \tag{10_1}$$

and

$$\mu_v=-4\phi_v\mu+\mu(\mu-\sigma), \quad \sigma_v=4\phi_v\sigma+\sigma(\sigma-\mu) \tag{10_2}$$

(10_1), (10_2) are completely integrable when ϕ satisfies sinh-Gordon equation (3). The new solution ϕ_1 is obtained from

$$e^{-4\phi_1}=\mu\sigma^{-1}e^{4\phi} \tag{11}$$

After solving (10_1), (10_2), we get μ, σ, α, the Darboux matrix S and the new solution ϕ_1. The new Su chain is obtained as

$$\begin{pmatrix}\widetilde{N}_1\\ \widetilde{N}_2\\ \widetilde{N}_3\\ \widetilde{N}_4\end{pmatrix}=\Phi_1(u, v, 1)=(I+\alpha)\Phi(u, v, 1)$$

$$=\begin{pmatrix}1 & 0 & -\mu & 0\\ 0 & 1 & 0 & \mu\\ 0 & \sigma & 1 & 0\\ \sigma & 0 & 0 & 1\end{pmatrix}\begin{pmatrix}N_1\\ N_2\\ N_3\\ N_4\end{pmatrix}=\begin{pmatrix}N_1-\mu N_3\\ N_2+\mu N_4\\ N_3+\sigma N_2\\ N_4+\sigma N_1\end{pmatrix}. \tag{12}$$

Thus we see $\widetilde{N}_1$, $\widetilde{N}_3$, $\widetilde{N}_2$, $\widetilde{N}_4$ lie on the lines N_1N_3, N_3N_2, N_2N_4, N_4N_1 respectively. We have:

Theorem 1. *We can get infinite new Su chains from a known Su chain by Darboux transformations. The tetrahedron of the new Su chains are correlated to the tetrahedron of the original Su chain.*

To obtain α explicitly, we establish the correspondence

$$\alpha=\begin{pmatrix}0 & \mu a\\ \sigma b & 0\end{pmatrix}\rightarrow\alpha_0=\begin{pmatrix}0 & \mu\\ \sigma & 0\end{pmatrix} \tag{13}$$

between the 4×4 matrix α and the 2×2 matrix α_0. With the aid of the Lax pair of 2×2 matrix

$$\underset{0}{\Phi}_u=\lambda\underset{0}{U}\underset{0}{\Phi}\qquad \underset{0}{\Phi}_v=\left(\underset{0}{C}+\frac{1}{\lambda}\underset{0}{D}\right)\underset{0}{\Phi} \tag{14}$$

we can construct the matrix α_0 as follows.

Let

$$h = \begin{pmatrix} h_1(\lambda_0) \\ h_2(\lambda_0) \end{pmatrix} \tag{15}$$

be a column solution to (14) with parameter $\lambda = \lambda_0 \neq 0$ and

$$H = \begin{pmatrix} h_1(\lambda_0) & h_1(\lambda_0) \\ h_2(\lambda_0) & -h_2(\lambda_0) \end{pmatrix} \tag{16}$$

with det $H \neq 0$. Then,

$$\alpha_0 = -H \begin{pmatrix} \frac{1}{\lambda_0} & 0 \\ 0 & -\frac{1}{\lambda_0} \end{pmatrix} H^{-1} = -\begin{pmatrix} 0 & \frac{h_1(\lambda_0)}{\lambda_0 h_2(\lambda_0)} \\ \frac{h_2(\lambda_0)}{\lambda_0 h_1(\lambda_0)} & 0 \end{pmatrix}. \tag{17}$$

Hence α is also obtained from (13).

Since the Darboux transformation for the 2×2 matrix case can be done successively with algebraic algorithm, we have:

Theorem 2. *The Darboux transformation for Su chain can be done successively with algebraic algorithm.*

The sequence of Darboux transformations and Su chains can be illustrated as

$$(\phi, \underset{0}{\Phi}, \Phi) \to (\phi_1, \underset{0}{\Phi}_1, \Phi_1) \to (\phi_2, \underset{0}{\Phi}_2, \Phi_2) \to \cdots \tag{18}$$

§ 3. Periodicity

The permutability theorem for the Bäcklund transformation of sine-Gordon equation is well known. The theorem was extended to many cases for the Darboux transformations. It is easy to prove that it holds for the 2×2 matrix Lax pair (14) and the sinh-Gordon equation. Consequently, we have

Theorem 3. *From a given Su-chain, there exist $2n$ ($n > 1$) consecutive Darboux*

transformation such that the final one is just the original one.

The theorem means the existence of the sequence of Su-chains of period $2n$.

§ 4. Examples

We take a trivial solution $\phi=0$ of sinh-Gordon equation (3). The equations (2) can be integrated explicitly. The result is

$$\begin{pmatrix} N_1 \\ N_2 \\ N_3 \\ N_4 \end{pmatrix} = \begin{pmatrix} e^{\tilde{u}}\cos\tilde{v} & -e^{\tilde{u}}\sin\tilde{v} & e^{-\tilde{u}}\cos\tilde{v} & -e^{-\tilde{u}}\sin\tilde{v} \\ e^{\tilde{u}}\sin\tilde{v} & e^{\tilde{u}}\cos\tilde{v} & -e^{-\tilde{u}}\sin\tilde{v} & -e^{-\tilde{u}}\cos\tilde{v} \\ -\frac{\sqrt{2}}{2}e^{\tilde{u}}B & \frac{\sqrt{2}}{2}e^{\tilde{u}}A & \frac{\sqrt{2}}{2}e^{-\tilde{u}}A & \frac{\sqrt{2}}{2}e^{-\tilde{u}}B \\ \frac{\sqrt{2}}{2}e^{\tilde{u}}A & \frac{\sqrt{2}}{2}e^{\tilde{u}}B & -\frac{\sqrt{2}}{2}e^{-\tilde{u}}B & \frac{\sqrt{2}}{2}e^{-\tilde{u}}A \end{pmatrix} \tag{19}$$

(here, $\tilde{u}=\frac{\sqrt{2}}{2}(u+v)$, $\tilde{v}=\frac{\sqrt{2}}{2}(u-v)$, $A=(\sin\tilde{v}+\cos\tilde{v})$, $B=(\cos\tilde{v}-\sin\tilde{v})$). From (19) it is seen that points N_1 and N_2 generated the same quadratic surface

$$x_1x_4=x_2x_3$$

while points N_3 and N_4 generate the same quadratic surface

$$x_1x_3+x_2x_4=0$$

Using the algorithm of Darboux transformation, we find

$$-\mu=\frac{\cosh\left(c_1v+\frac{u}{c_1}\right)-\coth\left(cv+\frac{u}{c}+\frac{1}{\lambda_0}\right)\sinh\left(c_1v+\frac{u}{c_1}\right)}{\cosh\left(c_1v+\frac{u}{c_1}\right)-\tanh\left(cv+\frac{u}{c}+\frac{1}{\lambda_0}\right)\sinh\left(c_1v+\frac{u}{c_1}\right)}.$$

The obtained Su chain is:

$$\begin{pmatrix} \widetilde{N}_1 \\ \widetilde{N}_2 \\ \widetilde{N}_3 \\ \widetilde{N}_4 \end{pmatrix} = \begin{pmatrix} N_1 - \mu N_3 \\ N_2 + \mu N_4 \\ N_3 + \dfrac{1}{\lambda_0^2 \mu} N_2 \\ N_4 + \dfrac{1}{\lambda_0^2 \mu} N_1 \end{pmatrix}.$$

References

[1] P. S. Finikoff, *Sur les suites de Laplace contenent des congruences de Wilczynski*, Comptes Rendus **189** (1929), 517 - 519.

[2] Su Buchin, *On certain periodic sequence of Laplace of period four in ordinary space*, Science Reports of the Töhoku Imp. Univ. (**1**) **25** (1936), 227 - 256, or Su Buchin Selected Mathematical Papers, Science Press, Beijing, China (1983) 147 - 176.

[3] Gu Chaohao & Hu Hesheng, *A unified explicit form of Bäcklund transformation for generalized hierarchies of KdV equations*, Letters in Mathematical physics **11** (1986), 325 - 335.

[4] Gu Chaohao, Hu Hesheng etc., *Soliton Theory and its Application*, (*Applied Mathematics series No. 1*), Zhejiang Publishing House of Science and Technology, China, 1990.

(本文曾发表于*Differential Geometry: Proceedings of the Symposium in Honour of Professor Su Buchin on His 90th Birthday*, World Scientific, 1993, 108 - 113)

Explicit Construction of Harmonic Maps From R^2 to $U(N)$ *

Gu Chaohao** Hu Hesheng**
Institute of Mathematics, Fudan University, Shanghai, China

Abstract Darboux transformation method is used for constructing harmonic maps from R^2 to $U(N)$. The explicit expressions for Darboux matrices are used to obtain new harmonic maps from a known one. The algorithm is purely algebraic and can be repeated successively to obtain an infinite sequence of harmonic maps. Single and multiple solitons are obtained with geometric characterizations and it is proved that the interaction between solitons is elastic. By introducing the singular Darboux transformations, an explicit method to construct new unitons is presented.

Keywords Harmonic map, Explicit construction, Darboux transformation method.

1991 MR Subject Classification 58E20.

§ 1. Introduction

In the present paper, we consider the explicit construction of harmonic maps from R^2 to $U(N)$. This is one of the most interesting problems in geometry and

---

* Project supported by the Chinese Tian Yuan Mathematical Foundation, the Scientific Foundation of State Education Commission of China, the Natural Science Foundation of Shanghai, the Department of Mathematical Sciences of the University of Tokyo and the Japanese Inoue Scientific Foundation.

** Manuscript received August 30, 1994.

mathematical physics. It is well-known that the equations of harmonic maps from R^{1+1} and R^2 to a Lie group admit a Lax pair with a spectral parameter. Hence the technique for solving integrable soliton equations is a powerful tool to study the harmonic maps form R^{1+1} and R^2 to Lie groups, especially, to $U(N)$. Among various methods the Darboux transformation method has the advantage that new solutions can be obtained explicitly by using purely algebraic algorithm and the same algorithm can be used successively to obtain an infinite sequence of explicit solutions. For the R^{1+1} case, the general method was introduced in [1]. By using this method, single and multiple soliton solutions were constructed with explicit formulas and the interaction of solitons has been proved to be elastic[2]. Harmonic maps from R^{1+1} to $U(N)$ have been studied also by Beggs[3], but the soliton solutions have not been expressed explicitly and the interaction of solitons has not been considered.

For the harmonic map from R^2 or S^2 to $U(N)$, there is a famous paper of K. Uhlenbeck[4]. She uses the loop action and Birkohoff factorization to construct new harmonic maps. Besides, the Bäcklund transformation and singular Bäcklund transformations are mentioned as tools for obtaining new harmonic maps and unitons. We find that Darboux transformation method, which we have used to study the harmonic maps from R^{1+1} to $U(N)$ in [1], is also valid for the case of R^2 to $U(N)$. Explicit solutions can be obtained too. The object of the present paper is to present this method and give the explicit formulas for new harmonic maps and unitons.

In § 2, we recall the generic formulation of harmonic maps from R^2 to $U(N)$ and the Lax pair. § 3 is devoted to the Darboux transformations. In particular, explicit formulas for new harmonic maps are established. In § 4, we construct single and multiple soliton-like solutions, they are the Darboux transformation of a kind of trivial solutions. Global behavior of these solitons is sketched. § 5 is devoted to obtain the explicit formula for obtaining unitons from a known one by introducing the singular

Darboux transformation.

§2. Harmonic Maps from R^2 to $U(N)$

A harmonic map $\phi(x, y)$ from $R^2=\{(x, y)\}$ to the group $U(N)$ is a critical point of the energy integral

$$S[\phi]=\int \operatorname{tr}(\phi_x\phi^{-1}\phi_x\phi^{-1}+\phi_y\phi^{-1}\phi_y\phi^{-1})dxdy \quad (\phi\in U(N)). \tag{2.1}$$

Let

$$U=\phi_x\phi^{-1},\ V=\phi_y\phi^{-1}. \tag{2.2}$$

We have

$$U_y-V_x+[U, V]=0, \tag{2.3}$$

$$U_x+V_y=0, \tag{2.4}$$

$$U+U^*=0,\ V+V^*=0. \tag{2.5}$$

Here (2.3) follows from (2.2), (2.5) is the condition for unitary group and (2.4) is the Euler equation for the variational problem. Thus, harmonic maps from R^2 to $U(N)$ are defined by (2.3), (2.4) and (2.5). If (2.3), (2.4) and (2.5) are satisfied, ϕ can be constructed by the integration of (2.2) provided that the initial data, say $\phi(0, 0)$, satisfy the condition $\phi(0, 0)\in U(N)$.

Introducing the complex coordinates $(\zeta, \bar\zeta)$ for R^2

$$\zeta=x+iy,\ \bar\zeta=x-iy; \tag{2.6}$$

$$\frac{\partial}{\partial\bar\zeta}=\frac{1}{2}\left(\frac{\partial}{\partial x}+i\frac{\partial}{\partial y}\right),\ \frac{\partial}{\partial\zeta}=\frac{1}{2}\left(\frac{\partial}{\partial x}-i\frac{\partial}{\partial y}\right). \tag{2.7}$$

Denoting

$$A = \frac{1}{2}(U + iV) = \phi_{\bar{\xi}}\phi^{-1}, \quad B = \frac{1}{2}(U - iV) = \phi_{\xi}\phi^{-1}, \tag{2.8}$$

the equations (2.3)-(2.5) become

$$A_{\xi} - B_{\bar{\xi}} + [A, B] = 0, \tag{2.9}$$

$$A_{\xi} + B_{\bar{\xi}} = 0, \tag{2.10}$$

$$A^{*} = -B, \tag{2.11}$$

respectively. Here $*$ denotes the conjugate and transpose of matrices. The Lax pair of the equations (2.9) and (2.10) is

$$\Phi_{\bar{\xi}} = \lambda A \Phi, \quad \Phi_{\xi} = \frac{\lambda}{2\lambda - 1} B \Phi, \tag{2.12}$$

i.e. the integrability condition (or zero-curvature condition) of (2.12)

$$\Phi_{\bar{\xi}\xi} = \Phi_{\xi\bar{\xi}}$$

for all λ ($\lambda \neq 1/2$) is equivalent to (2.9) and (2.10).

If $\Phi(\lambda)$ is a fundamental solution to the Lax pair (2.12) and valued in $U(N)$ at some point in R^2 for all λ with $|2\lambda - 1| = 1$, then $\Phi(\lambda)$ is valued in $U(N)$ for all $(x, y) \in R^2$ and all λ with $|2\lambda - 1| = 1$. Moreover, $\Phi(1)$ is a harmonic map. $\Phi(\lambda)$ is essentially the extended harmonic map in [4].

§ 3. Darboux Transformation

In our previous work [1, 2], we use Darboux transformation to obtain explicit formulas for the harmonic maps from $R^{1,1}$ to $U(N)$ and elucidate their behavior. The method is applicable to the case $R^2 \to U(N)$, with some modification.

Let $\Phi(\lambda)$ be an extended harmonic maps of $R^2 \to U(N)$, and A, B its potentials.

We want to construct an $N \times N$ matrix $\alpha(\zeta, \bar{\zeta})$, which is independent of λ, such that

$$\Phi_1(\lambda) = S\Phi = (I + \lambda\alpha)\Phi \tag{3.1}$$

satisfies

$$\Phi_{1\bar{\zeta}} = \lambda A_1 \Phi_1, \quad \Phi_{1\zeta} = \frac{\lambda}{2\lambda - 1} B_1 \Phi_1 \tag{3.2}$$

with some A_1, B_1 satisfying the $U(N)$ condition (2.11), then $S = I + \lambda\alpha$ is called a Darboux matrix and (3.1) the Darboux transformation. Substituting (3.1) to (3.2), it is seen that

$$A_1 = A + \alpha_{\bar{\zeta}}, \quad B_1 = B - \alpha_{\zeta}; \tag{3.3}$$

$$\alpha_{\bar{\zeta}}\alpha = \alpha A - A\alpha, \quad \alpha_{\zeta}\alpha + 2\alpha_{\zeta} = B\alpha - \alpha B. \tag{3.4}$$

(3.4) is a system of nonlinear equations of the matrix α. Explicit solutions α of (3.4) can be constructed by using $\Phi(\lambda)$ in the following way.

Let $\lambda_1, \lambda_2, \cdots, \lambda_N$ be N numbers such that at least two of them are unequal and $\lambda_\alpha \neq 0, \frac{1}{2}, 1$ ($\alpha = 1, 2, \cdots, N$). Choose N constant columns l_ρ ($\rho = 1, 2, \cdots, N$) and let

$$h_\rho = \Phi(\lambda_\rho) l_\rho (\rho = 1, 2, \cdots, N) \tag{3.5}$$

such that

$$H = [h_1, h_2, \cdots, h_N] \tag{3.6}$$

is a non-degenerate matrix. Note that h_ρ is a column solution to the Lax pair (2.12) with $\lambda = \lambda_\rho$. We have

Theorem 3.1. *The matrix*

$$\alpha = -H\Lambda^{-1}H^{-1} \ with \ \Lambda = \begin{bmatrix} \lambda_1 & & & \\ & \lambda_2 & & \\ & & \ddots & \\ & & & \lambda_N \end{bmatrix} \tag{3.7}$$

is a solution to (3.4).

Proof. From the definition, h_ρ satisfies

$$h_{\rho\bar{\xi}} = \lambda_\rho A h_\rho, \ h_{\rho\xi} = \frac{\lambda_\rho}{2\lambda_\rho - 1} B h_\rho. \tag{3.8}$$

Hence

$$H_{\bar{\xi}} = AH\Lambda, \ H_\xi = BH\Lambda'.$$

Here

$$\Lambda' = \begin{bmatrix} \frac{\lambda_1}{2\lambda_1 - 1} & & \\ & \ddots & \\ & & \frac{\lambda_N}{2\lambda_N - 1} \end{bmatrix}. \tag{3.9}$$

Thus

$$\begin{aligned} \alpha_{\bar{\xi}} &= -H_{\bar{\xi}}\Lambda^{-1}H^{-1} + H\Lambda^{-1}H^{-1}H_{\bar{\xi}}H^{-1} \\ &= -A + H\Lambda^{-1}H^{-1}AH\Lambda H^{-1} \end{aligned} \tag{3.10}$$

and further we have

$$\alpha_{\bar{\xi}}\alpha = -A\alpha + \alpha A. \tag{3.11}$$

The first equation of (3.4) is satisfied. The second one of (3.4) can be verified similarly.

Remark 3.1. It is easily seen that (3.4) is completely integrable and each solution α

can be determined by the value of α at a given point. Consequently, (3.7) gives all solutions of (3.4) which are similar to $-\Lambda^{-1}$ at the given point, and then at every point the property holds true.

From the expressions (3.3) of A_1, B_1 and the $U(N)$ condition (2.11), we should have

$$(\alpha_{\bar{\zeta}})^* = \alpha_\zeta \text{ or } (\alpha^* - \alpha)_\zeta = 0. \tag{3.12}$$

Moreover, in order that the solution obtained can be defined on the whole R^2 we should have $\det H \neq 0$ in R^2. The following choice of λ_ρ's and l_ρ's gives the explicit formula of α, which satisfies the above requirement.

We choose a complex number $\underset{1}{\lambda}$, and let

$$\underset{2}{\lambda} = \frac{\bar{\underset{1}{\lambda}}}{2\bar{\underset{1}{\lambda}} - 1},$$

$$\lambda_\rho = \begin{cases} \underset{1}{\lambda} & (\rho = 1, \cdots, k), \\ \underset{2}{\lambda} & (\rho = k+1, \cdots, N). \end{cases} \tag{3.13}$$

Here $\underset{1}{\lambda}$ satisfies

(i) $|2\underset{1}{\lambda} - 1| \neq 1$,

We choose l_ρ's such that

(ii) $\Phi(\underset{1}{\lambda})L_1$ and $\Phi(\underset{2}{\lambda})L_2$ are of rank k and $N-k$ respectively at some point (say (0, 0)), where

$$L_1 = [l_1, \cdots, l_k, 0, \cdots, 0],\ L_2 = [0, \cdots, 0, l_{k+1}, \cdots, l_N], \tag{3.14}$$

and

(iii) at a fixed point, say (0, 0),

$$h_p^* h_a = 0 \quad (a = 1, \cdots, k;\ p = k+1, \cdots, N). \tag{3.15}$$

We note that

$$(h_p^*)_{\bar{\zeta}} = (h_{p\zeta})^* = \left[\frac{\lambda_2}{2\lambda_2 - 1} B h_p\right]^* = -\lambda_1 h_p^* A$$

and hence

$$(h_p^* h_a)_{\bar{\zeta}} = 0. \tag{3.16}$$

Similarly, we have

$$(h_p^* h_a)_{\zeta} = 0. \tag{3.17}$$

Therefore (iii) holds on R^2 if it holds at a fixed point. Thus we have

Lemma. *Let h_a and h_p be two column solutions of (2.12) corresponding to the parameter λ_1 and λ_2 respectively. If $h_p^* h_a = 0$ at a point in R^2, then $h_p^* h_a = 0$ everywhere.*

Besides, it is easy to see that if h_a's (resp. h_p's) are linearly independent at some point, they should be linearly independent everywhere. From this construction, we have $\det H \neq 0$ on R^2.

Theorem 3.2. *If the λ_p's (given by (3.13)) and the constant columns l_p's satisfy the requirements (i) (ii) and (iii), then the potential*

$$A_1 = A + \alpha_{\bar{\zeta}}, \quad B_1 = B - \alpha_{\zeta},$$

where α is given by the explicit expression (3.7), satisfies the $U(N)$ condition $A_1^* = -B_1$ (i.e. α satisfies (3.12)), and thus defines a new harmonic map from R^2 to $U(N)$.

Proof. From (3.7), we see that $\alpha H = -H\Lambda^{-1}$, i.e.

$$\alpha h_a = -\frac{1}{\lambda_1} h_a, \quad \alpha h_p = -\frac{1}{\lambda_2} h_p, \quad (a = 1, \cdots, k;\ p = k+1, \cdots, N), \tag{3.18}$$

hence

$$h_a^*\alpha^* = -\frac{1}{\underset{1}{\bar\lambda}}h_a^*,\ h_p^*\alpha^* = -\frac{1}{\underset{2}{\bar\lambda}}h_p^*. \tag{3.19}$$

Consequently

$$\begin{aligned} h_a^*(\alpha^*-\alpha)h_b &= \left(-\frac{1}{\underset{1}{\bar\lambda}}+\frac{1}{\underset{1}{\lambda}}\right)h_a^*h_b, \\ h_p^*(\alpha^*-\alpha)h_q &= \left(-\frac{1}{\underset{2}{\bar\lambda}}+\frac{1}{\underset{2}{\lambda}}\right)h_p^*h_q, \\ h_p^*(\alpha^*-\alpha)h_a &= \left(-\frac{1}{\underset{2}{\bar\lambda}}+\frac{1}{\underset{1}{\lambda}}\right)h_p^*h_a = 0, \\ h_a^*(\alpha^*-\alpha)h_p &= \left(-\frac{1}{\underset{1}{\bar\lambda}}+\frac{1}{\underset{2}{\lambda}}\right)h_a^*h_p = 0, \end{aligned} \tag{3.20}$$

where $a, b = 1, \cdots, k$; $p, q = k+1, \cdots, N$. From the relation

$$2\underset{1}{\lambda}\,\underset{2}{\bar\lambda} = \underset{1}{\lambda} + \underset{2}{\bar\lambda}$$

for $\underset{1}{\lambda}$, $\underset{2}{\bar\lambda}$, we obtain

$$-\frac{1}{\underset{2}{\bar\lambda}}+\frac{1}{\underset{2}{\lambda}} = -\frac{2\underset{1}{\lambda}-1}{\underset{1}{\lambda}} + \frac{2\underset{1}{\bar\lambda}-1}{\underset{1}{\bar\lambda}} = -\frac{1}{\underset{1}{\bar\lambda}}+\frac{1}{\underset{1}{\lambda}}. \tag{3.21}$$

Thus, from (3.20) we have

$$h_\rho^*(\alpha^*-\alpha)h_\sigma = h_\rho^*\left(\frac{1}{\underset{1}{\lambda}}-\frac{1}{\underset{1}{\bar\lambda}}\right)I\,h_\sigma \quad (\rho, \sigma = 1, \cdots, N). \tag{3.22}$$

It follows that

$$\alpha^*-\alpha = \left(\frac{1}{\underset{1}{\lambda}}-\frac{1}{\underset{1}{\bar\lambda}}\right)I, \tag{3.23}$$

since h_ρ's are linearly independent. By differentiating (3.23) with respect to ζ, we see that the $U(N)$ condition (3.12) follows immediately. From Theorem 3.1, we know that A_1, B_1 satisfies (2.9), (2.10). The proof is completed.

In the following we first show that the Darboux matrix $S = I + \lambda\alpha$ can be expressed by an Hermitian projection π to $(N-k)$-dimensional subspaces of C^N.

In fact, from (3.23) we have

$$\alpha^* + \frac{1}{\lambda_1} I = \alpha + \frac{1}{\lambda_1} I.$$

So the matrix $\alpha + \frac{1}{\lambda_1} I$ is Hermitian. Since $\alpha = -H\Lambda^{-1}H^{-1}$, and the eigenvalues of α are $-\frac{1}{\lambda_1}$ and $-\frac{1}{\lambda_2}$, we have

$$\alpha + \frac{1}{\lambda_1} I = \left(\frac{1}{\lambda_1} - \frac{1}{\lambda_2}\right)\beta \begin{bmatrix} 0 & 0 \\ 0 & I_{N-k} \end{bmatrix} \beta^* = \left(\frac{1}{\lambda_1} - \frac{1}{\lambda_2}\right)\pi, \tag{3.24}$$

where $\beta \in U(N)$, I_{N-k} is the $(N-k)\times(N-k)$ unit matrix and π is an Hermitian projection.

Thus

$$\alpha = -\frac{1}{\lambda_1} I + \left(\frac{1}{\lambda_1} - \frac{1}{\lambda_2}\right)\pi = -\frac{1}{\lambda_2}\pi - \frac{1}{\lambda_1}\pi^{\perp}, \tag{3.25}$$

$$S = I + \lambda\alpha = \left(1 - \frac{\lambda}{\lambda_2}\right)\pi + \left(1 - \frac{\lambda}{\lambda_1}\right)\pi^{\perp}. \tag{3.26}$$

Moreover, by calculation we can prove that, for $|2\lambda - 1| = 1$,

$$\left|1 - \frac{\lambda}{\lambda_1}\right|^2 = \left|1 - \frac{\lambda}{\lambda_2}\right|^2. \tag{3.27}$$

Hence

$$S^* S = \left|1 - \frac{\lambda}{\lambda_2}\right|^2 \pi + \left|1 - \frac{\lambda}{\lambda_1}\right|^2 \pi^{\perp} = \left|1 - \frac{\lambda}{\lambda_1}\right|^2 I. \tag{3.28}$$

Thus we have

Theorem 3. 3. *A new extended solution* $\Phi^1(\lambda)$ *is obtained from the original extended solution* $\Phi(\lambda)$ *by the transformation*

$$\Phi^1(\lambda) = S\left(1 - \frac{\lambda}{\lambda_2}\right)^{-1} \Phi(\lambda) \tag{3.29}$$

and the corresponding new harmonic map is

$$\Phi^1(1) = S\left(1 - \frac{1}{\lambda_2}\right)^{-1} \Phi(1). \tag{3.30}$$

We turn to deduce the differential equations satisfied by π. From (3.25), we have

$$\alpha = -\frac{1}{\lambda_1} + \left(\frac{1}{\lambda_1} - \frac{1}{\lambda_2}\right)\pi.$$

Substitute it into (3.4), we obtain

$$\begin{aligned} \pi_{\bar{\zeta}} &= -\lambda_1 \pi A + \lambda_2 A \pi + (\lambda_1 - \lambda_2)\pi A \pi, \\ \pi_{\zeta} &= \bar{\lambda}_1 B \pi - \bar{\lambda}_2 \pi B + (\bar{\lambda}_2 - \bar{\lambda}_1)\pi B \pi. \end{aligned} \tag{3.31}$$

Thus we have

Theorem 3.4. *The projective operator*

$$\pi = \left(\alpha + \frac{1}{\lambda_1}\right) \Big/ \left(\frac{1}{\lambda_1} - \frac{1}{\lambda_2}\right) \tag{3.32}$$

is a solution to (3.31).

Remark 3.2. Equations in (3.31) are just the pair of equations (25) of Bäcklund transfor-mations for harmonic maps in Uhlenbeck's paper [4] with different notations. The differences of the notations between [4] and the present paper are as follows: (i) The order of multiplication of matrices in the present paper is different from that of [4]. (ii) Our A, B correspond to $2A_{\bar{\zeta}}$ and $2A_{\zeta}$ in [4], respectively. (iii) Our λ

corresponds to $\frac{1-\lambda}{2}$ in [4].

Remark 3.3. From (3.30) and (3.26), it is seen that the Darboux transformation gives new harmonic maps

$$\Phi^1(1)K=(\pi+\gamma\pi^{\perp})\Phi(1)\cdot K. \tag{3.33}$$

Here K is an arbitrary constant matrix in $U(N)$ and

$$\gamma=\left(1-\frac{1}{\lambda_1}\right)\Big/\left(1-\frac{1}{\lambda_2}\right). \tag{3.34}$$

From (3.27), we have $|\gamma|=1$.

(3.33) is actually the formula for new solution of Theorem 6.3 in [4]. But in our case, π can be constructed explicitly in terms of the extended solutions of the harmonic map $\Phi(1)$.

Remark 3.4. The system (3.31) is completely integrable. Hence each solution π of (3.31) is completely determined by the initial data $\pi(0,0)$. From our construction, for any fixed λ_1, we can choose l_α's such that

$$\alpha(0,0)=-\frac{1}{\lambda_1}+\left(\frac{1}{\lambda_1}-\frac{1}{\lambda_2}\right)\pi(0,0).$$

Thus our construction exhausts all solutions to the system (3.31).

§ 4. Soliton Solutions

In this section, we construct the single soliton solutions as applications of the Darboux matrix method. For simplifying the calculation, we take $N=2$ in the following. The results for general N are similar.

The elements in $SU(2)$ are matrices

$$\begin{bmatrix} \gamma & \beta \\ -\bar{\beta} & \bar{\gamma} \end{bmatrix} \tag{4.1}$$

with $|\beta|^2+|\gamma|^2=1$. Take the seed solution in the following form

$$g_0=\begin{bmatrix} e^{\tau\bar{\zeta}-\bar{\tau}\zeta} & 0 \\ 0 & e^{-(\tau\bar{\zeta}-\bar{\tau}\zeta)} \end{bmatrix}, \tag{4.2}$$

where τ is a constant and

$$A=\begin{bmatrix} \tau & 0 \\ 0 & -\tau \end{bmatrix},\ B=\begin{bmatrix} -\bar{\tau} & 0 \\ 0 & \bar{\tau} \end{bmatrix}. \tag{4.3}$$

Substituting (4.3) into the Lax pair (2.12) and integrating, we obtain

$$\Phi_0=\begin{bmatrix} \exp\left(\lambda\tau\bar{\zeta}-\frac{\lambda}{2\lambda-1}\bar{\tau}\zeta\right) & 0 \\ 0 & \exp\left(-\lambda\tau\bar{\zeta}+\frac{\lambda}{2\lambda-1}\bar{\tau}\zeta\right) \end{bmatrix}$$

$$=\begin{bmatrix} l(\lambda) & 0 \\ 0 & l^{-1}(\lambda) \end{bmatrix} \tag{4.4}$$

with

$$l(\lambda)=\exp\left(\lambda\tau\bar{\zeta}-\frac{\lambda}{2\lambda-1}\bar{\tau}\zeta\right). \tag{4.5}$$

Let λ_1, λ_2 be two distinct constants related by

$$2\lambda_1\bar{\lambda}_2=\lambda_1+\bar{\lambda}_2. \tag{4.6}$$

As in §3, we take

$$H=\begin{bmatrix} l(\lambda_1) & -\bar{a}l(\lambda_2) \\ a l^{-1}(\lambda_1) & l^{-1}(\lambda_2) \end{bmatrix}. \tag{4.7}$$

It is seen that

$$l^{-1}(\lambda_1)=\overline{l(\lambda_2)},\ l^{-1}(\lambda_2)=\overline{l(\lambda_1)} \tag{4.8}$$

and

$$h_2^* h_1=0. \tag{4.9}$$

Moreover,

$$\det H=\left|l(\lambda_1)\right|^2+|a|^2\left|l(\lambda_1)\right|^{-2}. \tag{4.10}$$

From (3.7), we obtain

$$\alpha=\frac{-1}{e^p+|a|^2e^{-p}}\begin{bmatrix}\frac{e^p}{\lambda_1}+|a|^2\frac{e^{-p}}{\lambda_2} & \left(\frac{1}{\lambda_1}-\frac{1}{\lambda_2}\right)\bar{a}e^{iq} \\ \left(\frac{1}{\lambda_1}-\frac{1}{\lambda_2}\right)ae^{-iq} & \frac{e^p}{\lambda_2}+|a|^2\frac{e^{-p}}{\lambda_1}\end{bmatrix}, \tag{4.11}$$

where

$$p=(\lambda_1-\lambda_2)\tau\bar{\zeta}+(\bar{\lambda}_1-\bar{\lambda}_2)\bar{\tau}\zeta, \tag{4.12}$$

$$iq=(\lambda_1+\lambda_2)\tau\bar{\zeta}-(\bar{\lambda}_1+\bar{\lambda}_2)\bar{\tau}\zeta. \tag{4.13}$$

Note that p, q are real, and linear with respect to x and y. From (3.30), the corresponding new harmonic map is

$$\Phi^1(1)=(I+\alpha)\left(1-\frac{1}{\lambda_2}\right)^{-1}\Phi_0(1)=\frac{-1}{e^p+|a|^2e^{-p}}\left(1-\frac{1}{\lambda_2}\right)^{-1}\times$$
$$\begin{bmatrix}e^p\left(\frac{1}{\lambda_1}-1\right)+|a|^2e^{-p}\left(\frac{1}{\lambda_2}-1\right)e^{\tau\bar{\zeta}-\bar{\tau}\zeta} & \left(\frac{1}{\lambda_1}-\frac{1}{\lambda_2}\right)\bar{a}\,e^{-iq}e^{-(\tau\bar{\zeta}-\bar{\tau}\zeta)} \\ \left(\frac{1}{\lambda_1}-\frac{1}{\lambda_2}\right)ae^{-iq}e^{\tau\bar{\zeta}-\bar{\tau}\zeta} & e^p\left(\frac{1}{\lambda_2}-1\right)+|a|^2e^{-p}\left(\frac{1}{\lambda_1}-1\right)e^{-(\tau\bar{\zeta}-\bar{\tau}\zeta)}\end{bmatrix}. \tag{4.14}$$

We write

$$\Phi^1(1)=\begin{bmatrix}\gamma & \beta \\ -\bar{\beta} & \bar{\gamma}\end{bmatrix}$$

and

$$\beta=\rho_1 e^{i\theta_1},\ \gamma=\rho_2 e^{i\theta_2}, \tag{4.15}$$

then

$$\rho_1=\frac{k\,|\,\bar{a}\,|}{e^p+|\,a\,|^2 e^{-p}}=\frac{k}{2}\operatorname{sech}(p-\ln|\,a\,|)\quad \left(k=\left|1-\frac{1}{\underset{2}{\lambda}}\right|^{-1}\left|\frac{1}{\underset{1}{\lambda}}-\frac{1}{\underset{2}{\lambda}}\right|\right), \tag{4.16}$$

$$\rho_2=(1-\rho_1^2)^{\frac{1}{2}}. \tag{4.17}$$

We see that $\rho_1\to 0$ and $\rho_2\to 1$ when $p\to\pm\infty$. Hence we call $\Phi^1(1)$ a single soliton solution.

Let $\gamma=x_1+ix_2$, $\beta=x_3+ix_4$, $\Phi^1(1)$ can be considered as a harmonic map from R^2 to S^3.

We describe the geometric character of the harmonic map $\Phi^1(1)$. Let l be a straight line in R^2, which is not parallel to the line $p=$const. It is easy to see that if (x, y) approaches to infinity along the line l, then p approaches to ∞ and $\rho_1\to 0$. Hence the image of l approaches to the equator $x_3=x_4=0$ of S^3. Such kind of straight lines are called generic lines. The straight lines $p=$ const., are called special lines and their images do not approach to the equator. Thus, for the single soliton solution $\Phi^1(1)$, there is one family of special lines $p=$const. whose images are some curves with $\rho_1=$ const.

We define the kth Darboux transformation of the trivial solution (4.2) to be k-soliton solution and describe their asymptotic behavior as we have done in [2]. Their explicit expressions for the extended solutions can be obtained recursively, i.e.

$$\Phi^k(\lambda) = (I + \lambda\alpha_{k-1})\cdots(I + \lambda\alpha_0)\Phi_0(\lambda) \cdot \left(1 - \frac{\lambda}{\lambda_2^{(0)}}\right)^{-1}\left(1 - \frac{\lambda}{\lambda_2^{(1)}}\right)^{-1}\cdots\left(1 - \frac{\lambda}{\lambda_2^{(k-1)}}\right)^{-1} \quad (4.18)$$

and the harmonic maps are

$$\Phi^k(1) = (I + \alpha_{k-1})\cdots(I + \alpha_0)\Phi_0(1) \cdot \left(1 - \frac{1}{\lambda_2^{(0)}}\right)^{-1}\left(1 - \frac{1}{\lambda_2^{(1)}}\right)^{-1}\cdots\left(1 - \frac{1}{\lambda_2^{(k-1)}}\right)^{-1}. \quad (4.19)$$

Here $(\lambda_1^{(i)}, \lambda_2^{(i)})$ for $i = 0, 1, \cdots, k-1$ are the parameters which are used in the successive Darboux transformations and satisfy the condition: all the $\left|2\lambda_2^{(i)} - 1\right|$ $(i=0, 1, \cdots, k-1)$ are distinct.

Moreover, in (4.18) and (4.19), α_i $(i = 0, 1, \cdots, k-1)$ are constructed from $\Phi^i(\lambda)$ by using Theorems 3.1 - 3.3. Define

$$p_i = \left(\lambda_1^{(i)} - \lambda_2^{(i)}\right)\tau\bar{\zeta} + \left(\bar{\lambda}_1^{(i)} - \bar{\lambda}_2^{(i)}\right)\bar{\tau}\zeta.$$

A straight line l in R^2 is called generic if it is not parallel to the lines $p_i =$ const. $(i=0, 1, \cdots, k-1)$, and the k families of lines $p_i =$ const. are called special lines. We have

Theorem 4.1. *The k-soliton solution* $\Phi^k(1) = \begin{pmatrix} \gamma^{(k)} & \beta^{(k)} \\ -\bar{\beta}^{(k)} & \bar{\gamma}^{(k)} \end{pmatrix} \in SU(2)$ *which can be considered as a harmonic map from* R^2 *to* S^3 *has the following properties:*

(i) $\beta^{(k)}$ *approaches to* 0 *when* (x, y) *approaches to infinity along a generic line* l, *i.e. the image of the line* l *approaches to the equator* $x_3 = x_4 = 0$ *asymptotically.*

(ii) There are k *families of special lines* $p_i = const.$ $(i = 0, 1, \cdots, k-1)$, *and* $\Phi^k(1)$ *behaves asymptotically as a single soliton when* (x, y) *approaches to infinity along each special line.*

The proof is similar to the proof of the main theorem for the R^{1+1} case[2].

If we consider y as the time coordinate and $y=$ const. are generic lines, Theorem 4.1 implies that when $y\to\pm\infty$, a k-soliton is splitting up into k single solitons asymptotically, and the interaction of solitons is elastic if we consider the magnitudes ρ_1, ρ_2 only.

§ 5. Transformation of Unitons

From now on, we write the parameter $\lambda=\frac{1-\mu}{2}$, and the Lax pair (2.12) can be written as

$$\frac{\partial\Psi(\mu)}{\partial\bar{\zeta}}=\frac{1-\mu}{2}A\Psi,\ \frac{\partial\Psi(\mu)}{\partial\zeta}=\frac{1-\mu^{-1}}{2}B\Psi. \tag{5.1}$$

Here

$$\Psi(\mu)=\Phi\left(\frac{1-\mu}{2}\right). \tag{5.2}$$

The parameter μ is used in [4] where it is denoted by λ.

The concept of uniton was introduced by Uhlenbeck in [4]. Let g be a harmonic map. If there is an extended solution $\Psi(\mu)$ which satisfies the following conditions

(a) $\Psi(\mu)=\sum_{a=0}^{n}T_a\mu^a$ (a polynomial of μ),

(b) $\Psi(1)=I$,

(c) $\Psi(-1)=g^{-1}Q$ ($Q\in U(N)$, a constant matrix),

(d) $(\Psi(\bar{\mu}))^*=(\Psi(\mu^{-1}))^{-1}$ $(\mu\neq 0)$.

g is called a uniton, and $\Psi(\mu)$ is called extended solution of a uniton or simply extended uniton.

From the above definition, we see that the soliton solutions which we obtained in the above section are not unitons.

By using the parameter μ

$$\underset{1}{\lambda}=\frac{1-\mu_1}{2},\ \underset{2}{\lambda}=\frac{1-\mu_2}{2}=\frac{1-\bar{\mu}_1^{-1}}{2}. \tag{5.3}$$

From (3.29), the Darboux transformation of an extended uniton is

$$\Psi^1(\mu)=\left(I+\frac{1-\mu}{2}\alpha\right)\left(1-\frac{1-\mu}{1-\mu_2}\right)^{-1}\Psi(\mu). \tag{5.4}$$

In general, it cannot be an extended uniton, since $\Psi^1(\mu)$ is not a polynomial of μ. Uhlenbeck introduced the singular Bäcklund transformation to obtain a uniton from a known uniton. Here we introduce the singular Darboux transformation as the limit of a sequence of Darboux transformations. We treat this problem as follows.

Let

$$\begin{aligned} L_1&=[l_1,\ l_2,\ \cdots,\ l_k,\ 0,\ \cdots,\ 0],\\ L_2&=[0,\ \cdots,\ 0,\ l_{k+1},\ \cdots,\ l_N] \end{aligned} \tag{5.5}$$

be two constant matrices of rank k and $N-k$ $(0<k<N)$ respectively. Here l_α $(\alpha=1,\ \cdots,\ N)$ are N constant columns satisfying

$$l_p^* l_a=0 \quad (p=k+1,\ \cdots,\ N;\ a=1,\ 2,\ \cdots,\ k). \tag{5.6}$$

We take $\mu_1=\varepsilon$, $\mu_2=\bar{\varepsilon}^{-1}$ $(\varepsilon\neq 0)$ and apply Darboux transformation to the extended uniton $\Psi(\mu)$. Let

$$H_\varepsilon=[\underset{1}{h},\ \cdots,\ \underset{N}{h}]=[\Psi(\varepsilon)l_1,\ \cdots,\ \Psi(\varepsilon)l_k,\ \Psi(\bar{\varepsilon}^{-1})l_{k+1},\ \cdots,\ \Psi(\bar{\varepsilon}^{-1})l_N]. \tag{5.7}$$

From the condition (d), we have

$$\begin{aligned} h_p^* h_a&=(\Psi(\bar{\varepsilon}^{-1})l_p)^*\,\Psi(\varepsilon)l_a\\ &=l_p^*\,\Psi^*(\bar{\varepsilon}^{-1})\Psi(\varepsilon)l_a \end{aligned}$$

$$= l_p^* l_a = 0. \tag{5.8}$$

We take ε such that $\det \Psi(\varepsilon) \neq 0$, $\det \Psi(\bar{\varepsilon}^{-1}) \neq 0$, hence

$$\det H_\varepsilon \neq 0. \tag{5.9}$$

Let

$$\alpha_\varepsilon = -H_\varepsilon \Lambda_\varepsilon^{-1} H_\varepsilon^{-1},$$
$$\pi_\varepsilon = \frac{1}{2(1-\varepsilon\bar{\varepsilon})}[(1-\varepsilon)(1-\bar{\varepsilon})\alpha_\varepsilon + 2(1-\bar{\varepsilon})I]. \tag{5.10}$$

The new extended solutions of the harmonic maps obtained by Darboux transformation are

$$\Psi_\varepsilon^{(1)}(\mu) = (\pi_\varepsilon + \mu\pi_\varepsilon^\perp)\left(1 - \frac{1-\mu}{1-\bar{\varepsilon}^{-1}}\right)^{-1} \Psi(\mu). \tag{5.11}$$

In order to elucidate the limiting process of $\varepsilon \to 0$, we need the following lemma.

Lemma. *Let $\mathcal{L}_1$ (resp. $\mathcal{L}_2$) be the subspace of C^N spanned by $l_1, l_2, \cdots, l_k$ (resp. $l_{k+1}, \cdots, l_N$). Then α_ε depends only on $\mathcal{L}_1$ and $\mathcal{L}_2$, and is independent of the choice of the basis of $\mathcal{L}_1$ and $\mathcal{L}_2$.*

Proof. We note at first that $\mathcal{L}_2 = \mathcal{L}_1^\perp$ is determined by $\mathcal{L}_1$. The change of basis of $\mathcal{L}_1$ and $\mathcal{L}_2$ means that we use

$$\widetilde{L}_1 = L_1 \begin{bmatrix} K_1 & 0 \\ 0 & 0 \end{bmatrix}, \widetilde{L}_2 = L_2 \begin{bmatrix} 0 & 0 \\ 0 & K_2 \end{bmatrix} \tag{5.12}$$

to replace L_1 and L_2. Here K_1 (resp. K_2) is a regular square matrix of order k (resp. $N-k$). We write H_ε and H_ε^{-1} by

$$H_\varepsilon = \begin{bmatrix} A_{11} & A_{12} \\ A_{21} & A_{22} \end{bmatrix}, H_\varepsilon^{-1} = \begin{bmatrix} B_{11} & B_{12} \\ B_{21} & B_{22} \end{bmatrix} \tag{5.13}$$

respectively, here the blocks of H and H^{-1} are $k \times k$, $k \times (N-k)$, $(N-k) \times k$, $(N-$

$k) \times (N-k)$ matrices.

The matrices $\widetilde{H}_\varepsilon$, $\widetilde{H}_\varepsilon^{-1}$ which are constructed by using the column vectors of $\widetilde{L}_1$ and $\widetilde{L}_2$ are:

$$\widetilde{H}_\varepsilon = \begin{bmatrix} A_{11}K_1 & A_{12}K_2 \\ A_{21}K_1 & A_{22}K_2 \end{bmatrix} \tag{5.14}$$

and

$$\widetilde{H}_\varepsilon^{-1} = \begin{bmatrix} K_1^{-1}B_{11} & K_1^{-1}B_{12} \\ K_2^{-1}B_{21} & K_2^{-1}B_{22} \end{bmatrix}. \tag{5.15}$$

By calculation, it is easily seen that

$$\widetilde{\alpha}_\varepsilon = -\widetilde{H}_\varepsilon \Lambda_\varepsilon^{-1} \widetilde{H}_\varepsilon^{-1} = -H_\varepsilon \Lambda_\varepsilon^{-1} H_\varepsilon^{-1} = \alpha_\varepsilon. \tag{5.16}$$

The lemma is proved.

We choose a special base of $\mathcal{L}_1$ such that

$$L_1 = L_1^{(0)} + L_1^{(1)} + \cdots + L_1^{(n)} \tag{5.17}$$

with

$$\begin{aligned} L_1^{(0)} &= [\tilde{l}_1, \cdots, \tilde{l}_{a_0}, \ 0, \cdots, 0, \ 0, \cdots, 0, \cdots, \ 0, \cdots, 0; \ 0, \cdots, 0], \\ L_1^{(1)} &= [0, \cdots, 0, \ \tilde{l}_{a_0+1}, \cdots, \tilde{l}_{a_1}, \ 0, \cdots, 0, \cdots, \ 0, \cdots, 0; \ 0, \cdots, 0], \\ &\cdots\cdots \\ L_1^{(n)} &= [0, \cdots, 0, \ 0, \cdots, 0, \ 0, \cdots, 0, \cdots, \ \tilde{l}_{a_{n-1}+1}, \cdots, \tilde{l}_{a_n}; \ 0, \cdots, 0]; \end{aligned} \tag{5.18}$$

$$\begin{gathered} T_i L_1^{(j)} = 0 \ (j > i), \\ \operatorname{rank}\{T_j L_1^{(j)}\} = \operatorname{rank}\{L_1^{(j)}\} = a_j - a_{j-1}, \\ (k = a_n \geqslant a_{n-1} \geqslant \cdots a_0 \geqslant 0, \ a_{-1} = 0). \end{gathered} \tag{5.19}$$

Define

$$\widetilde{L}_1 = L_1^{(0)} + L_1^{(1)}\varepsilon^{-1} + \cdots + L_1^{(n)}\varepsilon^{-n}(\varepsilon \neq 0). \tag{5.20}$$

Then

$$\Psi(\varepsilon)\widetilde{L}_1 = T_0 L_1^{(0)} + T_1 L_1^{(1)} + \cdots + T_n L_1^{(n)} + \varepsilon F_1, \tag{5.21}$$

where F_1 is a polynomial of ε. Similarly, we choose a special base of $\mathcal{L}_2$ and define

$$\widetilde{L}_2 = L_2^{(n)}\bar{\varepsilon}^{n} + L_2^{(n-1)}\bar{\varepsilon}^{n-1} + \cdots + L_2^{(0)}. \tag{5.22}$$

Here $L_2^{(i)}$ satisfy

$$T_j L_2^{(i)} = 0\ (j > i), \tag{5.23}$$

$$\mathrm{rank}\{T_j L_2^{(j)}\} = \mathrm{rank}\{L_2^{(j)}\}.$$

Then

$$\Psi(\bar{\varepsilon})\widetilde{L}_2 = T_0 L_2^{(0)} + T_1 L_2^{(1)} + \cdots + T_n L_2^{(n)} + \bar{\varepsilon} F_2, \tag{5.24}$$

where F_2 is a polynomial of $\bar{\varepsilon}$. Moreover, we take

$$\widetilde{H}_\varepsilon = \Psi(\varepsilon)\widetilde{L}_1 + \Psi(\bar{\varepsilon}^{-1})\widetilde{L}_2. \tag{5.25}$$

From the lemma, we have

$$\begin{aligned} \alpha_\varepsilon &= -\widetilde{H}_\varepsilon \Lambda_\varepsilon^{-1} \widetilde{H}_\varepsilon^{-1} \\ &= -\frac{2}{1-\bar{\varepsilon}^{-1}}\pi_\varepsilon - \frac{2}{1-\varepsilon}\pi_\varepsilon^{\perp}. \end{aligned} \tag{5.26}$$

Here

$$\Lambda_\varepsilon^{-1} = \begin{bmatrix} \frac{2}{1-\varepsilon} I_k & 0 \\ 0 & \frac{2}{1-\bar{\varepsilon}^{-1}} I_{N-k} \end{bmatrix} \tag{5.27}$$

and π_ε is a projection on $(N-k)$-dimensional subspaces of C^N.

Let $\varepsilon \to 0$. We have

$$\lim_{\varepsilon \to 0} H_\varepsilon = \sum_{a=0}^{n} T_a L_1^{(a)} + \sum_{a=0}^{n} T_a L_2^{(a)} = H. \tag{5.28}$$

Evidently, H is a regular matrix. Let

$$\lim_{\varepsilon \to 0} \alpha_\varepsilon = -H \begin{bmatrix} 2I_k & 0 \\ 0 & 0 \end{bmatrix} H^{-1} = \alpha, \tag{5.29}$$

$$\lim_{\varepsilon \to 0} \pi_\varepsilon^\perp = H \begin{bmatrix} I_k & 0 \\ 0 & 0 \end{bmatrix} H^{-1} = \pi^\perp, \tag{5.30}$$

$$\lim_{\varepsilon \to 0} \pi_\varepsilon = H \begin{bmatrix} 0 & 0 \\ 0 & I_{N-k} \end{bmatrix} H^{-1} = \pi. \tag{5.31}$$

Using (3.29), we have

Theorem 5.1. *From an extended uniton $\Phi(\mu)$, we can construct the extended uniton*

$$\Phi^1(\mu) = \left(I + \frac{1-\mu}{2}\alpha\right)\Phi(\mu). \tag{5.32}$$

Here

$$\alpha = -H \begin{bmatrix} 2I_k & 0 \\ 0 & 0 \end{bmatrix} H^{-1} \tag{5.33}$$

and H is defined by (5.28).

Remark 5.1. Since

$$\begin{aligned} I + \frac{1-\mu}{2}\alpha &= \pi + \pi^\perp + (\mu-1)\pi^\perp \\ &= \pi + \mu\pi^\perp, \end{aligned} \tag{5.34}$$

(5.32) can be written in

$$\Phi^1(\mu)=(\pi+\mu\pi^{\perp})\Phi(\mu). \tag{5.35}$$

This is just the main formula in Theorem 12.1 of [4]. In our case, the projection π has explicit expression (5.31). We will discuss the properties and concrete applications of the singular Darboux transformation in a forthcoming paper.

Acknowledgements. A part of results in the present paper was obtained in Tokyo when the authors were visiting the Dept. of Mathematical Sciences of the Univ. of Tokyo, July 1994. The results were announced there. The authors are grateful to Prof. H. Komatsu and the Dept. of Mathematical Sciences, Univ. of Tokyo for the hospitality and support. The first author is indebted to the Inoue Scientific Foundation for the support.

References

[1] Gu, C.H. & Zhou, Z.X., Darboux transformations for principal chiral fields, in the Proc. of the International Conference "Nonlinear Evolution Equations, Como, 1988", Manchester Univ. Press, 1990.

[2] Gu, C. H. & Hu, H. S., The soliton behavior of the principal chiral fields, Lectured in one of the main courses of the School on Qualitative Aspects and Applications of Nonlinear Evolution Equations, Trieste, 1990; in *Int. J. Mod. Phys. A* (Proc. of International Conference on Differential Geometrical Methods in Theoretical Physics, Tianjin, 1992), **3** (1993), 501－510.

[3] Beggs, E. J., Solitons in the chiral equations, *Commun. Math. Phys.*, **128** (1990), 131－139.

[4] Uhlenbeck, K., Harmonic maps into Lie groups (classical solutions of the chiral model), *J. Diff. Geom.*, **38** (1990), 1－50.

(本文曾发表于 *Chinese Annals of Mathematics*, 1995, 16B(2), 139－152)

Darboux Transformations Between $\Delta\alpha = \sinh\alpha$ and $\Delta\alpha = \sin\alpha$ and the Application to Pseudo-Spherical Congruences in $R^{2,1}$

Dedicated to the memory of Moshé Flato

Hu Hesheng (H.S. HU) *

Institute of Mathematics, Fudan University, Shanghai, China

(Received: 5 April 1999)

Abstract In this Letter, it is proved that by using Darboux transformation, from a known solution of the sinh-Laplace equation (resp. sine-Laplace equation), new solutions of sine-Laplace equation (resp. sinh-Laplace equation) are obtained explicitly. The corresponding geometrical configuration is the space-like pseudo-spherical line congruence with two focal surfaces of negative constant curvature being space-like and time-like, respectively.

Mathematics Subject Classifications (1991): 53C20, 35Q58, 58F07.

Key words: Darboux transformation, sine-Laplace equation, sinh-Laplace equation, pseudo-spherical congruence.

---

* Supported by the Chinese National Foundation of Natural Science, the Scientific Foundation of the National Education Department of China for Educating Doctors, and the Research Foundation of Education Commission of Shanghai.

1. Introduction

The sine-Gordon equation (SG) is one of the most important soliton equations. The study of SG started in the 19th century and was related to the construction of pseudo-spherical surfaces (surfaces of negative constant curvature) in Euclidean space R^3 [2]. Now SG is considered as one of the important roots of modern soliton theory [4].

The Bäcklund transformations (BT) between solutions of sine-Gordon and the pseudo-spherical surfaces play a very important role. On the other hand, the elliptic versions of SG, named the sine-Laplace equation, together with the sinh-Laplace equation are also very important in physics and geometry [5 - 10]. The BTs of such equations have been considered in [8, 10, 12].

The Bäcklund transformations have been used in soliton theory to construct new solutions from a known solution for some nonlinear PDEs [10, 8]. For SG and many other soliton equations, the BTs can be replaced by Darboux transformations (DT) [1, 3, 11]. The advantage of DT method is that the new solutions are expressed explicitly, the algorithm is purely algebraic and can be used successively.

In this Letter, it is shown that the DT method can be successfully applied to both the sinh-Laplace and sine-Laplace equations. Firstly, we consider the DT method for the complexified sinh-Laplace equation (Section 2). Afterwards, we prove that, from a known real solution of the sinh-Laplace equation (resp. sine-Laplace equation), new real solutions of sine-Laplace (resp. sinh-Laplace equation) can be obtained explicitly by using DT (Section 3). It is interesting to see that the corresponding geometrical configuration is the space-like pseudo-spherical congruence in Minkowski space $R^{2,1}$ in which one focal surface is space-like and the other focal surface is time-like, both are of negative constant curvature. Thus, DT also gives a method to construct these surfaces with an algebraic algorithm (Section 4). At the end of Letter, a complete classification of pseudo-spherical congruences is given (Section 5).

2. Complexified sinh-Laplace Equation

The sine-Laplace and sinh-Laplace equations

$$\Delta\alpha = \frac{\partial^2 \alpha}{\partial u^2} + \frac{\partial^2 \alpha}{\partial v^2} = \sin\alpha, \tag{1}$$

$$\Delta\alpha = \sinh\alpha \tag{2}$$

are important nonlinear partial differential equations. They have interesting applications in physics and geometry.

We introduce complex coordinates of the (u, v) plane

$$\zeta = \frac{u + iv}{2}, \quad \bar{\zeta} = \frac{u - iv}{2}. \tag{3}$$

Then

$$\partial_\zeta = \frac{\partial}{\partial u} - i\frac{\partial}{\partial v}, \quad \partial_{\bar{\zeta}} = \frac{\partial}{\partial u} + i\frac{\partial}{\partial v} \tag{4}$$

and Equations (1), (2) can be written as

$$\partial_{\zeta\bar{\zeta}}\alpha = \sin\alpha, \tag{1'}$$

$$\partial_{\zeta\bar{\zeta}}\alpha = \sinh\alpha. \tag{2'}$$

If α is a complex function, Equations (1) and (2) can be contained in the complexified sinh-Laplace equation

$$\Delta\alpha = \sinh\alpha. \tag{5}$$

In fact, when α is real, it is reduced to a real sinh-Laplace equation, when α is purely imaginary and $\alpha = i\beta$ (β real), β satisfies the sine-Laplace equation.

The complexified sinh-Laplace equation has a Lax pair

$$\Phi_\xi = \frac{\lambda}{2}\begin{bmatrix} 0 & -e^{-\alpha} \\ e^{\alpha} & 0 \end{bmatrix}\Phi, \quad \Phi_{\bar{\xi}} = \frac{1}{2}\begin{bmatrix} -\alpha_{\bar{\xi}} & \frac{1}{\lambda} \\ -\frac{1}{\lambda} & \alpha_{\bar{\xi}} \end{bmatrix}\Phi, \tag{6}$$

i.e., (5) is the integrability condition of (6). Let $\Phi(\lambda)$ be a solution of (6) with det $\Phi \neq 0$. If $(\alpha, \Phi(\lambda))$ is known, we use Darboux transformation to get new solutions of the complexified sinh-Laplace equation and its Lax pair. Let $\begin{bmatrix} h_1 \\ h_2 \end{bmatrix}$ be a column solution of the Lax pair with $\lambda = \lambda_1 \neq 0$, i.e.,

$$\begin{bmatrix} h_1 \\ h_2 \end{bmatrix} = \Phi(\lambda_1) l \quad (l \neq 0, \text{ a constant column}). \tag{7}$$

Let

$$S = \frac{1}{\lambda_1}\begin{bmatrix} 0 & \frac{h_1}{h_2} \\ \frac{h_2}{h_1} & 0 \end{bmatrix}, \tag{8}$$

we have the following theorem:

Theorem A. *Let α be a solution of the complexified sinh-Laplace equation, Φ be the solution of its Lax pair, then the transformation $(\alpha, \Phi) \to (\alpha_1, \Phi_1)$ with*

$$\Phi_1 = [I - \lambda S]\Phi, \tag{9}$$

$$e^{\alpha_1} = -\left(\frac{h_2}{h_1}\right)^2 e^{-\alpha}, \tag{10}$$

gives a new solution (α_1, Φ_1) to the sinh-Laplace equation and its Lax pair.

Proof. Substituting (9) into the Lax pair (6) in which α is replaced by α_1, we will see that it is satisfied if (10) holds. α_1 should satisfy the complexified sinh-Laplace equation(5), since (5) is the integrability condition of (6). □

It should be noted that we can do Darboux transformation

$$(\alpha, \Phi) \to (\alpha_1, \Phi_1) \to (\alpha_2, \Phi_2) \to (\alpha_3, \Phi_3) \to \cdots \tag{11}$$

successively by using the same algebraic algorithm.

3. Darboux Transformations Between the sine-Laplace Equation and the sinh-Laplace Equation

In general, the DTs in Section 2 give complex solutions. However, we have the following theorem:

Theorem B. *If α is a real solution of the sinh-Laplace equation, there is a DT such that $\alpha_1 = i\beta_1$ is purely imaginary, and β_1 is a real solution of the sine-Laplace equation. If α is purely imaginary, there is a DT such that α_1 is a real solution of the sinh-Laplace equation.*

Proof. If α is real, from (5), α_1 is purely imaginary iff

$$\left|\frac{h_2}{h_1}\right|^2 e^{-\alpha} = 1. \tag{12}$$

Let

$$A = \left|\frac{h_2}{h_1}\right|^2 e^{-\alpha} - 1. \tag{13}$$

From the Lax pair(6), it is seen that

$$
\begin{aligned}
h_{1\zeta} &= -\frac{\lambda_1}{2} e^{-\alpha} h_2, & h_{2\zeta} &= \frac{\lambda_1}{2} e^{\alpha} h_1, \\
h_{1\bar{\zeta}} &= -\frac{\alpha_{\bar{\zeta}}}{2} h_1 + \frac{1}{2\lambda_1} h_2, & h_{2\bar{\zeta}} &= -\frac{1}{2\lambda} h_1 + \frac{\alpha_{\bar{\zeta}}}{2} h_2.
\end{aligned} \tag{14}
$$

If $|\lambda_1| = 1$, for real α from (14) it follows

$$
A_\zeta = \frac{\lambda_1}{2}\left(\frac{h_2}{h_1} e^{-\alpha} - \frac{\bar{h}_2}{\bar{h}_1}\right) A, \quad A_{\bar{\zeta}} = \frac{\lambda_1}{2}\left(\frac{h_2}{h_1} + \frac{\bar{h}_2}{\bar{h}_1} e^{-\alpha}\right) A. \tag{15}
$$

Hence, if $A = 0$ at one point, then $A = 0$ everywhere.

We choose $\begin{bmatrix} h_1 \\ h_2 \end{bmatrix}$ such that $|h_2/h_1|^2 e^{-\alpha} = 1$ at one point, then α_1 is purely imaginary. Let $\alpha_1 = i\beta_1$. We get the real solution β_1 of the sine-Laplace equation.

Similarly, if α is purely imaginary, α_1 is real iff

$$
\overline{\left(\frac{h_2}{h_1}\right)^2 e^{-\alpha}} = \left(\frac{h_2}{h_1}\right)^2 e^{-\alpha} < 0 \quad (\overline{e^{-\alpha}} = e^{\alpha}) \tag{16}
$$

or

$$
B = \frac{\bar{h}_2 h_1}{\bar{h}_1 h_2} e^{\alpha} + 1 = 0. \tag{17}
$$

If $|\lambda_1| = 1$, for pure imaginary α, by using (14) we see that

$$
B_\zeta = \frac{\lambda_1}{2}\left(\frac{\bar{h}_1}{\bar{h}_2} + \frac{h_2}{h_1} e^{-\alpha}\right) B, \quad B_{\bar{\zeta}} = -\frac{1}{2\lambda_1}\left(\frac{h_2}{h_1} + \frac{\bar{h}_1}{\bar{h}_2} e^{\alpha}\right) B. \tag{18}
$$

Hence, we choose $\begin{bmatrix} h_1 \\ h_2 \end{bmatrix}$ such that $B = 0$ at one point, then $B = 0$ everywhere. Thus we obtain a solution α_1 of real sinh-Laplce equation. □

We can use DT successively and then obtain solutions of the sine-Laplace and sinh-

Laplace equations alternatively.

$$\alpha_0 \to i\beta_1 \to \alpha_2 \to i\beta_3 \to \alpha_4 \to \cdots \quad \Delta\alpha_j = \sinh\alpha_j,$$

$$i\beta_0 \to \alpha_1 \to i\beta_2 \to \alpha_3 \to i\beta_4 \to \cdots \quad \Delta\beta_j = \sin\beta_j.$$

Examples. Starting from the trivial solution $\alpha = 0$ of (1) and (2), we can get two sequences of solutions of the sinh-Laplace and sine-Laplace equations, respectively.

For $\alpha = 0$, the Lax pair (6) is reduced to

$$\Phi_\zeta = \frac{\lambda}{2}\begin{bmatrix} 0 & -1 \\ 1 & 0 \end{bmatrix}\Phi, \quad \Phi_{\bar\zeta} = \frac{1}{2}\begin{bmatrix} 0 & \frac{1}{\lambda} \\ -\frac{1}{\lambda} & 0 \end{bmatrix}\Phi, \tag{19}$$

Hence

$$\Phi(\lambda) = \begin{bmatrix} e^{\gamma} & e^{-\gamma} \\ -i\,e^{\gamma} & i\,e^{-\gamma} \end{bmatrix} \quad \left(\gamma = \left(\frac{\lambda}{2}\zeta - \frac{1}{2\lambda}\bar\zeta\right)i\right)$$

is a fundamental solution of (6).

Because $\alpha = 0$ can be considered as real or imaginary, we first consider the latter case and want to obtain real α_1 through DT. From (10), we have

$$e^{\alpha_1} = -\left(\frac{h_2}{h_1}\right)^2.$$

Let $\lambda = \lambda_1 (|\lambda_1| = 1)$, then

$$\gamma_1 = \left(\frac{\lambda_1}{2}\zeta - \frac{1}{2\lambda_1}\bar\zeta\right)i$$

is real and we choose

$$h_1 = e^{\gamma_1} + b\,e^{-\gamma_1}, \quad h_2 = -i\,e^{\gamma_1} + bi\,e^{-\gamma_1} \tag{20}$$

as the column solution.

For real b, we have

$$\frac{h_2 \bar{h}_1}{h_1 \bar{h}_2} = -1. \tag{21}$$

Taking $b = 1$, we have

$$e^{\frac{\alpha_1}{2}} = -\frac{1}{i}\frac{h_2}{h_1} = \frac{e^{\gamma_1} - e^{-\gamma_1}}{e^{\gamma_1} + e^{-\gamma_1}} = \tanh \gamma_1 \quad (\text{in the region } \gamma_1 > 0),$$

i.e.,

$$\alpha_1 = 2\ln \tanh \gamma_1. \tag{22}$$

They are solutions of the sinh-Laplace equation.

In order to get purely imaginary $\alpha_1 = i\beta_1$, we should have $h_2 \bar{h}_2 / h_1 \bar{h}_1 = 1$. Let $b = \pm i$, then

$$\beta_1 = \mp 2\mathrm{tg}^{-1}(\mathrm{csch}\, 2\gamma_1). \tag{23}$$

Thus, we get solutions to the sine-Laplace equation.

From α_1, β_1 together with the corresponding Φ_1 which can be written down explicitly by using (9), we can successively construct further solutions.

4. A Class of Pseudo-Spherical Congruence in $R^{2,1}$

In Euclidean space, a pseudo-spherical congruence consists of straight lines which are common tangents of two surfaces of constant negative curvature. These congruences define the Bäcklund transformations between surfaces of negative constant curvature. But the systems of partial differential equations for the Bäcklund transformation is not easy to solve. On the other hand, by using Darboux transformations, one can obtain explicit solutions. This is an advantage of Darboux transformations. In this section we consider a class of pseudo-spherical congruences in Minkowski space $R^{2,1}$ which are related to the Darboux transformations between the sine-Laplace and sinh-Laplace equations.

In Minkowski space $R^{2,1}$, a surface S may be expressed as $\mathbf{r}=\mathbf{r}(u, v)$. Let $\mathbf{e}_1$, $\mathbf{e}_2$ be two unit tangential vectors and $\mathbf{n}$ the normal, the fundamental equations of S are

$$\begin{aligned} &\mathrm{d}\mathbf{r}=\omega^a\mathbf{e}_a \quad (a=1,2), \\ &\mathrm{d}\mathbf{e}_a=\omega_a^b\mathbf{e}_b+\omega_a^3\mathbf{n}, \quad \mathrm{d}\mathbf{n}=\omega_3^a\mathbf{e}_a. \end{aligned} \tag{24}$$

If $\mathbf{n}$ is time-like, the surface is called space-like and we have

$$\mathbf{e}_1^2=\mathbf{e}_2^2=1, \quad \mathbf{n}^2=-1. \tag{25}$$

If $\mathbf{n}$ is space-like, the surface is called time-like and we have

$$\mathbf{e}_1^2=1, \quad \mathbf{e}_2^2=-1, \quad \mathbf{n}^2=1. \tag{26}$$

The Chebyshev coordinates for a surface of negative constant curvature in R^3 was generalized to the space-like and time-like surfaces of constant curvature in Minkowski space $R^{2,1}$ [5, 6].

For a time-like surface with $K=-1$, we have

$$\begin{aligned} &\omega^1=\cos\frac{\beta}{2}\mathrm{d}u, \qquad \omega^2=\sin\frac{\beta}{2}\mathrm{d}v, \\ &\omega_2^1=\omega_1^2=\frac{1}{2}(-\beta_v\mathrm{d}u+\beta_u\mathrm{d}v), \\ &\omega_1^3=-\omega_3^1=\sin\frac{\beta}{2}, \qquad \omega_2^3=\omega_3^2=\cos\frac{\beta}{2} \end{aligned} \tag{27}$$

and β is a solution of the sine-Laplace equation.

For a space-like surface of constant curvature $K=-1$, we have

$$\begin{aligned} &\omega^1=\cosh\frac{\alpha}{2}\mathrm{d}u, \qquad \omega^2=\sinh\frac{\alpha}{2}\mathrm{d}v, \\ &\omega_2^1=-\omega_1^2=\frac{1}{2}(\alpha_v\mathrm{d}u-\alpha_u\mathrm{d}v), \\ &\omega_1^3=\omega_3^1=\sinh\frac{\alpha}{2}\mathrm{d}u, \quad \omega_2^3=\omega_3^2=\cosh\frac{\alpha}{2}\mathrm{d}v \end{aligned} \tag{28}$$

and α is a solution of the sinh-Laplace equation.

We want to construct a congruence whose two focal surfaces are of $K=-1$ and are time-like and space-like, respectively.

We start from a time-like surface S with $K=-1$: $\mathbf{r}=\mathbf{r}(u, v)$. Let

$$S^* = \mathbf{r}^*(u, v) = \mathbf{r} + l(\cosh\theta \mathbf{e}_1 + \sinh\theta \mathbf{e}_2). \tag{29}$$

Here $l=\cosh\tau$ is a constant and θ is to be determined. It is required that S^* be another focal surface, $\overline{rr^*}$ be the common tangent to S and S^*, and $\mathbf{n}^* \cdot \mathbf{n} = \sinh\tau$. Here

$$\mathbf{n}^* = \cosh\tau(\sinh\theta\mathbf{e}_1 + \cosh\theta\mathbf{e}_2) + \sinh\tau\mathbf{n} \tag{30}$$

is the normal of S^*. The congruence $\{\overline{rr^*}\}$ is the pseudo-spherical congruence to be constructed and S, S^* are its focal surfaces.

Since $\mathbf{n}^*$ is the normal of S^*, we have $\mathbf{n}^* \cdot d\mathbf{r}^* = 0$. From (29) and (30) we have

$$\begin{aligned} l\left(\frac{\alpha_{1u}}{2} - \frac{\beta_v}{2}\right) &= \sinh\frac{\alpha_1}{2}\cos\frac{\beta}{2} + \sinh\tau\cosh\frac{\alpha_1}{2}\sin\frac{\beta}{2}, \\ l\left(\frac{\alpha_{1v}}{2} + \frac{\beta_u}{2}\right) &= -\cosh\frac{\alpha_1}{2}\sin\frac{\beta}{2} + \sinh\tau\sinh\frac{\alpha_1}{2}\cos\frac{\beta}{2}. \end{aligned} \tag{31}$$

Here $\alpha_1 = \theta/2$. This is a system of equations for α_1 and so the problem is reduced to solving (31) for α_1. The condition of integrability of (31) is $\Delta\alpha = \sin\alpha$ which is satisfied since S is time-like and of constant curvature -1, so the solution α_1 exists. Moreover, from (31) we can prove that α_1 satisfies the sinh-Laplace equation. Equation (31) has appeared in [9, 12]. We are going to prove that the Darboux transformation in Section 3 gives an explicit solution of (31).

Write Equation (10) of the Darboux transformation in the form

$$e^{\alpha_1} = -e^{-i\beta}\left(\frac{h_2}{h_1}\right)^2. \tag{32}$$

Then we have

$$e^{(\alpha_1+i\beta)/2}=i\frac{h_2}{h_1}.$$

Differentiate it, by using (14), and we obtain

$$\begin{aligned}(\alpha_1+i\beta)_\xi &= 2i\lambda_1\sinh\frac{i\beta-\alpha_1}{2},\\ (\alpha_1-i\beta)_{\bar\xi} &= \frac{2i}{\lambda_1}\sinh\frac{i\beta+\alpha_1}{2}.\end{aligned} \tag{33}$$

Let $\lambda_1=(i+\sinh\tau)/\cosh\tau$ and it can be shown that the DT (10) gives exact solutions to Equations (31). Thus, the DTs give explicit solutions of the system (31). We have proved the following theorem:

Theorem. *If* $(i\beta, \Phi) \to (\alpha_1, \Phi_1)$ *is a Darboux transformation with* $\lambda_1 = (i + \sinh\tau)/\cosh\tau$, *then* (29) *is a transformation of a time-like surface with* $K=-1$ *to a space-like surface with* $K=-1$. *Alternatively, a time-like surface of* $K=-1$ *can be obtained from a space-like surface of* $K=-1$ *by a Darboux transformation.*

Hence, we can use Darboux transformations to construct the pseudo-spherical congruences, space-like surfaces and time-like surfaces of negative constant curvatures. The examples corresponding to the solution (22) and (23) are easily constructed.

5. Complete List of Pseudo-spherical Congruences in R^3 and $R^{2,1}$

We have mentioned the pseudo-spherical congruence in R^3 in Section 4, and have discussed a class of pseudo-spherical congruence in $R^{2,1}$. Besides, there are two other classes of pseudo-spherical congruences in $R^{2,1}$ too. For completeness, we give a full classification of these pseudo-spherical congruences (without proof) (see Table I).

Table Ⅰ A complete classification of pseudo-spherical congruences in R^3 and $R^{2,1}$

Space	Surface	Congruence	DT	Equation
R^3	$\mathbf{r}$ $\mathbf{r}^*$ $K=-1$	$\mathbf{r}^*=\mathbf{r}+l\left(\cos\frac{\alpha_1}{2}\mathbf{e}_1+\sin\frac{\alpha_1}{2}\mathbf{e}_2\right)$ pseudo spherical congruence	α $\downarrow\uparrow$ α_1	$\alpha_{uu}-\alpha_{vv}=\sin\alpha$ $\alpha_{1uu}-\alpha_{1vv}=\sin\alpha_1$
$R^{2,1}$	$\mathbf{r}$ $\mathbf{r}^*$ $K=1$ space-like	$\mathbf{r}^*=\mathbf{r}+l\left(\cos\frac{\alpha_1}{2}\mathbf{e}_1+\sin\frac{\alpha_1}{2}\mathbf{e}_2\right)$ space-like congruence	α $\downarrow\uparrow$ α_1	$\alpha_{uu}-\alpha_{vv}=-\sin\alpha$ $\alpha_{1uu}-\alpha_{1vv}=-\sin\alpha_1$
$R^{2,1}$	$\mathbf{r}$ $\mathbf{r}^*$ $K=1$ time-like	$\mathbf{r}^*=\mathbf{r}+l\left(\cosh\frac{\alpha_1}{2}\mathbf{e}_1+\sinh\frac{\alpha_1}{2}\mathbf{e}_2\right)$ space-like congruence	α $\downarrow\uparrow$ α_1	$\alpha_{uu}-\alpha_{vv}=\sinh\alpha$ $\alpha_{1uu}-\alpha_{1vv}=\sinh\alpha_1$
$R^{2,1}$	$\mathbf{r}$ $K=-1$ time-like $\mathbf{r}^*$ $K=-1$ space-like	$\mathbf{r}^*=\mathbf{r}+l\left(\cosh\frac{\alpha_1}{2}\mathbf{e}_1+\sinh\frac{\alpha_1}{2}\mathbf{e}_2\right)$ space-like congruence	β $\downarrow\uparrow$ α_1	$\Delta\beta=\sin\beta$ $\Delta\alpha_1=\sinh\alpha_1$

All the cases have been treated by Darboux transformations. By complicated and careful calculations, we can prove that time-like pseudo-spherical congruence in $R^{2,1}$ does not exist. It is noted that in Euclidean space there is only one kind of pseudo-spherical congruence with two focal surfaces of negative constant curvature, and there is no pseudo-spherical congruence with focal surfaces of positive constant curvature. On the contrary, in Minkowski space, there are three kinds of pseudo-spherical congruences with all four kinds of surfaces of constant curvature (time-like, space-like and $K=\pm 1$) as focal surfaces.

References

[1] Darboux, G.: Sur une proposition relative aux équations linéaires, *C.R. Hebdomadairs des Seances de l'Academie des Sciences*, *Paris* **94**(1882), 1456.

[2] Eisenhart, L.P.: *A Treatise on the Differential Geometry of Curves and Surfaces*, Dover, New

York, 1960.

[3] Gu, C.H.: On the Bäcklund transformations for the generalized hierarchies of compound MKdV-SG equation, *Lett. Math. Phys.* **11**(1986), 31.

[4] Heyerhoff, M.: The history of the early period of soliton theory, In: D. Wójcik and J. Cieśliński (eds), *Nonlinearity and Geometry*, Polish Scientific Publishers, Warsaw, 1998, pp. 13–23.

[5] Hu, H.S.: The construction of hyperbolic surfaces in 3-dimensional Minkowski space and sinh-Laplace equation, *Acta Math. Sinica New Ser.* **1**(1985), 79–86.

[6] Hu, H.S.: Soliton and differential geometry, In: Gu Chaohao (ed.), *Soliton Theory and Its Applications*, Springer-Verlag, Berlin and Zhejiang Science and Technology Publishing House, 1995, pp.297–336.

[7] Hu, H. S.: Sine-Laplace equation, sinh-Laplace equation and harmonic maps, *Manuscripta Math.* **40**(1982), 205.

[8] Hu, H.S.: The geometry of sine-Laplace, sinh-Laplace equations, In: N. Hu (ed.), *Proc. Marcel Grossmann Meeting on General Relativity*, Science Press, Beijing, 1983, pp.1073–1076.

[9] Huang, Y. Z.: Bäcklund theorems in 3-dimensional Minkowski space and their higher-dimensional generalization, *Acta Math. Sinica* **29**(1986), 684 (in Chinese).

[10] Leibbrandt, G.: Exact solutions of the elliptic sine equation in two space dimensions with the application to the Josephson effect, *Phys. Rev. B* **15**(1977), 3353–3361.

[11] Matveev, V.B. and Salle, M.A.: *Darboux Transformations and Solitons*, Springer-Verlag, Berlin, 1991.

[12] Rogers, C. and Shadwick, W.R.: *Bäcklund Transformations and their Application*, Academic Press, New York, 1982.

（本文曾发表于 *Letters in Mathematical Physics*, 1999, **48**, 187–195）

Laplace Sequences of Surfaces in Projective Space and Two-Dimensional Toda Equations*

Hu Hesheng (H. S. HU)

Institute of Mathematics, Fudan University, Shanghai, China

(Received: 31 January 2001)

Abstract We find that the Laplace sequences of surfaces of period n in projective space $\mathbf{P}_{n-1}$ have two types, while type II occurs only for even n. The integrability condition of the fundamental equations of these two types have the same form

$$\frac{\partial^2 \omega_i}{\partial x \partial t} = -\alpha_{i-1} e^{\omega_{i-1}} + 2\alpha_i e^{\omega_i} - \alpha_{i+1} e^{\omega_{i+1}}, \quad \alpha_i = \pm 1 (i = 1, 2, \cdots, n).$$

When all $\alpha_i = 1$, the above equations become two-dimensional Toda equations. Darboux transformations are used to obtain explicit solutions to the above equations and the Laplace sequences of surfaces. Two examples in $\mathbf{P}_3$ of types I and II are constructed.

Mathematics Subject Classifications (2000) 53A20, 35Q58.

Key words Laplace sequence of surfaces, Toda equations, Darboux transformation.

1. Introduction

Toda equations are a class of important integrable systems. The two-dimensional Toda

* Supported by the Chinese National Foundation of Natural Science, the Scientific Foundation of National Education Department of China for Educating Doctors, and the Research Foundation of Science and Technology of Shanghai.

equations attract many authors [1 - 4] in recent years. In particular, the elliptic version of two-dimensional Toda equations has many applications in differential geometry, such as minimal surfaces, surfaces of constant mean curvature, harmonic maps, etc.

The hyperbolic two-dimensional Toda equations can be traced back to G. Darboux [5]. Starting from Laplace sequences of the following hyperbolic equations of second order:

$$z_{xt} + a_i z_x + b_i z_t + c_i z = 0, \tag{1}$$

he derived the equations

$$\frac{\partial^2 \log h_i}{\partial x \partial t} = -h_{i-1} + 2h_i - h_{i+1} \quad \left(h_i = \frac{\partial a_i}{\partial t} + a_i b_i - c_i\right). \tag{2}$$

If $h_i = e^{\omega_i}$, they are just the Toda equations

$$\frac{\partial^2 \omega_i}{\partial x \partial t} = -\mathrm{e}^{\omega_{i-1}} + 2\mathrm{e}^{\omega_i} - \mathrm{e}^{\omega_{i+1}} \tag{3}$$

of hyperbolic version.

The periodic Laplace sequences of surfaces (LSS) as an important subject in classical projective differential geometry, have been studied extensively [6 - 9]. In this Letter, we elucidate the relationship between LSS of period n in $\mathbf{P}_{n-1}$ and the Toda equations of period n. Different from Darboux, we consider this problem from the point of view of integrable system.

We start from the fundamental equations of periodic LSS which are written in the form of first-order partial differential equations. Multiplying suitable factors to the homogeneous coordinates of the points of the surfaces and changing an independent variable of the fundamental equations, we can simplify the fundamental equations of the periodic LSS quite significantly. We find that there are two types of n periodic LSS in $\mathbf{P}_{n-1}$. It is noted that type II occurs only for even n and was not mentioned by

Darboux and other researchers in this field. Moreover, both types have integrability conditions of the same form

$$\frac{\partial^2 \omega_i}{\partial x \partial t} = -\alpha_{i-1} e^{\omega_{i-1}} + 2\alpha_i e^{\omega_i} - \alpha_{i+1} e^{\omega_{i+1}} \quad (i = 1, 2, \cdots, n;\ \omega_{n+1} = \omega_1), \tag{4}$$

where $\alpha_i = \pm 1$. When all the $\alpha_i (i = 1, \cdots, n)$ are equal to 1, Equation (4) becomes the original two-dimensional Toda Equation (3). Hence (4) may be called two-dimensional signed Toda equation. In fact, it is noted that if some h_i are negative, then (3) should be replaced by (4).

Furthermore, we should point out here that the fundamental equations of the periodic LSS are actually the Lax pairs of Equations (4), and the spectral parameter λ can be easily introduced.

The theory of integrable systems contains many useful methods for constructing explicit solutions of the systems. Between them, the Darboux transformation (DT) method is an important one [10]. V. Matveev successfully used the DT method for the hyperbolic two-dimensional Toda equation (3). In this Letter we show that the method and its modification can be used to explicitly construct periodic LSS. Two examples are constructed for each type.

2. Two-Dimensional Toda Equations

The original Toda equations (periodic case) are

$$\frac{d^2 \omega_i}{dt^2} = e^{\omega_{i-1}} - 2e^{\omega_i} + e^{\omega_{i+1}} \quad (i = 1, 2, \cdots, n;\ \omega_{n+1} = \omega_1). \tag{5}$$

The two-dimensional Toda equations in $\mathbf{R}^{1+1}$ are a system of hyperbolic equations

$$\frac{\partial^2 \omega_i}{\partial x \partial t} = e^{\omega_{i-1}} - 2e^{\omega_i} + e^{\omega_{i+1}} \quad (i = 1, 2, \cdots, n) \tag{6}$$

or, equivalently,

$$\frac{\partial^2 \omega_i}{\partial x \partial t} = -e^{\omega_{i-1}} + 2e^{\omega_i} - e^{\omega_{i+1}} \quad (i=1, 2, \cdots, n), \tag{6'}$$

since we can replace t by $-t$ in (6).

The two-dimensional Toda equations can be deduced from their Lax pair

$$\begin{pmatrix} \psi_1 \\ \vdots \\ \psi_n \end{pmatrix}_t = \lambda \begin{pmatrix} 0 & 0 & \cdots & 0 & p_1 \\ p_2 & 0 & \cdots & 0 & 0 \\ 0 & p_3 & \cdots & 0 & 0 \\ \vdots & \vdots & \ddots & \vdots & \vdots \\ 0 & 0 & \cdots & p_n & 0 \end{pmatrix} \begin{pmatrix} \psi_1 \\ \vdots \\ \psi_n \end{pmatrix}, \tag{7}$$

$$\begin{pmatrix} \psi_1 \\ \vdots \\ \psi_n \end{pmatrix}_x = \begin{pmatrix} \sigma_1 & \frac{1}{\lambda} & 0 & \cdots & 0 \\ 0 & \sigma_2 & \frac{1}{\lambda} & \cdots & 0 \\ 0 & 0 & \sigma_3 & \cdots & 0 \\ \vdots & \vdots & \vdots & \ddots & \frac{1}{\lambda} \\ \frac{1}{\lambda} & 0 & 0 & \cdots & \sigma_n \end{pmatrix} \begin{pmatrix} \psi_1 \\ \vdots \\ \psi_n \end{pmatrix}, \tag{8}$$

i.e. the integrability condition of (7) and (8) are

$$\frac{\partial p_i}{\partial x} = p_i(\sigma_i - \sigma_{i-1}), \quad \frac{\partial \sigma_i}{\partial t} = p_i - p_{i+1}. \tag{9}$$

As is well known, in case of $p_i > 0$, we can write $p_i = e^{\omega_i}$ and obtain the two-dimensional Toda equation (6′)

$$\frac{\partial^2 \omega_i}{\partial x \partial t} = -e^{\omega_{i-1}} + 2e^{\omega_i} - e^{\omega_{i+1}}.$$

In particular, if

$$p_1 = p_3 = \cdots = p_{2k-1} = e^{\omega}, \quad p_2 = p_4 = \cdots = p_{2k} = e^{-\omega}, \quad n = 2k, \tag{10}$$

Equations (6′) are reduced to the sinh-Gordon equation

$$\omega_{xt} = 4\sinh\omega. \tag{11}$$

If x, t are replaced by ζ, $\bar{\zeta}$, the complex coordinates of the Euclidean plane, then (6) becomes an elliptic version of the Toda equations

$$\frac{\partial^2 \omega_i}{\partial\zeta\partial\bar{\zeta}} = e^{\omega_{i-1}} - 2e^{\omega_i} + e^{\omega_{i+1}}. \tag{12}$$

Two-dimensional Toda equations have attracted much attention in recent years (see [2, 3]).

In the research of the Laplace sequences of surfaces in projective space (see Section 3), we find that the case for some $p_i < 0$ in (9) is interesting and also important. Hence, we should put

$$p_i = \alpha_i e^{\omega_i}. \tag{13}$$

Here $\alpha_i = -1$ (resp. $\alpha_i = 1$) when $p_i < 0$ (resp. $p_i > 0$) and Equation (6′) becomes

$$\frac{\partial^2 \omega_i}{\partial t\partial x} = -\alpha_{i-1} e^{\omega_{i-1}} + 2\alpha_i e^{\omega_i} - \alpha_{i+1} e^{\omega_{i+1}}. \tag{14}$$

As to the one-dimensional case, we have

$$\frac{d^2 \omega_i}{dt^2} = -\alpha_{i-1} e^{\omega_{i-1}} + 2\alpha_i e^{\omega_i} - \alpha_{i+1} e^{\omega_{i+1}}. \tag{15}$$

The physical interpretation of Equation (15) is that the system consists of two kinds of particles and the interaction forces between them are either attraction or repulsion. We will discuss Equation (15) in detail in a forthcoming paper.

In particular, when

$$p_1 = p_3 = \cdots = p_{2k-1} = e^{-\omega}, \quad p_2 = p_4 = \cdots = p_{2k} = -e^{\omega}, \tag{16}$$

we have the cosh-Gordon equation

$$\omega_{xt} = 4\cosh\omega. \tag{17}$$

3. Laplace Sequences of Surfaces in Projective Space $\mathbf{P}_{n-1}$

Let $\mathbf{P}_{n-1}$ be $n-1$-dimensional projective space, $(x_1, x_2, \cdots, x_n)$ be the homogeneous coordinates of a point $N \in \mathbf{P}_{n-1}$, and

$$N = N(t, x) \quad (\text{i.e. } x_a = x_a(t, x),\ a = 1, 2, \cdots, n) \tag{18}$$

be the equations of a surface of $\mathbf{P}_{n-1}$ in homogeneous coordinates.

If a system of surfaces

$$N_i = N_i(t, x) \quad (i = 1, 2, \cdots, n) \tag{19}$$

satisfying

$$\begin{aligned} \frac{\partial N_i}{\partial t} &= \mu_i N_i + p_i N_{i-1} \quad (p_i \neq 0), \\ \frac{\partial N_i}{\partial x} &= \sigma_i N_i + q_i N_{i+1} \quad (q_i \neq 0), \end{aligned} \tag{20}$$

$(N_{n+1} = N_1,\ N_0 = N_n)$, then the system of surfaces constitutes a Laplace sequence of period n. In fact, the line $\overline{N_i N_{i+1}}$ is the common tangent of surfaces N_i and N_{i+1}. In other words, the surfaces N_i and N_{i+1} are the two focal surfaces of the line congruence $\{\overline{N_i N_{i+1}}\}$, or N_{i+1} (resp. N_i) is the Laplace transform of N_i (resp. N_{i+1}).

In this section, we will first simplify the fundamental equation (20) of the Laplace sequences by multiplying a suitable factor on the homogeneous coordinates of each surface $N_i = N_i(t, x)$.

Let

$$\widetilde{N}_i = k_i(x, t) N_i, \quad (k_i(x, t) \neq 0), \tag{21}$$

we have

$$\frac{\partial \widetilde{N}_i}{\partial t} = \widetilde{\mu}_i \widetilde{N}_i + \widetilde{p}_i \widetilde{N}_{i-1}, \quad \frac{\partial \widetilde{N}_i}{\partial x} = \widetilde{\sigma}_i \widetilde{N}_i + \widetilde{q}_i \widetilde{N}_{i+1}, \tag{20'}$$

where

$$\begin{aligned} \widetilde{\mu}_i &= \frac{k_{i,t}}{k_i} + \mu_i, \quad \widetilde{p}_i = \frac{k_i}{k_{i-1}} p_i, \\ \widetilde{\sigma}_i &= \frac{k_{i,x}}{k_i} + \sigma_i, \quad \widetilde{q}_i = \frac{k_i}{k_{i+1}} q_i. \end{aligned} \tag{22}$$

Using the expression $\widetilde{\mu}_i$, we choose

$$k_i = k_i^0(x) \mathrm{e}^{-\int \mu_i \mathrm{d}t}, \tag{23}$$

then $\widetilde{\mu}_i = 0$ holds true.

From the integrability condition of (20′), we get $\widetilde{q}_{i,t} = \widetilde{q}_i(\widetilde{\mu}_i - \widetilde{\mu}_{i+1}) = 0$. Hence, $\widetilde{q}_i$ depends on x only.

Furthermore, let $N'_i = m_i(x) \widetilde{N}_i (m_i(x) \neq 0)$ and we have

$$\partial_t N'_i = p'_i N'_{i-1}, \quad \partial_x N'_i = \sigma'_i N'_i + q'_i N'_{i+1}, \tag{20''}$$

where

$$p'_i = \frac{m_i}{m_{i-1}} \widetilde{p}_i, \quad \sigma'_i = \frac{m_{i,x}}{m_i} + \widetilde{\sigma}_i, \quad q'_i = \frac{m_i}{m_{i+1}} \widetilde{q}_i, \tag{24}$$

$$(m_{n+1} = m_1, \ m_0 = m_n).$$

From the last equation, we see that

$$q'_1 q'_2 \cdots q'_n = \widetilde{q}_1 \widetilde{q}_2 \cdots \widetilde{q}_n \equiv Q(x). \tag{25}$$

When $Q > 0$, or $Q < 0$ and n is odd, we define

$$q = Q^{1/n} \tag{26}$$

and take

$$m_1 \neq 0,\ m_{i+1} = m_i \tilde{q}_i q^{-1} \quad (i = 1, 2, \cdots, n), \tag{27}$$

then from (24) we have

$$q_i' = q \quad (i = 1, 2, \cdots, n). \tag{28}$$

We should note that the choice of m_{n+1} is consistent with $m_{n+1} = m_1$, since

$$m_{n+1} = m_n \tilde{q}_n q^{-1} = m_{n-1} \tilde{q}_{n-1} \tilde{q}_n q^{-2} = \cdots = m_1 \tilde{q}_n \tilde{q}_{n-1} \cdots \tilde{q}_1 q^{-n} = m_1. \tag{29}$$

When $Q < 0$ and n is even, we define

$$q = (-Q)^{1/n} \tag{30}$$

and take

$$m_1 \neq 0, \quad m_{i+1} = m_i \tilde{q}_i q^{-1} \quad (i = 1, 2, \cdots, n-1), \quad m_{n+1} = -m_n \tilde{q}_n q^{-1}. \tag{31}$$

Then we have

$$q_i'(x) = q(x) \quad (i = 1, 2, \cdots, n-1), \quad q_n' = -q(x) \tag{32}$$

and the choice of m_{n+1} is still consistent with $m_{n+1} = m_1$. Thus we have the following theorem.

Theorem 1. *There are two types of Laplace sequences of surface with period n in projective space $\mathbf{P}_{n-1}$. Their fundamental equations can be written as*

$$\textit{Type I}: \begin{cases} \dfrac{\partial N_i}{\partial t} = p_i N_{i-1} \quad (p_i \neq 0), \\[2ex] \dfrac{\partial N_i}{\partial x} = \sigma_i N_i + q N_{i+1} \quad (q \neq 0) \end{cases} \tag{33}$$

and

$$\text{Type II:}\begin{cases}\dfrac{\partial N_i}{\partial t}=p_iN_{i-1}\quad(p_i\neq 0),\\[2ex]\dfrac{\partial N_i}{\partial x}=\sigma_iN_i\pm qN_{i+1}\quad(q\neq 0),\end{cases}\tag{34}$$

respectively. Here q is a function of x only. The minus sign '—' in (34) appears only for $i=n$. However, the Laplace sequence of Type II can occur only for even n.

By a transformation of the variable x, (33) and (34) can be reduced to

$$\text{Type I:}\begin{cases}\dfrac{\partial N_i}{\partial t}=p_iN_{i-1},\\[2ex]\dfrac{\partial N_i}{\partial x}=\sigma_iN_i+N_{i+1}\end{cases}\tag{35}$$

and

$$\text{Type II:}\begin{cases}\dfrac{\partial N_i}{\partial t}=p_iN_{i-1},\\[2ex]\dfrac{\partial N_i}{\partial x}=\sigma_iN_i\pm N_{i+1}.\end{cases}\tag{36}$$

'—' is only taken for $i=n$.

Remark. In the case where $n=$even number, sgn (Q) is a projective invariant of LSS. This is the reason why there are two types of n periodic LSS for even n.

We write

$$\Psi=\begin{pmatrix}N_1\\N_2\\\vdots\\N_n\end{pmatrix}.\tag{37}$$

By the rescaling $(t, x, \sigma_1, p_i)\rightarrow(\lambda t, x/\lambda, \lambda\sigma_i, p_i)$, we can introduce the spectral parameter in (33) and (34). Thus we obtain the following theorem:

Theorem 2. *The Lax pair of two-dimensional Toda equations is just the fundamental equation* (35) *of the n periodic Laplace sequence of type I in* $\mathbf{P}_{n-1}$.

The Laplace sequence of surface of type II should correspond to the Lax pair

$$\begin{pmatrix} \psi_1 \\ \vdots \\ \psi_n \end{pmatrix}_t = \lambda \begin{pmatrix} 0 & 0 & \cdots & 0 & p_1 \\ p_2 & 0 & \cdots & 0 & 0 \\ 0 & p_3 & \cdots & 0 & 0 \\ \vdots & \vdots & \ddots & \vdots & \vdots \\ 0 & 0 & \cdots & p_n & 0 \end{pmatrix} \begin{pmatrix} \psi_1 \\ \vdots \\ \psi_n \end{pmatrix}, \tag{38}$$

$$\begin{pmatrix} \psi_1 \\ \vdots \\ \psi_n \end{pmatrix}_x = \begin{pmatrix} \sigma_1 & \frac{1}{\lambda} & 0 & \cdots & 0 \\ 0 & \sigma_2 & \frac{1}{\lambda} & \cdots & 0 \\ 0 & 0 & \sigma_3 & \cdots & 0 \\ \vdots & \vdots & \vdots & \ddots & \frac{1}{\lambda} \\ -\frac{1}{\lambda} & 0 & 0 & \cdots & \sigma_n \end{pmatrix} \begin{pmatrix} \psi_1 \\ \vdots \\ \psi_n \end{pmatrix}, \tag{39}$$

Equation (39) differs slightly from (8).

In the following, we do not use the parameter λ and put $\lambda = 1$. However, the integrability condition of (38), (39) is the signed Toda equations (14) if we set

$$p_a = \alpha_a e^{\omega_a} \quad (a = 2, \cdots, n) \tag{40}$$

and

$$-p_1 = \alpha_1 e^{\omega_1}, \tag{41}$$

where

$$\alpha_a = \mathrm{sgn}(p_a), \quad \alpha_1 = \mathrm{sgn}(-p_1). \tag{42}$$

Remark. If the LSS is nonperiodic, there is an infinite number of surfaces. The nonperiodic LSS correspond to the infinite number of Toda equations and the Lax pair contains matrices of infinite order.

4. The Construction of Laplace Sequence of Type I

We use Darboux transformation to construct a Laplace sequence of type I.

Let us recall the Darboux transformation of a two-dimensional Toda lattice [2]. Let (Ψ, p, σ) be a known solution of (7)-(9) where Ψ is a column. If Ψ^0 is a known solution (with same p, σ) too, then

$$\Psi'_a = \Psi_a - \frac{\Psi^0_a}{\Psi^0_{a-1}} \Psi_{a-1}, \tag{43}$$

$$p'_a = p_a - \left(\frac{\Psi^0_a}{\Psi^0_{a-1}}\right)_t, \tag{44}$$

$$\sigma'_a = \sigma_a + \frac{\Psi^0_{a-1}}{\Psi^0_a} - \frac{\Psi^0_a}{\Psi^0_{a-1}} \tag{45}$$

is a solution of (7)-(9) too. The transformation $(\Psi, p, \sigma) \to (\Psi', p', \sigma')$ is called a Darboux transformation (DT). This was obtained by V. B. Matveev [2].

Let

$$\Psi = (N_1, \cdots, N_n) \tag{46}$$

be a known LSS of period n in $\mathbf{P}_{n-1}$ and (35) is satisfied. Applying DT, we have

$$\Psi' = (N'_1, \cdots, N'_n) \tag{47}$$

with

$$N'_a = N_a - \frac{\Psi^0_a}{\Psi^0_{a-1}} N_{a-1}, \tag{48}$$

where

$$\Psi_a^0 = N_a l^a \quad (l^a \text{ is a constant column}). \tag{49}$$

Then Ψ' is a new Laplace sequence of surfaces of period n. It is noted that N'_a lies on the line $\overline{N_a N_{a-1}}$.

By using this method, we can construct a series of periodic LSS explicitly. We write down the procedure for $n=4$.

Take the trivial solution of two-dimensional Toda equation

$$\sigma_a = 0, \quad p_a = 1. \tag{50}$$

The fundamental equations (35) are reduced to

$$\begin{aligned} &N_{1,t} = N_4, \quad N_{2,t} = N_1, \quad N_{3,t} = N_2, \quad N_{4,t} = N_3, \\ &N_{1,x} = N_2, \quad N_{2,x} = N_3, \quad N_{3,x} = N_4, \quad N_{4,x} = N_1. \end{aligned} \tag{51}$$

Solving the equations, we obtain the Laplace sequence of period 4:

$$\begin{array}{lllll} N_1 & \cosh u & \sinh u & \cos v & \sin v \\ N_2 & \sinh u & \cosh u & \sin v & -\cos v \\ N_3 & \cosh u & \sinh u & -\cos v & -\sin v \\ N_4 & \sinh u & \cosh u & -\sin v & \cos v \end{array} \tag{52}$$

Here $u = t + x$, $v = t - x$.

In (52), the rows are homogeneous coordinates of the point N_a, and each column is a solution of the Lax pair

$$\frac{\partial \Psi_a}{\partial t} = \Psi_{a-1}, \quad \frac{\partial \Psi_a}{\partial x} = \Psi_{a+1} \quad (a = 1, 2, 3, 4;\ \Psi_0 = \Psi_4,\ \Psi_5 = \Psi_1). \tag{53}$$

It is seen that N_1 and N_3 generate the surface

$$S_1:\ x_1^2 - x_2^2 = x_3^2 + x_4^2 \quad \text{or} \quad x^2 + y^2 + z^2 = 1 \tag{54}$$

if we use the inhomogeneous coordinates $x = x_4/x_1$, $y = x_2/x_1$, $z = x_3/x_1$. Similarly,

N_2 and N_4 generate the surface

$$S_2: x_2^2 - x_1^2 = x_3^2 + x_4^2 \quad \text{or}$$

$$y^2 - x^2 - z^2 = 1. \tag{55}$$

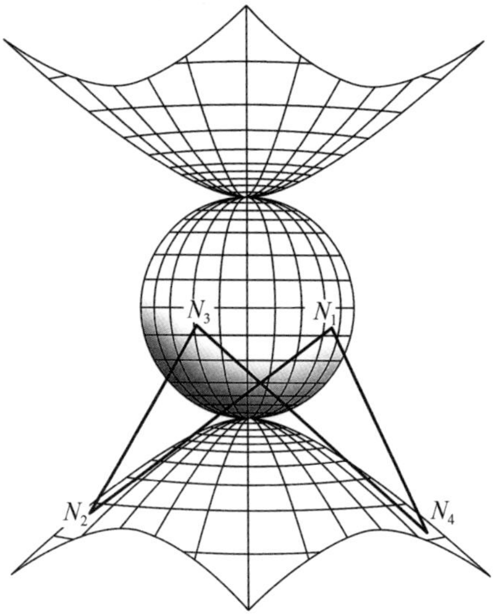

Figure 1. Example of Laplace sequences of surfaces of period 4 (Type I):

N_1 and N_3 are on the surface $x^2+y^2+z^2=1$, N_2 and N_4 are on the surface $y^2-x^2-z^2=1$

A typical quadrilateral together with S_1 and S_2 is shown in Figure 1.

By using DT, we obtain a new Laplace sequence of period 4

$$N'_1 = N_1 - \frac{\Psi_1^0}{\Psi_4^0} N_4, \quad N'_2 = N_2 - \frac{\Psi_2^0}{\Psi_1^0} N_1,$$

$$N'_3 = N_3 - \frac{\Psi_3^0}{\Psi_2^0} N_2, \quad N'_4 = N_4 - \frac{\Psi_4^0}{\Psi_3^0} N_3. \tag{56}$$

Here Ψ_a^0 is a linear combination of the four columns in (52), i.e.

$$\begin{aligned} \Psi_1^0 &= a\cosh u + b\sinh u + c\cos v + d\sin v, \\ \Psi_2^0 &= a\sinh u + b\cosh u + c\sin v - d\cos v, \\ \Psi_3^0 &= a\cosh u + b\sinh u - c\cos v - d\sin v, \\ \Psi_4^0 &= a\sinh u + b\cosh u - c\sin v + d\cos v. \end{aligned} \tag{57}$$

By a long calculation, we obtain

$$\begin{array}{lllll} N'_1 & b - cz_3 + dz_2 & -a - cz_1 - dz_4 & az_3 + bz_1 + d & -az_2 + bz_4 - c \\ N'_2 & -b - cz_4 + dz_1 & a + cz_2 + dz_3 & az_4 - bz_2 + d & -az_1 - bz_3 - c \\ N'_3 & b + cz_3 - dz_2 & -a + cz_1 + dz_4 & -az_3 - bz_1 + d & az_2 - bz_4 - c \\ N'_4 & -b + cz_4 - dz_1 & a - cz_2 - dz_3 & -az_4 + bz_2 + d & az_1 + bz_3 - c. \end{array} \tag{58}$$

Here

$$\begin{aligned} z_1 &= \cosh u \cos v + \sinh u \sin v, \\ z_2 &= \cosh u \cos v - \sinh u \sin v, \\ z_3 &= \cosh u \sin v + \sinh u \cos v, \\ z_4 &= \cosh u \sin v - \sinh u \cos v. \end{aligned} \tag{59}$$

It is not difficult to prove that N_1', N_2', N_3', N_4' generate four algebraic varieties in $\mathbf{P}_3$. In fact, between the parameters z_1, z_2, z_3, z_4, there are algebraic relations

$$z_1 z_2 + z_3 z_4 = 0, \quad z_1^2 - z_2^2 - z_3^2 + z_4^2 = 0. \tag{60}$$

From these relations and the parametric representation of N_i', we can eliminate these parameters and find that $x = x_4/x_1$, $y = x_2/x_1$, $z = x_3/x_1$ satisfy an algebraic equation $F_i(x, y, z) = 0$. Hence N_i' lies on an algebraic variety.

One can apply DTs successively to get an infinite sequence of 4-periodic Laplace sequences of surfaces in $\mathbf{P}_3$.

5. Laplace Sequences of Type II

The Lax equations for the Laplace sequences of surfaces in $\mathbf{P}_{n-1}$ of type II are slightly different from those of type I. We should find the formulas of the DT.

Let (Ψ, p, σ) be a solution of the Lax equations

$$\begin{pmatrix} \psi_1 \\ \vdots \\ \psi_n \end{pmatrix}_t = \begin{pmatrix} 0 & 0 & \cdots & 0 & p_1 \\ p_2 & 0 & \cdots & 0 & 0 \\ 0 & p_3 & \cdots & 0 & 0 \\ \vdots & \vdots & \ddots & \vdots & \vdots \\ 0 & 0 & \cdots & p_n & 0 \end{pmatrix} \begin{pmatrix} \psi_1 \\ \vdots \\ \psi_n \end{pmatrix},$$

$$\begin{pmatrix}\psi_1\\ \vdots\\ \psi_n\end{pmatrix}_x=\begin{pmatrix}\sigma_1 & 1 & 0 & \cdots & 0\\ 0 & \sigma_2 & 1 & \cdots & 0\\ 0 & 0 & \sigma_3 & \cdots & 0\\ \vdots & \vdots & \vdots & \ddots & 1\\ -1 & 0 & 0 & \cdots & \sigma_n\end{pmatrix}\begin{pmatrix}\psi_1\\ \vdots\\ \psi_n\end{pmatrix},\tag{61}$$

In order to apply DT, we try to use the same formula (43) for Ψ'_a, i.e. let

$$\Psi'_a=\Psi_a-\frac{\Psi^0_a}{\Psi^0_{a-1}}\Psi_{a-1},\quad (a=1,\cdots,n).\tag{62}$$

Here Ψ^0_a is a known column solution of (61). If $a\neq 1, n-1, n$, the situation is the same as the case Type I, i.e. (44) and (45) hold as well. By careful calculation for $a=1, n-1, n$, it is found that Ψ'_a is actually a new solution and the only change is

$$\begin{aligned}\sigma'_1&=\sigma_1+\frac{\Psi^0_2}{\Psi^0_1}+\frac{\Psi^0_1}{\Psi^0_n},\\ \sigma'_{n-1}&=\sigma_{n-1}+\frac{\Psi^0_n}{\Psi^0_{n-1}}-\frac{\Psi^0_{n-1}}{\Psi^0_{n-2}},\\ \sigma'_n&=\sigma_n-\frac{\Psi^0_1}{\Psi^0_n}-\frac{\Psi^0_n}{\Psi^0_{n-1}}.\end{aligned}\tag{63}$$

The integrability conditions of (61) are

$$\begin{aligned}&p_{a,x}=p_a(\sigma_a-\sigma_{a-1}),\quad \sigma_{a,t}=p_a-p_{a+1}\quad (a=2,\cdots,n-1),\\ &p_{1,x}=p_1(\sigma_1-\sigma_n),\qquad \sigma_{1,t}=-p_1-p_2,\\ &p_{n,x}=p_n(\sigma_n-\sigma_{n-1}),\quad \sigma_{n,t}=p_1+p_n.\end{aligned}\tag{64}$$

Comparing with (9), it is seen that the only change is the expressions for $\sigma_{1,x}$ and $\sigma_{n,x}$.

Now we turn to the explicit construction of periodic 4 Laplace sequences of surfaces of type II in $\mathbf{P}_3$.

For the trivial solution of (64), we can take

$$\sigma_1 = \sigma_2 = \sigma_3 = \sigma_4 = 0, \quad p_1 = -1, \quad p_2 = p_3 = p_4 = 1. \tag{65}$$

So the fundamental equations for the Laplace sequences are

$$\begin{aligned} &\frac{\partial N_1}{\partial t} = -N_4, \quad \frac{\partial N_2}{\partial t} = N_1, \quad \frac{\partial N_3}{\partial t} = N_2, \quad \frac{\partial N_4}{\partial t} = N_3, \\ &\frac{\partial N_1}{\partial x} = N_2, \quad \frac{\partial N_2}{\partial x} = N_3, \quad \frac{\partial N_3}{\partial x} = N_4, \quad \frac{\partial N_4}{\partial x} = -N_1. \end{aligned} \tag{66}$$

Solving these equations we obtain

$$\begin{array}{lllll} N_1 & e^v \cos u & e^v \sin u & e^{-v} \cos v & e^{-v} \sin v \\ N_2 & \frac{1}{\sqrt{2}} e^v (\cos u - \sin u) & \frac{1}{\sqrt{2}} e^v (\cos u + \sin u) & -\frac{1}{\sqrt{2}} e^{-v} (\cos u + \sin u) & \frac{1}{\sqrt{2}} e^{-v} (\cos u - \sin u) \\ N_3 & -e^v \sin u & e^v \cos u & e^{-v} \sin u & -e^{-v} \cos u \\ N_4 & -\frac{1}{\sqrt{2}} e^v (\sin u + \cos u) & \frac{1}{\sqrt{2}} e^v (\cos u - \sin u) & \frac{1}{\sqrt{2}} e^{-v} (\cos u - \sin u) & \frac{1}{\sqrt{2}} e^{-v} (\cos u + \sin u) \end{array} \tag{67}$$

Here $u = (x - t)/\sqrt{2}$, $v = (x + t)/\sqrt{2}$, N_1 and N_3 generate the surface S_1: $x_1 x_4 = x_2 x_3$ or $x = yz$, N_2 and N_4 generate the surface S_2: $x_1 x_3 = -x_2 x_4$ or $z = -xy$.

A typical quadralateral N_1, N_2, N_3, N_4 together with the surfaces S_1 and S_2 is shown in Figure 2.

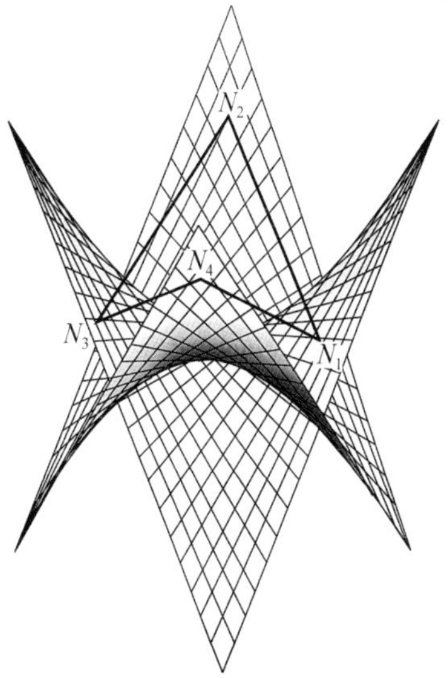

Figure 2. Example of Laplace sequences of surfaces of period 4 (Type II):
N_1 and N_3 are on the surface $x = yz$, N_2 and N_4 are on the surface $z = -xy$

By using the formula (62), we can get

$$N'_a = N_a - \frac{\Psi_a^0}{\Psi_{a-1}^0} N_{a-1} \tag{68}$$

as in the case of Type I. Hence, the DT gives a

new Laplace sequence of period 4. The four surfaces are algebraic surfaces too.

Remark. The surfaces (54), (55) do not contain straight lines, while the surfaces S_1 and S_2 appearing in LSS of type II contain two systems of straight lines. This fact reflects that the LSS of type I and those of type II are nonequivalent in the real projective geometry. The projective invariant sgn (Q) of type I and type II are $+1$ and -1, respectively.

The results were reported at the Moshé Flato Memorial Conference (Dijon) in September 2000 and at the W. L. Chow and K. T. Chen Memorial Conference (Tianjin) in October 2000.

References

[1] Toda, M.: *Nonlinear Waves and Solitons*, KTK Scientific Publishers, Tokyo and Kluwer Acad. Publ., Dordrecht, 1989.

[2] Matveev, V. B. and Salle, M. A.: *Darboux Transformations and Solitons*, Springer-Verlag, Berlin, 1991.

[3] Fordy, A. P. and Wood, J. C. (eds): *Harmonic Maps and Integrable Systems*, Vieweg, 1994.

[4] Guest, M. A.: *Harmonic Maps, Loop Groups and Integrable Systems*, London Math. Soc. Student Text 38, Cambridge Univ. Press, 1997.

[5] Darboux, G.: *Leçons sur la theorie générale des surfaces*, Vol. 2, 2nd edn, Gauthier-Villars, Paris, 1915.

[6] Finikoff, S.: Transformation T des congruences de droites, *Ann. Scuola Norm. Sup. Pisa, Sci. Fishche Mat.* (2) **2** (1933), 59 - 88.

[7] Su, B.: On certain Periodic sequences of Laplace of period four in ordinary space, *Sci. Rep. Töhoku Imp. Univ.* (1) **25** (1956), 227 - 256.

[8] Hu, H. S.: Darboux transformations between $\Delta\alpha=\sinh\alpha$ and $\Delta\alpha=\sin\alpha$ and the application to pseudo-spherical congruences in $\mathbf{R}^{2,1}$, *Lett. Math. Phys.* **48** (1999), 187 - 195.

[9] Hu, H. S.: Darboux transformations of Su-chain, In: *Proc. Sympos. in Honor of Prof. Su Buchin*, World Scientific, Singapore, 1993, pp.108 - 113.

[10] Gu, C. H., Hu, H. S. and Zhou, Z. X.: *Darboux Transformation in Soliton Theory and its Geometric Applications*, Shanghai Sci. Tech. Publ., 1999.

（本文曾发表于 *Letters in Mathematical Physics*, 2001, **57**, 19 - 32）

On Time-Like Surfaces of Positive Constant Gaussian Curvature and Imaginary Principal Curvatures

C.H. Gu[a,*], H.S. Hu[a], Jun-Ichi Inoguchi[b]

[a] Institute of Mathematics, Fudan University, Shanghai, China

[b] Department of Applied Mathematics, Fukuoka University, Nanakuma, Fukuoka 814-0180, Japan

Received 6 June 2001

Abstract We establish the Bäcklund transformation for the construction of time-like surfaces with positive Gaussian curvature and imaginary principal curvatures. The construction can be realized by algebraic algorithm via Darboux transformations. © 2002 Elsevier Science B.V. All rights reserved.

Subj. Class.: Differential geometry; General relativity

MSC: 53A30; 53A35; 37K35; 37K25

Keywords: Time-like surfaces; Bäcklund transformations; Darboux transformations; cosh-Gordon equation

0. Introduction

In 19th century, the construction of surfaces of negative constant curvature in Euclidean space $\boldsymbol{R}^3$ was one of the most important problems in differential geometry

* Corresponding author.

E-mail addresses: guch@fudan.ac.cn (C.H. Gu), inoguchi@bach.sm.fukuoka.ac.jp (J.-I. Inoguchi).

PII: S0393-0440(01)00066-3

[7, 26]. The related topics, such as sinh-Gordon equation, Bäcklund transformations [1] and Darboux transformations [5] have been developed extensively in the second half of 20th century and constitute an essential part in the modern soliton theory [10, 19, 23, 26]. In theory of relativity, geometry of indefinite metric is very crucial. Hence, the theory of surfaces in Minkowski space $\boldsymbol{R}^{2,1}$ which has the metric $\mathrm{d}s^2 = \mathrm{d}x^2 + \mathrm{d}y^2 - \mathrm{d}z^2$ attracted much attention. A series of papers are devoted to the construction of surfaces of constant mean curvature [6, 14, 17, 18] or constant Gaussian curvature [4, 12-15, 20, 22, 25]. The situation is much more complicated than the Euclidean case, since the surfaces may have a definite metric (space-like surfaces), Lorentz metric (time-like surfaces) or mixed metric.

Recently, the time-like surface of constant mean curvature in $\boldsymbol{R}^{2,1}$ have been studied systematically [6, 14, 17, 18]. In the meantime, the four kinds of surfaces (space-like and time-like surfaces of positive and negative constant curvature) have been considered in [11] by a unified approach — *Darboux transformation*. It should be mentioned that a time-like surface of positive constant Gaussian curvature is a parallel surface of a time-like surface of constant mean curvature. In [18], it is pointed that there is a class of surfaces of constant mean curvature whose principal curvatures are imaginary (see also [21]).

This class of surfaces has not been studied in [18] and has been missed in [11]. In [16,20], the time-like pseudo-spherical congruences whose two focal surfaces are time-like and of positive constant Gaussian curvature have been considered. However, the case of imaginary principal curvatures has not been studied either. The purpose of the present paper is to study the class of time-like surfaces with positive constant Gaussian curvature and imaginary principal curvatures. It is seen that these surfaces have real asymptotic lines, and hence, we can used some special asymptotic coordinates. The surfaces can be determined by a solution of cosh-Gordon equation

$$\omega_{uv} = \cosh\omega.$$

By using the time-like and space-like pseudo-spherical congruences the Bäcklund transformation approach for the cosh-Gordon equation is established. Moreover, it is proved that the system of partial differential equations can be solved explicitly by using the Darboux transformation. The construction can be continued successively via some algebraic algorithm.

We add a list of surfaces of constant Gaussian curvature and pseudo-spherical congruences in the Appendix A.

Moreover, we would like to mention another motivation (from theoretical physics) to the study of time-like surfaces of constant positive curvature with imaginary principal curvatures.

The relativistic string is the one-dimensional relativistic object whose time evolution extrimizes the Nambu-Goto action

$$S(\boldsymbol{r}) = -\kappa\int \mathrm{d}A,$$

where $\mathrm{d}A$ is the area element of the world sheet and κ is a constant.

From the mathematical point of view, world sheet of the relativistic string is a *time-like minimal* surface in the spacetime (cf. [2,8]).

In [3], Barbashov et al. suggested a generalization of relativistic string model in three-dimensional spacetimes with the action

$$S_H(\boldsymbol{r}) := S(\boldsymbol{r}) - 2\kappa H\int \mathrm{d}V$$

called *relativistic string with an external field*. Here, $\mathrm{d}V$ is the volume element of the region bounded by the world sheet and H is a nonzero constant.

One can easily to see that critical points of this action integral are time-like surfaces with *nonzero* constant mean curvature H in the spacetime.

Moreover, they mentioned a relation between this generalized string model and soliton theory. More precisely they claimed that the field equation of this generalized model in three-dimensional Minkowski space $\boldsymbol{R}^{2,1}$ coincides with the following sinh-Gordon equation:

$$\omega_{uv} = \sinh \omega.$$

In their argument, they took isothermal-curvature line coordinates for the world sheet. Namely, they assumed implicitly that the world sheet has two real distinct principal curvatures everywhere.

However, as we explained above, time-like surfaces of constant mean curvature may have imaginary real principal curvatures.

This observation also motivates us to study the soliton theory and differential geometry of cosh-Gordon equation and the corresponding time-like surfaces.

1. Asymptotic Chebyshev coordinates and cosh-Gordon equation

Let $\boldsymbol{R}^{2,1}$ be Minkowski three-space with Lorentzian metric $\mathrm{d}s^2 = \mathrm{d}x^2 + \mathrm{d}y^2 - \mathrm{d}z^2$, and S be a connected orientable two-manifold and $\boldsymbol{r}: \mathrm{S} \to \boldsymbol{R}^{2,1}$ an immersion. The immersion $\boldsymbol{r}$ is said to be *time-like* if the induced metric $\mathrm{I} := \mathrm{d}\boldsymbol{r} \cdot \mathrm{d}\boldsymbol{r}$ of S is Lorentzian. We use the abbreviation " • " for the scalar product of vectors in Lorentzian metric $\mathrm{d}s^2$.

The unit normal vector field $\boldsymbol{n}$ of S can be regarded as a smooth map $\boldsymbol{n}: S \to \boldsymbol{S}^{1,1}$ into the unit pseudosphere

$$\boldsymbol{S}^{1,1} := \{(x, y, z) \in \boldsymbol{R}^{2,1} \mid x^2 + y^2 - z^2 = 1\}$$

and called *Gauss map* of S.

Let S be a time-like surface in $\boldsymbol{R}^{2,1}$ with unit normal vector field $\boldsymbol{n}$. We can introduce a system of frames $\{\boldsymbol{r}; \boldsymbol{e}_1, \boldsymbol{e}_2, \boldsymbol{n}\}$ such that $\boldsymbol{e}_1$ and $\boldsymbol{e}_2$ are tangential null vector fields. Thus,

$$e_a^2=0,\quad n^2=1,\quad n\cdot e_a=0,\ (a=1,\ 2). \tag{1.1}$$

The fundamental equations of the surface are

$$\mathrm{d}r=\omega^a e_a,\quad \mathrm{d}e_a=\omega_a^b e_b+\omega_a^3 n,\quad \mathrm{d}n=\omega_3^a e_a,\quad (a=1,\ 2). \tag{1.2}$$

From Eq. (1.1) it follows

$$\omega_1^2=\omega_2^1=0, \tag{1.3}$$

and

$$\omega_1^3+\frac{1}{2}\mathrm{e}^{\omega}\omega_3^2=0,\quad \omega_2^3+\frac{1}{2}\mathrm{e}^{\omega}\omega_3^1=0. \tag{1.4}$$

Here, we assume $e_1\cdot e_2$ to be positive and equal to $\mathrm{e}^{\omega}/2$. We suppose that there exist special coordinates $(u,\ v)$ such that

$$\omega^1=\mathrm{d}u-\mathrm{e}^{-\omega}\mathrm{d}v,\quad \omega^2=-\mathrm{e}^{-\omega}\mathrm{d}u-\mathrm{d}v \tag{1.5}$$

and

$$\omega_3^1=-\mathrm{d}u-\mathrm{e}^{-\omega}\mathrm{d}v,\quad \omega_3^2=-\mathrm{e}^{-\omega}\mathrm{d}u+\mathrm{d}v. \tag{1.6}$$

Afterward we will see that this is equivalent to say that the surface is of $K=1$ and the principal curvatures are imaginary. From Eq. (1.4), we have

$$\omega_1^3=\frac{1}{2}(\mathrm{d}u-\mathrm{e}^{\omega}\mathrm{d}v),\quad \omega_2^3=\frac{1}{2}(\mathrm{e}^{\omega}\mathrm{d}u+\mathrm{d}v)$$

From $\mathrm{d}^2 r=0$, we obtain

$$\omega_1^1=\omega_u\mathrm{d}u,\quad \omega_2^2=\omega_v\mathrm{d}v. \tag{1.7}$$

The first and second fundamental forms are, respectively

$$\mathrm{I}=\mathrm{d}r^2=-\mathrm{d}u^2-2\sinh\omega\,\mathrm{d}u\,\mathrm{d}v+\mathrm{d}v^2, \tag{1.8}$$

$$\mathrm{II} = -\mathrm{d}\boldsymbol{n} \cdot \mathrm{d}\boldsymbol{r} = -2\cosh\omega \mathrm{d}u \mathrm{d}v \tag{1.9}$$

It is easily seen that $K=1$ and the principal curvatures of S are imaginary. The *Gauss equation* $\mathrm{d}\omega_b^a + \omega_c^a \wedge \omega_b^c = -\omega_3^a \wedge \omega_b^3$ is

$$\omega_{uv} = \cosh\omega. \tag{1.10}$$

It is seen that *Codazzi equations*

$$\mathrm{d}\omega_3^1 + \omega_1^1 \wedge \omega_3^1 = 0, \quad \mathrm{d}\omega_3^2 + \omega_2^2 \wedge \omega_3^2 = 0 \tag{1.11}$$

or

$$\mathrm{d}\omega_1^3 + \omega_1^3 \wedge \omega_1^1 = 0, \quad \mathrm{d}\omega_2^3 + \omega_2^3 \wedge \omega_2^2 = 0 \tag{1.12}$$

hold true. Thus, from the fundamental theorem of surfaces, we have

Theorem 1.1. *From a solution ω of the cosh-Gordon* Eq. (1.10) *there is a time-like surface S with $K=1$ and imaginary principal curvatures.*

According to the expressions (1.8) and (1.9) of the fundamental forms I and II, we call the coordinates (u, v) *asymptotic Chebyshev coordinates*.

Remark 1.2. Similar as that in Euclidean case, the surfaces S of $K=1$ in $\boldsymbol{R}^{2,1}$ is a parallel surface of a surface of constant mean curvature $H=1/2$ with distance 1. From the conformal flat coordinates of time-like surface of constant mean curvature [17], we see the existence of asymptotic Chebyshev coordinates of S.

Remark 1.3. The cosh-Gordon equation has been appeared in the study of the Cauchy problem of the harmonic maps from $\boldsymbol{R}^{1,1}$ into $\boldsymbol{S}^{1,1}$[9].

2. Bäcklund transformations

Let S be a known time-like surface with $K=1$ and two imaginary principal curvatures (u, v) be the asymptotic Chebyshev coordinates. We use the pseudo-spherical congruences to construct surfaces of the same characters. The congruences can be time-like and space-like either. Let

$$\boldsymbol{r}' = \boldsymbol{r} + l(a\boldsymbol{e}_1 + b\boldsymbol{e}_2). \tag{2.1}$$

Here l is a real constant and $ab\,\mathrm{e}^{\omega} = -1$. The lines $\boldsymbol{r}\boldsymbol{r}'$ generate a *time-like line congruence* Σ_- and for the time-like vector $\overrightarrow{\boldsymbol{r}\boldsymbol{r}'}$

$$(\overrightarrow{\boldsymbol{r}\boldsymbol{r}'})^2 = -l^2. \tag{2.2}$$

Let S' be the surface generated by $\boldsymbol{r}'$. The surface S is a focal surface of Σ_-. We demand that S' is another focal surface of the congruence Σ_- and the unit normal vector field of S' should be

$$\boldsymbol{n}' = \cos\tau(a\boldsymbol{e}_1 - b\boldsymbol{e}_2) + \sin\tau\boldsymbol{n} \tag{2.3}$$

which is evidently perpendicular to the line $\overline{\boldsymbol{r}\boldsymbol{r}'}$. Here, $\sin\tau = \boldsymbol{n}\cdot\boldsymbol{n}'$ is a constant. Thus, $\Sigma_-$is a time-like pseudo-spherical congruence in $\boldsymbol{R}^{2,1}$. In order that S' is another focal surface of Σ_-, we should have $\boldsymbol{n}'\cdot \mathrm{d}\boldsymbol{r}' = 0$. From

$$\mathrm{d}\boldsymbol{r}' = \mathrm{d}\boldsymbol{r} + l(\mathrm{d}a\boldsymbol{e}_1 + a\,\mathrm{d}\boldsymbol{e}_1 + \mathrm{d}b\boldsymbol{e}_2 + b\,\mathrm{d}\boldsymbol{e}_2) \tag{2.4}$$

and the fundamental equations, we have

$$\begin{aligned}\mathrm{d}\boldsymbol{r}' = {} & [(\mathrm{d}u - \mathrm{e}^{-\omega}\mathrm{d}v) + l(a_u\mathrm{d}u + a_v\mathrm{d}v) + la\omega_u\mathrm{d}u]\boldsymbol{e}_1 + \\ & [(-\mathrm{e}^{-\omega}\mathrm{d}u - \mathrm{d}v) + l(b_u\mathrm{d}u + b_v\mathrm{d}v) + lb\omega_v\mathrm{d}v]\boldsymbol{e}_2 + \\ & \frac{l}{2}[a(\mathrm{d}u - \mathrm{e}^{\omega}\mathrm{d}v) + b(\mathrm{e}^{\omega}\mathrm{d}u + \mathrm{d}v)]\boldsymbol{n},\end{aligned} \tag{2.5}$$

or

$$\boldsymbol{r}'_u = (1 + la_u + la\omega_u)\boldsymbol{e}_1 + (-\mathrm{e}^{-\omega} + lb_u)\boldsymbol{e}_2 + \frac{l}{2}(a + b\,\mathrm{e}^{\omega})\boldsymbol{n},$$

$$\boldsymbol{r}'_v = (-e^{-\omega} + l a_v)\boldsymbol{e}_1 + (-1 + l b_v + l b \omega_v)\boldsymbol{e}_2 + \frac{l}{2}(-a e^{\omega} + b)\boldsymbol{n}. \tag{2.6}$$

By using $a b e^{\omega} = -1$ and

$$a b_u + a_u b + a b \omega_u = 0, \quad a b_v + a_v b + a b \omega_v = 0 \tag{2.7}$$

we can write

$$\boldsymbol{r}'_u = \left(1 - l a \frac{b_u}{b}\right)\boldsymbol{e}_1 + (-e^{-\omega} + l b_u)\boldsymbol{e}_2 + \frac{l}{2}(a - a^{-1})\boldsymbol{n},$$

$$\boldsymbol{r}'_v = (-e^{-\omega} + l a_v)\boldsymbol{e}_1 + \left(-1 - l b \frac{a_v}{a}\right)\boldsymbol{e}_2 + \frac{l}{2}(b + b^{-1})\boldsymbol{n}. \tag{2.8}$$

Take $l = \cos\tau$ and define $\mu = \sec\tau - \operatorname{tg}\tau$, the equations $\boldsymbol{n}' \cdot d\boldsymbol{r}' = 0$ can be written as

$$2b^{-1} b_u = -\mu(a - a^{-1}), \quad 2a^{-1} a_v = -\frac{1}{\mu}(b + b^{-1}). \tag{2.9}$$

Without loss of generalities, we may assume that a is positive and put

$$a := \exp\frac{\omega' - \omega}{2}, \quad b := -\exp\frac{-\omega' - \omega}{2}. \tag{2.10}$$

Then, we obtain the following system of equations:

$$(\omega' + \omega)_u = 2\mu \sinh\frac{\omega' - \omega}{2}, \quad (\omega' - \omega)_v = \frac{2}{\mu}\cosh\frac{\omega' + \omega}{2}. \tag{2.11}$$

By differentiation, it is seen that

$$(\omega' + \omega)_{uv} = 2\cosh\frac{\omega' - \omega}{2}\cosh\frac{\omega' + \omega}{2},$$

$$(\omega' - \omega)_{uv} = 2\sinh\frac{\omega' - \omega}{2}\sinh\frac{\omega' + \omega}{2}. \tag{2.12}$$

The integrability condition of (2.11) for ω' is just the cosh-Gordon equation (1.10) and the solution ω' satisfies (1.10) too. Thus, Eq. (2.11) describes the *Bäcklund transformation* $\omega \mapsto \omega'$ for the cosh-Gordon equation. We may use the Bäcklund's theorem to confirm that S' is a time-like surface with $K=1$. However, for the purpose of applying the Bäcklund transformation successively and showing that S' has two imaginary principal curvatures, we have to prove that (u, v) are also the asymptotic Chebyshev coordinates of S'. By direct but long calculation we obtain

$$\boldsymbol{r}_u'^2=-1,\quad \boldsymbol{r}_v'^2=1,\quad \boldsymbol{r}_u'\cdot\boldsymbol{r}_v'=-\sinh\omega'. \tag{2.13}$$

Hence, the first fundamental form of S' is

$$\mathrm{I}'=-\mathrm{d}u^2-2\sinh\omega'\mathrm{d}u\,\mathrm{d}v+\mathrm{d}v^2. \tag{2.14}$$

Moreover, by differenting (2.3), we obtain

$$\boldsymbol{n}_u'=\left(-a\frac{b_u}{b}\cos\tau-\sin\tau\right)\boldsymbol{e}_1+(-b_u\cos\tau-\mathrm{e}^{-\omega}\sin\tau)\boldsymbol{e}_2+\frac{\cos\tau}{2}(a+a^{-1})\boldsymbol{n},$$

$$\boldsymbol{n}_v'=(a_v\cos\tau-\mathrm{e}^{-\omega}\sin\tau)\boldsymbol{e}_1+\left(b\frac{a_v}{a}+\sin\tau\right)\boldsymbol{e}_2-\frac{\cos\tau}{2}(b-b^{-1})\boldsymbol{n}. \tag{2.15}$$

By calculation, we have

$$-\boldsymbol{n}_u\cdot\boldsymbol{r}_u=0,\quad -\boldsymbol{n}_v\cdot\boldsymbol{r}_v=0,\quad -(\boldsymbol{n}_u\cdot\boldsymbol{r}_v+\boldsymbol{n}_v\cdot\boldsymbol{r}_u)=-2\cosh\omega'. \tag{2.16}$$

Thus, the second fundamental for S' is

$$\mathrm{II}'=-2\cosh\omega'\mathrm{d}u\,\mathrm{d}v. \tag{2.17}$$

From Eqs. (2.15) and (2.17) it is seen that S' is a surface of $K=1$ and has two imaginary principal curvatures. Moreover, (u, v) are the asymptotic Chebyshev coordinates of S'. Consequently, the Bäcklund transformation can be done successively, if the system (2.12) can be solved successively.

The above results can be summarized as the following theorem:

Theorem 2.1 (Time-like Bäcklund transformation). *Let* S *be a time-like surface with* $K=1$ *and two imaginary principal curvature and* (u, v) *be the asymptotic*

Chebyshev coordinates, then

1. *the system of* Eq. (2.11) *is completely integrable;*

2. *if* ω' *is a solution of* (2.11), *then* Eq. (2.1) *in which,* a, b *have the expression* (2.10) *defines a time-like surface* S' *with* $K=1$ *and two imaginary principal curvatures;*

3. *the coordinates* (u, v) *are asymptotic Chebyshev coordinates of* S' *too.*

Now, we use a space-like congruence to construct S' from S. We choose $ab\,e^{\omega}=1$ in Eq. (2.1), then a *space-like congruence* Σ_+ is obtained. In this case, we should have

$$\boldsymbol{n}'=\sinh\tau(a\boldsymbol{e}_1-b\boldsymbol{e}_2)+\cosh\tau\boldsymbol{n} \tag{2.18}$$

which is still a unit space-like vector field. The Eq. (2.6) holds as well and (2.8) is replaced by

$$\boldsymbol{r}'_u=\left(1-la\frac{b_u}{b}\right)\boldsymbol{e}_1+(-e^{-\omega}+lb_u)\boldsymbol{e}_2+\frac{l}{2}(a+a^{-1})\boldsymbol{n},$$

$$\boldsymbol{r}'_v=(-e^{-\omega}+la_v)\boldsymbol{e}_1+\left(-1-lb\frac{a_v}{a}\right)\boldsymbol{e}_2-\frac{l}{2}(b^{-1}-b)\boldsymbol{n}. \tag{2.19}$$

Let $l=\sinh\tau$, the condition $\boldsymbol{n}'\cdot \mathrm{d}\boldsymbol{r}'=0$ becomes

$$\frac{2b_u}{b}=\mu(a+a^{-1}),\quad \frac{2a_v}{a}=-\frac{1}{\mu}(b-b^{-1}). \tag{2.20}$$

Here, $\mu=\operatorname{csch}\tau-\operatorname{cth}\tau$. Let

$$a=\exp\frac{-\omega'-\omega}{2},\quad b=\exp\frac{\omega'-\omega}{2}. \tag{2.21}$$

Then, we obtain

$$(\omega'-\omega)_u=2\mu\cosh\frac{\omega'+\omega}{2},\quad (\omega'+\omega)_v=\frac{2}{\mu}\sinh\frac{\omega'-\omega}{2}. \tag{2.22}$$

This is the equations for Bäcklund transformation of the cosh-Gordon equations too. It is seen that (2.22) becomes (2.11), if we interchange (u, v) and use $(1/\mu)$ instead of μ. Further calculations imply the anologue of Theorem 2.1.

Theorem 2.2 (Space-like Bäcklund transformation). *Let S be a time-like surface with $K=1$ and two imaginary principal curvatures and (u, v) be the asymptotic Chebyshev coordinates, then*

1. *the system (2.22) is completely integrable;*
2. *if ω' is a solution of (2.22), then (2.1), in which, a, b have the expressions (2.21), defines a time-like surface S' with $K=1$ and two imaginary principal curvatures;*
3. *the coordinates (u, v) are the asymptotic Chebyshev coordinates of S'* too.

3. Darboux transformations

We use Darboux transformations to realize the Bäcklund transformations explicitly. At first, we have

Lemma 3.1. *The cosh-Gordon equation is the zero curvature condition of the Lax pair*

$$U := \Phi_u \Phi^{-1} = \frac{\lambda}{2}\begin{bmatrix} 0 & -e^{-\omega} \\ e^{\omega} & 0 \end{bmatrix}, \quad V := \Phi_v \Phi^{-1} = \frac{1}{2}\begin{bmatrix} -\omega_v & \dfrac{1}{\lambda} \\ \dfrac{1}{\lambda} & \omega_v \end{bmatrix}. \tag{3.1}$$

Here λ $(\neq 0)$ is the spectral parameter and Φ is a matrix valued function, depending on the real variable (u, v) and the complex variable $\lambda (\lambda \neq 0)$.

Proof. By direct calculation, we see that the zero curvature condition

$$U_v - V_u + [U, V] = 0 \tag{3.2}$$

is equivalent to the cosh-Gordon equation. □

Lemma 3.2. *If $(h_1, h_2)^t$ is a column solution to the Lax pair for $\lambda = \lambda_0 (h_1 \neq 0, h_2 \neq$*

0), *then* $(-h_1, h_2)^t$ *is a column solution to the Lax pair for* $\lambda=-\lambda_0$.

Proof. Substitute $(-h_1, h_2)^t$ in the Lax pair (3.1) for $\lambda=-\lambda_0$, it is easily seen that the Lax pair is satisfied. □

Lemma 3.3. *Let* ω *be a real solution of cosh-Gordon equation,* λ_0 *pure-imaginary number and* $(h_1, h_2)^t$ *a column solution of the Lax pair for* $\lambda=\lambda_0$. *If* h_2/h_1 *is pure-imaginary at one point* (u_0, v_0), *then it is pure-imaginary on any connected region containing* (u_0, v_0), *where the solution* $(h_1, h_2)^t$ *makes sense and* $h_1\neq 0$, $h_2\neq 0$.

Proof. From the Lax equation (3.1), we see that

$$\left(\frac{h_2}{h_1}\right)_u=\frac{\lambda_0}{2}\left[e^{\omega}+e^{-\omega}\left(\frac{h_2}{h_1}\right)^2\right], \tag{3.3}$$

$$\left(\frac{h_2}{h_1}\right)_v=\frac{1}{2\lambda_0}+\omega_v\left(\frac{h_2}{h_1}\right)-\frac{1}{2\lambda_0}\left(\frac{h_2}{h_1}\right)^2. \tag{3.4}$$

Let $A=h_2/h_1+\overline{h_2/h_1}$. Then

$$A_u=\frac{\bar{\lambda}_0}{2}e^{-\omega}\left(\frac{\bar{h}_2}{\bar{h}_1}-\frac{h_2}{h_1}\right)A,\quad A_v=\omega_v A+\frac{1}{2\bar{\lambda}_0}\left(\frac{h_2}{h_1}-\frac{\bar{h}_2}{\bar{h}_1}\right)A. \tag{3.5}$$

Consequently, if $A=0$ at (u_0, v_0), then $A=0$ on any connected region containing (u_0, v_0). Hence, h_2/h_1 is pure-imaginary. □

Theorem 3.4. (Darboux transformation). *If* ω *is a real solution of cosh-Gordon equation, then the function* ω', *defined by*

$$e^{\omega'}:=-\left(\frac{h_2}{h_1}\right)^2 e^{-\omega} \tag{3.6}$$

is a real solution of the cosh-Gordon equation too. Here, $(h_1, h_2)^t$ *is a column solution of the Lax* Eq. (3.1) *corresponding to a pure-imaginary spectral parameter* λ_0 *and* h_2/h_1 *is pure-imaginary.*

Proof. First, we notice that the preceding Lemma implies that it is possible to have pure-imaginary h_2/h_1.

From the general theory of Darboux transformation in matrix form [10,11]

$$\Phi^1 = D(\lambda)\Phi, \quad D(\lambda) = I + \lambda A \tag{3.7}$$

is a solution of the Lax equation (3.1) for some ω'. Here,

$$A = -H \begin{bmatrix} \frac{1}{\lambda_0} & 0 \\ 0 & -\frac{1}{\lambda_0} \end{bmatrix} H^{-1} = -\frac{1}{\lambda_0} \begin{bmatrix} 0 & \frac{h_1}{h_2} \\ \frac{h_2}{h_1} & 0 \end{bmatrix} \tag{3.8}$$

with $H = \begin{bmatrix} h_1 & -h_1 \\ h_2 & h_2 \end{bmatrix}$. Hence,

$$\Phi^1 = \left(I - \frac{\lambda}{\lambda_0} \begin{bmatrix} 0 & \frac{h_1}{h_2} \\ \frac{h_2}{h_1} & 0 \end{bmatrix} \right) \Phi. \tag{3.9}$$

We should have

$$\Phi^1_u = \frac{\lambda}{2} \begin{bmatrix} 0 & -e^{-\omega'} \\ e^{\omega'} & 0 \end{bmatrix} \Phi^1, \quad \Phi^1_v = \frac{1}{2} \begin{bmatrix} -\omega'_v & \frac{1}{\lambda} \\ \frac{1}{\lambda} & \omega'_v \end{bmatrix} \Phi^1. \tag{3.10}$$

Substituting (3.9) into (3.10), we obtain (3.6). Besides, the zero-curvature condition of (3.10) implies that ω' is a solution of cosh-Gordon equation. □

The transformation $(\omega, \Phi) \mapsto (\omega', \Phi^1)$ is called the *Darboux transformation* (DT) and D the *Darboux matrix*. The new solution (ω', Φ^1) is called the *Darboux transform* of (ω, Φ).

It is noted that the conclusion of Theorem 3.4 can be deduced by somewhat tedious and lengthy but straightforward calculations. However, the explicit formula (3.7) of the matrix-function Φ^1 is necessary for applying DT successively.

Theorem 3.5. *The solution ω', obtained by the Darboux transformation, and the seed solution ω to the cosh-Gordon equation are related by the Bäcklund transformation (2.11).*

Proof. From Eq. (3.6), we see that

$$e^{\omega'+\omega} = -\left(\frac{h_2}{h_1}\right)^2. \tag{3.11}$$

Differentiating (3.11) and using (3.3), we obtain

$$(\omega' + \omega)_u = 2\mu \sinh \frac{\omega' - \omega}{2}. \tag{3.12}$$

Here μ is a nonzero real number defined by $\mu = -\sigma\sqrt{-1}\lambda_0 \in \boldsymbol{R}^* = \boldsymbol{R}\backslash\{0\}$, where $\sigma = \pm 1$ such that

$$\exp\left\{\frac{\omega' + \omega}{2}\right\} = \sigma\sqrt{-1}\,\frac{h_2}{h_1} > 0.$$

Similarly, we have

$$(\omega' - \omega)_v = \frac{2}{\mu}\cosh(\omega' + \omega). \tag{3.13}$$

Consequently, Eq. (2.11) holds. This completes the proof. □

Theorem 3.5 implies the following result:

Corollary 3.6. *The (time-like) Bäcklund transformation $\omega \mapsto \omega'$ of cosh-Gordon equation can be solved explicitly by* Eq. (3.6), *provided a general solution Φ of the Lax equation* (3.1) *is known.*

Remark 3.7. The time-like surface of $K = 1$ with principal curvature $\kappa_1 = \kappa_2$ and free of umblics can be construct through Bäcklund transformation and Darboux transformation as well.

We sketch the procedure. Take

$$\omega^1 = du - e^{-\omega} dv, \quad \omega^2 = -dv. \tag{3.14}$$

$$\omega_3^1 = -du - e^{-\omega} dv, \quad \omega_3^2 = dv. \tag{3.15}$$

instead of Eqs. (1.4) and (1.6). We still take $\boldsymbol{e}_1^2 = \boldsymbol{e}_2^2 = 0$ and $\boldsymbol{e}_1 \cdot \boldsymbol{e}_2 = e^{\omega}/2$. Then

$$\mathrm{I} = -e^{\omega} du\, dv + dv^2, \quad \mathrm{II} = -e^{\omega} du\, dv. \tag{3.16}$$

The Gauss equation becomes the Liouville equation

$$\omega_{uv} = \frac{1}{2} e^{\omega}. \tag{3.17}$$

The time-like Bäcklund transformation takes the form

$$\boldsymbol{r}' = \boldsymbol{r} + l(a\boldsymbol{e}_1 + b\boldsymbol{e}_2), \quad ab\, e^{\omega} = -1, \quad \boldsymbol{n}' = \cos\tau (a\boldsymbol{e}_1 - b\boldsymbol{e}_2) + \sin\tau \boldsymbol{n}. \tag{3.18}$$

Eqs. (2.11) become

$$(\omega' + \omega)_u = -\mu \exp\frac{\omega - \omega'}{2}, \quad (\omega' - \omega)_v = \frac{2}{\mu}\cosh\frac{\omega + \omega'}{2}. \tag{3.19}$$

The integrability condition for ω' is the Liouville equation (3.17). By using the Lax pair

$$\Phi_u = \frac{\lambda}{2}\begin{bmatrix} 0 & 0 \\ e^{\omega} & 0 \end{bmatrix}\Phi, \quad \Phi_v = \frac{1}{2}\begin{bmatrix} -\omega_v & \dfrac{1}{\lambda} \\ \dfrac{1}{\lambda} & \omega_v \end{bmatrix}\Phi, \tag{3.20}$$

we can apply Darboux transformation to construct explicit solutions. The space-like Bäcklund transformation takes the form

$$\boldsymbol{r}' = \boldsymbol{r} + l(a\boldsymbol{e}_1 + b\boldsymbol{e}_2), \quad ab\, e^{\omega} = 1, \quad \boldsymbol{n}' = \sinh\tau (a\boldsymbol{e}_1 - b\boldsymbol{e}_2) + \cosh\tau \boldsymbol{n}, \tag{3.21}$$

which can be treated in the same way.

4. Geometrical meaning of Lax pair

In this section, we give the geometrical meaning of the Lax pair (3.1) together with a *Sym formula* for time-like $K=1$ surfaces with imaginary principal curvatures. To this end, we identify Minkowski three-space $\boldsymbol{R}^{2,1}$ with the Lie algebra $\underline{g}=\underline{sl}_2\boldsymbol{R}$ (cf. [17]).

We take the following basis $\{\underline{e}_1, \underline{e}_2, \underline{e}_3\}$ of $\underline{g}$:

$$\underline{e}_1=\begin{bmatrix}-1 & 0\\ 0 & 1\end{bmatrix},\quad \underline{e}_2=\begin{bmatrix}0 & 1\\ 1 & 0\end{bmatrix},\quad \underline{e}_3=\begin{bmatrix}0 & -1\\ 1 & 0\end{bmatrix}. \tag{4.1}$$

Hereafter, we identify $\boldsymbol{R}^{2,1}$ with $\underline{g}$ via this basis

$$(x, y, z)\leftrightarrow x\underline{e}_1+y\underline{e}_2+z\underline{e}_3. \tag{4.2}$$

By the linear isomorphism (4.2) the Lorentzian metric ds^2 corresponds to the scalar product

$$\langle X, Y\rangle=\frac{1}{2}\operatorname{tr}(XY),\quad X, Y\in \underline{g}. \tag{4.3}$$

The special linear group $G=SL_2\boldsymbol{R}$ acts isometrically on $\underline{g}$ via the Ad^*-action Ad^* : $\underline{g}\times G\to\underline{g}$

$$\mathrm{Ad}^*(\sigma)X=\mathrm{Ad}(\sigma^{-1})X=\sigma^{-1}X\sigma,\quad (\sigma\in G). \tag{4.4}$$

The Ad^*-action induces a double covering $G\to SO^+(2, 1)$. Here, $SO^+(2, 1)$ denotes the identity component of the Lorentz group $O(2, 1)$.

The Ad^*-action of G on $\boldsymbol{S}^{1,1}$ is transitive and isometric. The pseudosphere $\boldsymbol{S}^{1,1}$ is represented as

$$\boldsymbol{S}^{1,1}=\mathrm{Ad}^*(G)\underline{e}_1=\{\mathrm{Ad}(\sigma^{-1})\underline{e}_1 \mid \sigma\in G\}. \tag{4.5}$$

Compare with [17], we have

Proposition 4.1. *Let* $\Phi: D\times \boldsymbol{R}^* \to SL_2\boldsymbol{R}$ *be a solution to the Lax pair* (3.1) *over a simply connected region* D *of* S *and* $\lambda\in\boldsymbol{R}^*$. *Then*

$$\boldsymbol{r}_\lambda := 2\Phi^{-1}\frac{\partial}{\partial t}\Phi,\quad \lambda=\pm \mathrm{e}^{-t} \tag{4.6}$$

describes a real loop of time-like $K=1$ *surfaces in* $\boldsymbol{R}^{2,1}$ *with imaginary principal curvatures. The unit normal vector field of each* $\boldsymbol{r}_\lambda$ *is given by*

$$\boldsymbol{n}_\lambda = \mathrm{Ad}(\Phi^{-1})\,\underline{e}_1. \tag{4.7}$$

The first and second fundamental forms of each $\boldsymbol{r}_\lambda$ *are given by*

$$\mathrm{I}_\lambda = -\lambda^2 \mathrm{d}u^2 - 2\sinh\omega\, \mathrm{d}u\,\mathrm{d}v + \lambda^{-2}\,\mathrm{d}v^2,\quad \mathrm{II}_\lambda = -2\cosh\omega\,\mathrm{d}u\,\mathrm{d}v. \tag{4.8}$$

Proof. Under the identification (4.2), differentiating (4.6), we have

$$\frac{\partial}{\partial u}\boldsymbol{r}_\lambda = -\lambda\,\mathrm{Ad}(\Phi^{-1})\{\sinh\omega\,\underline{e}_2+\cosh\omega\,\underline{e}_3\},\quad \frac{\partial}{\partial v}\boldsymbol{r}_\lambda = \lambda^{-1}\mathrm{Ad}(\Phi^{-1})\,\underline{e}_2, \tag{4.9}$$

$$\frac{\partial}{\partial u}\boldsymbol{n}_\lambda = \lambda\,\mathrm{Ad}(\Phi^{-1})\{\cosh\omega\,\underline{e}_2+\sinh\omega\,\underline{e}_3\},\quad \frac{\partial}{\partial v}\boldsymbol{n}_\lambda = \lambda^{-1}\mathrm{Ad}(\Phi^{-1})\,\underline{e}_3. \tag{4.10}$$

From these formulas, we get the required result. □

Note that Eq. (4.6) is a formula of Sym's type [24]. If we take

$$\boldsymbol{e}_1 := \mathrm{Ad}(\Phi^{-1})\left\{-\frac{\mathrm{e}^{\omega}}{2}(\underline{e}_2+\underline{e}_3)\right\},\quad \boldsymbol{e}_2 := \mathrm{Ad}(\Phi^{-1})\left\{\frac{\mathrm{e}^{\omega}}{2}(\underline{e}_2-\underline{e}_3)\right\} \tag{4.11}$$

and $\lambda=1$, we obtain (1.2) with Eqs. (1.5) and (1.6), i.e. $\boldsymbol{r}:=\boldsymbol{r}_1$ is just the surface considered in Section 2 and (u, v) are asymptotic Chebyshev coordinates.

Let $u_1=\lambda u$, $v_1=v/\lambda$, the fundamental forms of $\boldsymbol{r}_\lambda$ can be written as

$$\mathrm{I}_\lambda = -\mathrm{d}u_1^2 - 2\sinh\omega\left(\frac{u_1}{\lambda}, \lambda v_1\right)\mathrm{d}u_1\mathrm{d}v_1 + \mathrm{d}v_1^2. \tag{4.12}$$

$$\mathrm{II}_\lambda = -2\cosh\omega\left(\frac{u_1}{\lambda},\ \lambda v_1\right)\mathrm{d}u_1\mathrm{d}v_1. \tag{4.13}$$

Thus, $\omega^\lambda(u_1, v_1) := \omega(\lambda^{-1}u_1, \lambda v_1)$ is a solution to

$$\omega^\lambda_{u_1 v_1} = \cosh\omega^\lambda, \tag{4.14}$$

and hence, $\boldsymbol{r}_\lambda$ is a surface corresponding to ω^λ in the sense of Theorem 1.1. In other words, $\boldsymbol{r}_\lambda$ is the Lie transformation of $\boldsymbol{r}_1$[7]. Thus, we have

Theorem 4.2. *The loop of time-like surfaces $\boldsymbol{r}_\lambda$ with $K=1$ and two imaginary principal curvatures are the Lie transformations of $\boldsymbol{r}_1$.*

Similar results for other kinds of surfaces of constant curvature have been obtained in [11].

Remark 4.3. The quadruple $\{\boldsymbol{r}_\lambda, \mathrm{Ad}(\Phi^{-1})\underline{e}_2, \mathrm{Ad}(\Phi^{-1})\underline{e}_3, \boldsymbol{n}_\lambda\}$ constitutes a system of moving orthonormal frames on $\boldsymbol{r}_\lambda$. This is the geometrical meaning of the Lax pair (3.1). The Darboux transform Φ^1 induces a new loop $\{\boldsymbol{r}^1_\lambda\}$ of time-like surfaces of constant curvature $K=1$ with two imaginary principal curvatures. We call each $\boldsymbol{r}^1_\lambda$ the Darboux transform of each $\boldsymbol{r}_\lambda$.

Theorem 4.4. *Let $\boldsymbol{r}: S \to \boldsymbol{R}^{2,1}$ be a time-like surface of constant curvature $K=1$ with two imaginary principal curvatures. Let Φ be a solution to the Lax equation (3.1) with $\det\Phi = 1$. Take a pure-imaginary number $\lambda_0 \neq 0$ and Darboux matrix $D(\lambda)$ determined by λ_0. Put $\widetilde{\Phi}^1 := \Phi^1/\sqrt{\det D}$. Then the Darboux transform $\boldsymbol{r}^1_\lambda$ of $\boldsymbol{r}_\lambda$ is*

$$\boldsymbol{r}^1_\lambda = 2(\widetilde{\Phi}^1)^{-1}\frac{\partial}{\partial t}\widetilde{\Phi}^1,\quad \lambda = \pm\mathrm{e}^{-t},\quad t\in\boldsymbol{R}. \tag{4.15}$$

The pseudo-spherical line congruence correponding to D is given by the following formula

$$\boldsymbol{r}^1_\lambda = \boldsymbol{r}_\lambda + 2\mathrm{Ad}(\Phi^{-1})\left\{\left(\frac{D}{\sqrt{\det D}}\right)^{-1}\frac{\partial}{\partial t}\left(\frac{D}{\det D}\right)\right\}$$

$$=\boldsymbol{r}_{\lambda}+2\mathrm{Ad}(\Phi^{-1})\left\{\frac{\pm\lambda^{2}}{\lambda^{2}-\lambda_{0}^{2}}\underline{1}+D^{-1}\frac{\partial}{\partial t}D\right\}. \tag{4.16}$$

Here $\underline{1}$ *denotes the identity matrix.*

Proof. First, we notice that det $D(\lambda)=1-(\lambda^{2}/\lambda_{0}^{2})>0$, since, λ_{0} is pure-imaginary. The normalized matrix $\widetilde{\Phi}^{1}$ is a solution to (3.1) with det $\widetilde{\Phi}^{1}=1$. Thus, by Proposition 4.1, the Darboux transform $\boldsymbol{r}_{\lambda}^{1}$ is given by Eq. (4.12). Since $D/\sqrt{\det D}\in SL_{2}\boldsymbol{R}$, $(D/\sqrt{\det D})^{-1}(\partial/\partial t)(D/\sqrt{\det D})\in\underline{g}$. Thus, Eq. (4.13) gives the pseudo-spherical line congruence corresponding to D. □

Thus, if we know a time-like surface of constant curvature $K=1$ with imaginary principal curvatures and the solution Φ of Lax pair, then an infinite series of surfaces of the same characters together with the solutions of Lax pair can be obtained successively by Eqs. (2.1), (3.6) and (4.15). The algorithm consists of *elementary operations* (algebraic operation and substitution) only, if we note that, in the algorithm, we need only $e^{\omega'}$ rather than ω'. However, the seed solution with explicit expressions is to be found.

The above discussions is mainly of local character. It is interesting to develope those results to a global theory.

Acknowledgements

The first author is supported by the Chinese special funds for major state basic research project ("Nonlinear Science"). The second author is supported by the Chinese National Fundation of Natural Sciences and the Commission of Science and Technology of Shanghai. The third author is partially supported by Grant-in-Aid for Encouragement of Young Scientists no. 12740051, Japan Society for Promotion of Science.

Most of this work was done when the third named author visited Institute of Mathematics, Fudan University. He would like to express his sincere thanks to Institute of Mathematics for the warm hospitality.

Appendix

A complete list of surfaces of constant Gaussian curvature and pseudo-spherical congruences in $\boldsymbol{R}^3$ and $\boldsymbol{R}^{2,1}$

Space	Surface	Congruence and BT	DT	Equation
$\boldsymbol{R}^3$	$\boldsymbol{r}$ $\boldsymbol{r}'$ $K=-1$ $\mathrm{I}=\cos^2(\alpha/2)\mathrm{d}u^2+\sin^2(\alpha/2)\mathrm{d}v^2$ $\mathrm{II}=\cos(\alpha/2)\sin(\alpha/2)(\mathrm{d}u^2-\mathrm{d}v^2)$	$\boldsymbol{r}'=\boldsymbol{r}+l(\cos(\alpha'/2)\boldsymbol{e}_1+\sin(\alpha'/2)\boldsymbol{e}_2)$ $\boldsymbol{e}_1^2=\boldsymbol{e}_2^2=1,\ \boldsymbol{e}_1\cdot\boldsymbol{e}_2=0$	α $\downarrow\uparrow$ α'	$\alpha_{uu}-\alpha_{vv}=\sin\alpha$ $\alpha'_{uu}-\alpha'_{vv}=\sin\alpha'$
$\boldsymbol{R}^{2,1}$	$\boldsymbol{r}$ $\boldsymbol{r}'$ $K=1$ space-like $\mathrm{I}=\cos^2(\alpha/2)\mathrm{d}u^2+\sin^2(\alpha/2)\mathrm{d}v^2$ $\mathrm{II}=-\cos(\alpha/2)\sin(\alpha/2)(\mathrm{d}u^2-\mathrm{d}v^2)$	$\boldsymbol{r}'=\boldsymbol{r}+l(\cos(\alpha'/2)\boldsymbol{e}_1+\sin(\alpha'/2)\boldsymbol{e}_2)$ Space-like congruence $\boldsymbol{e}_1^2=\boldsymbol{e}_2^2=1,\ \boldsymbol{e}_1\cdot\boldsymbol{e}_2=0$	α $\downarrow\uparrow$ α'	$\alpha_{uu}-\alpha_{vv}=-\sin\alpha$ $\alpha'_{uu}-\alpha'_{vv}=-\sin\alpha'$
$\boldsymbol{R}^{2,1}$	$\boldsymbol{r}$ $\boldsymbol{r}'$ $K=1$ time-like Principal curvatures $\kappa_1\neq\kappa_2$, real $\mathrm{I}=\cosh^2(\alpha/2)\mathrm{d}u^2-\sinh^2(\alpha/2)\mathrm{d}v^2$ or $\sinh^2(\alpha/2)\mathrm{d}u^2-\cosh^2(\alpha/2)\mathrm{d}v^2$ $\mathrm{II}=\cosh(\alpha/2)\sinh(\alpha/2)(\mathrm{d}u^2-\mathrm{d}v^2)$	$\boldsymbol{r}'=\boldsymbol{r}+l(\sinh(\alpha'/2)\boldsymbol{e}_1+$ $\cosh(\alpha'/2)\boldsymbol{e}_2)$ Time-like congruence $\boldsymbol{r}'=\boldsymbol{r}+l(\cosh(\alpha'/2)\boldsymbol{e}_1+$ $\sinh(\alpha'/2)\boldsymbol{e}_2)$ Space-like congruence $\boldsymbol{e}_1^2=-\boldsymbol{e}_2^2=1,\ \boldsymbol{e}_1\cdot\boldsymbol{e}_2=0$	α $\downarrow\uparrow$ α'	$\alpha_{uu}-\alpha_{vv}=\sinh\alpha$ $\alpha'_{uu}-\alpha'_{vv}=\sinh\alpha'$

(continued)

Space	Surface	Congruence and BT	DT	Equation
$\boldsymbol{R}^{2,1}$	$\boldsymbol{r}$ $\boldsymbol{r}'$ $K=1$ time-like $\kappa_1=\kappa_2$ $\mathrm{I}=-e^{\omega}du\,dv+dv^2$ $\mathrm{II}=-e^{\omega}du\,dv$	$\boldsymbol{r}'=\boldsymbol{r}+l(a\boldsymbol{e}_1+b\boldsymbol{e}_2)$ $abe^{\omega}=1$ Space-like congruence $abe^{\omega}=-1$ Time-like congruence $\boldsymbol{e}_1^2=\boldsymbol{e}_2^2=0$, $\boldsymbol{e}_1\cdot\boldsymbol{e}_2=e^{\omega}/2$	ω ↓↑ ω'	$\omega_{uv}=(1/2)e^{\omega}$ $\omega'_{uv}=-(1/2)e^{\omega'}$
$\boldsymbol{R}^{2,1}$	$\boldsymbol{r}$ $\boldsymbol{r}'$ $K=1$ time-like κ_1, κ_2 imaginary $\mathrm{I}=-du^2-2\sinh\omega\,du\,dv+dv^2$ $\mathrm{II}=-2\cosh\omega\,du\,dv$	$\boldsymbol{r}'=\boldsymbol{r}+l(a\boldsymbol{e}_1+b\boldsymbol{e}_2)$ $abe^{\omega}=1$ Space-like congruence $abe^{\omega}=-1$ Time-like congruence $\boldsymbol{e}_1^2=\boldsymbol{e}_2^2=0$, $\boldsymbol{e}_1\cdot\boldsymbol{e}_2=e^{\omega}/2$	ω ↓↑ ω'	$\omega_{uv}=\cosh\omega$ $\omega'_{uv}=\cosh\omega'$
$\boldsymbol{R}^{2,1}$	$\boldsymbol{r}$ $K=-1$ time-like $\mathrm{I}=\cos^2(\alpha/2)du^2-\sin^2(\alpha/2)dv^2$ $\mathrm{II}=\cos(\alpha/2)\sin(\alpha/2)(du^2\quad dv^2)$ $\boldsymbol{r}'$ $K=-1$ space-like $\mathrm{I}=\cosh^2(\alpha'/2)du^2+\sinh^2(\alpha/2)dv^2$ $\mathrm{II}=\cosh(\alpha'/2)\sinh(\alpha'/2)(du^2+dv^2)$	$\boldsymbol{r}'=\boldsymbol{r}+l(\cosh(\alpha'/2)\boldsymbol{e}_1+\sinh(\alpha'/2)\boldsymbol{e}_2)$ $\boldsymbol{e}_1^2=-\boldsymbol{e}_2^2=1$, $\boldsymbol{e}_1\cdot\boldsymbol{e}_2=0$ $\boldsymbol{r}=\boldsymbol{r}'+l(\cos(\alpha/2)\boldsymbol{e}'_1+\sin(\alpha/2\boldsymbol{e}'_2)$ Space-like congruence $\boldsymbol{e}'^2_1=\boldsymbol{e}'^2_2=1$, $\boldsymbol{e}'_2\cdot\boldsymbol{e}'_2=0$	α ↓↑ α'	$\Delta\alpha=\sin\alpha$ $\Delta\alpha'=\sinh\alpha'$

References

[1] A.V. Bäcklund, Einiges über Curve und Flächentransformationen, Lund Universitets Arsskrift, Vol. 10, 1885.

[2] B.M. Barbashov, V. V. Nesterenko, Differential geometry and nonlinear field models, Fortschritte Phys. 28(1980) 427－464.

[3] B.M. Barbashov, V.V. Nesterenko, A.M. Chervjakov, Generalization of the relativistic string model in the scope of the geometric approach (in Russian), Theret. Mat Fiz. 45 (1980), 365－376, English translation: Theoret. and Math. Phys. 45 (1980), 1082－1089.

[4] S.S. Chern, Geometrical interpretation of sinh-Gordon equation, Ann. Polon. Math. 39 (1980)

74 - 80.

[5] G. Darboux, Leçons sur la théorie des surfaces, Vol. 2, 2nd Edition, Gauthier-Villars, Paris, 1915.

[6] J. Dorfmeister, J.-I. Inoguchi, M. Toda, Weierstraß type representation of timelike surfaces with constant mean curvature, preprint, 2001.

[7] L.P. Eisenhart, A Treatise on the Differential Geometry on Curves and Surfaces, Ginn and Company, 1909, reprinted by Dover, 1960.

[8] C.H. Gu, On the motion of a string in curved spacetime, in: N. Hu (Ed.), Proceedings of the third Marcel Grossmann Meeting on General Relativity (1982, Shanghai), Science Press and North-Holland Publishing, 1983, pp.139 - 142.

[9] C.H. Gu, On the harmonic maps from $\boldsymbol{R}^{1,1}$ to $S^{1,1}$, J. Reine Angew. Math. 346 (1984) 101 - 109.

[10] C.H. Gu (Ed.), Soliton Theory and its Applications, Springer-Verlag and Zhejiang Science and Technology Publishing House, 1995.

[11] C.H. Gu, H.S. Hu, Z.X. Zhou, Darboux Transformations in Solition Theory and its Geometric Applications (in Chinese), Modern Mathematical Series, Shanghai Sci. Tech. Publ., 1999.

[12] H.S. Hu, The geometry of sine-Laplace, sinh-Laplace equations, in: N. Hu(Ed.), Proceedings of the 3rd Marcel Grossman Meeting on General Relativity (1982, Shanghai), Science Press and North-Holland Publishing, 1983, pp. 1073 - 1076.

[13] H.S. Hu, The construction of hyperbolic surfaces in three-dimensional Minkowski space and sinh-Laplace equations, Acta Math. Sinica 1 (1985) 79 - 86.

[14] H.S. Hu, On the geometry of sinh-Gordon equation, in: Beirao da Veiga, et al. (Eds.), Qualitative Aspects and Applications of Nonlinear Evolution Equations (Trieste, 1993) World Sci. Publishing, River Edge, NJ, 1994, pp. 35 - 47.

[15] H.S. Hu, Darboux transformations between $\Delta\alpha = \sinh\alpha$ and $\Delta\alpha = \sin\alpha$ and the application to pseudo-spherical congruence in $\boldsymbol{R}^{2,1}$, Letts. Math. Phys. 48 (1999) 187 - 195.

[16] Y.Z. Huang, Bäcklund theorem in three-dimensional Minkowski space and their higher-dimensional generalization (in Chinese), Acta Math. Sinica 29 (1986) 684 - 690.

[17] J.-I. Inoguchi, Timelike surfaces of constant mean curvature in Minkowski 3-surfaces, Tokyo J. Math. 21(1998) 141 - 152.

[18] J.-I. Inoguchi, Darboux transformations on timelike constant mean curvature surfaces, J. Geom. Phys. 32(1999) 57 - 78.

[19] V.B. Matveev, M.A. Salle, Darboux Transformations and Solitons, Springer Series in Nonlinear Dynamics, Springer, Heidelberg, 1991.

[20] L. McNertney, One-parameter families of surfaces with constant curvature in Lorentz three-space, Ph.D. Thesis, Brown University, 1980.

[21] T.K. Milnor, Harmonic maps and classical surface theory in Minkowski three-space, Trans. Am. Math. Soc. 280(1983) 161 - 185.

[22] B. Palmer, Bäcklund transformation for surfaces in Minkowski space, J. Math. Phys. 31 (1990) 2872 - 2875.

[23] C. Rogers, W.R. Shadwick, Bäcklund Transformations and their Applications, Academic Press, New York, 1982.

[24] A. Sym, Soliton surfaces and their applications-soliton geometry from spectral problems, in: R. Martini (Ed.), Geometric Aspects of the Einstein Equations and Integrable Systems, Lecture Notes in Phys. 239 (1985), Springer, Berlin, pp.154 - 231.

[25] C. Tian, Bäcklund transformations on surface with $K = -1$ in $R^{2,1}$, J. Geom. Phys. 22 (1997) 212 - 218.

[26] D. Wójcik, J. Cieśliński (Eds.), Nonlinearity and Geometry, Polish Scientific Publishers PWN, Warszawa, 1988.

（本文曾发表于 *Journal of Geometry and Physics*, 2002, **41**, 296 - 311）

The Emmi Noether Lecture at ICM 2002: Two-Dimensional Toda Equations and Laplace Sequences of Surfaces in Projective Space

Hu Hesheng

Institute of Mathematics, Fudan University, Shanghai, China

In this lecture we consider the 2-dimensional signed Toda equations of period n

$$\frac{\partial^2 \omega_i}{\partial x \partial t} = -\alpha_{i-1} e^{\omega_{i-1}} + 2\alpha_i e^{\omega_i} - \alpha_{i+1} e^{\omega_{i+1}},$$

$$\alpha_i = \pm 1, \quad \omega_{n+i} = \omega_i \quad (i = 1, 2, \cdots, n).$$

If all $\alpha_i = 1$, they are reduced to the original one, which can be traced back to G. Darboux. The equations are closely related to the n-periodic Laplace sequences of surfaces (LSS) in the projective space P_{n-1}. According to the projective invariant normal forms of the fundamental equations, we find that the n-periodic LSS have two types.

It is shown that the Lax pair fo (1) is just the fundamental equations of the n-periodic LSS in P_{n-1}. The Darboux transformation method is applied to the n-periodic LSS so as to construct a new one explicitly from a known one. Examples in P_3 of types Ⅰ and Ⅱ are constructed.

We also solve the following problem: To extend a one-parametric family of n-polygons in P_{n-1} into an n-periodic LSS which may be considered as a two-parametric family of n-polygons with elegant geometrical structures. We prove that the elliptic version of 2-

dimensional signed Toda equations may be deduced from the harmonic sequences of the harmonic maps from a Riemann surface to the Grassmannian $G_{1,n}$ of indefinite metric. Besides, the corresponding fact for harmonic sequences from the Minkowski plane is in consideration.

(本文选自于 *International Congress of Mathematicians: Abstracts of Short Communications and Poster Sessions*, *Beijing*, *August 20 - 28*, *2002*, Higher Education Press)

附录一 数学：超越国界和性别

——中法女院士南师附中深情话数学

沙国祥

2003年11月18日下午，法国科学院数学女院士、巴黎第六大学 Yvonne Choquet-Bruhat 教授与中国科学院数学女院士、复旦大学胡和生教授，应南京师范大学附属中学江宁分校的邀请，与该校师生欢聚在一起，畅谈她们对数学的理解、挚爱以及学习数学的方法，并特别强调：女同学在数学上与男同学相比毫不逊色，数学之门对女性是敞开的. 两位女院士还愉快地回答了同学们提出的问题.

Choquet：工作是一种快乐的源泉，而并不仅仅是一种必需，使得我们可以面对好多困难. 对我来讲，最大的快乐是理解、认识和工作. 我个人就是一个例子，数学对女性是不关门的. 数学与女性、与日常生活都是密切相关的. 现在的男士们找妻子，不仅仅是找一个生活中的伴侣，也不仅仅是让她管理家务.（Choquet 院士的丈夫也是法国科学院院士）

尽管我本人是一个数学家，我也不觉得数学是生活中唯一必须学习的学问. 在这个世界上每个人都可以尽他的所能做他自己的事情，不仅仅是数学. 在所有的工作中都有创造.

胡和生：我是南京人，所以对南京是挺有感情的. 小学时曾经在南京崇文小学学习过.

数学是一门非常重要的学科，这一点现在越发显示出来. 数学是各门学科的基础和工具. 数学的基础非但自然科学需要，而且社会科学也非常需要. 在其他科学的研究中，譬如在医学、经济学、生命科学等方面，数学都有很大的作用，最近我国载人宇宙飞船上天与数学是非常相关的.

数学思维与数学能力是现代公民与专业人才所必需的，从公民的素质来讲，数学应该学得更好一些.现在搞经济的，有些人数学很好.大家知道，获得经济学诺贝尔奖的数学家已有好多位了，去年来中国访问过的纳什就是一个例子，他在基础数学中的贡献也是很大的.要想在其他学科上取得非常大的成就，就必须学好数学，打好数学基础.

我们中国这20多年来，数学有了长足的进步.去年在北京召开的国际数学家大会上，我们看到有十多位年轻的中国数学家做1小时或45分钟邀请报告.中国现在搞数学的人数大大超过从前了，水平有了很大的提高.但是，如果跟美国、法国、德国、俄罗斯相比，还有很大的距离.所以，我们成为数学大国，就是说有很多数学家出现比较容易些，但是要成为数学强国就很难.如果成为数学强国，就要有相当多的水平非常高、研究成果非常突出的大数学家，这时数学也可以带动其他科学发展.在座的同学都还很年轻，这个任务就落在你们肩上.

现在学数学的条件与以前大大不同了.我小的时候，社会是不安定的，国家受侵略，基本生活、甚至生命安全都没有保障.新中国成立以后，情况不同了，我到了浙江大学做研究生.浙大是一所非常好的学校，提倡“求是”精神，学校的校风很好.我到浙大以后发愤学习，师从苏步青教授，他认为我是一个可以造就的人才，十分重视对我的培养.苏步青教授对我的要求非常严格，也非常爱护我，给我做的事情很多，而且放手让我去做.下面，我讲几点对学习的看法.

第一点是要对数学发生兴趣.有些学生对数学没兴趣，实际上兴趣是在不断克服困难的过程中培养起来的.当你解决了一个个困难问题以后，你的兴趣就一步步地提高了.所以要钻进去，不要怕难.

第二点是要肯吃苦.深入钻研的确是非常苦的.要掌握数学中的思想方法跟一般的学习要求就不同，不仅要会代公式、会算题，更重要的是要自己会思考，抓住最根本的要点.我早上在东南大学做报告，虽然是我自己做的东西，前一天我还要反复去想，当时做的时候是什么想法，现在再进一步想哪些东西是最实质性的，这是要很刻苦的.有时候做一个研究工作，往往3个月或半年都没有进展，这个时候是非常艰苦的阶段，要不怕困难，要坚持下去，想尽办法去克服困难.

第三点，我觉得要自信，要有勇气，要坚强，要有进取精神.不要看到困难就退缩了.将来攀登科学高峰，会碰到一个又一个困难，在克服困难中，走向成功之路.从前遇到一些同学，他们缺乏自信，中途就放弃了.要相信自己一定有能力，一定能够克服困难，这样你就能全力以赴.

还有一点，是我自己学习的经验，就是看到自己的弱点或是最需要改进的地方，立一个座右铭.我在做研究生的时候，写了16个字的座右铭，自己天天要看一遍.这16个字的座右铭是："专心致志，刻苦钻研，持之以恒，不受干扰."因为总有形形色色的干扰，如有时候考得不好，闹情绪等，都是干扰.现在经济情况比较好，玩的东西也多，如游戏机等，还有赚钱的事情，当然你们还小，大学生中有些人课外赚钱去了，这样一来考虑学问就少了.所以要排除这些干扰，专心致志，持之以恒.当你受到委屈了，受到批评了，或者不顺利了，自己能很好地控制，仍然发愤努力，这样你就会成功.我曾经受过很多干扰，总是设法排除.我读的是基础理论，当年往往要受到批评、批判，说是不联系实际，有些人受到批评、批判就不干了.我觉得批评中有对的就吸收过来，其他我就不管了，拼命用功，后来我就成功了.这一点非常重要，所以我今天讲给同学们听.

刚才Choquet院士特别讲了女子学科学的情况.我现在也特别对女同学讲一些.当然我对女同学讲的话，有些对男同学也有用.

女同学一样能在科学上做出杰出的贡献，在数学上也是如此.过去因为条件限制，女性的大数学家少一些.但是也有一批有突出贡献的女性数学家，其中有名的如埃米·诺特，是德国人；还有柯瓦列夫斯卡娅，是俄国人；还有在你们面前的Y. Choquet.埃米·诺特是近代代数学的奠基人之一，她建立了一个学派，培养了许多很好的学生.这个学派在国际上非常有名，所以她不仅是有名的女数学家，也是一位伟大的数学家.柯瓦列夫斯卡娅是研究偏微分方程、动力系统的，她在偏微分方程中有的定理是我们现在经常要用的. Choquet 院士是法国科学院第一个女院士.我们知道，法国有个居里夫人，大家很佩服.但居里夫人没有做成院士，因为那时候法国科学院还是排斥女性的.Choquet 成为法国第一个女院士，是非常不容易的，这是因为她在科学上的成就，她主要研究偏微分方程、广义相对论.今天上午在东南大学 Choquet 院士做了一个广义相对论的报告.她今年已经 80 岁了，今天做的报告是她最近的结果，是跟美国耶鲁大学的一位数学家一起做的.虽然她年纪这么大了，但她一直走在最前沿，得出了非常好的成果，她是我的榜样.

刚才介绍了世界上几位女性大数学家，她们的共同点是非常坚强.虽然受到歧视，有的大学不让她们进去，但她们以顽强的钻研精神，做出了那么多非常好的成绩，这些都常常鼓舞我.她们还有一个共同点是非常热爱自己的事业，对数学事业有献身精神.埃米·诺特没结婚，大家可能认为她这样有点怪，其实并不怪.她有很多学生，她常与学生一起从事研究，以此为最大的乐趣.

对女性来讲，要自信，更要自强.我们从小要有自己的理想，还要有本事.我小时候，父亲就教育我们要有本事，不做寄生虫，不做花瓶.我们姐妹几个都学得蛮好的，大家都自强不息.现在的生活中，看到电视上时装、化妆品很多，这些都是诱惑.当然女孩子稍微爱打扮一点，弄得漂亮点，这也是很好的.我自己也是爱漂亮的，想来 Choquet 院士也是如此.但是如果你花很多功夫在这个上面，那你一定学不好.所以我觉得要集中精力好好学习，年轻的时候，这点非常重要.

另外，我想女同学要“不娇不骄”，这样才能永远向上.

（部分对话摘录）

学生：中国学生学习数学的方式方法与法国学生有什么不同？你们对数学奥林匹克有什么看法？

Choquet：我不知道中国学生学习数学的方法是怎样的，但觉得在法国同样是有很多问题的.譬如说，有人喜欢数学，有人不喜欢数学，在同一个班上对这两种人同时教，就有一定的困难.

胡和生：我对法国学生学数学也不太了解.但我觉得外国人，如法国人、美国人，他们学的时候是相当灵活的，这一点是他们比我们优秀的地方.我们的同学题目做得很多，有时在数学竞赛中取得很好的成绩，但是有许多人是经过训练的.我希望通过训练以后，思维能灵活，而不仅仅是掌握了一些方法——到时候相当于运用一下这类题目现成的方法，而对临时解决问题的能力注意得不够.所以同学们训练自己的能力是非常重要的.要多思考，而且要思考得深入一些.

学生：我特别喜欢物理，想做一个物理学家.但我的物理不是很好，是不是您在中学时就有非常出色的表现，是不是您天生就有这种素质?

Choquet：肯定地，你要成为一个职业的数学家或物理学家，一定要有相当的才能.如同成为一个音乐家，必须具有音乐方面的某些才能；成为一个画家，在绘画方面应当有一定的才能.但我想补充的是，我觉得数学对大家都是非常有用的，因为数学很有逻辑，需要去分析、推理.

胡和生：我觉得物理可以学得很活，不仅仅是掌握几个定律，要会应用，要与现实生活结合.物理也是一门非常重要的科学，如果你的数学好，最好物理也能学好.

学生：请问您认为我们中学时怎样培养数学素养?

胡和生：数学素养我在高中时是得到了培养，当然不是有意识的.但是我跟别人有一点不大相同，就是我喜欢理论的东西，喜欢思考，喜欢了解新鲜事物.数学素养有时跟教师也很有关系，教师的启发性的讲课能引起学生的兴趣，能鼓励和激发你去努力、去学习.我在中学的时候，有几位数学老师曾说我的几何题目做得非常之好，这样我就很有信心.

另外，自己有时候稍微看一点课外读物，像“三角”，那时候有比较好的书，能够看看其中的思想方法，还有高次方程，觉得也是很有趣的.我们的家庭是个艺术世家，我的父亲、祖父虽然在科学方面没有帮助过我，但都鼓励我们勤奋学习，要有一技之长，我们姐妹几个都有个特点，就是独立.我中学里对科学很有兴趣，觉得什么事情都要问个为什么，弄得清清楚楚.

那时不分第一名、第二名的.当然同学交流,题目你做得来,他做不来,就知道谁的功课好一些,班上老师并没有对大家排出谁是第一名、第二名.人家讲女同学之间有嫉妒,但我们同学之间没有,大家好像姐妹一样,互相帮助,我觉得非常之好.现在分数卡得太厉害了.我想得第一名还是第二名有什么关系呢,实际上可能第五名、第六名比第一名、第二名还要好.主要的是你掌握知识的能力,思考的能力,想问题的深度,这是中学时应该培养起来的.

学生:在平时的数学学习中,怎样启发我们的兴趣?怎样学好数学?

Choquet:主要要有一个好的数学老师.他会给你练习,给你一些作业在家里做.对数学来讲,个人的独立工作是非常必要的.当然老师的讲解也是必需的,但特别要求个人独立思考.对学生来说,如果能独自解决一个问题,当然对学生来说不是特别难的问题,自己独立解决了,就是一种乐趣.

学生:在数学学习或人生道路上遇到困难的时候,驱使你们克服困难的,是你们对科学的热爱,还是一种使命感?

Choquet:生活与工作是两样不同的事情.工作对我来讲,永远是一种快乐.

胡和生:我平常做工作有一种性格:做事情就要做得好,一定要做得非常好.不做得很好,我心里就不安,这种性格是从小养成的.一方面是责任感,另一方面是自己有这样的性格.我感兴趣的事情还是比较多的,但是到后来觉得一个人要想在某一方面做好,在专业中有重要建树,兴趣不能太广.如果太爱看电影,或者太喜欢看小说,或者女同志对家务太关心,对生活要求过高,就不能在自己的专业上花很多功夫.大学毕业工作以后,责任感、使命感就不断增强了,感到自己的工作对国家、对社会都是很有意义的.我周围的许多同志都很有责任感,答应的事就好好去做.想到自己受到较高的培养,就更加感到应该很好地从事自己的事业,培养学生.

最后,南师附中江宁分校的一位老师也向两位女院士提了一个问题:"作为一名教师,最重要的是要教给学生什么?"

Choquet 院士的回答出人意料,令人感动:"最重要的是喜欢教."

胡和生院士的回答同样令人感动：一是教给学生知识和培养他们的能力，二是培养学生的品德.中学时学生的思想不太稳定，要在品德上加以培养，使他们成为全面发展的人.

附录二　胡和生院士速写

张咏晴

见到胡先生,她穿了一件暗红色的格子外套,领口露出大红色毛衣,非常精神.她长得秀气而娇小,在言谈中不时显露出她独立和坚强的气质.她说:“在任何时候,我都要进取,特别是女同志,更要有自强不息的精神,刻苦钻研的精神.1958 年,有人批判我,说我理论脱离实际,我不服气,学了弹性力学、相对论、核物理,视野和研究能力又前进了一步.”在科研中胡先生强调了思想要有深度,要有冲击困难问题的勇气和拼劲.她时常在科学研究中提出新的问题并努力解决它.她是苏步青教授在微分几何方面的接班人,1991 年当选为中国科学院院士,是我国数学领域中迄今唯一的女院士.

最近,她收到一份通知,邀请她今年 8 月份在北京召开的世界数学家大会(四年一度)作 Noether 讲座报告,这是专门为女数学家设置的一个讲座,国际上只选一位女数学家作为报告人,纪念近代最杰出的德国伟大女数学家 Noether.近年来,国际上女数学家增加得相当快,胡先生能被选中意味着她的研究工作得到国际数学界的重视.

她比丈夫谷超豪小二岁,夫妻之间在数学工作上既有各自研究的方向和问题,又有相互的合作,保持着青春活力.在生活上也相互照顾,这一对七十多岁的伉俪之间,相互极为尊重和体贴.采访结束时,两位先生送我们下楼,他们搀扶着走过一级级楼梯,那种默契是只能用意会而无法言传的.

(本文曾发表于《文汇报》2002 年 3 月 7 日第 6 版《实验室内外的独特风景——上海 3 位女院士速写》)